Natural Time Analysis: The New View of Time

Precursory Seismic Electric Signals, Earthquakes and other Complex Time Series

Panayiotis A. Varotsos, Nicholas V. Sarlis and
Efthimios S. Skordas

Natural Time Analysis: The New View of Time

Precursory Seismic Electric Signals, Earthquakes and other Complex Time Series

Springer

Published in association with
Praxis Publishing
Chichester, UK

PRAXIS

Professor Panayiotis A. Varotsos
University of Athens
Athens
Greece

Dr. Nicholas V. Sarlis
University of Athens
Athens
Greece

Dr. Efthimios S. Skordas
University of Athens
Athens
Greece

SPRINGER–PRAXIS BOOKS IN GEOPHYSICAL SCIENCES
SUBJECT *ADVISORY EDITOR*: Philippe Blondel, C.Geol., F.G.S., Ph.D., M.Sc., F.I.O.A., Senior Scientist, Department of Physics, University of Bath, Bath, U.K.

ISBN 978-3-642-16448-4 e-ISBN 978-3-642-16449-1
DOI 10.1007/978-3-642-16449-1
Springer Heidelberg Dordrecht London New York

Library of Congress Control Number: 2011925062

Cover design: Jim Wilkie
Project copy editor: Mike Shardlow
Author-generated LaTex, processed by EDV-Beratung Herweg, Germany

Printed on acid-free paper

Springer is part of Springer Science+Business Media (www.springer.com)

To the memory of my parents Antonios and Efrosyni
and of my Professor Kessar D. Alexopoulos

P.V.

Contents

II. Natural Time Foundations

Preface

There is a widespread belief that it is not space but time that in the end poses the greatest challenge in science. It is according to Gödel "that mysterious and seemingly self-contradictory being which, on the other hand, seems to form the basis of the world's and our own existence".

The basic scope of this monograph is to present the new view of time, termed natural time χ (from the Greek word "χρόνος" which means "time"), introduced by the authors in 2001. In this new view, time is *not continuous*, thus being in sharp contrast with the hitherto used conventional time t which is modeled as the one-dimensional *continuum* $\mathscr{R}$ of the real numbers.

The results deduced to date on the basis of this new domain reveal that novel dynamical features hidden behind time series in complex systems can emerge upon analyzing them in natural time, but cannot when the analysis is carried out within the frame of conventional time. Furthermore, the analysis in natural time enables the study of the dynamical evolution of a complex system and identifies when the system enters a critical stage. Hence, it seems that natural time plays a key role in predicting impending catastrophic events in general.

The present monograph comprises three Parts:

Part I (Chapter 1) provides a review of the so-called seismic electric signals (SES) which are low-frequency ($\lesssim$1 Hz) electric signals that precede earthquakes. A sequence of such signals, termed SES activity, constitutes the first example of a time series emitted from a complex system, like the Earth, to which natural time analysis has been applied.

Part II, consisting of the Chapters 2 and 3, sheds light on the foundations of natural time by providing the necessary mathematical background in each step. Furthermore, this Part describes how the analysis of a time series is made in the frame of natural time and explains how the entropy in natural time is introduced and calculated.

Part III, consisting of the six Chapters 4 to 9, presents examples of data analysis in natural time that have appeared to date (mainly in *Physical Review* and *Physical Review Letters*) in diverse fields, including Biology, Earth Sciences (Geophysics, Seismology), Environmental Sciences, Physics (Condensed Matter, Statistical Physics, Physics of Complex Systems) and Cardiology.

The contents of the nine Chapters can be summarized as follows.

Chapter 1 summarizes the mechanisms suggested to date for the SES generation as well as the physical properties of SES including those that SES are observed only at certain points of the Earth's surface called "sensitive points" and that their amplitude is interrelated with the magnitude of the impending earthquake. It is explained that these physical properties can be theoretically understood on the basis of Maxwell equations if we just consider that the earthquakes occur by slip on pre-existing faults, which constitute conductive paths in the solid Earth's crust. In addition, general background in Statistical Physics is provided on the basis of which we show that the observed SES activities exhibit *scale invariance* over four orders of magnitude. This is consistent with the pressure stimulated currents SES generation model proposed by Varotsos and Alexopoulos in the early 1980s based on thermodynamical grounds which (motivated the SES research in general and) suggests that SES are emitted when the stress in the focal region reaches a *critical* value, thus SES should be governed by *critical* dynamics.

In Chapter 2, we first present aspects advanced by such giants as A. Einstein, E. Schrödinger, W. Pauli, J. von Neumann, H. Weyl, and K. Gödel, in order to shed light on the crucial point that the continuity of conventional time is not demanded from *any* fundamental principle. We then introduce the natural time χ, which is *not* continuous, and indicate that its values, as well as those of the energy, form *countable* sets by using the set theory developed by Cantor. Furthermore, we demonstrate that natural time analysis is optimal for enhancing the signals in time-frequency space when employing the Wigner function and measuring its localization property. In other words, natural time analysis conforms to the desire to reduce uncertainty and extract signal information as much as possible. The normalized power spectrum $\Pi(\omega)$ is introduced in natural time, and its Taylor expansion leads, at low natural (cyclic) frequencies ω ($\omega \to 0$), to the expression $\Pi(\omega) \approx 1 - \kappa_1 \omega^2$ where the coefficient κ_1 is just the variance of natural time, i.e., $\kappa_1 = \langle \chi^2 \rangle - \langle \chi \rangle^2$. This quantity is useful in identifying the approach to a critical point as in the case of SES whose κ_1 value is shown to be 0.070. In addition, natural time analysis enables the distinction between the two origins of self-similarity, which is a challenging point when analyzing data from complex systems, i.e., whether self-similarity solely results from long-range temporal correlations (the process's memory only) or solely from the process's increments' infinite variance (heavy tails in their distribution).

In Chapter 3, we define the entropy S in natural time and show that the entropy S_- deduced from the natural time analysis of the time series obtained upon time reversal is in general different from S, thus the entropy in natural time does satisfy the condition to be "*causal*". Furthermore, the physical meaning of the change $\Delta S \equiv S - S_-$ of the entropy in natural time under time reversal, which is of profound importance for the study of the dynamical evolution of a complex system, is discussed. In addition, complexity measures are introduced that quantify the fluctuations of the entropy S and of the quantity ΔS upon changing the length scale as well as the extent to which they are affected when shuffling randomly the consecutive events.

Chapter 4 deals with the natural time analysis of all the measured SES activities which are shown to be characterized by very strong memory and their normalized power spectra $\Pi(\omega)$ versus ω fall on a *universal* curve having κ_1 value equal to 0.070. This curve coincides with the one obtained on theoretical grounds when assuming that SES are governed by *critical* dynamics. Concerning the distinction of SES activities from similar-looking "artificial" (man-made) noise, we find that modern techniques of Statistical Physics, e.g., detrended fluctuation analysis (DFA), multifractal DFA, wavelet transform, *can* achieve such a distinction, but when they are *applied* in natural time. The entropy S in natural time as well as $\Pi(\omega)$ can also achieve such a distinction. Finally, the study of the fluctuations $\Delta\chi_l$ of the average value of natural time under time reversal versus the window length l, also achieves a distinction between SES activities and "artificial" noises.

In Chapter 5, we investigate the effect of significant data loss on the identification of a SES activity. In particular, the following two cases are treated. First, the effect of the random removal of data segments of fixed length on the scaling properties of SES activities. Second, the appearance of a periodic noise like in Japan, where the electric field measurements at some sites are seriously contaminated by high noise from 06:00 to 22:00 LT every day, i.e., around 70% data loss. We show that in both cases, when combining natural time analysis with DFA, the identification of a long duration SES activity becomes possible with probability around 70% even after severe data loss (e.g., 70–80%).

Chapter 6 is focused on the natural time analysis of the seismicity, a careful inspection of which reveals that the quantity κ_1 may be considered as an order parameter. This allows the determination of the constant b in the Gutenberg–Richter law for earthquakes, $N(\geq M) = 10^{a-bM}$, by applying the Maximum Entropy Principle. It leads to $b \approx 1$, which agrees with real seismic data. Studying the order parameter fluctuations relative to the standard deviation of its distribution, the scaled distributions of different seismic areas as well as that of the worldwide seismicity fall on a *universal* curve which also exhibits features similar to those in several critical phenomena. This curve changes upon randomly shuffling, which reflects that temporal correlations exist between the earthquake magnitudes (M). This is confirmed by applying DFA to the earthquake magnitude time series and upon employing either multifractal cascades in natural time or nonextensive statistical mechanics combined with natural time analysis. Finally, the probability density function $P(\kappa_1)$ versus κ_1 plot *before* mainshocks exhibits a significant bimodal feature. This is strikingly similar to the bimodal feature of the order parameter when approaching (from below) T_c in equilibrium critical phenomena.

In Chapter 7, assuming that the mainshock is a new phase, we show how natural time analysis enables the determination of the occurrence time of an impending major earthquake. Considering that the detection of a SES activity signifies that the system enters in the critical regime, the time series of the small earthquakes that occur in the candidate region to suffer the mainshock after the SES detection are analyzed in natural time. It is found that the variance κ_1 becomes equal to 0.070 a few days to around one week *before* the mainshock. This behavior, which exhibits spatial as well as magnitude threshold invariance, has been observed to date for *all* major earthquakes in Greece since 2001. For example, the occurrence time of the $M_w6.9$ earthquake on February 14, 2008, which was the strongest earthquake to occurr in Greece during the last 27 years, was publicly announced as imminent on February 10, 2008. The procedure has been also ascertained in the case of the volcanic-seismic swarm activity in 2000 in the Izu island region in Japan as well as in the $M_s7.1$ Loma Prieta earthquake in California in 1989.

In Chapter 8, we apply natural time analysis to the time series of the avalanches in several self-organized criticality (SOC) models as well as to other dynamical models including a simple deterministic version of the "train" model for earthquakes introduced by Burridge and Knopoff, the Olami–Feder–Christensen earthquake model, the 2D Ising model when quenching at temperatures close to but below T_c, which is qualitatively similar with the pressure stimulated currents SES generation model, a deterministic version of the original Bak–Tang–Wiesenfeld sandpile model and a generalized stochastic SOC model. In all these dynamical models, we find that the value $\kappa_1 = 0.070$ can be considered as quantifying the extent of the organization of the system at the onset of the critical stage. In addition, in this Chapter, we present the natural time analysis of the avalanches observed in laboratory experiments on three-dimensional piles of rice getting progressively closer to the critical state and on the penetration of the magnetic flux into thin films of high T_c superconductors. The results reveal κ_1 values around $\kappa_1 = 0.070$.

Chapter 9 deals with the natural time analysis of electrocardiograms and basically aims at identifying the risk of sudden cardiac death, which is a frequent cause of death and may occur even if the electrocardiogram seems to be strikingly similar to that of a healthy individual. Upon employing the fluctuations of the entropy in natural time, we find that sudden cardiac death individuals (SD) can be clearly distinguished from the truly healthy ones. Furthermore, by using complexity measures that quantify the change of the natural entropy fluctuations either by changing the time window length scale or by shuffling the "pulses" (heartbeats) randomly, we can achieve the classification of individuals into three categories: healthy, heart disease patients and SD. In addition, when considering the entropy change under time reversal, not only the SD risk can be identified, but also an estimate of the time of the impending cardiac arrest can be provided. Finally, a $1/f$ model in natural time is presented which is consistent with the progressive modification of heart rate variability in healthy children and adolescents. The model also results in complexity measures that separate healthy dynamics from heart disease patients as well as from SD.

For the reader's convenience, the figures and the tables that refer to others than those included in this monograph, begin with small "f" and "t". As for the Supplemental Material, cited as EPAPS document, it is freely available from `www.aip.org/pubservs/epaps.html`. Furthermore, bold face symbols correspond to vectors, as usual.

We would like to express our gratitude to the leading contemporary figure in Earth Sciences, i.e., Professor Seiya Uyeda (Member of the Japan Academy), who has gone through the text in several of our papers focused on natural time with meticulous care and generously offered his advice by making very useful suggestions.

Athens *P.A. Varotsos, N.V. Sarlis and E.S. Skordas*

Acronyms

AE	Approximate Entropy
AF	Atrial Fibrillation
AN	"Artificial" Noise
BK	Burridge & Knopoff
BTW	Bak–Tang–Wiesenfeld
CDFT	Cumulative Distribution Function Transformation
CH	Continuum Hypothesis
CHF	Congestive Heart Failure
CP	Critical Point
DCS	Deterministic Cantor Set
DFA	Detrended Fluctuation Analysis
DMM	DANSK (coil) Magnetometers
ECG	Electrocardiogram(s)
EQ	Earthquake
FA	Fluctuation Analysis
FBC	Free Boundary Conditions
fBm	fractional Brownian motion
fGn	fractional Gaussian noise
fLsm	fractional Lévy stable motion
GI-NOA	Geodynamics Institute National Observatory of Athens
G-R	Gutenberg–Richter
HRV	Heart Rate Variability
ICFMC	Ionic Current Fluctuations in Membrane Channels
iff	if and only if
i.i.d.	independent and identically distributed
INAGS	Ivanov–Nunes Amaral–Goldberger–Stanley
JMA	Japan Meteorological Agency
LT	Local Time
MF DFA	Multifractal Detrended Fluctuation Analysis
MSE	Mean Squared Errors
MT	Magnetotellurics

MW	Maxwell–Wagner
OBC	Open Boundary Conditions
OFC	Olami–Feder–Christensen
pdf	probability density function
p.i.i.d.	positive, independent and identically distributed
PSC	Pressure Stimulated Currents
PSDC	Pressure Stimulated Depolarization Currents
PSPC	Pressure Stimulated Polarization Currents
QRS	QRS complex in ECG
QT	QT interval in ECG
PBC	Periodic Boundary Conditions
PVC	Premature Ventricular Contraction
ROC	Receiver Operating Characteristics
RTS	Random Telegraph Signal
RR	beat to beat interval in ECG
SCEC	Southern California Earthquake Catalog
SCK	Smoluchowski–Chapman–Kolmogorov
SCS	Stochastic Cantor Set
SD	Sudden cardiac Death individual
SDNN	Standard Deviation of all Normal sinus RR intervals over 24 h
SE	Sample Entropy
SES	Seismic Electric Signals
STAI	Short-Term Aftershock Incompleteness
STD	Standard Deviation
SOC	Self-Organized Criticality
TIP	Time Increased Probability
USGS	United States Geological Survey
UT	Universal Time
VAN	Varotsos–Alexopoulos–Nomicos
WTMM	Wavelet Transform Modulus Maxima method
WWS	Worldwide Seismicity

Symbols

$\Gamma(x)$	the gamma function, i.e., $\Gamma(x) = \int_0^\infty t^{x-1}\,\mathrm{e}^{-t}\,\mathrm{d}t$
$\mathrm{Cov}(x,y)$	the covariance of the random variables x and y, i.e., $\mathrm{Cov}(x,y) = \mathscr{E}\{[x-\mathscr{E}(x)]\,[y-\mathscr{E}(y)]\}$.
$\delta(x)$	the usual Dirac's delta distribution function
δS	the standard deviation of the time series of the entropy in natural time
ΔS	the change of the entropy in natural time under time reversal
$\Delta\chi$	the change of the average value of natural time under time reversal, see Eq. (2.12)
$\Delta V/L$	SES amplitude
-e	the electron charge
e	the base of natural logarithms, i.e., $e = 2.71828\ldots$
ε_0	the dielectric permittivity of vacuum ($\varepsilon_0 = 8.85 \times 10^{12}$ F/m)
$\mathscr{E}(x)$	the expectation value of the random variable x
f_{exp}	the experimental sampling frequency or rate
$\Theta(x)$	the Heaviside unit-step function, i.e., $\Theta(x) = 0$ for $x < 0$ and $\Theta(x) = 1$ for $x > 0$
I	current intensity
$\Im(z)$	the Imaginary part of the complex number z
κ_1	the variance of natural time, see Eq. (2.39)
κ_u	the variance of natural time for a "uniform" distribution, $\kappa_u = 1/12$, see Subsection 2.4
$\kappa_{1,p}$	the most probable value of κ_1
k_B	Boltzmann constant
M_D	"duration" magnitude of an earthquake
M_L	Local magnitude of an earthquake
M_s(ATH)	$\equiv M_L + 0.5$
M_w	moment magnitude of an earthquake
$N(\mu,\sigma)$	the normal (Gaussian) distribution with mean μ and standard deviation σ
p_k	stands for the point probabilities in natural time, see Eq. (2.2)
p_n	stands for the n-th cosine Fourier coefficient of $p(\chi)$, see Eq. (2.44)
$\Pi(\omega)$ or $\Pi(\phi)$	normalized power spectrum in natural time

$\Re(z)$	the Real part of the complex number z
S	entropy in natural time
S_-	entropy in natural time under time reversal
S_u	entropy in natural time for a "uniform" distribution, $S_u \approx 0.0966$.
S_D	the surface area of the unit sphere in D dimensions
$\mathscr{N}$	the set of natural numbers
$\mathscr{Q}$	the set of rational numbers
Q_k	a quantity proportional to the energy released during the k-th event
$\mathscr{R}$	the set of real numbers
$\mathrm{Var}(x)$	the variance of the random variable x, i.e., $\mathrm{Var}(x) = \mathscr{E}\{[x - \mathscr{E}(x)]^2\}$
χ	the natural time
$\aleph_0$	the cardinality of $\mathscr{N}$
$\mathscr{Z}$	the set of integer numbers

Part I
Seismic Electric Signals

1. Introduction to Seismic Electric Signals

Abstract. In the early 1980s, Varotsos and Alexopoulos showed that when the pressure (stress) on an ionic solid reaches a *critical* value, a *cooperative* orientation of the electric dipoles (that anyhow exist due to lattice defects) may occur, which results in the emission of a transient electric signal. This may happen before an earthquake since the stress gradually increases in the focal region before rupture. Thus, a detailed experimentation started in Greece in 1981, which showed that actually transient variations of the electric field of the Earth, termed seismic electric signals (SES), are observed before major earthquakes. In the meantime, several other SES generation mechanisms have been proposed. The field experiments revealed the physical properties of SES including those that SES can be observed only at certain points of the Earth's surface called "sensitive points" and that their amplitude is interrelated with the magnitude of the impending earthquake. Each sensitive station enables the detection of SES only from a restricted number of seismic areas, a phenomenon termed "selectivity effect", which provides the basis for the determination of the epicenter of an impending earthquake. These physical properties can be theoretically explained on the basis of Maxwell equations, if we take into account that the earthquakes occur by slip on pre-existing faults and that the faults constitute conductive paths (electrical inhomogeneities) in the solid Earth's crust. Finally, the observed SES activities, which are series of SES observed within a short time, as well as the associated magnetic field variations, exhibit *scale invariance* over four orders of magnitude. This is consistent with the original model suggested by Varotsos and Alexopoulos for the SES generation according to which SES should be characterized by *critical* dynamics which is always accompanied by scale invariance.

1.1 Data collection and the telemetric network

The experimental study of seismic electric signals (SES) which are low frequency ($\lesssim 1$ Hz) electric field variations that precede earthquakes (EQs), started almost thirty years ago in Greece and was motivated by solid state physics aspects, namely the thermodynamics of defects in solids. These have been developed by Varotsos and Alexopoulos during the

1970s in a series of publications (see Ref. [129] and references therein) and led to the theoretical possibility that the earthquake rupture could be preceded by transient electric signals(§ 1.6.2). In order to check its validity, continuous measurements of the electric field of the Earth have been carried out in Greece since 1982 (note that temporary field experiments had already started in 1981 by Varotsos, Alexopoulos and Nomicos [132], from the initials of which this study is also known as the VAN method). A telemetric network of eighteen measuring stations (via leased telephone lines) was completed in the end of 1983 [127, 128]. Data were transmitted to the central station located at an Athens suburb, labeled GLY, on a real-time basis. Technical details on this data transmission system can be found in Refs. [74, 75, 125]. The sites of the stations operating at that time are shown in Fig. 1.1, while the present configuration of the telemetric network consists of the stations depicted in Fig. 1.2.

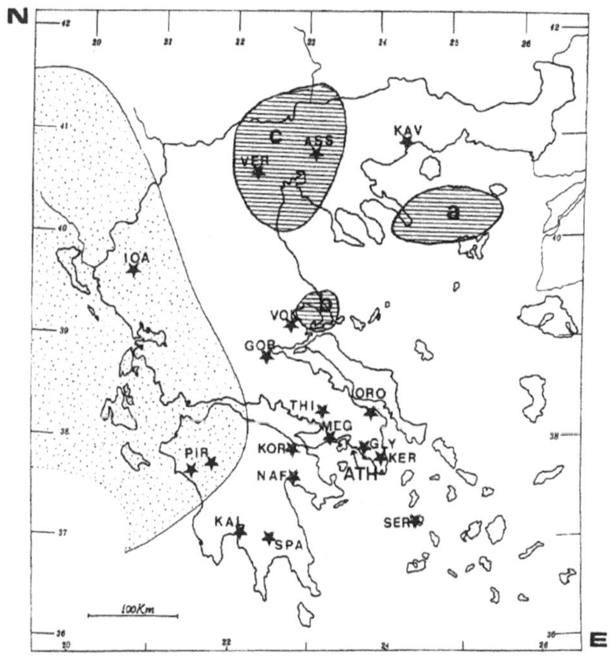

Fig. 1.1 The sites of the 18 telemetric stations in Greece installed in 1983. The western shaded area corresponds to the selectivity map of IOA, but a more updated version is given in Ref. [137]. The shaded regions a, b and c (that may be joined to a single larger area) refer to seismic areas from which SES have been recorded at ASS. These selectivity maps need gradual completion (see the text). The boundaries of the IOA selectivity map are yet uncertain. Taken from Ref. [138].

Experimental details were published in Varotsos and Lazaridou [140] and in Varotsos, Alexopoulos and Lazaridou [131] which are reviewed in Varotsos [125]. Therefore, only a brief description will be given here.

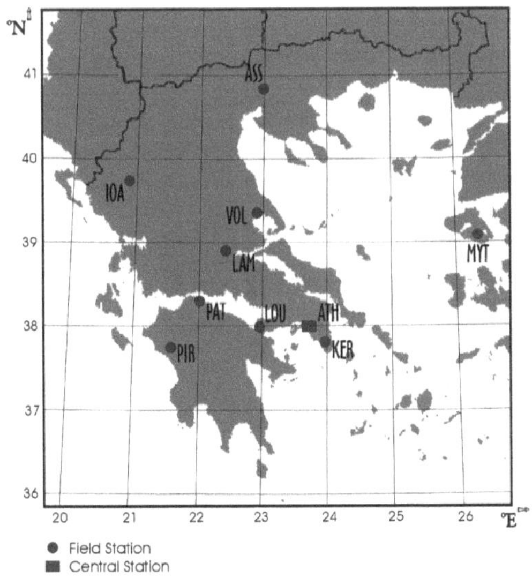

Fig. 1.2 Map showing the sites of the stations of the real-time telemetric network currently operating in Greece.

Non polarizable electrodes $Pb/PbCl_2$ at a depth of 2 m are used and the potential difference between two of them (that constitute a measuring electric dipole) is measured. A minimum of eight (horizontal) measuring electric dipoles was initially installed at each station; some of these dipoles have lengths (L) between 50 m and 400 m and are called *"short dipoles"*, while others have appreciably longer lengths (usually between 2 and 20 km) and are called *"long dipoles"*.

A minimum of four short dipoles was installed in perpendicular directions (usually along EW and NS), e.g., two parallel short dipoles with unequal lengths in the EW direction and two others in the NS direction. No common electrode should be used (§ 1.2.2). As for the long dipoles, their sites were carefully selected (see Varotsos and Lazaridou [140] and Appendix II of Varotsos, Alexopoulos and Lazaridou [131]), so as to allow the distinction between true SES and *"artificial"* noises coming from man-made sources lying up to several kilometers away from the station (see Section 1.2 and § 1.2.3).

In 1990, beyond the aforementioned real-time data collecting system, dataloggers (Campbell 21X connected to a portable PC) were installed at several stations to collect data with sampling rate $f_{exp} = 1$ sample/sec. These data were finally stored only during SES collection, and during the period extending from several minutes before a significant earthquake (EQ), until a few minutes after. The averages, taken every 20 seconds (note that initially it was 1 sample/10 sec), were transmitted to the central station (once or twice per day) through dial-up. Typical example of the configuration of the electrodes is given in Fig. 1.3, for the station IOA in northwestern Greece. Furthermore, at each of the four stations ZAK (which is located at Zakymthos island in western Greece, the closest island to PIR), IOA, KER and ASS, the vertical electrotelluric component was also measured

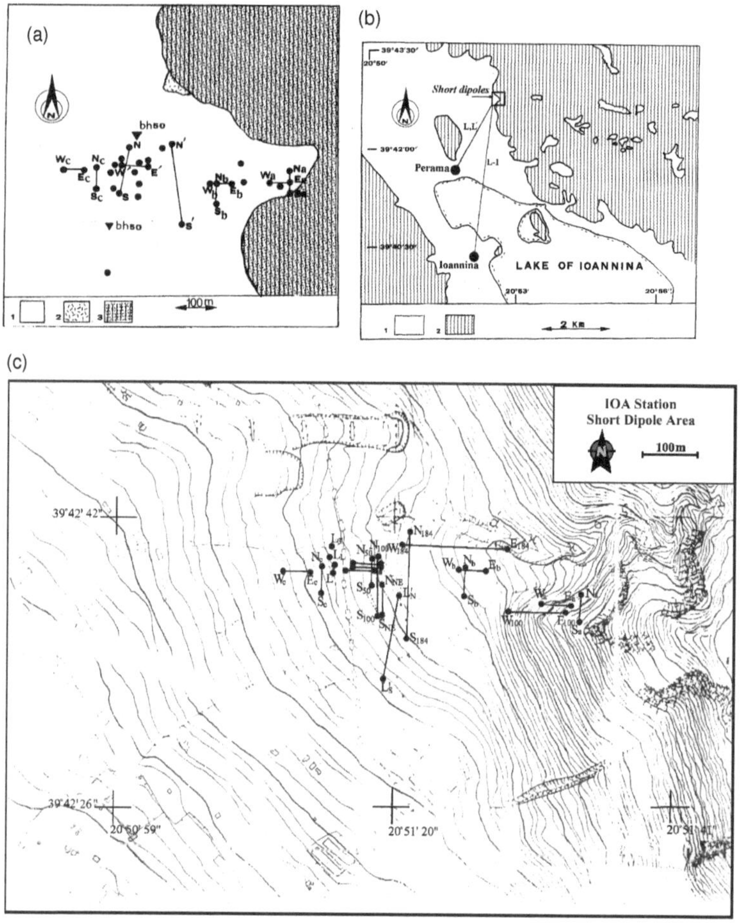

Fig. 1.3 Current configuration of the short (a) and long (b) dipoles at IOA, while (c) depicts a more detailed view of the short dipoles array. In (a), the two solid triangles (bh50) stand for the two boreholes (with depth $h = 50$ m). In (a) and (c), the short dipoles with subscripts a, b, c are those located at the sites termed "A", "B" and "C", respectively. In the real-time telemetric network the data of the following 8 (horizontal) electric dipoles are collected: the three long dipoles L, L′ and L-I depicted in (b) as well as the five short dipoles lying between the sites "B" and "C" in (c), i.e., the two EW dipoles (with lengths ~50 m) and the three NS dipoles $N_{50}S_{50}$, $N_{100}S_{100}$ and $N_{184}S_{184}$ (with lengths around 50 m, 100 m, and 184 m, respectively). In the datalogger, the data of the 3 coil magnetometers (DMM) along EW, NS and Z (vertical) together with the following 13 electric dipoles are collected: two short dipoles along EW and NS at the sites "A", "B" and "C" (labeled with subscripts a, b and c, respectively), two additional short dipoles located at site "B" but rotated by 22° counterclockwise with respect to the EW and NS directions, respectively, two vertical dipoles at the boreholes, and three long dipoles L, L′ and L-I (occasionally labelled L'_s-I). Concerning the geology in (a): (1) alluvial deposits, (2) flysch of the Ionian unit, (3) limestones; in (b): (1) quaternary sedimentary deposits, (2) alpine formation of the Ionian unit (mainly limestones). Reprinted from Ref. [125], Copyright (2005), with permission from TerraPub.

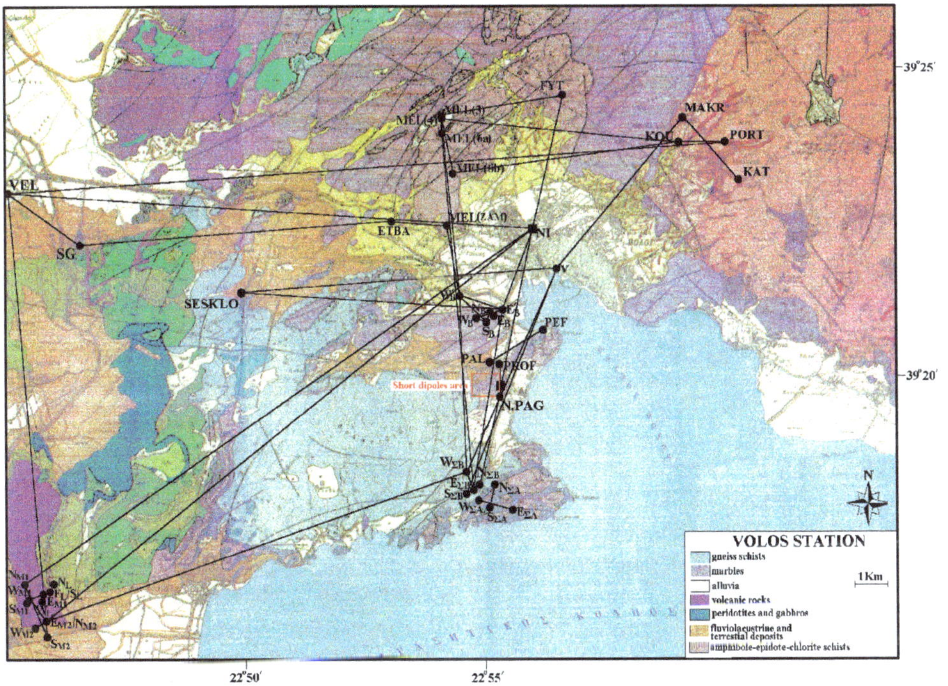

Fig. 1.4 Current configuration of the measuring short and long electric dipoles at the station VOL. Taken from Ref. [151].

by using pairs of electrodes installed in each of two independent boreholes with depths of around 50 m (e.g., see Refs. [1, 135, 125]). The variations of *both* fields, electric and magnetic, are studied since the magnetic field was also continuously measured at four stations, i.e., at IOA, LOU, ASS and VOL, using coil magnetometers. In 1995, beyond the aforementioned two independent measuring systems (i.e., the real-time telemetric network with 8 channels per station, and the dataloggers with 8 or 16 channels per station) instrumentation was increased more (see Section 1.4 of Ref. [125]).

In particular, since 1995, beyond the coil magnetometers, several tens of short and long measuring electric dipoles have been operating around each of the stations depicted in Fig. 1.2.

As an example, Fig. 1.4 depicts the current configuration of the measuring short and long dipoles at VOL (central Greece). Details for the low pass filters used for the electric field measurements as well as for the calibration of the three component DANSK coil magnetometers (DMM) can be found in the Appendix, see also pp. 1–2 and 24–29 of Ref. [125], or in Refs. [149, 161, 157, 159].

1.2 Distinction of SES from noise

The simultaneous operation of multiple short and long measuring electric dipoles at each station is *necessary* for the SES recognition.

Frequent electric field variations, termed magnetotelluric (MT) variations appear (practically) simultaneously at all stations of the network because they are induced by small variations of the geomagnetic field.

An electric disturbance is classified as a SES after it has met *all* of the following four criteria [131, 140], which are currently applied *in addition to* those based on the natural time concept that are summarized in Section 4.10 (the same holds in the case of significant data loss, see Chapter 5):

(1) The SES can be distinguished from MT variations, because the latter appear at all stations (practically) simultaneously, while the former do *not* (due to the "selectivity" phenomenon that will be explained later in § 1.3.4).

(2) The SES must appear simultaneously on the short and long dipoles at the station(s) concerned.

(3) The SES should obey the criterion $\Delta V/L =$ constant for the short dipoles oriented in the same direction (if they are *not* located on locally inhomogeneous ground).

(4) The polarity and amplitude of the SES on the short and long dipoles must be compatible with a *distant source* assumption: the projection onto the long dipole of the $\Delta V/L$ vector calculated from the short dipoles must have the same polarity and comparable amplitude with the observed signal on the long dipole. Thus, the criterion $\Delta V/L \approx$ const. should (approximately) hold when considering two dipoles, a long dipole and a short one that are parallel.

An independent confirmation of the objectivity of these four criteria was made by Nagao, Uyeshima and Uyeda [73].

Furthermore, note that beyond the aforementioned four criteria, there are two additional ways of discriminating SES from "artificial" noise which make use of: (a) the time-difference between the "arrivals" of the electric field and the magnetic field (see § 1.3.6.1), (b) the difference of the time evolutions (rise times) of the signal recorded at two directions (see § 1.3.5; see also section 13.2 in Ref. [125]).

We now describe below how the SES are discriminated from noise of various sources.

1.2.1 Distinction of SES from magnetotelluric (MT) changes

When a network of several electrotelluric stations is operating, it is difficult for the MT changes (which is alternatively called here "MT noise") to be misinterpreted as SES. This is so because, as mentioned, MT appear simultaneously at all the stations, in contrast to the SES which are recorded only at a restricted number of them [127] (in most cases, in a non-dense network, the SES are recorded *only* at one station). The following procedures for the distinction between SES and MT are also applied:

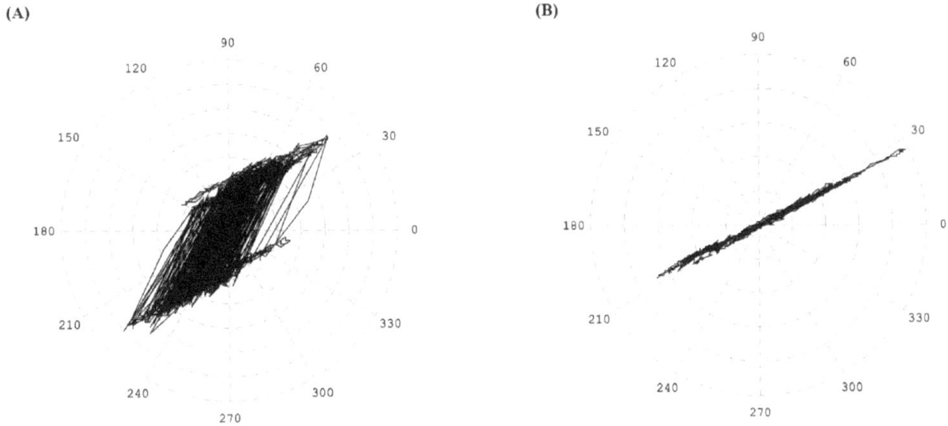

Fig. 1.5 The electric field polarization of the SES activity on April 19, 1995 (A), and that of the MT variations (B) at the site "B" of IOA; the latter is almost directed along EW, as confirmed by long time recordings as well. Attention is drawn to the point that the origin of the axes has been displaced to the average value of the electric field. Reprinted from Ref. [125], Copyright (2005), with permission from TerraPub.

(a) *Conventional MT method*: After having determined for each site, the impedance tensor that interconnects the variations of the magnetic and electric field, a "real-time" subtraction of the MT noise from the VAN records has been developed and tested with satisfactory results (see chapter 2 of Ref. [125]).

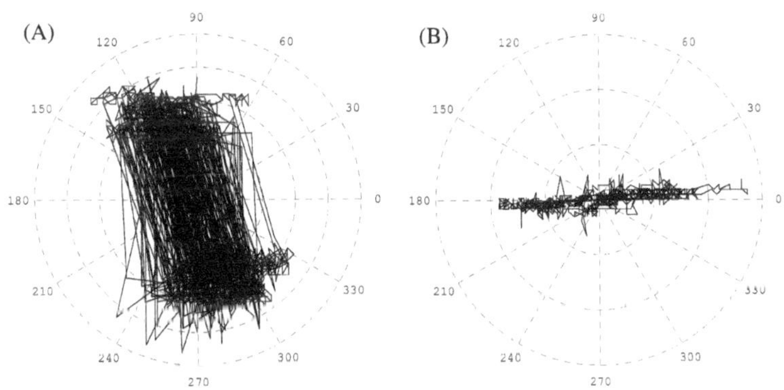

Fig. 1.6 The same as Fig. 1.5, but for the site "C" of IOA. Reprinted from Ref. [125], Copyright (2005), with permission from TerraPub.

(b) *Detection of a SES in the direction in which the MT noise is minimum*: Let us consider for example the case of the station IOA, at which Figs. 1.3(a) and (c) depict the areas termed "A" ,"B" and "C" of the short dipoles arrays. At "B", measurements of the electric and magnetic fields (see Fig. 1.5(B)) led to the conclusion that the MT variations are least along the direction ~N22°W (cf. a highly anisotropic 2D regional geoelectrical structure with a strike angle N40°W is found in Ioannina area, see section 2.5 of Ref. [125]). Thus, two more dipoles (labeled $N_{b'}$ $S_{b'}$ and $E_{b'}W_{b'}$) have been installed at "B", which were oriented along to NS(−22°) and EW(−22°), respectively. (It will be hereafter called site "B′".) This reflects that an SES can be easily recognized at site "B" on NS(−22°), in which direction the MT variations are least (provided of course, that the polarization of the MT significantly differs from that of the SES under consideration; see also § 1.3.3). By the same token two additional short dipoles had been installed in the neighboring area "C" (see Figs. 1.3(a) and (c)), where the NS direction was found experimentally (Fig. 1.6(B)) to be almost the direction at which the MT disturbances are appreciably small.

Thus, in short, the SES recognition becomes easier in a direction in which the MT noise is minimum.

1.2.2 Distinction of SES from noise of electrochemical origin

This type of noise, which is usually ascribed to a change in the contact potential between the electrodes and the ground, e.g., due to rain, can easily be recognized when parallel dipoles for each measuring direction are installed [127]. This noise is usually not recorded simultaneously at the parallel dipoles, because, as mentioned, care is taken to have independent electrodes. In the rare case that it is simultaneously recorded, the variations ΔV of the potential difference do not scale (in a *homogeneous* ground) with the length of the short dipoles in a given direction, i.e., with $\Delta V/L \neq$ constant. The long dipoles are less affected by the electrochemical disturbances, when compared to the short dipoles (e.g., see Ref. [129]).

1.2.3 Distinction of SES from "artificial" (man-made) noise. The $\Delta V/L$ criterion

The proper use of the $\Delta V/L$ criterion (see Section 1.2) can lead to the noise recognition. In particular, depending on the location of the emitting noise source, the application of this criterion can be made as follows:

(1) A noise source lying in the immediate vicinity of the short dipoles can be easily recognized, because a cultural signal N does not generate the same field strength $\Delta V/L$ in neighboring parallel (short) dipoles of different lengths. This also excludes any electrochemical noise, arising from electrode-instabilities, e.g., due to rain.

Multiple neighboring short dipoles exclude the electrochemical and the local "artificial" noise.

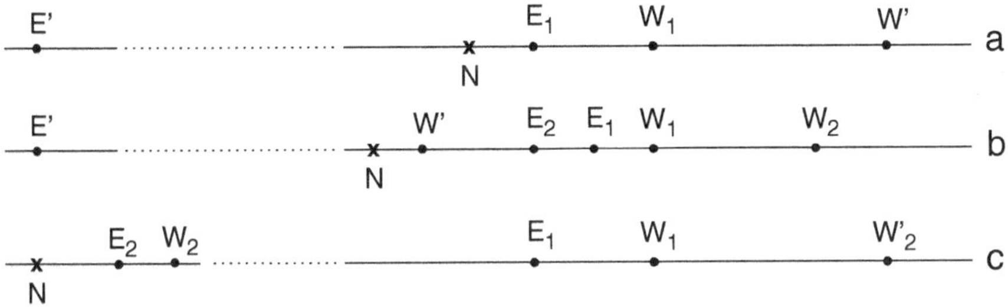

Fig. 1.7 Configuration of short (unprimed) and long (primed) dipoles suggested by Varotsos and Lazaridou [140] for an easy recognition of noise. N = noise source.

(2) In the case of a noise source the location of which is known and lies a few km (up to several km) away from the measuring site, the installation of a single long dipole in combination with a short parallel dipole is usually sufficient for the noise recognition, under the following condition: one of the two electrodes of the long dipole should be located close to the measuring site, while the other, i.e., the remote one, should lie on the same side (with respect to the measuring site) with the noise source. As for the exact location of this remote electrode, the following two configurations are recommended:

(i) If the noise source N lies closer to the measuring site than the remote electrode as in Figs. 1.7(a),(b), or case II of Fig. 1.8, the noise is easily recognized, because it gives signals with opposite polarities on the long dipole and the parallel short dipole.

(ii) If the noise source lies close to the remote electrode, the noise is again recognized, because the ratio ($\Delta V/L$ of the long dipole)/($\Delta V/L$ of the short dipole) of the $\Delta V/L$ values differs drastically from unity (e.g., we compare the long dipole E_2W_2' and the short dipole E_1W_1 in Fig. 1.7(c)).

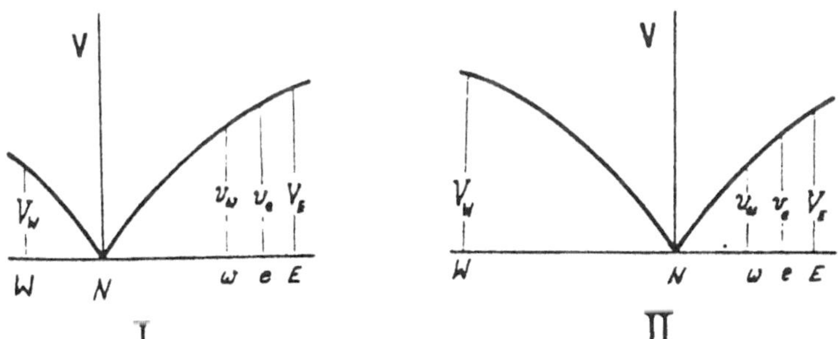

Fig. 1.8 Configuration of the signal strength of short and long dipoles as a function of distance suggested by Varotsos et al. [131] for an easy recognition of noise. The electrodes e and w correspond to the short dipole, while E and W to the long dipole. In configuration I, a disturbance from noise source N (or a SES) is recorded on both dipoles ew and EW, with the same polarity. In II, a disturbance from the noise source N give signals of different polarities on the dipoles ew and EW, in contrast to SES. Taken from [131].

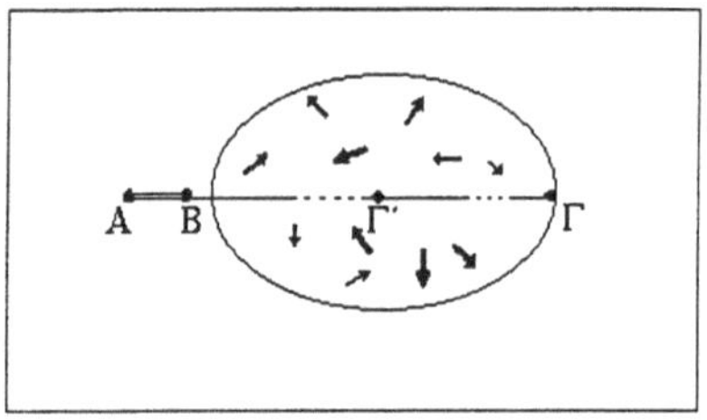

Fig. 1.9 The "optimal" configuration of a short dipole AB and two long dipoles $A\Gamma'$ and $A\Gamma$, for an easy recognition of noise; the emitting noise source is assumed to lie within the elliptical area. Taken from Ref. [145].

(3) In the case of "artificial" sources (up to several km away from the measuring site), the exact location of which is not known, two (parallel) long dipoles with non-equal lengths should be installed (in addition to the short dipole array). One of the short dipoles should be parallel to the long ones of which one of their electrodes has to be close to the measuring site. If the candidate area of "artificial" sources can be roughly estimated, say as an ellipse, care should be taken in order its major axis to coincide, if possible, with one of the long dipoles (the other long dipole may have a smaller length, e.g., by a factor of 2, or so, see Fig. 1.9). For such a configuration, any noise originating from the candidate area, cannot give equal $\Delta V/L$ values (within 50% or so) at both the long and the short dipole(s).

Detailed calculations which explain the aforementioned recommended configurations of the short and long dipoles are presented in chapter 7 of Varotsos [125], a review of which is given here in § 1.7.7.

In the case of lateral inhomogeneities, however, the $\Delta V/L$ criterion may not be applicable at all [140]. In this case, as well as in all other cases where the eventual inhomogeneities are not known beforehand, the SES activities (i.e., a sequence of SES within a relatively short time, see below) can be distinguished from "artificial" (man-made) noises by analyzing the signal in *natural time* and then applying the criteria summarized in Section 4.10. These criteria stem from the fact that solely SES activities are characterized by *critical dynamics (infinitely ranged temporal correlations)*, while the "artificial" noises do not.

We also note that *if* a noise source remains geographically fixed, the relevant noise on the records can be permanently eliminated by applying a technique developed in Ref. [140] (see also pp. 6, 7 in Varotsos [125]).

1.3 SES physical properties

Seismic electric signals are classified into two types: (a) single SES, i.e., a single transient change of the electric field, (b) SES activity, i.e., many *one-sided* SES within a relatively small time. Examples are given in Figs. 1.10 and 1.11. These examples, show SES activities of duration up to a few hours. During recent years, however, SES activities of appreciably longer-duration, i.e., from several hours to 1 week have been recorded in Greece (see §1.4.3 and Section 4.11). These will be hereafter termed long duration SES activities. In Japan, beyond several SES of duration comparable to that recorded in Greece, an SES activity lasting for around 2 months has been also reported [119, 120] that was related with volcanic-swarm activity; see Section 7.4. In addition, an SES activity lasting for more than 2 months [83] has been also recorded before the Guerrero-Oaxaca EQ ($M_s = 7.4$) that occurred on September 14, 1995, in southern Mexico at a measuring station lying close to Acapulco, i.e., at a distance around 110 km from the EQ epicenter [83, 36, 84].

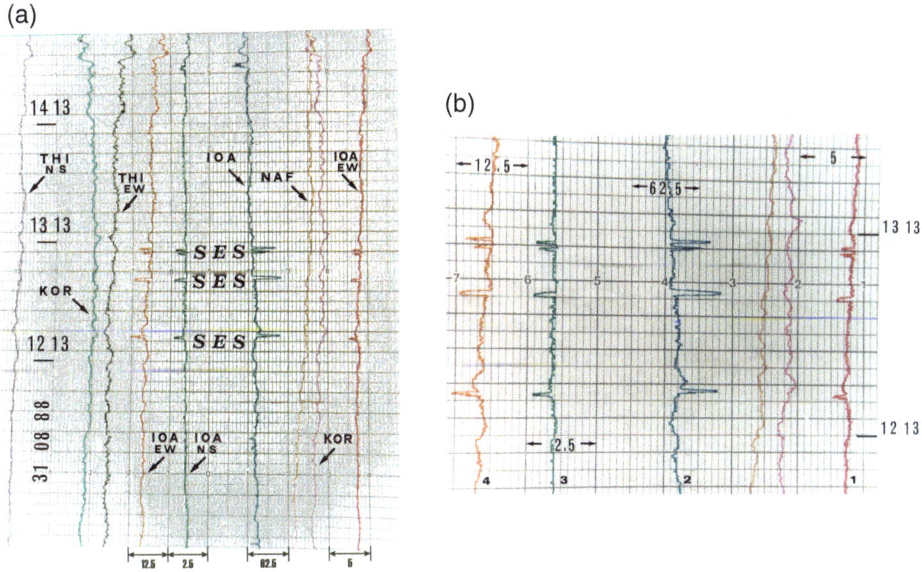

Fig. 1.10 (a): SES activity at IOA (records from the real-time telemetric network) on Aug. 31, 1988, which preceded the destructive EQs at Killini-Vartholomio (i.e., approximately at 38.0°N 21.0°E) that occurred on Sep. 22 and Oct. 16, 1988. This SES activity provided the basis for the public warning issued by Haroun Tazieff on September 3, 1988, the social impact of which has been studied in detail in Ref. [164]. Taken from Ref. [140]. (b): Excerpt of (a) taken from Ref. [131]. The corresponding directions and lengths of the dipoles are as follows: red (EW, $L = 47.5$ m), green (NS, $L = 48$ m), orange (EW, $L = 181$ m). The dipole $N^{\underline{o}}2$ labeled IOA (blue) is the long dipole labeled "L" in Fig. 1.3(b); the apparently reversed polarity of the SES on the latter dipole is due to the connection of this dipole to the recorder, having been intentionally reversed, as explained by Varotsos and Lazaridou [140]. The horizontal scale is in mV, while the vertical is time in UT.

1.3.1 Lead time of SES. Other electrical precursors

The lead time of SES lies, in general, between several hours and several weeks. In particular:

(a) For single SES, the lead time is usually around 11 days [123] or smaller [127].

(b) For SES activities: the study of several cases observed until 2000 (compiled in Ref. [125]) led to the following empirical rule [141]:

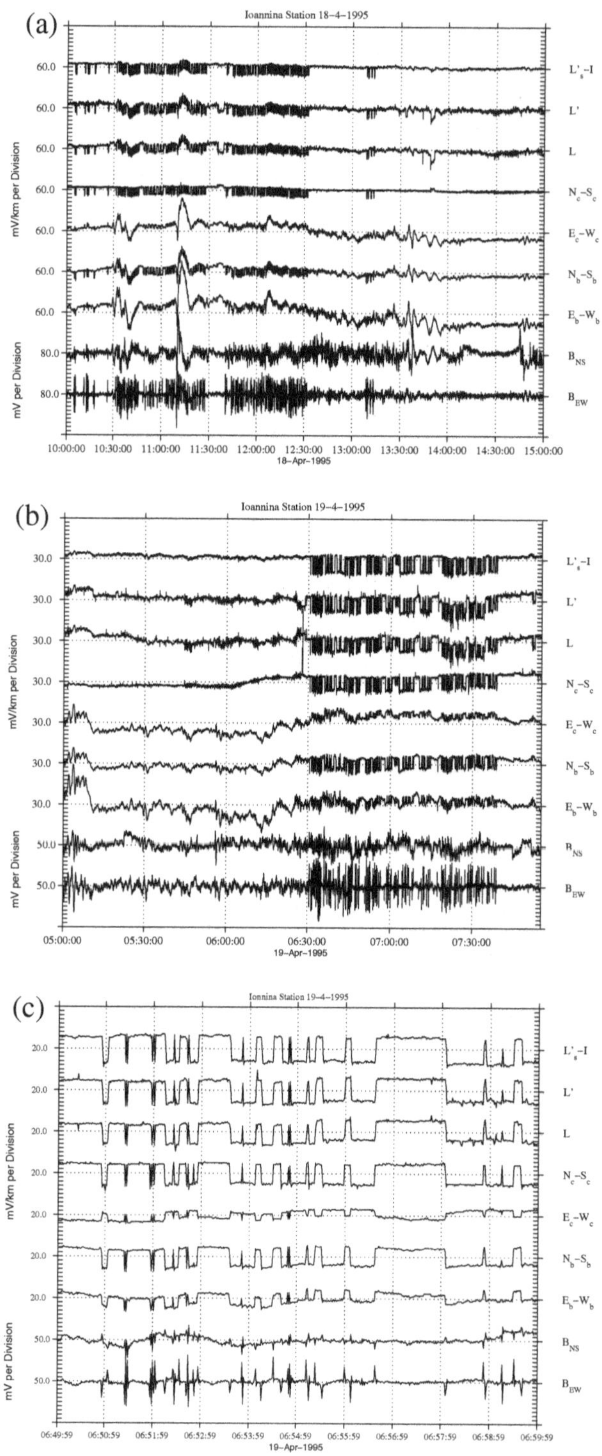

Fig. 1.11 The two SES activities recorded by datalogger at IOA with a sampling rate $f_{exp} = 1$ sample/sec before the M_w6.6 EQ at Grevena-Kozani that occurred on May 13, 1995 (see § 7.2.1). In § 1.3.6, a short description of the recordings at various channels is given. (a): April 18, 1995. (b): April 19, 1995. Figure (c) is an excerpt of (b), showing the variations in an expanded time-scale. The range (10 ticks) of 80 mV or 50 mV in the lower two channels of (a) and (b), (c) correspond to 0.4 nT/sec or 0.25 nT/sec, respectively, for a constantly increasing magnetic variation (see Appendix). Taken from Ref. [150].

A time period of at least 3 weeks elapses between the initiation of the SES activity and the occurrence of a strong EQ (small shocks may start earlier), see Fig. 7.15. The strongest EQ usually occurs during the fourth week, see the middle case in Fig. 7.15; otherwise smaller EQ(s) with magnitude around 5-units appear during this week and the strongest EQ occurs after an additional period of 2–3 weeks, see the upper and the lowest case in Fig. 7.15.

Some later examples, however, e.g., the SES activity that preceded the M_w6.5 EQ in the Aegean sea on July 26, 2001 (see § 7.2.3), revealed that the lead time Δt *may occasionally be longer*. Upon the introduction of natural time in 2001, however, a methodology has been developed which enables the determination of the time-window of an impending mainshock with an accuracy usually around a few days to 1 week. Chapter 7 is reserved to explain this methodology along with several examples.

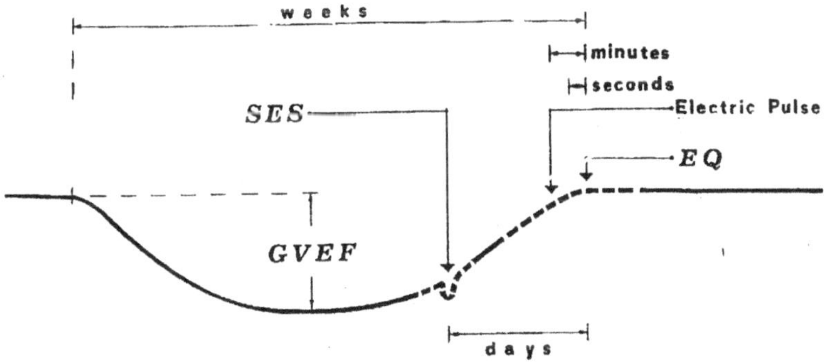

Fig. 1.12 Schematic representation (not to scale) of the three types of electrical precursors (GVEF, SES and electric pulses) observed in Greece. Modified from Ref. [131].

We clarify that, beyond the SES, two other types of electrical precursors have been detected in Greece as summarized in Fig. 1.12. First, a gradual variation of the electric field of the earth (GVEF) which is a transient anomaly of long duration (of the order of one month) with an amplitude one order of magnitude larger than that of the SES [128, 129]. Second, electric pulses of short duration, labeled "electric pulse" in Fig. 1.12, which appear some minutes before the earthquake occurrence [159, 129]; see also § 1.3.7.

1.3.2 Interrelation between SES amplitude and EQ magnitude

For SES registered at a given station and originating from a given seismic area, their amplitude expressed as $\Delta V/L$ (for a dipole with given orientation) scales with the magnitude M according to the relation [127, 129, 140]

$$\log_{10}(\Delta V/L) = (0.32-0.37)\mathrm{M}+\beta. \tag{1.1}$$

Equation (1.1) reflects that when the EQ magnitude increases by 1-unit, the SES amplitude becomes almost double.

The plots for the dipoles of two different orientations have the same slope but different intercepts. This is schematically shown in Fig. 1.13.

The following three comments are in order.

First, the $\Delta V/L$ value is of the order of 10 mV/km, for EQs of magnitude $\mathrm{M} \approx 6.0$–6.5, when measured at epicentral distances of around 100 km.

Second, the aforementioned relation, which was found empirically by Varotsos and Alexopoulos [127] has the form:

$$\log_{10}(\Delta V/L) = a\mathrm{M} + \beta, \tag{1.2}$$

where the constant a is approximately $a \approx 0.3$–0.4 for *all* measuring cites (*universal*).

Third, let us now consider that the seismic energy E released during the EQ is proportional to the seismic moment M_0 and that M_0 is interrelated with the moment magnitude M_w through [35] (see also Section 6.1)

$$M_0 \propto 10^{cM_w} \tag{1.3}$$

where the constant $c = 1.5$. Thus, the empirical relation (1.2), in view of Eq. (1.3), is in fact a power law relation that interconnects $\Delta V/L$ with the energy E, thus pointing to the conclusion that the future focal region has a *fractal geometry* (§1.5.1). This is consistent with the early suggestion by Varotsos and Alexopoulos [127] (see p.92) that the *universal a-value stems from the geometry of the SES emitting source.*

In addition, when considering (§ 1.5.2) that *criticality* is always associated with *fractality*, Eq. (1.2) is consistent with the SES generation model of pressure stimulated currents (§ 1.6.2) that motivated the SES research and assumes that SES is emitted when the stress reaches a *critical* value in the future focal region. Note that a number of independent workers, e.g. Sornette and Sornette [102] (see § 1.6.9) and Surkov, Uyeda, Tanaka and Hayakawa [109] (see § 1.6.8) have also provided a theoretical justification of Eq. (1.2) based on concepts of critical phenomena.

1.3.3 SES polarity and the ratio of the two SES components

For a given short dipole (e.g., EW or NS) a SES from a given seismic area always has the same algebraic sign. For each pair "seismic area-station", the ratio of the amplitudes in the EW and NS directions, i.e. $(\Delta V/L)_{EW}/(\Delta V/L)_{NS}$, remains the same for EQs of different magnitudes but from the same seismic area (Fig. 1.14).

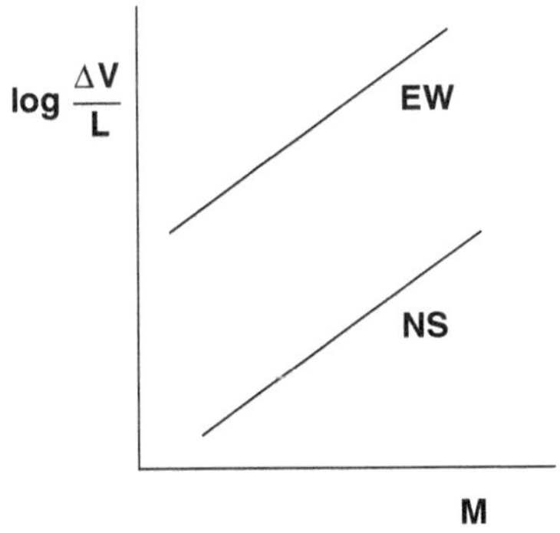

Fig. 1.13 Schematic representation of the variation of $\log_{10}(\Delta V/L)$ versus the magnitude M, for SES coming from the same seismic area and recorded at a given sensitive station. Taken from Ref. [140].

The above property seems to hold when the EQ source parameters from the seismic region under discussion remain the same. In the case that the EQ mechanism changes, a reversal of the polarity may be observed [151].

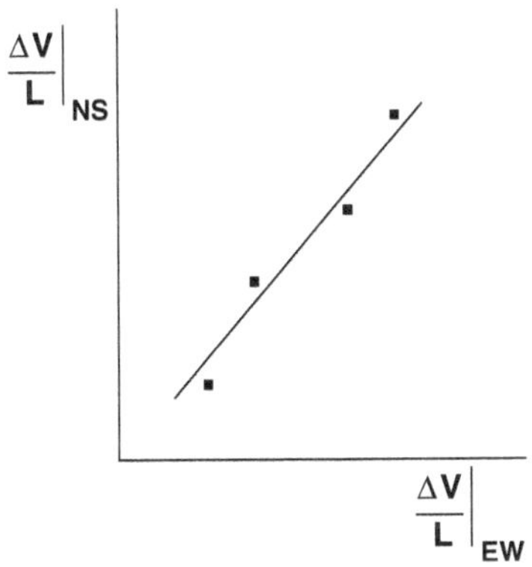

Fig. 1.14 Schematic representation of the ratio of the two SES components, for a given pair "seismic area – (SES sensitive) station". It remains almost the same. Taken from Ref. [140].

The ratio of the two SES components is, in general, different from that of the MT-polarization. We present two characteristic examples. Let us consider the case for the SES activities that preceded the M_w6.6 EQ of Grevena-Kozani. In Fig. 1.11(a), we give the recordings of April 18, 1995, at IOA, at the sites termed "B" and "C" (see the maps in Figs. 1.3(a) and (c)). An inspection of this figure reveals the difference between MT disturbances and SES recordings. First, the MT disturbances in the NS direction (e.g., 09:30 UT on April 18), measured at site "C", are appreciably smaller than those in NS measured at site "B" (the opposite holds for the EW direction), but the SES amplitude at N_cS_c is larger than that at N_bS_b. Second, when comparing the recordings at the long dipoles L'_s-I and L' (or L), we find that they respond differently to MT disturbances than to SES activities.

1.3.4 SES sensitive sites. Selectivity effect

SES cannot be observed at all points on the Earth's surface but only at certain points [127, 128, 130] called "sensitive sites" [140].

Therefore the installation of a station appropriate for SES collection *should necessarily be preceded by a tedious experimentation* to find a sensitive site [131]. In short, a number (e.g., 10) of temporary low-noise stations are installed and only after a long period, i.e., after the occurrence of several significant EQs from a given seismic area, can we select the most appropriate site (if any). Although general rules cannot be drawn, it seems that [131] the areas which are more likely to be sensitive are: (a) the vicinity of a major fault, (b) regions of crystalline rocks close to large heterogeneities, such as geological contacts with significantly different conductivities and (c) areas with strong local inhomogeneities. The stations depicted in Fig. 1.2 have been selected to be SES sensitive after long experimentation. As an example, the procedure we followed for the selection of IOA sensitive site has been described in detail by Kondo, Uyeda and Nagao [53].

Selectivity effect. It refers to the experimental fact that a sensitive station is capable of collecting SES *only* from a restricted number of seismic areas [140, 131].

The following may happen: a given station S_A situated at A may be able to record SES originating from a seismic area B but not from an area C, even when the distance between A and C is significantly smaller than that between A and B [127, 128, 130]. A map showing the seismic areas that emit SES detectable (for EQs above a magnitude threshold) at a given station is called "*selectivity map* of this station". In Fig. 1.1 the western large dotted area corresponds to the selectivity map of IOA, while the shaded regions a, b and c (which are likely to be joined to a single larger area) refer to the selectivity map of ASS.

The construction of the selectivity map of a given station takes a long time. There are, in principle, three categories of seismic regions with respect to a given SES station. The first two categories refer to those regions at which strong EQs occurred during the operation of the station: the one category refers to the regions which gave detectable SES at the station (thus belonging to the selectivity map) and the other which did not. There is, however, a third category, referring to those seismic regions that were not seriously activated during the operation of the station. Thus, the completion of the selectivity map of a station needs gradual improvement as more data are collected. For example, the seismic region of Kozani-Grevena until 1995 was *not* initially included in the selectivity map of IOA depicted in Fig. 1.1; however, after the M_w6.6 EQ on May 13, 1995, with an epicenter at 40.2°N 21.7°E, it became included (see Refs. [141, 137]). Thus, the construction of the selectivity map requires a long time operation of a (sensitive) station. The procedure is described in detail in Ref. [131] (see also pp. 13–14 of Ref. [125]).

An explanation for the selectivity effect has been suggested by Varotsos and Alexopoulos [129] (see Section 1.7 and Ref. [131]).

This explanation is supported by analytical solutions of Maxwell equations (§ 1.7.4), as well as by numerical ones (§ 1.7.5), and suggests that selectivity is a natural consequence of the fact that EQs occur by slip on faults, which are appreciably more conductive than the surrounding medium.

Moreover, a detailed experimentation summarized in section 1.4 of Ref. [125] has been carried out in Greece aiming at a better understanding of the SES selectivity effect. This experimentation included: (a) the additional installation of several tens of short and long dipoles in order to determine the spatial extent of each of the SES sensitive sites, and (b) a detailed magnetotelluric investigation around each SES sensitive station. For example, the results showed [137] that the SES sensitive site at IOA extends to a zone with a width of about a few km and a length of the order of $\approx$10 km. These dimensions seem to be compatible with the results obtained from analytical and numerical solutions of Maxwell equations, e.g. see Fig. 1.38(c) and § 1.7.5.3.

1.3.5 Determination of the epicenter and magnitude of an impending mainshock from the SES data

In a non-dense network, the SES of an impending EQ is recorded at one station *only*. In this case, the following procedure is applied [140]:

The determination of the epicenter and magnitude of an impending EQ needs the knowledge of the selectivity map of the station that recorded the SES. Then by considering the SES polarity and the value of the ratio $(\Delta V/L)_{EW}/(\Delta V/L)_{NS}$ (see § 1.3.3) obtained from the short dipoles array, we select the region(s) of the selectivity map which might have emitted the observed SES. Subsequently, from the $\log_{10}\Delta V/L$ vs M plot (see § 1.3.2) that corresponds to this concrete pair: "candidate seismic area-station", we estimate the magnitude.

We clarify that the additional use of long dipole information may significantly improve the epicenter estimation. Let us consider, as an example, one of the two (independent) long dipoles (L or L′) installed between the IOA station and the town of Perama, which are oriented along an angle $\hat{\theta} = $ N30°E (see Fig. 1.3(b)). Let us denote the electric field and potential difference by E and ΔV with relevant suffices. If we denote the NS and EW components of electric field variation measured by the short dipole by $E_{NS} = \Delta V_{NS}/L_{short}$ and $E_{EW} = \Delta V_{EW}/L_{short}$, the component of the electric field variation in the direction of the long dipole, $E_{long} = \Delta V_{long}/L_{long}$, would be: $E_{long} \approx E_{NS}\cos\hat{\theta} + E_{EW}\sin\hat{\theta}$, in the case that the resistivity structure is assumed homogeneous and isotropic. Then:

$$(\Delta V_{long}/\Delta V_{NS}) = (L_{long}/L_{short})[\cos\hat{\theta} + (E_{EW}/E_{NS})\sin\hat{\theta}]. \quad (1.4)$$

As $\hat{\theta}$ and (L_{long}/L_{short}) are constant, $(\Delta V_{long}/\Delta V_{NS})$ can be considered as a *sensitive measure* of (E_{EW}/E_{NS}), because it "amplifies" the latter almost by the ratio $(L_{long}/L_{short} \approx 20) \times (\sin\hat{\theta} = 1/2) \approx 10$. Since the underground structure is neither homogeneous nor isotropic, the observed ratio $(\Delta V_{long}/\Delta V_{NS})$ is considered to be merely a parameter sensitive to (E_{EW}/E_{NS}). (This ratio is termed "directional parameter" by Kondo, Uyeda and Nagao [53], who independently checked this procedure.)

By summarizing, since the ratio $\Delta V_{long}/\Delta V_{NS}$ is a sensitive measure for the ratio of the two SES components, the use of the ratio E_{EW}/E_{NS} deduced from the short dipoles array together with the ratio $\Delta V_{long}/\Delta V_{NS}$ improves the epicentral determination.

By the same token, in some cases, the ratio $\Delta V_{long}/\Delta V'_{long}$ of two long dipoles with unequal lengths may help towards distinguishing one among different candidate epicentral areas. This has been used for example in the text of the prediction of the EQ at Grevena-Kozani (May 13, 1995), that will be discussed in § 7.2.1.

We also emphasize the following. *If* the two components of the electric field detected on the Earth's surface are measured along appropriate directions, i.e., by two short dipoles, oriented along and perpendicular to the (local) current channeling [136, 152], they exhibit markedly *different* time evolutions. This stems from the fact that the electric field obeys *diffusion* type equations (in the low frequency range, see §1.8.1 and §1.8.3).

An analysis of the difference between the time evolutions of these two SES components can then reveal the *distance of the emitting source from the measuring station*(see Refs. [157, 151, 152]; see also pp. 275–280 of Ref. [125]).

1.3.6 Magnetic field variations associated with SES

The experiments show that small amplitude SES, e.g., those corresponding to magnitude $M \sim 5.5$ EQs at epicentral distances of the order of 100 km, are not accompanied by easily observable variations of the horizontal components of the magnetic field [141]. This does not imply that SES are not accompanied at all by magnetic field (B) variations, the existence of which is *obligatory* from Maxwell equations, but does imply that they are very small compared to those which produce (comparable) magnetotelluric electric signals.

On the other hand, *both* theoretical calculations (see § 1.7.6 and chapter 10 of Ref. [125]) and experiments (e.g., see Figs. 1.11(a),(b)) indicate that strong SES activities and in particular those related with EQs of magnitude M $\gtrsim$ 6.5 at epicentral distances ~100 km, should be accompanied by detectable magnetic field variations, e.g., $B \sim 10^{-1}$ nT.

This was actually observed (details are given in subsection 1.6.3 of Ref. [125]) for the SES activities detected at IOA before the occurrence of the M_w6.6 EQ at Grevena-Kozani on May 13, 1995 [138, 150, 161], to which we now turn.

Figures 1.11(a),(b) depict the SES activities recorded at IOA on April 18 and April 19, 1995, respectively. All variations were collected with the Campbell X21 datalogger by 20 ms integration to avoid 50 Hz noise with $f_{exp} = 1$ sample/s. The following nine (out of 16) channels are shown in the figures: two for the horizontal coil magnetometers DMM (see Section 1.1), labeled B_{EW} and B_{NS}, four for the short electric dipoles at sites "B" and "C" and three for the long electric dipoles L, L′ and L'_s-I (see Fig. 1.3). Note that "1 Hz" low pass filters were used in the measurements of the electric field variations of the three long dipoles and "10 Hz" (see Appendix) for the short dipoles at sites "B" and "C". An inspection of this figure reveals that there are obvious magnetic disturbances at DMM, which accompanied the SES activities. A part of the SES activities and the accompanied DMM recordings is presented, but in a more expanded time scale, in Fig. 1.11(c). These SES activities meet [138] the criteria for distinguishing true SES from noise (Section 1.2).

1.3.6.1 The time difference between the variations of the electric and magnetic fields

Varotsos et al. [161] studied in the aforementioned example of Figs. 1.11(a),(b) the cross-correlation values, calculated between the following two time series versus their time-difference Δt: one component of the electric field and one component of the horizontal DMM. The results, either for the SES activity on April 18, 1995 (Fig. 1.11(a)), or for that on April 19, 1995 (Fig. 1.11(b)), showed that [161]: the electric field variations *preceded* the magnetic ones (i.e., dB/dt since we used coil magnetometers) by a Δt of around 1–2 sec.

The fact that the SES electric field variations precede those of the magnetic field by a measurable time difference, i.e., of the order of 1 sec (for epicentral distances of the order of ∼100 km), *may be critical in distinguishing SES from "artificial" noises.* This is so, because for noises emitted from nearby "artificial" (man-made) sources, the "arrivals" of the variations of the two fields are simultaneous at the measuring station within experimental accuracy.

Such a distinction can be achieved, of course, when the epicentral distance is ≈ several tens of km. The value of the time-difference may also be used for the estimation of the epicentral distance of the impending EQ, which when combined with the selectivity map can significantly improve the epicenter determination.

1.3.6.2 The polarization of the electric and magnetic field variations of SES

The SES polarization differs [127, 128] from that of the MT as already mentioned in § 1.3.3. Let us recall, for example, the SES activity of April 19, 1995: the SES recordings at site "C" (Fig. 1.6(A)) indicate an almost linear polarization of the electric field variations along NS, in contrast to that of the MT, which is almost directed along EW

(Fig. 1.6(B)). A smaller directional difference was observed at site "B" (Fig. 1.5). Furthermore, the magnetic field recordings associated with this SES activity also show an almost linear polarization as can be seen in fig. 1.6.17 of Ref. [125].

1.3.7 Magnetic field variations associated with the precursory short-duration electric pulses

Figure 1.15(a) shows the electric and magnetic field variations associated with the short-duration pulses (§ 1.3.1 and Fig. 1.12) that have been observed at the station IOA *shortly* before the aforementioned M_w6.6 EQ which occurred at 08:47:13 UT on May 13, 1995. Here, we present the electric field recordings of two horizontal short dipoles of length 50 m installed at site "C" (see Fig. 1.3) that are oriented along EW and NS. The magnetic field variations were measured by two horizontal DMM, also oriented along EW and NS, which act as dB/dt detectors. The details of the instrumentation are the same as described in § 1.3.6.

The bottom panel of Fig. 1.15(a) reveals that five magnetic pulses marked "a" to "e" are detected before the EQ occurrence. They started at 16 min before the EQ and the last one was at 3 min before. They were identified also in the electric records, with varying definitude. Note also that simultaneously with the arrival of the seismic waves, several seconds after the origin time (OT), disturbances reminiscent of seismograms were recorded by both, the electric field and magnetic field sensors. No "true" coseismic, i.e., cofracture signal, was observed at OT. Observation of such signals but no cofracture signal was reported for several other cases [125, 98, 72].

Example pulses "c" and "e" of Fig. 1.15(a) are shown in Figs. 1.15(b) and 1.15(c) in a more expanded time scale. It can be noted that the magnetic field disturbance was recorded around 1 s after the electric field variation. These pulses cannot be attributed to noise from a nearby source or to a magnetotelluric origin for the following reasons: (i) the time difference between the arrivals of the electric field variations and the magnetic field recordings should not be in the observable range in the case of emission from nearby sources [161] (see also § 1.3.6) and (ii) in the case of MT disturbances, the magnetic variations (dB/dt) are recorded before [161, 125] (not after) the electric field variations, see the examples marked MT in Fig. 1.15(d).

In other words, the time difference of 1 s observed between the electric and magnetic field recordings of the short duration electric pulses detected shortly before the aforementioned M_w6.6 EQ, supports the view that they were not from a local man-made source but from the same origin as the SES activities of Fig. 1.11 (see also § 1.3.6), i.e., the EQ epicentral area lying $\sim$100 km away from the measuring site.

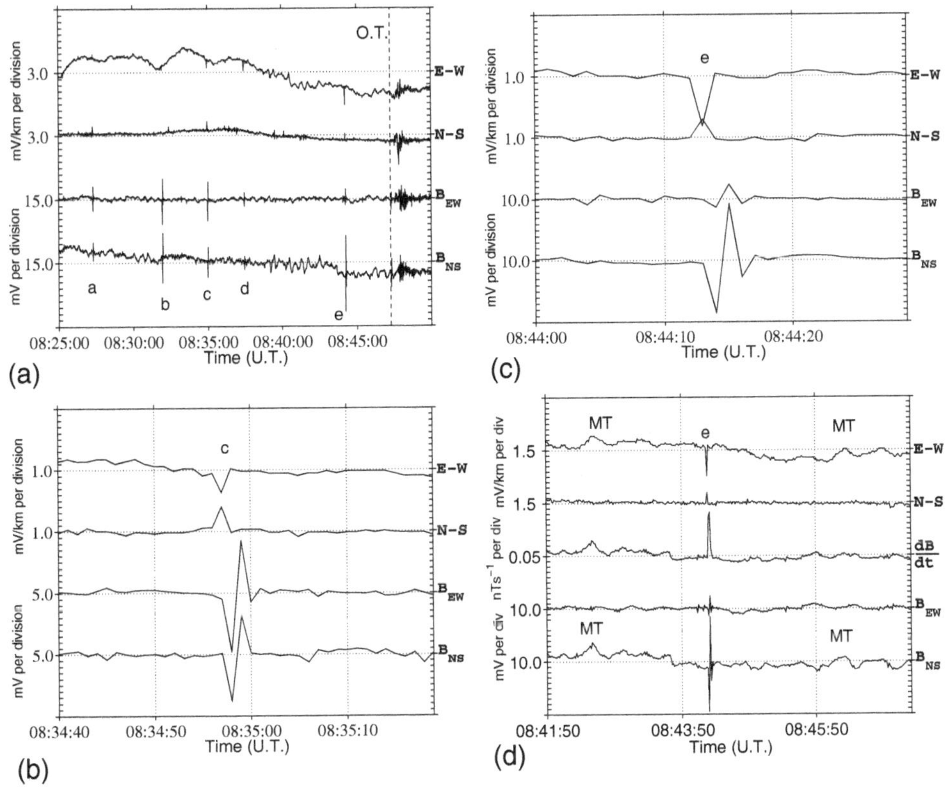

Fig. 1.15 Variations of the electric field (the upper two channels) and the associated magnetic field recordings (the lowest two channels) at IOA station. (a): For the time period from 22 min before, until 3 min after the occurrence of the M_w6.6 mainshock on May 13, 1995. The symbols "a" to "e" mark the five pulses observed before the earthquake occurrence, while the vertical broken line shows the earthquake origin time (OT). (b) and (c): The pulses "c" and "e", respectively, in an expanded time scale. (d): Excerpt from (a) to show MT disturbances in an expanded time scale. For the scale in the vertical axis for the magnetometers: 20 mV correspond to a constantly increasing magnetic field of 0.1 nT/s. In the middle channel of (d) the amplitude of dB/dt in nT/s. Reprinted with permission from Ref. [159]. Copyright (2007), American Institute of Physics.

1.4 Scale invariance of SES activities and their associated magnetic field variations

Complex systems exhibit scale-invariant features characterized by long-range power law correlations, which are often difficult to quantify due to the presence of erratic fluctuations, heterogeneity and nonstationarity embedded in the emitted signals. This also happens when monitoring geoelectric (and geomagnetic) field changes aiming at detecting SES activities and their associated magnetic field variations. Different types of nonstationarities appear in these measurements, such as random spikes and pseudo-sinusoidal trends, that may affect the long-range correlation properties of signals. Since these nonstationarities may either be epiphenomena of external conditions or may arise from the intrinsic dynamics of the system, it is crucial to distinguish their origin. This is attempted in the present Section for both the magnetic and the electric field variations associated with SES activities by employing the detrended fluctuation analysis (DFA) as a scaling method to quantify long-range temporal correlations.

DFA, originally introduced by Peng et al. [80], has been established as an important method to reliably detect long-range correlations in data effected by trends. Before proceeding to its description, some comments on the notion of long-range correlations are forwarded below (e.g. Ref. [6] and references therein).

1.4.1 Long-Range Correlations. Background

Let us consider a record (x_i) of $i = 1, \ldots, N$ equidistant measurements in which we are interested in the correlation of the values x_i and x_{i+s} for different time lags s, i.e., correlations over different time scales . In order to remove the constant offset in the data, the mean $\bar{x} = \frac{1}{N}\sum_{i=1}^{N} x_i$ is usually subtracted, i.e.,

$$\tilde{x}_i \equiv x_i - \bar{x}\,. \tag{1.5}$$

Correlations between x values separated by "s" steps are usually defined by the autocorrelation function:

$$C(s) = \frac{\langle \tilde{x}_i \tilde{x}_{i+s} \rangle}{\langle \tilde{x}_i^2 \rangle} = \frac{1}{(N-s)\langle \tilde{x}_i^2 \rangle} \sum_{i=1}^{N-s} \tilde{x}_i \tilde{x}_{i+s}\,. \tag{1.6}$$

If the x_i are *uncorrelated*, $C(s)$ is zero for $s > 0$. *Short-range* correlations of the x_i are described by an *exponentially* decreasing $C(s)$, i.e.,

$$C(s) \sim \exp(-s/\tau) \tag{1.7}$$

with a decay time τ (cf. a convenient way to exclude short-range correlations up to a scale s_1 is downsampling the original data by the same factor s_1). In the case of the so-called *long-range* correlations, $\tau = \int_0^\infty C(s)\, \mathrm{d}s$ diverges and the decay time τ cannot be defined. For example, in this case $C(s)$ may decrease as a power law

$$C(s) \sim s^{-\gamma} \tag{1.8}$$

with an exponent [51] in the range $0 < \gamma < 1$.

The determination of the *correlation exponent* γ cannot be made by a direct computation of $C(s)$ in view of the following:

The autocorrelation function $C(s)$ strongly fluctuates around zero on large scales s, making it extremely difficult to find the potential scaling (fractal) behavior of Eq. (1.8).

In addition, a direct calculation of $C(s)$ is usually not suitable due to underlying nonstationarities and trends of unknown origin. This also holds for the traditional method of power spectrum, because a power spectrum calculation assumes (e.g. see Ref. [8]) that the signal is stationary and hence when applied to nonstationary time series it can lead to misleading results.

Thus, a power spectrum analysis should be necessarily preceded by a test for the stationarity of the (portions of the) data analyzed.

1.4.2 Detrended fluctuation analysis (DFA)

Here, we summarize the detrended fluctuation analysis (DFA) [80, 111].

DFA is established, as mentioned, as a robust method suitable for detecting long-range power law correlations embedded in nonstationary signals.

It has been applied to diverse fields such as DNA [79, 107], heart dynamics [81, 3], human locomotion [2, 5, 45], circadian rhythms [39, 44], meteorology [47, 110], MT during magnetic storms [156], economics [58, 121, 4, 46] etc. A major advantage of DFA is the systematic elimination of polynomial trends of different order [12, 51, 38, 6]. In addition, as it will be discussed in § 4.6.1.1, DFA is the best estimator (compared to wavelet-based estimators) of self-similarity or long-range dependence for short time series.

DFA is based on random walk theory and its brief description is as follows. We first calculate the 'profile':

$$y(n) = \sum_{i=1}^{n} (x_i - \overline{x}) \tag{1.9}$$

of a time series (x_i), $i = 1, 2, ..., N$ with mean $\overline{x}$:

$$\overline{x} = \frac{1}{N} \sum_{i=1}^{N} x_i \tag{1.10}$$

where N is the length of the signal. In Eq. (1.9), $y(n)$ can be considered as the position of a random walker on a linear chain after n steps.

Second, the profile $y(n)$ is divided into $N_s \equiv [N/s]$ (where the symbol [ ] here stands for the integer part) non-overlapping segments of equal length ("scale") s. Third, we estimate a piecewise polynomial trend $y_s^{(l)}(n)$ within each segment by least-squares fitting, i.e., $y_s^{(l)}(n)$ consists of concatenated polynomials of order l which are calculated separately for each of the segments. The degree of the polynomial can be varied in order to eliminate linear ($l = 1$), quadratic ($l = 2$), or higher-order trends [12] of the profile function. DFA is named after the order of the fitting polynomial, i.e., DFA-1 if $l = 1$, DFA-2 if $l = 2, \ldots$. Note that, due to the integration procedure in the first step, DFA-l removes polynomial trends of order $l - 1$ in the original signal (x_i). Fourth, the detrended profile function $\tilde{y}_s(n)$ on scale s is determined by

$$\tilde{y}_s(n) = y(n) - y_s^{(l)}(n) \tag{1.11}$$

which, in other words, means that the profile $y(n)$ is detrended by subtracting the local trend in each segment. Fifth, the variance of $\tilde{y}_s(n)$ yields the fluctuation function on scale s

$$F(s) = \sqrt{\frac{1}{N}\sum_{n=1}^{N}[\tilde{y}_s(n)]^2} \tag{1.12}$$

which corresponds to the trend-eliminated root mean square displacement of the random walker mentioned above. Sixth, the above computation is repeated for a broad number of scales s to provide a relationship between $F(s)$ and s.

A power law relation between $F(s)$ and s, i.e.,

$$F(s) \sim s^{\alpha} \tag{1.13}$$

indicates the presence of scale-invariant (fractal) behavior embedded in the fluctuations of the signal. The fluctuations can be characterized by the scaling exponent α, a self-similarity parameter (see § 1.5.1):

If $\alpha = 0.5$, there are no correlations in the data and the signal is uncorrelated (white noise); the case $\alpha < 0.5$ corresponds to anti-correlations, meaning that large values are most likely to be followed by small values and vice versa. If $\alpha > 0.5$, there are long-range correlations, which are stronger [6] for higher α. Note that $\alpha > 1$ indicates a nonstationary local average of the data and the value $\alpha = 1.5$ indicates Brownian motion (integrated white noise).

For stationary signals with long-range power law correlations the value of the scaling exponent α is interconnected with the exponent β characterizing the power spectrum $S(f) \sim f^{-\beta}$ (f =frequency) through [79]

$$\beta = 2\alpha - 1\,. \tag{1.14}$$

Thus, the widely observed case of the so-called $1/f$ noise, where $\beta = 1$, corresponds to $\alpha = 1$. Furthermore, if $0.5 < \alpha < 1$, the scaling exponent α is interconnected with the correlation exponent γ (see Eq. (1.8)) through the relation:

$$\gamma = 2 - 2\alpha\,. \tag{1.15}$$

If the type of trends in a given data set is not known beforehand, the fluctuation function $F(s)$ should be calculated for several orders l of the fitting polynomial.

If l is too low, $F(s)$ will show a pronounced cross-over to a regime with larger slope for large scales [51, 38]. The cross-over will move to larger scales s or disappear upon increasing l, unless it is a real cross-over associated with intrinsic fluctuations and not due to trends [38].

1.4.3 DFA of long duration SES activities

In several applications of DFA, long-range correlations have been revealed in SES activities of duration up to a few hours [154, 156, 155, 163] (these will be summarized later in Section 4.4 along with the corresponding analysis in natural time). During the last few years, however, as already mentioned in Section 1.3, SES activities of appreciably longer duration, i.e., from several hours to a couple of days, have been collected (see § 7.2.4 to § 7.2.6). Since such long duration data enable the investigation of scaling in a wider range of scales, we focus below on such a case.

Here, we analyze as an example the long duration SES activity at PIR that lasted from February 29 until March 2, 2008. It was followed [93] by a $M_w6.4$ earthquake at 38.0°N 21.5°E on June 8, 2008 (see § 7.2.6). The time series of this electrical disturbance, which is not of obvious dichotomous nature, is reproduced in the channel "a" of Fig. 1.16. The signal, comprising a number of pulses, is superimposed on a background which exhibits frequent small MT variations.

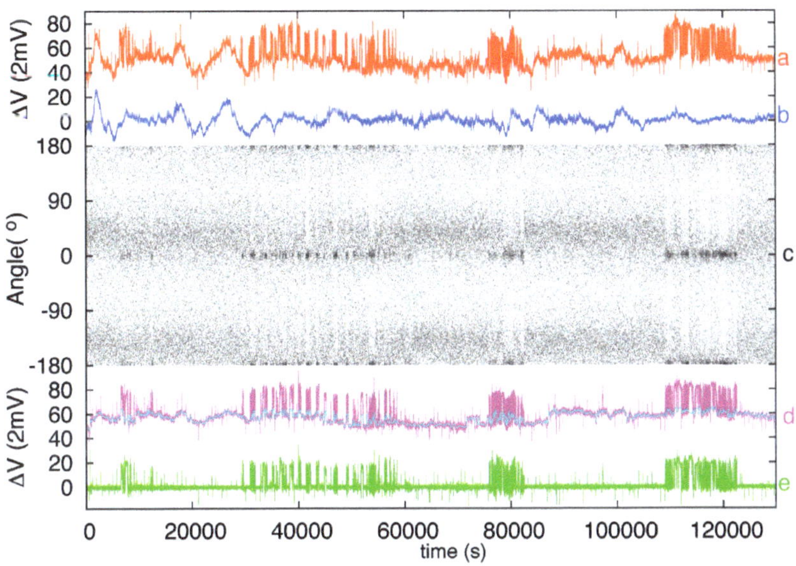

Fig. 1.16 The long duration SES activity from February 29 to March 2, 2008, recorded at the station PIR. Channel "a" original time series, channel "b": recordings at a measuring dipole which did not record the SES activity, but does show MT variations, "c": the angle of the resulting vector upon assuming that the 1 s increments of channel "a" lie along the x axis and those of channel "b" along the y-axis. Channel "d": the residual of a linear least-squares fit of channel "a" with respect to channel "b"; channel "e": the same as "d" but after eliminating the slight variations of the MT background. For the sake of clarity, channels "a", "b" and "d" have been shifted vertically. Reprinted with permission from Ref. [158]. Copyright (2009), American Institute of Physics.

1.4.3.1 The procedure to subtract the magnetotelluric background variations

In order to separate the MT background, the following steps are taken: First, we look into the simultaneous data of another measuring dipole of the same station (PIR), i.e., the data shown in channel "b" of Fig. 1.16, which has not recorded the signal but does show the MT pseudo-sinusoidal variations. Second, since the sampling rate of the measurements f_{exp} is 1 sample/sec, we now read the increments every 1 s of the two time series of channels "a" and "b". Placing the "1 s increments" of channel "a" along the x-axis and those of "b" along the y-axis, we obtain increment vectors and plot in the middle panel "c" of Fig. 1.16 their angles with dots. When a non-MT variation (e.g. a dichotomous pulse) appears (disappears) in channel "a", the angle in "c" abruptly changes to 0° ($\pm 180^\circ$). Thus, the dots in panel "c" mark such changes. In other words, an increased density of dots (dark regions) around 0° or $\pm 180^\circ$ marks the occurrence of these pulses, on which we should focus. To this end, we plot in channel "d" of Fig. 1.16 the residual of a linear least-squares fit of channel "a" with respect to channel "b". Comparing channel "d" with channel "a", we notice a significant reduction of the MT background but not of the signal. The small variations of the MT background that still remain in "d" are now marked by the light blue line. When this is removed, channel "d" results in channel "e". Hence, channel "e" provides the time series which should now be further analyzed.

1.4.3.2 The results of DFA

The DFA plot of the time series of channel "e" of Fig. 1.16 is shown in Fig. 1.17. It reveals an almost linear $\log F(s)$ vs $\log s$ plot with an exponent $\alpha \approx 1$ practically over all scales available (approximately four orders of magnitude). Note, that this value of the exponent remains the same irrespective of whether we apply DFA-1, DFA-2 or DFA-3 (see § 1.4.2). This result is in agreement with the exponents obtained [154, 156, 155, 163] for SES activities of *shorter* duration (see Section 4.4). Furthermore, note that in order to distinguish whether the signal "e" in Fig. 1.16 is a true SES activity or a man-made electric signal, its analysis in natural time has been employed as described in Section 4.11.

By summarizing, when DFA is applied to recent SES activities of long duration (several hours to a couple of days) it reveals a scale-invariant feature with an exponent $\alpha \approx 1$ over all scales available (around four orders of magnitude)

1.4.4 DFA of the magnetic field variations that accompany SES activities

For major earthquakes, i.e., with magnitude $M_w \gtrsim 6.5$, the SES activities are accompanied [161], as mentioned above (§ 1.3.6), by detectable variations of the magnetic field B. These variations, when measured by coil magnetometers, have the form of 'spikes' of alternating sign. Here, we investigate whether these 'spikes' exhibit long-range temporal correlations. This investigation, which is of major importance since only magnetic field data are usually available in most countries [25, 52, 64] (because it is easier to conduct magnetic field

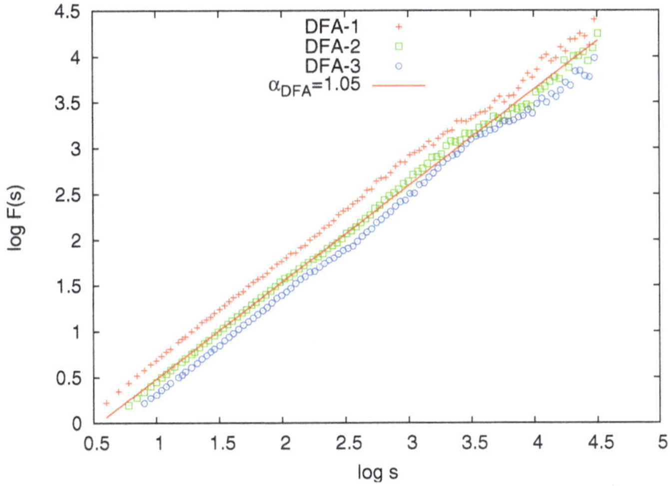

Fig. 1.17 The DFA$-l$ ($l = 1$, 2 and 3) for the lower channel labeled "e" of Fig. 1.16. Logarithm base 10 is used. Reprinted with permisson from Ref. [158]. Copyright (2009) American Institue of Physics.

measurements than electric field ones), was reported in Ref. [158] and is briefly described here.

Figures 1.11(a),(b) provide the simultaneous recordings of the electric and magnetic variations on April 18, 1995, and April 19, 1995, respectively, at the station IOA. As mentioned, a magnitude M_w6.6 earthquake with an epicenter at 40.2°N 21.7°E occurred almost three weeks later, i.e., on May 13, 1995 (see also § 7.2.1). The recordings of the two horizontal magnetometers oriented along the EW- and NS-directions labeled B_{EW} and B_{NS}, are shown in the lower two channels. They consist of a series of 'spikes' of alternating sign as more clearly seen in Fig. 1.11(c) which provides a 10 min excerpt of the recordings in Fig. 1.11(b), but in a more expanded time scale. The 'spikes' are superimposed on a background which exhibits almost pseudo-sinusoidal MT variations of duration much larger than 1 s. We now apply DFA to the original time series of the magnetic field variations and focus our attention on the B_{EW} component where the intensity of 'spikes' is higher. We find the corresponding $\log F(s)$ versus $\log s$ plot, shown in Fig. 1.18. This refers to the data on April 18, 1995, i.e., Fig. 1.11(a). If we fit the data with two straight lines (which are also depicted in Fig. 1.18) the corresponding scaling exponents are $\alpha = 0.52 \pm 0.04$ and $\alpha = 0.89 \pm 0.03$ for the short and long time lags (i.e., smaller than ~12 sec and larger than ~12 sec), respectively. The cross-over occurs at a time-lag $\Delta t \sim 12$ sec which is roughly equal to the average duration $\langle T \rangle \approx 11.01 \pm 0.03$ sec of the electric pulses, corresponding to the interval between two consecutive alternating 'spikes'. Thus, Fig. 1.18 shows that, at time-lags Δt larger than $\langle T \rangle$, long-range power law correlations prevail (since $\alpha > 0.5$), while at shorter time-lags the α value is very close to that ($\alpha = 0.5$) of an uncorrelated signal (white noise). This is understood, if we consider the conclusions drawn in § 4.1.3 for the DFA of dichotomous time series.

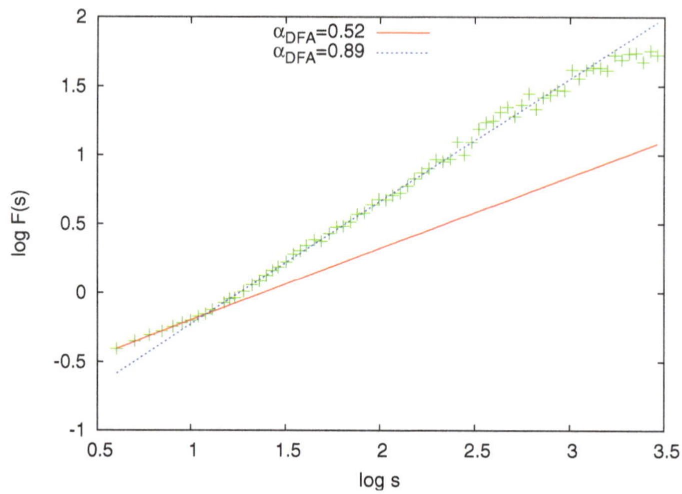

Fig. 1.18 The DFA for the B_{EW} channel of Fig. 1.11(a). Logarithm base 10 is used. Reprinted with permission from Ref. [158]. Copyright (2009), American Institute of Physics.

In summary, DFA was used here as a scaling analysis method to investigate long-range correlations in the original time series of the magnetic field variations. When using coil magnetometers, these magnetic field variations have the form of 'spikes' of alternating sign. We find a scaling exponent α close to 0.9 for time-lags larger than the average time interval $\langle T \rangle$ between consecutive 'spikes', while at shorter time-lags the exponent is close to 0.5.

Electric field variations are interconnected, of course, with the magnetic field ones through Maxwell equations. Thus, it is expected that when the former exhibit long-range correlations (§ 1.4.3.2) the same should hold for the latter.

1.5 Criticality, complexity and fractals. An introduction

1.5.1 Introductory note on fractal dimension and self-similarity. Fractional Brownian motion and fractional Gaussian noise

Fractals can be introduced in a simple way as follows [33]: Assume a reference volume V which consists of N smaller elementary volumes r^D, i.e., $N = V/r^D$. The smaller volume r^D is the reference volume used for measurement. For $D = 1$, a segment of unit length can be decomposed into N smaller segments of length r, i.e., $N = 1/r$. In two dimensions, i.e., $D = 2$, a surface of unit area can be decomposed into N smaller areas $1/r^2$. Generalizing this procedure, a dimension D can be defined through the relation:

$$D = \log N / \log(1/r) . \qquad (1.16)$$

This definition allows a noninteger dimension, or fractional dimension. N is the number of elementary elements necessary to cover the unit surface, curve or volume, $N = 1/r^D$.

The latter can be equivalently formulated as $r = N^{-(1/D)}$. Therefore, when the fractal curve has length L and is measured with a ruler of length ε:

$$D = \log(L/\varepsilon)/\log(1/\varepsilon) \tag{1.17}$$

and hence

$$L(\varepsilon) = \varepsilon^{1-D}. \tag{1.18}$$

Obviously, if $D = 1$ (Euclidean dimension), L is a constant independent of ε. On the other hand, if $D \neq 1$ the length of the curve depends on the choice of ε. An example is the so-called Von Koch's curve, which is obtained by a process of repeated dissection as follows:

A segment AB is dissected into four new segments, each being one-third of the original length (see Fig. 1.19, where note that the length at Fig. 1.19(b) is 4/3 the length at Fig. 1.19(a)). This is repeated at the next stage: each of the four segments obtained in the previous step is dissected into four new segments each being one-third the length at the preceding step, etc. Thus, if ε denotes the length of the elementary segment at a stage n, then the elementary segment at the stage $n+1$ will be of length $\varepsilon/3$. Therefore, if $L(\varepsilon)$ and $L(\varepsilon/3)$ denote the total lengths at the stages n and $n+1$ respectively, we have:

$$L(\varepsilon/3) = 4/3L(\varepsilon) \tag{1.19}$$

If $L(\varepsilon) = \varepsilon^{1-D}$, we obtain $(\varepsilon/3)^{1-D} = (4/3)\varepsilon^{1-D}$ with $D = \log 4/\log 3 \approx 1.26$. Von Koch's "curve" (which is *continuous* but *not* differentiable) is a fractal of dimension $D = 1.26$. It is a self-similar curve, i.e., a curve invariant on a change of scale. The ratio of *self-similarity* is 4/3.

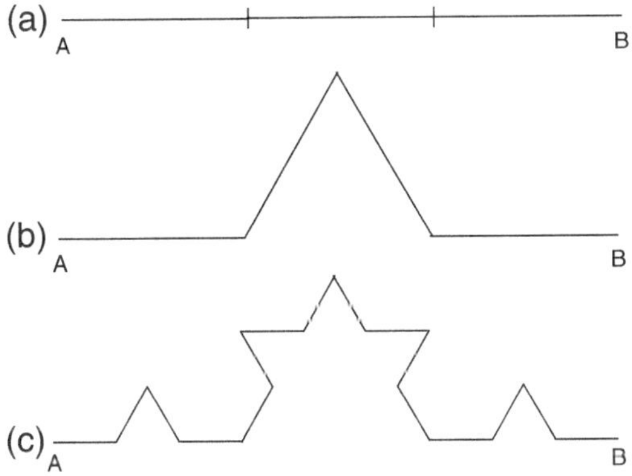

Fig. 1.19 Von Koch's curve. The segment AB is replaced by four segments each of length AB/3. This is repeated n fold. The fractal dimension is $D = \log 4/\log 3 \approx 1.26$ and the ratio of *self-similarity* is 4/3 (see the text). Reprinted from Ref. [125], Copyright (2005), with permission from TerraPub.

In general, we may say the following. *Self-similarity* of an object is equivalent to the invariance of its geometrical properties under *isotropic* rescaling of lengths.

Furthermore, we clarify that:

A stochastic process $X(t)$ is called self-similar with index H if it has the property

$$X(\lambda t) \stackrel{d}{=} \lambda^H X(t) \tag{1.20}$$

where the equality concerns the finite-dimensional distributions of the process $X(t)$ on the right- and the left-hand side of the equation (*not* the values of the process).

In many physically relevant cases, the structure of the objects is such that it is invariant under dilation transformation only if the lengths are rescaled by *direction dependent factors*. These anisotropic fractals are called *self-affine* (Vicsek [162]). Such examples are the fracture surfaces.

In summary, a fracture surface $z(x,y)$ is said to be a *self-affine* object in the sense that it remains invariant under the transformation $(x,y,z) \to (\alpha x, \alpha y, \alpha^{\zeta} z)$, where ζ is the so-called *roughness exponent*.

1.5.1.1 Fractional Brownian motion and fractional Gaussian noise

Among the simplest models that display self-similarity and long-range dependence, one can consider the example of the fractional Brownian motion (fBm), introduced by Mandelbrot and van Ness [60]. Let $B(t)$ be an ordinary Brownian motion (or *Wienner process*, in mathematical jargon), then the fractional Brownian motions of the exponent H $(0 < H < 1)$ denote [60] a family of Gaussian random functions (continuous-time stochastic processes) $B_H(t)$ which are the moving averages of $dB(s)$ in which the past increments of $B(t)$ are weighted by the kernel $(t-s)^{H-1/2}$.

More specifically, $B(t)$ is called ordinary Brownian motion (or Wienner process) when

- The increments $B(t) - B(s)$ are Gaussian.
- The increments $B(t) - B(s)$ have zero mean:

$$\mathscr{E}[B(t) - B(s)] = 0. \tag{1.21}$$

- The increments $B(t) - B(s)$ have variance:

$$\mathrm{Var}[B(t) - B(s)] = |t - s|. \tag{1.22}$$

- The increments $B(t) - B(s)$ are independent if they correspond to non-overlapping intervals, i.e., when $[t_1, t_2]$ does not overlap with $[t_3, t_4]$, i.e., ($t_2 < t_3$):

$$\mathscr{E}\{[B(t_2) - B(t_1)][B(t_4) - B(t_3)]\} = 0. \tag{1.23}$$

- $B(t)$ is almost surely continuous with $B(0) = 0$.

According to Mandelbrot and van Ness [60], the fBm $B_H(t)$ for $t > 0$ is then given by

$$B_H(t) - B_H(0) = \frac{\int_{-\infty}^{0} \left[(t-s)^{H-1/2} - (-s)^{H-1/2}\right] \mathrm{d}B(s) + \int_0^t (t-s)^{H-1/2}\, \mathrm{d}B(s)}{\Gamma(H + \frac{1}{2})} \tag{1.24}$$

where the integration is taken in the pointwise sense (as well as in the mean square sense) by using the usual methods involving integration by parts. The symbol $\Gamma(H + \frac{1}{2})$ is the gamma function of the real argument $H + 1/2$ and throughout this monograph we assume $B_H(0) = 0$. Note that $B_{\frac{1}{2}}(t) = B(t)$, thus fractional Brownian motions are divided into three very different families corresponding, respectively, to $0 < H < \frac{1}{2}$, $\frac{1}{2} < H < 1$ and $H = \frac{1}{2}$ (see also § 4.3.1).

The function $B_H(t)$ has [60] the following important properties (that generalize those of the Brownian motion stated above):

- The increments $B_H(t) - B_H(s)$ are Gaussian.
- The increments $B_H(t) - B_H(s)$ have zero mean:

$$\mathscr{E}[B_H(t) - B_H(s)] = 0. \tag{1.25}$$

- The increments $B_H(t) - B_H(s)$ have variance:

$$\mathrm{Var}[B_H(t) - B_H(s)] = \sigma^2 |t-s|^{2H}, \tag{1.26}$$

  where $\sigma^2 \equiv \mathrm{Var}[B_H(s+1) - B_H(s)]$ (e.g. see Ref. [57]).
- $B_H(t)$ is (mean square) continuous, has all sample paths continuous *but is not (mean square) differentiable* and it almost surely does not have differentiable paths.

The fBM process, Eq. (1.24), which is *not* stationary, has [60] two unique properties: *self-similarity* and *stationary increments* (cf. Eqs. (1.25) and (1.26)). As mentioned in Ref. [54], self-similarity can be understood in the sense that if a time segment is taken from the fBm trajectory, after proper normalization, the segment has the same behavior as any segments of other time scales (see also §1.5.1). Moreover, stationary increments means that the distribution of the increments $B_H(t) - B_H(s)$ depend only on the time lag $t - s$.

According to Eqs. (1.25), (1.26) and the fact that $B_H(0) = 0$, one has

$$\mathscr{E}[B_H(t)^2] = \mathrm{Var}[B_H(t)] = \sigma^2 |t|^{2H}. \tag{1.27}$$

as well as

$$\text{Var}[B_H(t) - B_H(s)] = \mathscr{E}\left\{[B_H(t) - B_H(s)]^2\right\} = \sigma^2 |t-s|^{2H}. \tag{1.28}$$

Expanding the square in Eq. (1.28) and using Eq. (1.27), one can obtain the autocorrelation function (§ 1.4.1) of the fBm [60] (e.g. see also Ref. [57])

$$\mathscr{E}[B_H(t)B_H(s)] = \frac{\sigma^2}{2}\left(|t|^{2H} + |s|^{2H} - |t-s|^{2H}\right). \tag{1.29}$$

The fact that fBm, like ordinary Brownian motion, has no derivative is inconvenient, but fractional Gaussian noise $G_H(t;\varepsilon)$ can be defined [60] through the random function ($\varepsilon > 0$)

$$\mathscr{B}_H(t) = \frac{1}{\varepsilon}\int_t^{t+\varepsilon} B_H(s)\mathrm{d}s \tag{1.30}$$

as

$$G_H(t;\varepsilon) = \frac{\mathscr{B}_H(t+\varepsilon) - \mathscr{B}_H(t)}{\varepsilon}. \tag{1.31}$$

The fractional Gaussian noise (fGn) $G_H(t;\varepsilon)$ is Gaussian with zero mean, stationary, almost surely continuous, but surely nondifferentiable [60].

Note that it is not uncommon to call [57] fractional Gaussian noise simply the time series of the increments of fBm, i.e., $B_H(t+s) - B_H(t)$, but it is actually in the sense of Eq. (1.31) that fGn is applied [54] in physics.

Since fGn is stationary the autocorrelation function of fGn

$$C_H(t,s;\varepsilon) \equiv \mathscr{E}[G_H(t;\varepsilon)G_H(s;\varepsilon)] \tag{1.32}$$

depends solely in the time-lag $\tau = t - s$, i.e., $C_H(t,s;\varepsilon) = r(\tau;\varepsilon)$. Mandelbrot and Van Ness [60] showed that

$$C_H(t,s;\varepsilon) = \frac{\sigma^2\varepsilon^{2H-2}}{2}\left[\left(\frac{|t-s|}{\varepsilon}+1\right)^{2H} - 2\left|\frac{t-s}{\varepsilon}\right|^{2H} + \left|\frac{|t-s|}{\varepsilon}-1\right|^{2H}\right]. \tag{1.33}$$

The fGn is a generalization of the so-called "white Gaussian noise" and contains three subclasses of time series [60]:

- When $H = \frac{1}{2}$, the fGn reduces to white Gaussian noise whose "integral" is the Brownian motion.
- In the case of $H \in (0.5, 1)$, $r(\tau;\varepsilon)$ is positive and finite for all τ. Moreover, in this case fGn exhibits long-range dependence (see § 1.4.1)

$$\int_0^\infty r(\tau;\varepsilon)\,\mathrm{d}\tau = \infty. \tag{1.34}$$

- For $H \in (0,0.5)$ the integral of $r(\tau;\varepsilon)$ in Eq. (1.34) is zero and $r(0;\varepsilon)$ diverges as $\varepsilon \to 0$; $r(\tau;\varepsilon)$ changes sign once, from positive to negative, as τ approaches ε from below.

Mandelbrot and Van Ness [60] proposed that (verbatim) *"1:f noises be relabeled as fractional noises"* this is indeed fully justified since the Fourier transform (considered [56] in the sense of a generalized function over the Schwartz space of test functions, due to Eq. (1.34)) of the autocorrelation function of Eq. (1.33), which corresponds to the power spectral density $S(\omega;\varepsilon)$ of the fGn $G_H(t;\varepsilon)$, has the property [54, 56]

$$S_H(\omega) \equiv \lim_{\varepsilon \to 0} S(\omega;\varepsilon) \propto \frac{\sin(H\pi)\Gamma(2H+1)}{|\omega|^{2H-1}} . \tag{1.35}$$

The power spectrum $S_H(\omega)$ of fGn in Eq. (1.35), which has been rigorously proven in Ref. [56], explains why the suggestion by Mandelbrot and Van Ness [60] provides a unique Gaussian model for the ubiquitous $1/f^\beta$ ($\beta = 2H-1$) noise in nature.

1.5.2 Critical phenomena and fractality

Critical points are for example temperatures (T), densities (ρ), etc. above which some parameter describing, say, long-range order in a ferromagnetic, or the density change in liquid–gas transition, vanishes. For example, spontaneous magnetization, $M(T)$, is zero above some critical temperature T_C in a ferromagnet.

Order parameters are quantities, which are nonzero below T_C and zero above it, and are found to be a common feature associated with critical points in a large variety of physical systems (see also § 6.2.1). For example, $M(T)$, is the magnetic order parameter, whereas $\rho_L - \rho_G$ is the order parameter in the case of liquid–gas transition through a critical point (the subscripts L and G stand for liquid and gas, respectively).

Correlation length ξ is the distance over which specific thermodynamic variables in the system are correlated and is relevant in a system near a critical point.

As an example, we refer to the 2D Ising model (§ 8.4.1), where one can see correlations of spins over larger and larger distances, as T_C is approached. Above T_C, such correlations of spins show short-range order (correlations over short distances), whereas just below (and more precisely *at*) T_C the system exhibits (infinitely ranged, e.g., $\propto 1/r^{1/4}$) long-range correlations. In other words, the physical meaning of ξ is the following: it is the length scale up to which critical correlations have been established (see also § 1.5.3).

The power laws. Following Stanley [106], who hat an interesting discussion aimed at providing a qualitative justification of power laws characteristic of critical phenomena, let us consider the aforementioned example of a ferromagnet. The probability that a spin at the origin O is aligned with a spin at a distance r is unity only at $T = 0$. For $T > 0$, we may imagine intuitively that the spin correlation function $C(r)$, in the sense of Eq. (1.6), would decay exponentially with r. Thus, we might expect that $C(r) \propto \exp(-r/\xi)$, where ξ, the correlation length, is the characteristic length scale above which the correlation function is negligibly small. Calculations on mathematical models, confirmed by experiments, indicate that actually correlations decay exponentially. However, if the system is *at* its critical point $T = T_c$, then the rapid exponential decay "magically" turns into a long-range power law decay

$$C(r) \propto 1/r^{D-2+\eta}, \tag{1.36}$$

where D denotes the system dimensionality and η is a critical exponent.

Scale-free systems. Many systems in nature are scale-free, meaning that their structure or behavior is represented by *power laws* over several orders of magnitude. The best understood of these are critical systems, like thermodynamic systems at a critical phase transition, or dynamical systems on the verge of the period-doubling transition to chaos. Examples are the solid–liquid transition, which is a first-order (*abrupt*) transition (there exists latent heat), and those termed "dynamic phase transitions" (§ 1.5.4), which are second-order (*continuous*) phase transitions, without latent heat. See also § 6.2.1 on this point.

Concerning the usual statement in the literature that, once a power law is found it reveals a critical phenomenon, we emphasize that this should be considered with extreme care, because it is not always valid.

In other words, *criticality* is always accompanied with *fractality* (power laws), but the inverse is *not* always valid.

A characteristic example of such a case is the following: the SES activities and some "artificial" noises of RTS shape bear as a common signature the power law behavior (e.g., see Section 4.4), but *only* the SES activities exhibit *critical* dynamics [156, 155].

1.5.3 Non-equilibrium critical dynamics. The scaling hypothesis

The approach to equilibrium, from a non-equilibrium initial state, in a system *at* its critical point is usually described by a scaling theory with a single growing length scale, $\xi(t) \propto t^{1/z}$, where z is the dynamic exponent that governs the equilibrium dynamics. Such a case is called the single growth model.

The simplest scenario consists of a system evolving at its critical point from a non-equilibrium initial state in which the system was prepared at time $t = 0$ (e.g., see Bray et al. [11]). The system evolves toward equilibrium through a non-equilibrium scaling state. We consider, as an example of our discussion below, the equal-time pair correlation function:

$$C(r,t) = \langle \phi(\mathrm{x},t)\phi(\mathrm{x}+\mathrm{r},t) \rangle \tag{1.37}$$

where ϕ is the order-parameter field. In the non-equilibrium scaling state, this function has the following form:

$$C(r,t) = \frac{c}{r^{D-2+\eta}} f\left[\frac{r}{\xi(t)}\right] \tag{1.38}$$

which holds in the limit $r \gg a$, $\xi(t) \gg a$, with $r/\xi(t)$ arbitrary, where a is a microscopic cut-off, e.g., a lattice spacing. In Eq. (1.38), the factor $c/r^{D-2+\eta}$ (where $c =$ constant, D is the dimension of space and η the usual critical exponent of Eq. (1.36)) denotes the equilibrium correlation function (which holds for $a \ll r \ll \xi(t)$). Requiring that this be recovered for $t = \infty$, forces $f(0) = 1$.

The *physical meaning of* $\xi(t)$. This is, as mentioned, the length scale up to which critical correlations have been established at time t.

Dynamic scaling, for large t, suggests:

$$\xi(t) \propto t^{1/z} \tag{1.39}$$

where z is the usual *dynamic exponent* characterizing temporal correlations in *equilibrium.*

First, we emphasize that Eq. (1.39) shows that relaxation *to* equilibrium is governed by the same exponent as correlations *in* equilibrium. Secondly, we note that this relation holds independently of the non-equilibrium initial state, which can affect the scaling function, $f(x)$, but *not* the exponent z (since this is a property of the equilibrium renormalization group fixed point, e.g., Ref. [11]).

The point that domain *growth* is a scaling phenomenon could be simply understood [10] when considering the result of a Monte Carlo simulation of a 2D Ising model quenched from $T_{initial} = \infty$ to $T_{final} = 0$ and then visualizing the time sequence of domain growth at $T = 0$: We then see (e.g. fig. 2 of Ref. [10]) that the domain patterns at later times look statistically similar to those at earlier times, apart from a global change of scale.

As for the scaling hypothesis, it states that a single characteristic length scale $L(t)$ exists, such that at late times the domain structure is (in a statistical sense) independent of time when lengths are scaled by $L(t)$.

1.5.4 Current aspects on the non-equilibrium driven dynamics. Dynamic phase transitions

Several efforts have been devoted to understand how quenched impurities (i.e., time-independent) influence the transport of charge-density waves (CDW), pinning of magnetic flux in type-II superconductors, roughness of crystalline surfaces, propagation of invasion fronts, etc. Many aspects of these systems can be described either by the model of randomly-pinned CDWs or by the model of randomly pinned directed manifold (RM). The progress made in understanding the non-equilibrium driven dynamics of these systems revealed that, in the extreme limit where thermal fluctuations can be neglected, a driving force F exceeding a critical value F_c is necessary to depin the system.

A (continuous) *dynamic phase transition* occurs at the depinning threshold, where the dynamics exhibit complex stick–slip motion with "avalanches" of all sizes.

The current conclusion of thought on the dynamics of driven CDW/RM systems could be summarized as follows. For F below some threshold F_c, the average velocity v is zero. Upon approaching the threshold from below, the dynamics (e.g., response to perturbation) becomes very "jerky"; it consists of a series of "avalanches", whose (linear) size l obeys a power law distribution $\text{Prob}(l > s) = s^{-\kappa}\rho(s/\xi)$ where ξ is the correlation length of the system, $\rho(x)$ is a scaling function, which is constant for $x \ll 1$ and drops off sharply for $x \gg 1$. The correlation length diverges as $\xi \propto (F_c - F)^{-\nu}$ when $F \to F_c$ (for example, when studying the dynamics of planar crack fronts in heterogeneous media we have $\nu \approx 1.52$; the exponents ν and κ are interconnected through $\kappa = 1 - 1/\nu$, e.g., see Ref. [124]). For $F > F_c$ the motion becomes continuous due to overlapping avalanches. There, the interface (e.g., in planar crack fronts in solids) acquires a finite velocity $v \propto (F - F_c)^{\beta}$, similar to the emergence of the order parameter in a critical phenomenon.

1.6 Physical mechanisms suggested for the generation of SES

1.6.1 Introduction. Views on seismogenesis and classes of SES generation models

As already mentioned in Section 1.1, aspects related to defects motivated the SES research and the pressure-stimulated currents (PSC) model [129, 133, 127, 128] was proposed for the SES generation. Several others have been also proposed later. All these models can be grouped into various classes depicted in Fig. 1.20.

The fact that a plethora of models have been suggested to explain the SES generation is not surprising, especially if one considers that there is still a diversity of views on which physical phenomenon, e.g., friction, fracture, etc. plays the prominent role in seismogenesis. This diversity could be summarized as follows.

In the standard rebound theory of earthquakes, elastic deformation energy is gradually stored in the crust until a threshold is reached at which it is suddenly released in an earthquake. The traditional view of tectonics is that the lithosphere comprises a strong brittle layer overlying a weak ductile layer, which gives rise to two forms of deformation: brittle fracture, accompanied by EQs, in the upper layer, and aseismic ductile flow in the layer beneath; thus, earthquakes are associated with strength and brittleness. Scholz [95] noticed that although this view is not incorrect, it is imprecise and in ways this can lead to serious misunderstandings. A different point of view has been suggested stating that [95]:

An EQ is a frictional, rather than a fracture, phenomenon, with brittle fracture playing a secondary role in the lengthening of faults and frictional wear.

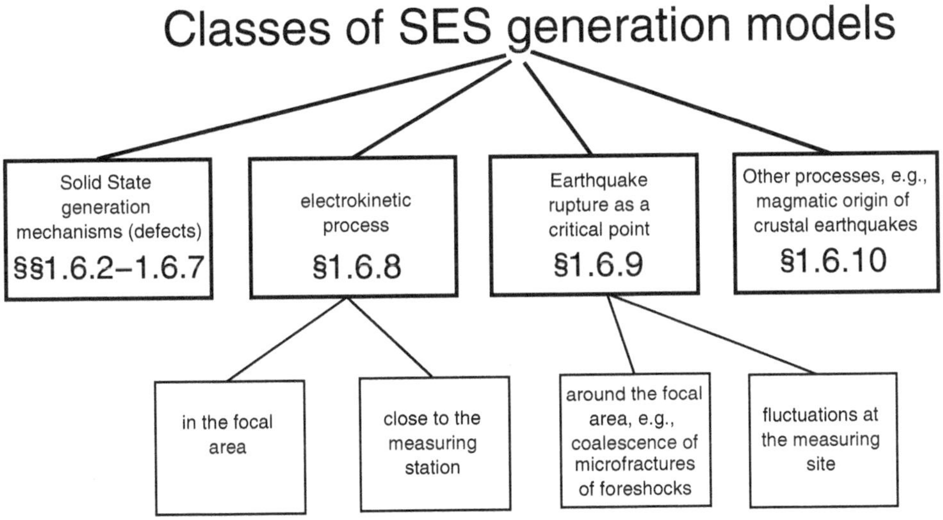

Fig. 1.20 Schematic classification of the more than ten models proposed to date for the explanation of the SES generation.

Within this frame, the seismogenic behavior of the fault is solely determined by its frictional stability, and *not* by its strength. Brace and Byerlee [9] suggested that EQs must be the result of a *"stick–slip frictional instability"* which could be interpreted as EQs in the "slip" and interseismic period of elastic strain accumulation in the "stick". This is consistent with the observation that tectonic EQs seldom, if ever, occur by the sudden appearance and propagation of a new shear crack (or "fault"), but instead they occur by sudden slippage along a pre-existing fault or plate interface. This point of view suggests that properties traditionally thought to control EQ processes – strength, brittleness and ductility – are subsumed within the overarching concept of frictional stability regimes. Furthermore, there are aspects, which consider that the earthquake rupture is a mixed process between frictional slip failure and a shear fracture of intact rock and conclude that a constitutive law for earthquake rupture can be formulated as a unifying law that governs both frictional slip failure and shear fracture of intact rock [76].

Another point of view focuses attention to the prominent role of water, both mechanically (pore pressure) and chemically (e.g., recrystallization, texture) and their probable interplay. Water was previously seen to have mainly the mechanical effect of decreasing the normal lithostatic stress in the fault core on one hand, and to weaken rock materials via hydrolytic weakening and stress corrosion on the other (e.g., see Refs. [103, 104]). Sornette [103] pointed out that nowdays there is a growing recognition that mineral structures, when in contact with water or in the presence of anisotropic strain and stress, can form and deform at much milder pressures and temperatures than their pure equilibrium phase diagram would suggest: water in the presence of finite localized strain within fault gouges may lead to the modification of mineral textures, involving dynamic recrystallization and

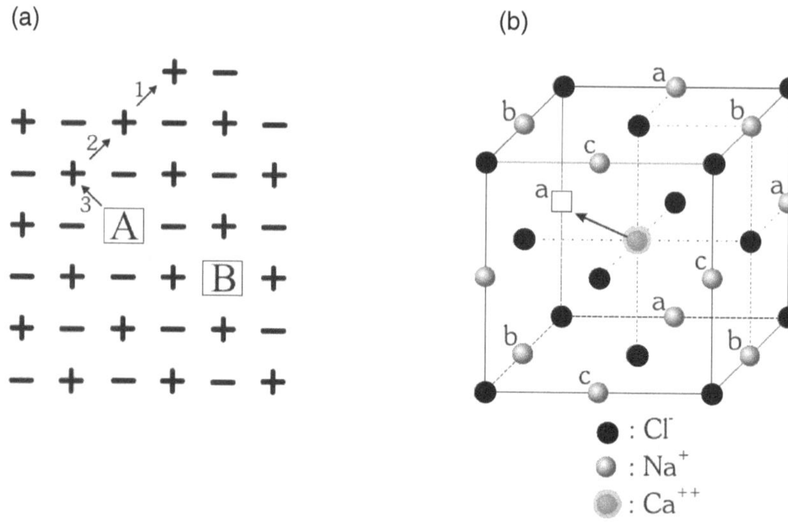

Fig. 1.21 (a) Schottky defect in an ionic crystal, e.g., Na^+Cl^-: one cation migrates to the surface through the consecutive jumps 3, 2, 1, thus leaving a cation vacancy at site A. For reasons of electrical neutrality an anion also migrates to the surface (the corresponding jumps are not shown in the figure), thus leaving the anion vacancy at site B in the interior of the crystal. These vacancies, cation and anion vacancies, are free vacancies, i.e., they do not lie close to each other, so that they can freely move into the crystal through jumps of the neighboring ions. (b) Production of a (bound) cation vacancy, when Ca^{+2} enter into Na^+Cl^-. This vacancy is attracted by the divalent cation, and hence it lies in its vicinity, thus forming an electric dipole: "Ca^{+2}-cation vacancy", which can change orientation in space. Reprinted from Ref. [125], Copyright (2005), with permission from TerraPub.

maybe phase transformations of stable minerals into metastable polymorphs of higher free energy density.

Finally, we note that concepts on dynamic phase transitions (critical phenomena) and on the physics of complex systems have been also forwarded for understanding the physics of earthquakes (for a review see Ref. [89]). In addition, a model which generalizes the Gutenberg-Richter law in the frame of nonextensive Statistical Mechanics has been proposed (§ 6.5.1).

Note that SES generation models have been developed in the frame of *all* the aforementioned streams.

1.6.2 Pressure (stress) stimulated currents (PSC) model

This, as mentioned, is the model that motivated the SES research. Its description can be found in Varotsos and Alexopoulos [129] (see also Varotsos et al. [147]) and its essence, which also explains that SES activities exhibit critical dynamics, could be summarized as follows:

In the focal area of an impending earthquake, which contains ionic materials, the stress gradually increases. In ionic solids extrinsic defects (see Fig. 1.21(b)) are always formed (beyond the usual intrinsic lattice defects, see Fig. 1.21(a)), because they contain aliovalent impurities. Some extrinsic defects are attracted by the nearby impurities and hence form electric dipoles (see Fig. 1.21(b)) the orientation of which can change through defect migration. The stress increase may result in a gradual decrease of the relaxation time of these dipoles and when the stress (pressure) σ reaches a *critical* value σ_{cr}, a *cooperative* orientation of these dipoles occurs. This reflects the emission of a transient electric signal which constitutes the SES. More details on this model are now given below.

We start with the electric dipoles caused by aliovalent impurities. As an example, we refer to the presence of M^{+2} impurities in a crystal $A^{+}B^{-}$, which introduces an equivalent number of cation vacancies in order to maintain charge neutrality in the bulk. A portion of these cation vacancies are attracted by the divalent cations and form electric dipoles that can change their orientation in space (the rest of cation vacancies remain "free", contributing to the d.c. conductivity of the crystal in a way shown by the arrows in Fig. 1.21(a)). For simplicity, we may assume that this change of orientation can be achieved only through jumps of the neighboring cations into the cation vacancies, which are usually called "bound" cation vacancies. This bound (b) cation vacancy motion (m) requires a migration Gibbs free energy $g^{m,b}$. For example, when adding $CaCl_2$ to NaCl, Fig. 1.21(b) shows that a Ca^{+2} enters the lattice in a normal Na^{+} site and charge neutrality results in a creation of a vacant cation site. This cation vacancy, being negatively charged, is attracted by the divalent Ca^{+2} and hence is located at one of its neighboring sites, usually at one of the 12 nearest neighboring sites, 4 in sites "a", 4 in "b", and 4 in "c", depicted in Fig. 1.21(b). Thus, an electric dipole "Ca^{+2}-vacancy" is formed, which can change its orientation through the vacancy jumps from one neighboring site to another (overcoming a potential barrier, which regulates the migration Gibbs energy). The time needed for such a dipole to change orientation is related to the so-called relaxation time $\tau^{m,b}$:

$$\tau^{m,b} = (\lambda \nu)^{-1} \exp\left(g^{m,b}/kT\right) \tag{1.40}$$

where λ denotes the number of jump paths accessible to the jumping species with an attempt frequency ν.

A solid containing electric dipoles, due to defects, can emit [129] PSC under isothermal conditions, as a result of either increasing or decreasing pressure. They can be classified into two categories [129]: "pressure-stimulated polarization currents" (PSPC) or "pressure-stimulated depolarization currents" (PSDC). PSPC refers to the polarization that arises under a gradual variation of pressure (pressure increase, if the migration volume $v^{m,b}$ is negative, or pressure decrease, if $v^{m,b} > 0$). In the PSDC category, the solid is initially brought into a "fully" polarized state under the action of an external field for a time appreciably longer than the relaxation time; if $v^{m,b} > 0$, the pressure is increased to a final value P_f, thus increasing $\tau^{m,b}$, so that the dipoles are practically immobilized and the electric field is then switched off; the pressure is then gradually decreased, and a depolarization current density j is liberated, the absolute value of which reaches a maximum j_M, signalling the *cooperative* reorientation of dipoles at a certain pressure P_M. If $v^{m,b} < 0$, the PSDC are emitted upon increasing pressure.

Thus, in short, pressure stimulated polarization or depolarization currents are emitted upon a gradual pressure *increase* if the migration volume $v^{m,b}$ is negative, or pressure *decrease* if $v^{m,b} > 0$.

A general condition for the appearance of the maximum in the absolute value of the current density j can be derived as follows (for the sake of convenience, the superscripts m, b in the relaxation time $\tau^{m,b}$ will be dropped hereafter).

Assuming the operation of a *single* relaxation time, the current density j is given by:

$$j = -\frac{d\Pi}{dt} = \frac{\Pi(t) - \Pi_E}{\tau(t)}, \tag{1.41}$$

$\Pi(t)$ being the polarization each time, and Π_E the isothermal saturation of polarization (cf. $\Pi_E = 0$ for PSDC, see Ref. [134]). This relation can be also written as: $j\tau(t) = \Pi(t) - \Pi_E$, which by differentiating with respect to time, for $T =$ constant, gives:

$$j\frac{d\tau}{dt} + \tau\frac{dj}{dt} = \frac{d\left[\Pi(t) - \Pi_E\right]}{dt}. \tag{1.42}$$

In the simple case of free-rotating dipoles, according to Langevin theory, we may write $\Pi_E = \mu^2 N_D E_{loc}/3kT$, where μ is the dipole moment, N_D is the dipole concentration and E_{loc} the local electric field. In materials for which E_{loc} varies only slightly with pressure [126], (and if μ, N_D are assumed independent of P) the quantity $d\Pi_E/dt$ can be disregarded, at least in the pressure region where $-d\Pi(t)/dt$ starts to become significant, and hence the previous relation turns to

$$j\frac{d\tau}{dt} + \tau\frac{dj}{dt} = -j. \tag{1.43}$$

The maximum value j_M occurs when $\frac{dj}{dt}\big|_{j=j_M} = 0$, and hence we find:

$$\left.\frac{d\tau}{dt}\right|_{j=j_M} = -1. \tag{1.44}$$

This is the condition for the (*cooperative* rotation of dipoles and hence the) appearance of a maximum in the absolute value of PSC.

It is exact for PSDC (because, in the above derivation, $\Pi_E = 0$, see also Ref. [134]), but also holds for PSPC as long as the pressure variation of Π_E is not significant.

The above relation, for *any* form of the pressure rate $b \equiv \frac{dP}{dt}\big|_T$, can be alternatively written as:

$$\left.\frac{dP}{dt}\right|_T \left.\frac{d\tau}{dP}\right|_{j=j_M} = -1, \tag{1.45}$$

or

$$b \left. \frac{d\tau}{dP} \right|_{j=j_M} = -1 \tag{1.46}$$

which, after considering that [129, 134]

$$\frac{1}{\tau} \left. \frac{\partial \tau}{\partial P} \right|_T = \frac{v^{m,b}}{kT} \tag{1.47}$$

becomes:

$$\frac{bv^{m,b}}{kT} = -\frac{1}{\tau(P_M)} \tag{1.48}$$

where $\tau(P_M)$ denotes the relaxation time at the pressure at which j_M is detected. This relation holds either for PSPC or PSDC, without assuming that the pressure rate b is constant.

In short, PSC maximizes when a gradual pressure variation *decreases* the relaxation time with a rate obeying the relation $bv^{m,b}\tau = -kT$. In particular: (a) a transient depolarization electric signal is emitted in absence of any external electric field (PSDC) in a previously polarized solid, which contains electric dipoles due to defects; (b) a transient polarization electric signal (PSPC) is emitted under either the action of an electric field or an inhomogeneous stress deformation (see below) if the solid is *not* initially polarized.

Let us now consider a material in which a pressure increase results in a decrease of the relaxation time (see Fig. 1.22) and assume that at the initial pressure P_0 the relaxation time τ is very long. In view of the long τ, the dipoles, which have initially random orientations, still retain their random orientation even after applying an external field. When the pressure starts increasing, the relaxation time becomes gradually smaller, and when the pressure reaches a critical value P_{cr}, in the neighborhood of which the relaxation time becomes sufficiently small so that the dipoles align from their initial random orientation into the direction of the continuously applied electric field, a transient electric signal is emitted (because the crystal is brought from a non-polarized state to a polarized one within a very short time). When the pressure (stress) still continues to increase, it reaches at later times the failure stress P_f (see Fig. 1.22).

Therefore, the emission of this PSC signal can be considered as a *precursor* of the failure of the solid.

The following point should be also noticed. Let us assume, for simplicity, an ellipsoidal stressed volume in the EQ preparation area (cf. in p. 92 of Ref. [127], it was argued that this volume *cannot be spherical*, because the slope of $\log \Delta V/L$ versus M would then result around unity, which disagrees with the experimental data, see § 1.3.2). In addition, upon increasing stress, a simultaneous achievement of the stimulating stress σ_{cr} at *all* points of the stressed volume is *not* intuitively expected. The "points" obeying the condition $\sigma = \sigma_{cr}$ should lie on a surface A (which may be very complicated in view of the existing

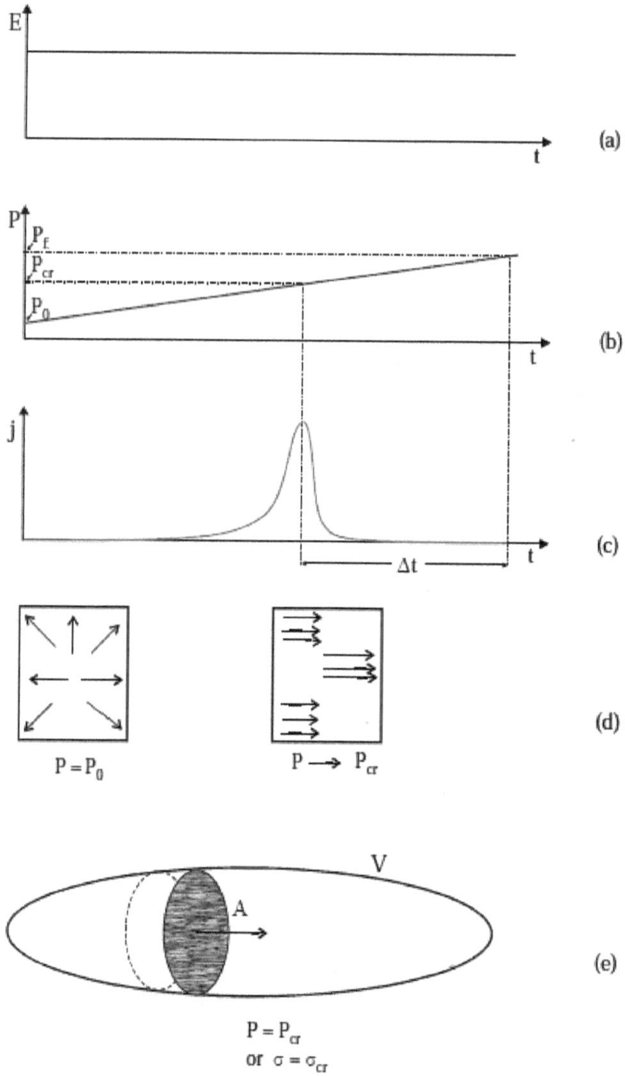

Fig. 1.22 The mechanism of the emission of a transient polarization current (PSPC) before the failure (f) at P_f when $v^{m,b} < 0$. The current is emitted when the pressure (stress) reaches a critical value P_{cr} (b) at which the relaxation time of the dipoles becomes sufficiently short and hence a *cooperative* orientation of the dipoles can occur within a very short time under the action either of an external electric field E shown in (a) or an inhomogeneous stress deformation (§ 1.6.2.1). The time Δt shown in (c) elapsed between the emission of this current and the failure is the lead time. The cartoon in (d) shows, that the electric dipoles at $P \approx P_0$ are randomly oriented, while at $P \rightarrow P_{cr}$ become oriented cooperatively(see the text). (e): the "points" obeying the condition $P = P_{cr}$ or $\sigma = \sigma_{cr}$ lie on a surface A which sweeps through the stressed volume V. Reprinted from Ref. [125], Copyright (2005), with permission from TerraPub.

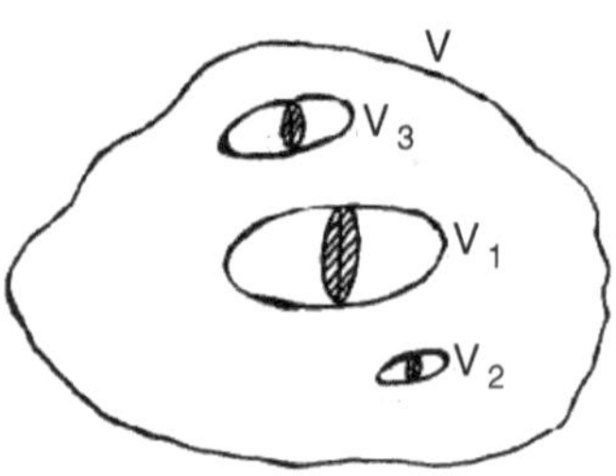

Fig. 1.23 During the last preparatory stage before a seismic event in the volume V, the stress field σ gradually changes. However the critical value σ_{cr} is not reached simultaneously in the various sub-volumes V_1, V_2, V_3 etc., so that the corresponding SES are not emitted from these sub-volumes at the same time. Taken from Ref. [140].

inhomogeneities) that sweeps through the stressed volume (Fig. 1.22(e), see p. 419 of Ref. [129]).

Note that sweeping the whole stressed volume (presumably comprising several sub-volumes, see Fig. 1.23) produces the SES activity, while each sub-volume corresponds to a SES transient "pulse" [140, 129].

1.6.2.1 Comments on the pressure (stress) stimulated currents model

We now proceed to a number of comments, which aim at clarifying a few misunderstandings of the model published in the literature:

(1) The negative sign in Eq. (1.44) comes from the following necessity: in order to approach the critical pressure/stress, the relaxation time τ should gradually decrease as the time progresses (increases).

In the above frame, we can also understand the negative sign in Eq. (1.48), which is equivalent to Eq. (1.44). Since the relaxation time $\tau(P_M)$ is always positive, the quantity $bv^{m,b}$ should always be negative. The latter means, for example ($b > 0$, $v^{m,b} < 0$), simply that, when the time progresses (increases) in order for a dipole to rotate more easily, the following possibility can be envisaged: upon increasing the stress (and hence $b > 0$) the migration volume should be negative ($v^{m,b} < 0$); an alternative possibility is, of course, upon decreasing the stress, if the volume $v^{m,b}$ is positive (note that Eq. (1.48) was derived in p. 403 of Ref. [129] with a positive sign since the absolute value of b was used there).

(2) Concerning the existence of an external electric field or of the piezoelectric inclusions in the rocks, which was postulated at an earlier stage [133], we clarify that neither of them are prerequisite for the PSC explanation of SES. This is so because in the case of *inhomogeneous* stress deformation (which happens during the EQ preparation stage), as emphasized long ago in Ref. [21] (see also § 1.6.4 and Ref. [139]): "*the effect of the applied stress gradient is similar to that of an electric field*".

Thus, neither external electric field nor piezoelectric inclusions in the rocks (to give rise to an internal field "driving" the dipoles) are a prerequisite for PSC and hence for the relevant explanation of the SES generation.

(3) The Maxwell–Wagner (MW) polarization mechanism, is very likely to occur in the EQ preparation process, in view of the heterogeneities in the structure. This is [118, 70] what usually appears in heterogeneous structures, in which the ratio of permittivity to conductivity is different in the different phases. It is the result of: (i) the formation of charge layers at the interfaces due to the different conductivity currents within the various phases, or (ii) the migration of the carriers over microscopic distances and the subsequent trapping. The relaxation time of this mechanism obeys again an equation of the form: $\tau \approx \text{constant} \times \exp(g^{act}/kT)$, which finally leads to Eq. (1.44) or, equivalently to Eq. (1.48), where $v^{m,b}$ should be simply replaced by $v^{act} = \frac{\mathrm{d}g^{act}}{\mathrm{d}P}|_T$ (where the superscript

"act" now refers to the corresponding quantities in the activation process of the MW mechanism). Thus, PSPC may well arise from the MW mechanism (without the necessity of any external electric field, in view of the clarifications given above in point 2). In other words, we emphasize that the PSPC model for the SES generation *should not be necessarily considered* as being connected solely with (re)orientation of electric dipoles of the simple type of Fig. 1.21(b) (discussed at an early stage in Ref. [133] to exemplify the PSPC concept), but may also include more complicated processes like that of the MW mechanism. The latter mechanism may lead a PSPC amplitude drastically larger than that resulting from the mechanism depicted in Fig. 1.21(b).

Putting the details aside, the basic spirit of the pressure stimulated currents (PSC) model can be summarized as follows. During the EQ preparation stage, the stress σ gradually changes; when σ reaches a critical value σ_{cr}, a transient current is emitted associated with a (cooperative) reorientation of electric dipoles taking place through motion of defects. This is a (second-order) dynamic phase transition (see also § 8.4.1). Later the rupture occurs.

Note that PSC model is solely based on thermodynamical concepts and hence can be applied to all cases that could be described by a single relaxation time.

1.6.2.2 Laboratory activation volumes in hydrated rocks

The basic prerequisite for the applicability of the PSC model to the SES generation, i.e., $v^{m,b} < 0$, has been recently confirmed in laboratory measurements [77]. In particular, the complex impedance spectra of as-received and hydrated rocks from Greece in the frequency range 10^{-2} Hz to 10^{7} Hz were investigated at various temperatures (from room temperature to 373 K) and *hydrostatic* pressures (from 1 bar to 3.5 kbar). In such a spectroscopic scheme, mechanisms with negative activation volume are shifted to higher frequencies (smaller relaxation times τ) on increasing pressure, i.e., the relaxation time of a mechanism with negative activation volume decreases on pressurization and hence can easily be detected at higher frequencies. So, a proper combination of temperature and pressure is likely to reveal slow relaxation processes, which – having long relaxation time – are activated in the low-frequency region (whereas dc conductivity and electrode polarization are strongly dominating masking any bulk relaxation process). Following this strategy, Papathanasiou et al. [77], among different rocks measured (such as pelite, sandstone, leukolite and limestone), were able to identify experimentally relaxation mechanisms with *negative activation volume* in hydrated leukolite and hydrated limestone. For example, for hydrated leukolite, at temperature $T = 349$ K, the relaxation time varies from 150 s at ambient pressure to 1 s at 3 kbar.

1.6.3 Charged dislocation mechanism

1.6.3.1 Dislocations. Background

The observed shear strength of crystalline materials is orders of magnitude smaller than what is expected theoretically for a perfect crystal lattice structure, and can only be explained by the presence of imperfections that can act as sources of mechanical weakness. These linear defects, called dislocations, are known to exist in almost all crystals.

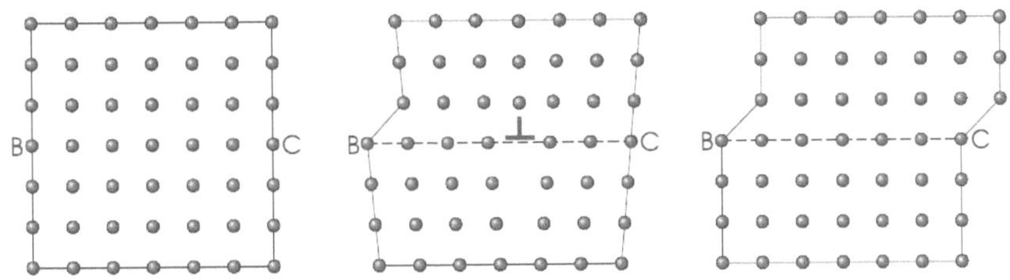

Fig. 1.24 Motion of an edge dislocation under a shear stress; the edge dislocation is indicated by an upside-down "T" with the base representing the slip plane and the stem the end of the extra half-plane; the upper surface moves to the right. Reprinted from Ref. [125], Copyright (2005), with permission from TerraPub.

The movement of dislocations is responsible for slip at very low applied stress. In slip, one part of the crystal slides as a unit across an adjacent part. There are several types of dislocations, but we only refer here to the so-called *edge* dislocation, which is simply the presence of an extra half-plane of atoms. The mechanism responsible for the mobility of an edge dislocation is depicted in Fig. 1.24, where its motion is shown under a shear, tending to move the upper surface of the specimen to the right. This reminds that a wrinkle moves more easily than a whole rug. Points of particular interest along the edge dislocation are the *"jogs"*, where the edge of the extra half-plane makes an abrupt step from one slip plane to an adjacent, parallel one.

Electrical effects of dislocations. In an ionic crystal, a dislocation line, being a region where the regular alternation of positive and negative ions is disturbed, may carry an electrical charge, or, if the line as a whole is uncharged, jogs in it may carry charges. A line, which is itself uncharged, may carry an atmosphere of charged point defects (e.g., see Fig. 1.25).

Jogs which are charged, attract charged point defects. In view of the fact that the charge of jogs is half-integral (see pp. 233–234 of Ref. [125]), they can never be neutralized. Screw dislocations are uncharged [99].

As mentioned, the individual types of intrinsic point defects, such as cation and anion vacancies, have effective electric charge ($-e$ or $+e$ respectively in a crystal like NaCl, where $-e$ stands for the electron charge). A jog may emit or absorb a vacancy; this causes the jog to change electrical sign and to move along the dislocation by one interatomic spacing. As the jogs must establish equilibrium with each of the species of point defects, this results in the presence of more jogs of one sign than the other, so that the dislocations carry a net charge.

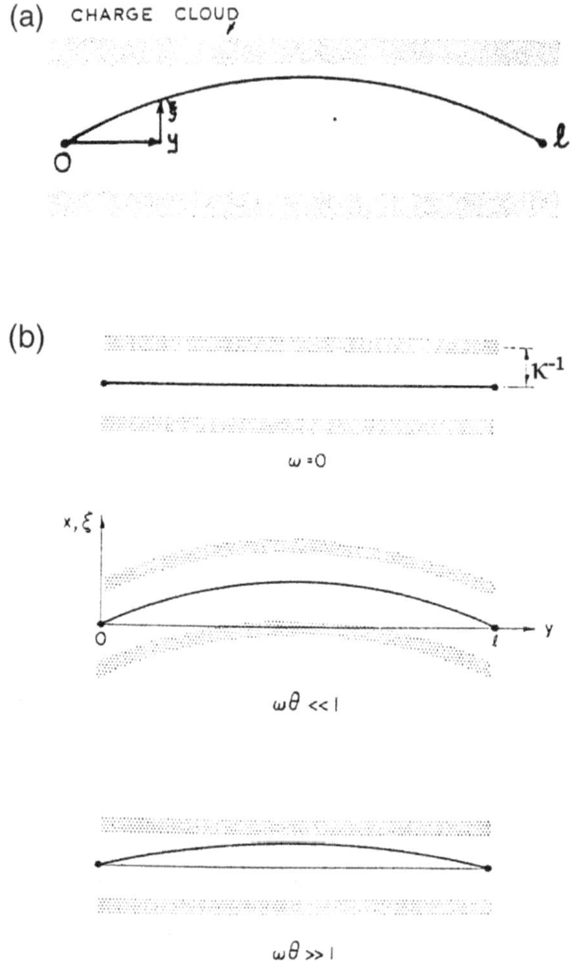

Fig. 1.25 (a): Schematic drawing of a charged edge dislocation, with loop length l, surrounded by a cylindrical charge-cloud of effective radius κ^{-1}. We consider the motion of the dislocation in the x-y slip plane. The displacement of the dislocation in the x-direction from its equilibrium position is given by $\xi(y,t)$, where y denotes the co-ordinate of the dislocation; this displacement ξ is zero at $y=0$ and $y=l$ as the result of pinning at dislocation nodes. (b): The upper figure schematically shows the effective radius κ^{-1} of the charge cloud and the next two depict the limiting cases of $\omega\theta$, i.e., $\omega\theta \ll 1$ and $\omega\theta \gg 1$ (see the text). Reprinted from Ref. [125], Copyright (2005), with permission from TerraPub.

The linear charge density on the dislocations is compensated by an almost cylindrical space charge around the dislocation, consisting of an excess of point defects – e.g., vacancies and heterovalent impurity ions – of opposite sign (Fig. 1.25).

Ionic surface charge. Lattice defects in a simple ionic crystal only occur in pairs because the overall electrical neutrality of the crystal must be maintained. For example, in NaCl vacancies of the Na^+ sublattice must be compensated by vacancies of the Cl^- sublattice and vice versa (see Fig. 1.21(a)). The equality of the concentrations of the two components of the appropriate defect pair holds *only* in the deep interior of a pure and otherwise perfect crystal (where the net charge density is zero), but *not* near the surface. The concentration of defects near the surface is governed by the energies of individual defects and one member of the defect pair can dominate. Thus, at steady state, there is *an ionic surface charge compensated by a space charge of opposite sign, which reflects strong electric field reaching* 10^7 *V/m near the surface* [99].

As for the spatial distribution of the potential within the sub-surface space charge, in a very rough approximation it resembles an exponential decay, but with the curvature suppressed for small depths. The decay distance, or Debye screening length, is given by:

$$\lambda = \left(\frac{\varepsilon\varepsilon_0 kT}{2ne^2}\right)^{1/2} \tag{1.49}$$

where ε is the (relative) dielectric constant and n is the number of Frenkel pairs (if we take as an example silver halides) per unit volume deep inside the crystal. In simple words, the screening length λ is a quantity demanding that essentially all the space charge is contained within a layer of thickness about 2λ. For example, in experiments in which thin layers of two simple ionic crystals BaF_2 and CaF_2 were investigated [94], the value of λ was found to be around 15 nm.

Ionic crystals containing charged dislocations exhibit a type of (dynamic) piezoelectric behavior (see below). This piezoelectric behavior has been attributed solely to the movement of charged dislocations.

In simple words the following happens: Let us consider an ionic crystal which in its undeformed state is *not* piezoelectric. In a crystal which becomes plastically bent, a production of an excess of dislocations of one mechanical sign takes place to accommodate the bend. These dislocations move away from the center of curvature of the specimen upon the application of a compressive stress. If a segment δy of a dislocation with a charge q per unit length is displaced a distance δh from the center of its immobile compensating charge cloud (Fig. 1.25(a)), the resulting polarization δP_h will be:

$$\delta P_h = \xi_h q \delta y\,. \tag{1.50}$$

If the crystal is mechanically loaded, the total strain is the sum of the elastic and inelastic components. The inelastic component is mainly governed by the mobility of structural defects contained in the crystal. The main difference between the elastic and inelastic strain components is the following:

Elastic deformation occurs "instantaneously", while inelastic deformation is a function of time. The latter is due to the existence of a certain *relaxation time* θ characterizing the mobility of structure defects. The relaxation time θ characterizes the rate at which the defect concentrations in the charge cloud can adjust to a perturbation caused by displacement of the dislocation.

Let us consider the motion in the *x-y* slip plane of a charged edge dislocation, which is acted on by an *oscillating* stress $\sigma = \sigma_0 \sin \omega t$. The displacement of the dislocation in the x-direction $\xi(y,t)$ is zero at $y=0$ and $y=l$ as a result of pinning at the dislocation nodes.

In Fig. 1.25(a) a schematic drawing of a charged dislocation, with loop length l, surrounded by a charge-cloud of radius κ^{-1} is depicted. The displacement in the x-direction from its equilibrium position is given by $\xi(y,t)$. Figure 1.25(b) shows the two limiting cases of $\omega\theta$. In the lowest case, the frequency ω of the applied stress is far greater than that associated with the relaxation of the charge cloud. This corresponds to a dislocation oscillating within an almost immobile charge cloud. On the other hand, in the middle figure, ω is far less than the θ^{-1}; the charge cloud, in this case, can remain in phase with the dislocation and thereby keep the dislocation at its center. The experimental results are usually obtained in the case of $\omega\theta \gg 1$, i.e., the dislocation is oscillating within an almost immobile charge cloud. The piezoelectric behavior (defect) therefore arises from the oscillating electric dipole formed by the separation of the dislocation from the center of its charge cloud. This is the basis for the mechanism that has been proposed in Ref. [125] as being responsible for the generation of the "oscillating" electric signals (reminiscent of seismograms, see Fig. 1.15(a)) upon the arrival of seismic waves at a measuring site.

1.6.3.2 Description of the charged dislocation mechanism for the SES generation

Slifkin [100, 101] suggested the following mechanism for the production of electric dipoles upon abrupt stress variation in materials with significant concentrations of impurities. Such is the case of geophysically interesting materials, as silicates, oxides and the like, in which the space charge around an electrically charged edge dislocation consists largely of aliovalent ions (note that vacancies and aliovalent impurity ions carry effective charges; see § 1.6.2 and § 1.6.3.1).

Assume that a shear stress is applied to a slab; although all the dislocations may not be able to move through large distances, segments of them can indeed bow out between the points at which the dislocations are pinned. For example, the pins may be impurity ions or points of intersection with other dislocations (see sections 11.3 to 11.5 of Ref. [125]). The dislocation loops between the "pinning" points respond to applied shear stress as if they were non-Hookean elastic bands (Fig. 1.25). Hence, at any abrupt change in stress, the bowed loops will quickly respond; the space charge around them, however, cannot respond so quickly, because it requires motion of ions. Thus, immediately after each abrupt change in stress, the space charge distribution remains practically unrelaxed and hence its center remains at the same site, no longer coinciding with the line of the dislocation; see Fig. 1.25(b)-bottom (recall that the relaxation time of the charge cloud, e.g., at 100°C is very long, i.e., lies roughly between 10^6 and 10^{12} hours; see subsection 11.5.4 of Ref. [125]).

In other words, an electric dipole has been produced, the dipole moment of which lies in the slip plane and is oriented perpendicular to the dislocation line. Note that these dipoles do *not* have random orientations.

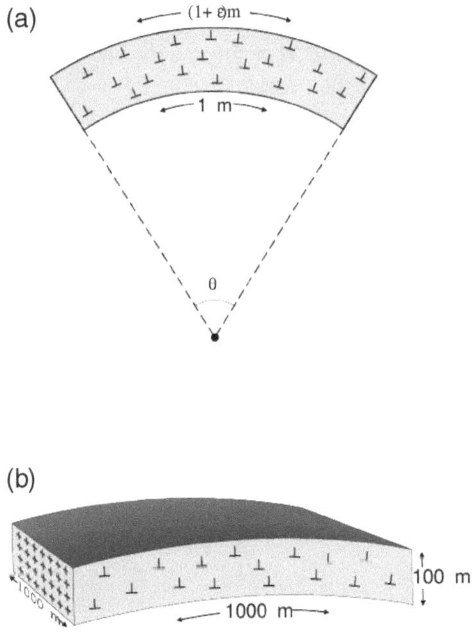

Fig. 1.26 (a) Introduction of edge dislocations (of the same mechanical sign) by plastic bending. If the angle $\theta = 1^\circ$, the upper edge of the block is longer by $\varepsilon = 20$ mm than the lower. (b) Mechanism of the production of a dipole moment in a specimen which was *previously* plastically bent (and hence extra edge dislocations have been introduced). A shear stress is applied parallel to the slip plane of the dislocations. Reprinted from Ref. [125], Copyright (2005), with permission from TerraPub.

This is due to the fact that the distribution of dislocations in most rock strata is not random, because of the deformation they have undergone in earlier history. This can be understood from the fact that plastic bending of crystalline material is accomplished through the introduction of a set of edge dislocations, all of the same "mechanical" sign, i.e., their extra half-planes all come from above or all from below. Consider, for example, the specimen shown in Fig. 1.26(a) (not in scale), a slab of edge length 1 m. If it has been bent through a very small angle of 1°, the upper edge must be longer than the lower one by 20 mm. If we assume that the lattice spacing is around 5 Å(1 Å $= 10^{-10}$ m), the bending must have resulted from the introduction of 20 mm/5 Å $= 4 \times 10^7$ new edge dislocations, all of the same mechanical sign, as shown in the figure. This excess density of dislocations (i.e., $4 \times 10^7/\text{m}^2$) is a quite modest value, as noticed by Slifkin, and probably greatly underestimates the dislocation densities in naturally occurring rocks. (For the sake of comparison, we note that most annealed simple crystals have dislocation densities in the range 10^8–$10^{10}/\text{m}^2$, while heavily deformed material contains 10^3–10^4 times larger densities.) Slifkin then proceeded to an estimate of the (total) electric dipole moment for a horizontal block 100 m thick and 1000 m wide on the other two sides, that has been folded around a horizontal axis perpendicular to one of the end faces (Fig. 1.26(b)). Assuming that the angle through which the block has been bent is 1°, the excess dislocations must have a density around $4 \times 10^7/\text{m}^2$ in order to achieve the aforementioned bending. This value led Slifkin to a dipole moment of around 8×10^{-4} Cb.m, after first estimating that the stress-induced dipole moment per unit length of dislocation is 2×10^{-19} Cb.m/m; in other words, Slifkin based his estimation on the relation: (stress induced dipole moment of a block) = (stress-induced dipole moment per unit length of dislocation) $\times$ (density of the excess dislocations) × (cross-sectional area) × (length of the dipole) (as will be discussed later in § 1.7.4.1, a typical seismic source of an EQ with magnitude M ≈ 5 has a dipole moment larger than that estimated by Slifkin [101] by a factor of the order of 10^2).

In this example, Slifkin [101] found an electric field value E of 7×10^{-6} V/m at a distance of 10 km from the dipole by assuming a $1/d^3$ decrease.

By summarizing, Slifkin's mechanism suggests that upon an abrupt change in stress, the bowed dislocation loops will quickly respond, but the space charge distribution remains unrelaxed. Thus, an electric dipole is produced, whose dipole moment lies in the slip plane and is oriented perpendicular to the dislocation line.

Slifkin [101] also proceeded to the following remark: "*The electric relaxation time (the RC time constant) of wet soil and minerals is much shorter than the duration of the observed signal. This suggests that the recorded signals may be due to the superposition of many rapid, closely spaced, unidirectional pulses such as could be propagation of a sequence of mechanical relaxations*". We shall return to this interesting point in § 1.8.4.

1.6.4 The deformation-induced charge flow mechanism

1.6.4.1 The phenomenon of deformation-induced charge flow. Background

Deformation-induced charge flow is the occurrence of a charge flow as a result of *inhomogeneous* deformation; this takes place even in the absence of an electric field and ceases abruptly when the stress *increment* producing it is removed. This phenomenon is a *current flow* and not a voltage effect.

Fischbach and Nowick [22, 21] reported that when a single crystal of NaCl is plastically deformed in an inhomogeneous fashion, a transient electric current flows through the crystal, even when no external field is applied. The effect was observed when cleaned single crystals of NaCl approximately $0.1 \times 1 \times 1$ cm were deformed in compression (note that the load was applied to a smaller area on the one face of the specimen than on the other) at temperatures between room temperature and about 100°C. The greatest flow was observed whenever a given stress was applied to the crystal for the *first* time.

The direction of the current flow was found to be governed solely from the deforming stress gradient. In particular, negative charge flows from the crystal face to which the higher compressive stress has been applied. In other words, *the effect of the applied stress gradient is similar to that of an electric field.*

The total charge flow produced by a stress increment of 1 kg/cm^2, was of the order of 10^{-12} Cb. Similar experiments in non-piezoelectric rock meterials have been performed by Varotsos and coworkers (see section 3.7 of Ref. [125]).

If the deforming load increment is removed before the current has decayed to zero, the residual current drops immediately to a value near zero. On reapplying the load increment, the charge flow is resumed. An applied electric field has little influence on the effect. Even under fields of the order of 10^5 V/m, the direction of the initial charge flow produced by deformation is determined by stress gradient rather than the direction of the applied electric field.

Fischbach and Nowick [21] concluded that the only suitable carrier to explain the aforementioned characteristics appears to be the electrically charged dislocations as follows. As mentioned in § 1.6.3.1, jogs on edge dislocation lines in NaCl may have an effective charge $\pm e/2$, the sign depending on whether the jog occurs at a positive or negative ion. When a dislocation, however, is in motion, vacancies may be "absorbed" at a jog, thereby changing the sign of the charge at the jog. Thus, when a negatively charged jog generates a cation vacancy, it becomes a positively charged jog. A dislocation in motion may be expected to have equal numbers of negative and positive jogs if cation and anion vacancies are formed with equal probability. This is not the case, however, because both theoretical and experimental results show that the formation energy for an anion vacancy is somewhat larger than that required to form a positive-ion vacancy. Therefore, a *moving edge dislocation may be expected to achieve a net positive charge* due to the preferential loss of positive-ion vacancies from jogs. Since these positively charged dislocations move into the crystal, from regions of stress concentration (§ 1.6.3.1), leaving behind a net excess of (negatively charged) cation vacancies, this model correctly predicts the sign of the observed charge flow. The model explains all the other observed characteristics of the phenomenon (see subsection 11.6.1 of Ref. [125]).

1.6.4.2 The model of the deformation-induced charge flow mechanism for the SES generation

This model, suggested by Varotsos et al. [139], is based on the "deformation-induced charge flow" phenomenon (hereafter called DICF) discussed above in § 1.6.4.1.

The DICF, as mentioned, is a current flow (as a result of inhomogeneous deformation) and not a voltage effect; thus, the RC time constant of the circuit does not have a significant influence on the phenomenon. This implies that ionic materials, subjected to inhomogeneous deformation, can still emit a current even when they are shorted through the surrounding conductive material. Such a situation would exist in the focal area, where the material undergoes inhomogeneous deformation before an EQ. In fact, the stress levels involved in the laboratory experiments by Varotsos and coworkers on rocks, presented in section 3.7 of Ref. [125], are of the same order of magnitude as those in the Earth at the shallow depths of the earthquakes for which SES measurements have been reported. Assuming that DICF is the likely mechanism for the SES generation, in what follows we will interpret a few properties of the SES (for the explanation of other SES properties see subsection 12.4.4 of Ref. [125]):

Concerning the form of the signal recorded at remote sites, when Fischbach and Nowick [21] compared their DICF measurements between 30 and 90°C, they reported the following difference: "A 'jerky' type of charge flow often occurred in the crystal at 90°C, i.e., instead of decaying smoothly with time, the current/time curve obtained at 90°C was interrupted at random intervals by sudden brief rises in the current."

These rises were in the *same* direction as the initial increase on application of the load increment. This agrees with the field observations, because in the latter a multi-peak occurrence (i.e., SES activity) was found to have the same direction (and hence is an one-sided signal; see Section 1.3).

Concerning the amplitude of the signal, considering that DICF is a volume effect (§ 1.6.4.1) and using the value 10^{-13} A/cm^3 found in the aforementioned laboratory experiments on rock materials (see section 3.7 of Ref. [125]), we expect a current intensity I of the order of $I = 10^2$–10^3 A to be emitted from a similarly stressed volume in the eventual focal area, which may be of the order of 1 to 10 km^3, e.g., a cross-sectional area of $\approx$1 km^2 and a length l of a few to several km. Such values of Il, when the current source is located inside a homogeneous medium with conductivity $\sigma_{\text{host}} \approx 10^{-3}$ S/m would generate at $d \approx 100$ km an electric field E_{host} of the order of $E_{\text{host}} \approx 10^{-1}$ to 1 mV/km (see § 1.7.2, § 1.7.4 and § 1.7.4.1). We now take into account that the current source lies close to a fault, which is a conductive path having a conductivity (σ) significantly larger than that (σ_{host}) of the surrounding medium. This, when the highly conductive path terminates (see § 1.7.4.1) within the host medium, gives rise to an electric field E_{outside} (measured within the host medium but close to the edge of the path) which is around $E_{\text{host}} \times (\sigma/\sigma_{\text{host}})$. Considering a reasonable conductivity ratio $\sigma/\sigma_{\text{host}} = 10^2$, we find $E_{\text{outside}} \approx 10$–$10^2$ mV/km, which agrees with the SES field experiments and in addition explains the existence of SES sensitive sites.

We stress again, that in the above calculation we used the value 10^{-13} A/cm^3 (per unit load increment, see § 1.6.4.1) measured in rock experiments at stress levels comparable to those occurred at faults. Interestingly this value is comparable to that obtained for NaCl when a relatively small load increment is removed and reapplied, while the major part of the total load still remains applied to the crystal.

The latter is reminiscent of the earthquake preparation process, and this is why the value 10^{-13} A/cm^3 was preferred in the calculation above instead of the values 10^{-11} to 10^{-12} A/cm^3 deduced from the experiments by Fischbach and Nowick [21] in NaCl upon the *initial* attainment of a 1 kg load increment (recall, see § 1.6.4.1, that the greatest flow was observed whenever a given stress was applied to the crystal for the *first* time).

We now turn to the following important point about the explanation of the SES generation. Fischbach and Nowick [21] do *not* report any critical stress value above which the charge flow is detectable. The following suggestion could be forwarded concerning the stress critical value. In subsection 11.3.3 of Ref. [125], it was discussed that there exists a *critical* stress τ, above which dislocation overcomes the second maximum of the so-called Peierls relief, and continues to overcome the succeeding maxima *dynamically*, thus moving forward a long distance. This could explain why the SES initiates when the stress reaches a critical value σ_{cr}.

1.6.5 Teisseyre's model on the precursory electric signals generation related with dislocation dynamics

Teisseyre [116] reviewed the theory of stress and dislocation evolution, which is based on the equation of motion for the self-stresses (dislocation stresses) on the slip-fault plane. Furthermore, the effects of dislocation dynamics on the generation of electrical precursors were reviewed [116].

Stress evolution. On a fault plane the fracturing processes produce a rapid increase of the self-stress fields: formation of dislocations, dislocation arrays and cracks [114]. The theory of the stress and dislocation evolution is developed on the basis of the interaction between the dislocations and dislocation arrays of opposite signs [115]. This interaction, which leads to stress accumulation and release, is governed by the so-called source/sink function Π which describes the nucleation of new dislocations and coalescence processes (mutual annihilation of dislocations with opposite signs, being equivalent to the coalescence of two neighboring dislocated elements or formation of a crack). The role of the source/sink function Π is similar to that of the body forces acting in the zone of seismic source and introduces an instability factor into the equation of motion; thus it is reminiscent of the friction weakening laws (see § 8.2.1) used in the fault-slip dynamics.

Based on the aforementioned concept on Π, Teisseyre [112, 113] suggested a model that combines the dipole polarization and the motion of the charged dislocations under the influence of the evolving field of stresses (this model was further discussed in Refs. [117, 71]). Finally, Teisseyre [116], after solving numerically the equations that govern the evolution of Π versus the time, found the electric field values generated at the preseismic zone.

The results show sharp increases of the electric current source intensities at the extrema or zeros of the corresponding derivatives of the function Π, which leads to the conclusion that electromagnetic phenomena precede the seismic event.

1.6.6 The peroxy defects model

Freund and coworkers studied the defects in oxide and silicate crystals and ran across rather peculiar defects (e.g., see Refs. [27, 28] for brief reviews). They call them "*peroxy defects*" because they arise from lattice oxygen that has been oxidized from its normal O^{2-} state to the O^- or $Si/^{OO}\backslash Si$. Valence changes between O^{2-} and O^- can introduce electronic charge carriers into oxides. An O^- represents a defect electron on the O^{2-} sublattice, which is called a positive hole. A similar concept has been earlier used in Ref.

[122] to explain that the dc conductivity in LiD and LiH decreases upon X-irradiation as follows: "molecular" H_2 is produced that (somehow) associates with a certain vacancy to form a charge-compensated complex (a cation vacancy traps a positive hole).

Peroxy defects derive from small amounts of H_2O incorporated as OH^- or Si-OH into the matrix of nominally anhydrous minerals when they crystallize in H_2O-laden environments [31, 29]. Upon cooling below 500°C, pairs of OH^- or Si-OH undergo [26] a redox conversion to molecular H_2 plus peroxy defects (peroxy moieties; the word 'moiety' meaning here one of the two parts into which something is divided). The peroxy moieties are inconspicuous and inactive as long as they are spin-coupled and localized. However, they are a dormant source of powerful charge carriers [30].

A tentative model for the generation of precursory electromagnetic signal goes as follows. We can conceive situations where mobile charge carriers might be generated at depth through physical changes in the stress state of rocks. This may happen because of the very short O^--O^- bond (which is [15] about 1.5 Å, and hence very short compared to the nearly 3 Å length of the regular O^{2-}-O^{2-} bond in MgO structure), thus the pressure increase stabilizes peroxy moieties and pressure release destabilizes them.

If O^- charge carriers are generated, the resulting surge of mobile O^- is expected to lead to an outflow of positive charges, and when the O^- outflow is asymmetrical, an electric dipole will emerge. This in turn will generate an electromagnetic signal, which propagates through the surrounding rock [30].

1.6.7 The model of the large-scale motion of lattice defects

It is alternatively called Lazarus's model for SES generation since it has been suggested by Lazarus [55]. This is based on a *phase transition* associated with large-scale motion of lattice defects. An EQ is presumed to occur through a *three-step process*, arising from a uniaxial stress across a pre-existing fault-plane. The phase transition region is confined to a thin region immediately adjacent to and parallel to the fault plane. A likely candidate for such a phase transition would be from a hydrous to an anhydrous form of a mineral. Nearly all terrestrial minerals are formed with varying amounts of water crystallization in the lattice; under sufficient compression, they must transform to anhydrous phases, expelling the water. The compression of the lattice at this phase transition would cause large-scale (deformation and hence) motion of lattice defects, which would result in a large electrostatic signal, roughly proportional to the length of the region transformed, in a similar fashion as in the Slifkin's model considered above in § 1.6.3.2. This phase transition is the first step in the EQ generation; it occurs when the orthogonal stress component, in combination with the overall hydrostatic stress caused by the overburden, reaches a threshold sufficient to bring about the aforementioned phase transition in the material lying in the vicinity of the fault plane. The second step may occur when the water (or any other low-density, low-friction substance), released during the phase transition in microscopic amounts, diffuses to the interface of the fault plane (where there is excess free volume). The third, and final, stage may occur when the tangential component of the stress reaches a threshold sufficient to cause macroscopic rapid slip along the (possibly now "lubricated") fault plane. In this model, there is naturally a time delay between the first threshold, at which SES is generated, and the third, at which the EQ occurs [55].

Thus, in short, this model suggests that SES are generated during a phase transition (i.e., from a hydrous to an anhydrous form of a mineral) associated with large-scale motion of lattice defects. This occurs within a thin region immediately adjacent and parallel to the fault plane.

1.6.8 SES generation mechanisms based on electrokinetic phenomena

Several publications suggested that electrokinetic phenomena can provide the basis for the generation of electrical precursors (e.g., see Refs. [61, 62, 23, 24, 42, 16, 65, 66, 32, 48, 49]). The electric field results from fluid flow through the crust in the presence of an electric double layer at the solid–liquid interfaces. This fluid flow transports the ions in the fluid in the direction of flow, thus leading to electric currents.

We present below some background material on the physics of the phenomenon, and then focus only on those suggestions that are relevant with the explanation of the SES generation.

These suggestions could be grouped broadly into two types: hypocentral and local. The hypocentral mechanisms, e.g. Refs. [65, 63], consider electrokinetic phenomena *in* the EQ preparation zone and transmission (diffusion) of electric field to the recording site. On the other hand, the local mechanisms, e.g., Refs. [16, 32, 7], require electrokinetic phenomena (in response to precursors of strain or stress) close to the SES measuring site.

1.6.8.1 The physical basis of the electrokinetic effect

Permeability. Darcy's law. From a physical point of view, the permeability of one substance in another is a transport property. In porous, permeable media, traversed by a fluid in the $+x$ direction, the permeability coefficient k (which, in reality, reflects an effective cross-section for flow, see below) relates the fluid flux to the force (the fluid pressure gradient):

$$J = -\frac{k}{\eta}\frac{\mathrm{d}P}{\mathrm{d}x} \tag{1.51}$$

where η denotes the viscosity of the fluid and J the fluid volume that crosses a section S (perpendicular to the x-axis) per unit area and per unit time; note that Eq. (1.51) is usually called Darcy's law. Thus, J is a volume flux and although it is measured in m/s does not reflect the real velocity v of the fluid. This can be understood on the basis of the simple case depicted in Fig. 1.27(a), in which the fluid flow is taking place through N capillaries

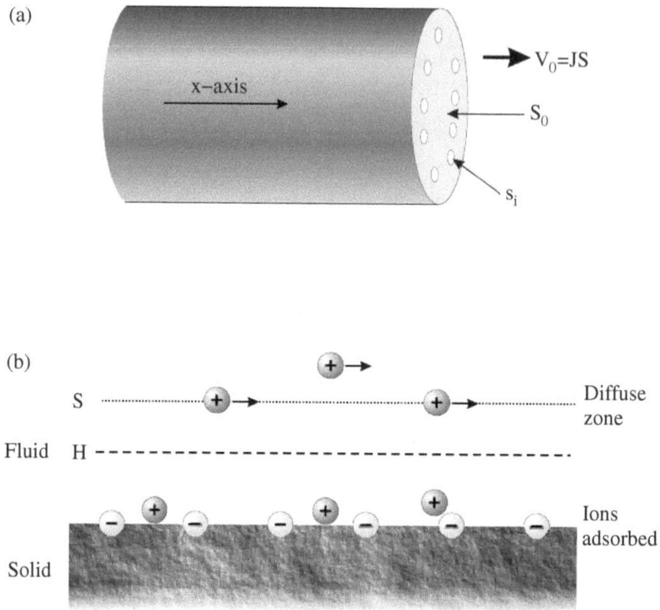

Fig. 1.27 Schematic diagram of (a): The fluid flow crossing the total section S_0 through capillaries of individual cross-section s_i (the fluid volume V_0 that crosses the section S_0 per unit time should be equal to the sum of the (volume) fluxes flowing through each channel, and hence $V_0 = JS_0$). (b): The electrochemical solid/solution interface; the absolute value of the ionic charge of the surface layer is equal to the sum of the ionic charges in the Helmholtz layer and in the diffuse zone, thus leading to an electrically neutral double layer, if the appropriate charge signs for each zone are considered. The plane H lies at a distance from the zero surface of the solid, and corresponds to the beginning of the diffuse zone. Reprinted from Ref. [125], Copyright (2005), with permission from TerraPub.

of individual cross-sections s_i. In this case, assuming isotropic distribution of porosity ϕ, thus $\phi = (1/S_0)\sum_{i=1}^{N} s_i$, and that the velocity v_i in every channel i is equal to v, we find that the quantities J and v are interconnected through: $J = \phi v$.

The origin of electrokinetic phenomena. An electric double layer is formed at a solid/liquid interface (Fig. 1.27(b)). This is made up of a layer of ions (*the Helmholtz layer*) adsorbed on the surface of the solid (e.g., the rock) and of a diffuse mobile layer of ions (*the Gouy–Chapman zone*) which extends into the liquid phase.

When a fluid flows through a porous medium, a potential will be developed across the (length of) sample, because of the relative motion between the solid and the liquid; this is the so called *streaming potential* (which may be thought of as the reverse of electro-osmosis). If S denotes the closest plane to the surface on which fluid motion takes place (see Fig. 1.27(b)), the potential of this plane is defined as the ζ potential. (Note that this is manifested in the streaming potential measurements. In other words, the *zeta potential* ζ is the potential at the slipping plane; the latter is the plane where the fluid velocity goes to zero.) In such cases, if V denotes the streaming potential, both forces ∇V and ∇P (while in Eq. (1.51) only the second force, i.e., ∇P, was present) must be considered and then the fluid flow becomes:

$$\mathbf{J} = -\frac{k}{\eta}\nabla P + \frac{\varepsilon\zeta}{\eta F^0}\nabla V \tag{1.52}$$

while the electric current density j is given by:

$$\mathbf{j} = \frac{\varepsilon\zeta}{\eta F^0}\nabla P - \frac{\sigma}{F}\nabla V \tag{1.53}$$

where σ, ε are the electrical conductivity and the dielectric constant of the fluid. F^0 is the so-called formation factor (the ratio of the conductivities of the fluid and the porous medium, e.g., σ/σ_r, the ratio of the fluid over the rock conductivity [33, 49]) when surface conduction is absent, while F is the formation factor for the fluid conductivity being studied (possibly with surface conductivity).

In a steady state, the absolute values of the two terms in the right-hand side of Eq. (1.53) (i.e., the first term represents current resulting from mechanical energy being offered to the system, while the second term corresponds to the conduction current) become equal; this equality leads to the so-called *Helmholtz–Smoluchowski equation* [17]:

$$\frac{\Delta V}{\Delta P} = \frac{\varepsilon\zeta}{\eta\sigma}\frac{F}{F^0} \tag{1.54}$$

(note that we again clarify that ΔP and ΔV are measured across the length of the sample).

The quantity $C \equiv \Delta V/\Delta P$, i.e.,

$$C \equiv \frac{\Delta V}{\Delta P} = \frac{\varepsilon\zeta}{\eta\sigma}\frac{F}{F^0} \tag{1.55}$$

is called the *streaming potential cross-coupling coefficient* or simply the *coupling coefficient*. For a circular cross-section pore of radius r, we have [68]

$$\frac{F}{F^0} = \frac{1}{\left(1+\frac{2S_s}{r\sigma}\right)} \tag{1.56}$$

where S_s denotes the surface conductance. If surface conductivity is absent, $F = F_0$ and Eq. (1.55) simplifies to:

$$C \equiv \frac{\Delta V}{\Delta P} = \frac{\varepsilon\zeta}{\eta\sigma} \tag{1.57}$$

In other words, streaming potentials occur in a fluid when there is relative motion between the fluid and a *charged surface*, for the latter recall § 1.6.3.1.

The electrical double layer, which as mentioned forms at the interface between the fluid and the charged surface, has a charge density ρ which decays exponentially away from the surface; the distance at which this charge density decays by $1/e$ is the so-called Debye length (see § 1.6.3.1; see also subsection 11.4.1 of Ref. [125]). As the fluid moves tangentially to the double layer, it pulls the ions of the double layer along a length l.

These moving ions near the surface give rise to convection current: $I_{conv} = \int v(r)\rho(r)\,dr = (\pi\varepsilon r^2\zeta\Delta P)/(\eta l)$, where $v(r)$ is the fluid velocity. The conduction current $I_{cond} = (\pi\sigma r^2/l)\Delta V$ flows through the resistive bulk fluid to generate a potential referred to as the streaming potential.

The electrokinetic current density averaged with respect to the cross-section is usually written as [24, 16]:

$$j = -\sigma_r C\nabla P,\ C \approx \frac{\varepsilon\varepsilon_0\zeta\phi}{\eta\sigma_r} \tag{1.58}$$

where σ_r the electric conductivity of the rock. Alternatively the following relation is given [109]:

$$j = -C_e\nabla P \tag{1.59}$$

where

$$C_e \approx \frac{\varepsilon\zeta\phi}{\eta} \tag{1.60}$$

The physical parameters (including the temperature and/or tri-axial stress) that influence the electrokinetic coupling coefficient (e.g., Refs. [68, 42, 48, 78]) and the frequency dependence of permeability (e.g., Refs. [82, 85, 87, 86]) have been discussed in pp. 267–268 of Ref. [125].

1.6.8.2 Electrokinetic phenomena in the hypocentral area of an impending earthquake

Morgan et al. [68] and Morgan and Nur [67] suggested the following mechanism that may have significant relevance to the generation of transient electric signals. Two-phase fluid flow enhances cross-coupling properties; this enhancement originates from an increase of resistivity at partial saturation which reflects an increase of C by a factor of 2 or more. Partial saturation may result at the earthquake source region either by dilatancy or boiling [59]. However, boiling is unlikely except as a coseismic phenomenon [78]. Morgan [65] first presented a calculation showing that such electrokinetic phenomena in the source area can account for the observed SES amplitudes. Beyond this suggestion, a few other more recent publications [63, 43] appeared, the results of which will be summarized below.

Molchanov's multifractures model [63]. The idea is that SES is generated during the enhanced fracturing that occurs a few weeks or a few days before the main shock. Molchanov considered for the source current the electrokinetic mechanism (in the focal area), by assuming a water-induced relaxation of stress-drop just after the appearance of microfractures. A similar process was considered in Ref. [20] for the explanation of ULF magnetic variations.

The following model was proposed [63]. Assume that many fractures with scale L_s and rate $\dot{N}(L_s)$ occur in a volume $V_0 \sim L_0^3$, where L_0 is a length of the order of the large-scale stress gradient. Before a large EQ with magnitude M, fracture occurrences are probably intensified; the latter can be represented as foreshocks obeying the Gutenberg–Richter law with the usual exponent b, e.g. see Eq. (6.1).

Considering that the electric field variations (due to the *largest* fractures from the region V_0) recorded at the same site probably retain the same polarity, the electric field amplitude E resulting from the overlapping (clustering) of a significant number of electric pulses, after assuming Poisson distribution (in time) for the occurrence of these electric pulses, is found to be [63]:

$$\log_{10} E = \left(\frac{3}{4} - \frac{b}{2}\right) \mathrm{M} + \mathrm{const}. \tag{1.61}$$

Molchanov [63] assumed $b = 0.89$, and found that the slope of $\log_{10} E$ versus M is 0.305 which is very close to the experimental values, see Eq. (1.1). The empirical b values are around unity (i.e., between 0.8 and 1.2, see Section 6.1), thus resulting in a slightly different slope which is still in remarkable agreement with the SES observations.

Numerical simulation of electric field produced by fluid flow within a fault. The electrokinetic potential on the ground surface, produced by the fluid flow within a vertical thin fault zone, has been computed in Ref. [43]. "Dilatancy" or "over-pressure" was assumed as a driving force of the fluid flow. It was found that, if the resistivity of the crust is of the order of 10^3–10^4 Ωm and a high conductivity (vertical) channel exists between the fault zone and the Earth's surface, the electric field values in the peripheral area of the outcrop of the channel reach values well above the detectability limit.

1.6.8.3 Electrokinetic phenomena close to the SES measuring station

Dobrovolsky et al. [16] and Gershenzon and Gokhberg [32] suggested that SES can be explained by the electrokinetic effect resulting from strain changes that affect fluid dynamics close to the measuring site or around it. Moreover, they suggested that SES can be better recorded on a vertical electric dipole as follows: at the surface the excess pressure is always zero, but at points within the crust strain changes will produce pressure changes and hence, a potential difference ΔV will result between an electrode located at the surface and another at a depth h (e.g., in a well, provided that the latter electrode lies below the water level and is covered by the Earth, making the pressure at the pore fluid around it approximately equal to what it would have been if there were no well at all). We clarify that ΔVdoes not have [16] to be proportional to h.

A model slightly different from that of Refs. [16, 32] for detecting electrical anomalies due to electrokinetic phenomena close to the measuring site has been suggested by other authors [48]. Based on empirical observations of abrupt upheaval of the underground water level measured postseismically (at distances of the order of 50 km from the epicenter), it is assumed that there is a ground-water recharge during interseismic periods; this recharge implies vertical fluid flows in shallow aquifers which could induce electrokinetic anomalies.

Within such a scheme, a precursory electrical anomaly can be observed irrespective of either the exact value of the coupling coefficient or whether a lateral heterogeneity exists or not.

Note that the measurements of Jouniaux and Pozzi [48] show that a streaming potential up to 30 mV could be produced by an underground water level change of 50 cm, for a fluid resistivity of 10^2 Ωm and a permeability of 10^{-12} m^2. In addition, Jouniax and Pozzi [49] observed transient streaming potential variations with pulses of amplitude 15–40 mV and frequency 0.1 to 0.5 Hz under geochemical changes; they suggest that such geochemically induced effects may possibly be responsible for the observation of electrical precursors in this frequency range.

1.6.8.4 Electrokinetic phenomena in porous media with fractal structure

Fractal critical exponents of electrokinetic current parameters. Surkov, Uyeda, Tanaka and Hayakawa [109] supposed that the EQ hypocenter is surrounded by water-saturated porous rock with fluid-filled pore channels. The pre-earthquake stage was assumed to be accompanied by the appearance of fresh cracks in the fracture zone. The scale of this zone may vary from hundreds of meters up to several kilometers. It was assumed that the pore space in the fracture zone exhibited fractal structure. Apparently, most of the fresh cracks are closed when formed. Because of the pressure release due to cracking, they are under lower pressure, so that water from the un-cracked outer region can penetrate into them as soon as a network of connected channels or fractal clusters is formed. The closed fresh cracks may be regarded as the sink of water from the surrounding higher-pressure areas.

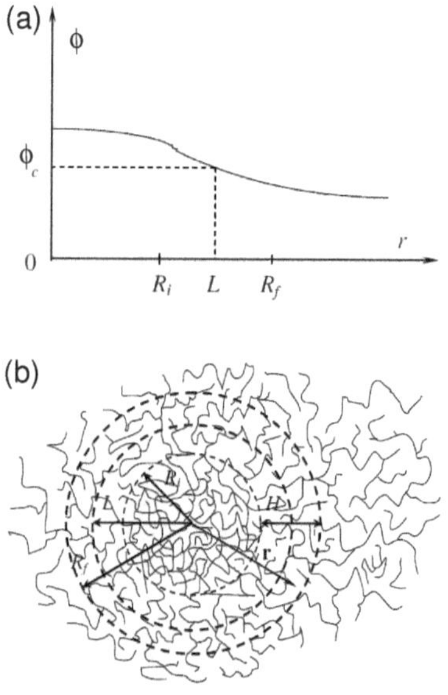

Fig. 1.28 (a) The porosity ϕ at various distances r from the fracture zone assumed in Ref. [109]; the percolation threshold ϕ_c is exceeded only in the internal area $r < L$, and non-fractal zone is situated at $r < R_i$. The fractal region is confined by the radii R_i and R_f. (b) Schematic picture of the fracture zone assumed in Ref. [109]. The internal high permeability area is restricted by the radius R_i. The fractal region occupies the field from $r = R_i$ to $r = R_f = R_i + H$. L denotes the distance where $\phi = \phi_c$. Reprinted from Ref. [109], Copyright (2002), with permission from Elsevier.

Surkov et al. [109] assumed that the porosity ϕ, after the cluster formation, decreases from the center of the fracture zone towards the periphery according to a certain law, see Fig. 1.28(a):

The percolation threshold ϕ_c is exceeded only in the internal area that has a typical size L.

Fractal properties, near the threshold, are governed by a correlation length(see § 1.5.2):

$$\xi \propto \frac{1}{|\phi(\mathbf{r}) - \phi_c|^{\nu}}, \tag{1.62}$$

where $\nu = 0.88$ is the correlation length critical exponent [108]. This equation indicates that an increase in the porosity in the internal area ($r < L$) is followed by a decrease of correlation length. Thus, the central part of the internal area may lose its fractal properties, because of a multiplex intersection of channels. This part with high porosity and high permeability was called the *non-fractal central zone*, in contrast to the *peripheral fractal region*. Let $R_i = L - H$ be the typical radius/size of the non-fractal central zone, as shown in Figs. 1.28(a) and 1.28(b). The fractal region occupies a spherical shell confined by the radii R_i and $R_f = L + H$. The typical size H of the fractal region was assumed to be of the order of the correlation length, i.e., $H \propto \xi(R_i) \propto \xi(R_f)$, and the porosity near $r = L$ was written as a power series of the parameter H:

$$\phi(r) \approx \phi_c + \frac{\mathrm{d}\phi}{\mathrm{d}r} H. \tag{1.63}$$

Substituting the latter expression into Eq. (1.62) and using the approximation $\mathrm{d}\phi/\mathrm{d}r \approx \Delta\phi/L$, at $r = L$, where $\Delta\phi$ is the porosity variation, at $r = 0$ and $r = L$, they obtained:

$$H \propto \left(\frac{\mathrm{d}\phi}{\mathrm{d}r}\right)^{-\frac{\nu}{\nu+1}} \propto L^{\frac{\nu}{\nu+1}} \tag{1.64}$$

which is valid if $H \ll L$. For this case, they supposed that the size of the fracture zone is of the same order as that of the internal area, i.e., L, and so the H dependence on the fracture zone size looks like Eq. (1.64). This dependence will be needed later to find the relation between the SES amplitude and the EQ magnitude.

In the fractal peripheral region, the coefficient C, as well as the diffusion coefficient, the permeability and other rock parameters, change with distance through power laws; the coefficient C_e (see Eq. (1.60)) was assumed to have the form:

$$C_e \approx \frac{G}{\rho^{\mu}}, \tag{1.65}$$

where G is a constant, ρ the distance between two points in the fractal region, and μ denotes an unknown critical exponent, which can be determined as follows. The coefficient C_e in the form of Eq. (1.60) *cannot* be applied, because the fractal geometry should be taken into account. In the latter case, the average conductivity σ_r of the rock containing pore channels and the permeability k depend on the porosity ϕ according to: $\sigma_r \propto k \propto (\phi - \phi_c)^t$, where $t \approx 1.6$ is the transport critical exponent [108]. Using these dependencies, Surkov et al. find $C_e \propto H^{-t/\nu}$, which when compared to Eq. (1.65), leads to: $\mu = t/\nu \approx 1.82$ (where they used the values $t = 1.6$ and $\nu = 0.88$ mentioned above).

The dependence of the current dipole source moment on the dimensions of the fracture zone. At remote distances from the fracture zone, the low-frequency electromagnetic field generated by a system of electrokinetic currents can be characterized by an effective short linear current element $I\Delta l$. Here I denotes the total source current and Δl is the effective current length. To replace an extrinsic current system by a point current element, they integrate the electrokinetic current density over the volume V of the fracture zone, i.e.,

$$\mathbf{p} = I\Delta\mathbf{l} = \int_V \mathbf{j}_e \, \mathrm{d}V = -\int_V C_e \nabla P \, \mathrm{d}V \tag{1.66}$$

According to Surkov et al. [109] there are two contributions to this integral: one (p_c) from the central non-fractal zone and a second one (p_f) from the external fractal region. The first one was shown [109] to be

$$p_c \propto L^2 \tag{1.67}$$

while the second contribution (p_f) is given by:

$$p_f \propto C_e \frac{\Delta P}{L} L^2 H \propto L^{1-\frac{t-\nu}{1+\nu}}. \tag{1.68}$$

Explanation of the empirical VAN relation of Eq. (1.1) or Eq. (1.2). The SES amplitude $E(=\Delta V/L)$ must be proportional to the amplitude $|\mathbf{p}|$ of the current dipole source moment. Furthermore, recall that the EQ focal dimension L is related to the EQ magnitude M, through the empirical equation [50]:

$$\log_{10} L = 0.5\mathrm{M} - 1.9 \tag{1.69}$$

where L is measured in km. Due to the *perfect* spherical symmetry of the *non-fractal* zone the corresponding dipole moment **p** should be vanishingly small, thus its contribution to the observed electric field value can be disregarded. Therefore, the leading contribution comes from the fractal region, i.e. $p_f \gg p_c$, with $E \propto p_f$, which when taking into account Eqs. (1.68) and (1.69), gives:

$$\log_{10} E = \alpha \mathrm{M} + b, \;\; \alpha = \frac{1}{2}\left(1 - \frac{t-\nu}{1+\nu}\right) \approx 0.31 \tag{1.70}$$

after using the critical exponents: $t = 1.6$ and $\nu = 0.88$.

Thus, when assuming fractal geometry in the fracture region, Surkov et al. [109] find a value of α in Eq. (1.2) very close to the one observed.

This is consistent with an early suggestion (see p. 92 of Ref. [127]), see also § 1.3.2, that the α value stems from the *geometry of the emitting source* (where it is clear that when the source enters the critical regime, it must be characterized by fractal geometry).

1.6.9 SES generation mechanisms when assuming the earthquake rupture as critical point

Sornette and Sornette [102], based on the concept that rupture in EQs could be considered as a critical point (CP), suggested a model for the SES generation. (Since percolation is a *critical* phenomenon, the model by Surkov et al. [109] discussed in § 1.6.8.4, as well as the PSC model (§ 1.6.2), could have been classified in the same framework as the model of Sornette and Sornette [102].) This model provides naturally an explanation of the SES detectability at certain cites at long distances (selectivity effect; see § 1.3.4). In addition, this model explains the experimental fact mentioned above that the logarithm of the SES amplitude scales linearly with the magnitude M for a given focal area and a given SES sensitive station, see Eq. (1.2), i.e., $\log_{10}(\Delta V/L) = \alpha \mathrm{M} + \beta$, where the experimental value of α is around $\alpha = 0.32$–0.37 for all measuring sites (*universal*).

The essence of the Sornette and Sornette [102] proposal is that SES is closely related to a *local* piezoelectric effect, which does not average to zero in some places and thus can be detected as the CP is approached.

In particular, they state [102] that as the CP is approached, the medium becomes more and more fragile, ending eventually in rupture at the CP. Then they assume that an effective control parameter "q" can be defined such that the CP is attained when q reaches some threshold q_c. As the CP is approached, the increasing fragility of the medium is reflected in the "critical" behavior of the effective deformation modulus Y, which should decrease eventually to zero at CP. As q approaches q_c, the material becomes very weak and hence a finite stress creates a locally large strain, resulting in a large polarization and electric field. Thus, one expects to measure an increasing potential difference between two points of a piezoelectric system as the CP is approached. The small value of the measured SES can be associated with the absence of a real macroscopic effect, but is more closely related to a local piezoelectric effect which does not average to zero in some places and thus can be detected [102]. In other words, the detectability of the potential difference variations is related with the strong spatial variations in the polarization of the medium.

Sornette and Sornette [102] made use of the theory of critical piezoelectricity in percolation [105] and derived an interconnection between its critical exponents and Eq. (1.2). They finally obtained an α value comparable with the aforementioned one obtained experimentally.

1.6.10 Other SES generation mechanisms

Beyond the mechanisms mentioned above, several others have been suggested. Two of them are mentioned below:

(a) *Pulsed charge model*: Ikeya et al. [41, 40] suggested a model, based on the piezoelectricity of quartz, and raise the possibility that SES may come from the time averaged evanescent ULF waves. In simple words, their model assumes an ensemble of emitting dipoles in the fault area (due to the piezoelectricity of quartz-bearing rocks) and therefrom the SES are explained as being the envelopes of these electromagnetic pulse waves.

This model seems to satisfactorily explain [40] the empirical relation found for SES: $\log_{10}(\Delta V/L) = \alpha M + \beta$, where α is around 0.35.

(b) *SES generation based on the magmatic mechanism of shallow crustal earthquake preparation*: Rokityansky [88] forwarded a model, which is based on a suggestion of Guterman and Khazan [34] that crustal seismic activity (or some parts of it) is a manifestation of contemporary magmatic activity. The model assumes a mantle chamber (that can be a mantle plume head, for instance) and a crustal magma chamber. The latter is connected to the former by a magma channel. Based on giant radiating dike swarms data [19], Rokityansky also assumes secondary crustal chambers. Such a model can qualitatively explain several SES features: a SES-sensitive station is situated near a magma channel or crustal magma chamber. Long-distance SES-sensitivity is observed when the station and the focal area are situated at opposite ends of a (quasi-)radial chain of crustal chambers, the station being near the first chamber, and the EQ near a peripheral one. The SES generation is attributed to magma flow or, more probably, to the beginning of magma flow, with the opening of magma channel(s).

1.7 Explanation of the selectivity effect and other SES properties

When a current-emitting solid is surrounded by a medium of conductivity σ', the question arises whether the signal can be observed at long distances (i.e., at distances appreciably larger than the dimensions of the emitting source). Of special interest, for practical applications, is the case when the emitting source is located at a small distance from a "path", which has conductivity σ, orders of magnitude larger than the conductivity σ' of the surrounding medium. The study of this case is closely related to the SES transmission in the Earth and the question of SES detectability at long distances. This is so, because, as will become clear in this Section:

Maxwell equations can naturally explain the *selectivity effect*, if we consider that the SES generation source lies in the future focal region and the earthquakes occur by slip on faults, which are highly conductive paths in the Earth's crust. This is the essence of the model described below and was proposed by Varotsos and Alexopoulos [129] for the explanation of the SES selectivity.

1.7.1 The model for the explanation of the selectivity effect

Varotsos and Alexopoulos [129] (see also Ref. [131]) suggested the following model for the SES transmission from its source to the measuring station. When the SES is emitted, the current follows the most conductive (neighboring) path through which most of this current travels and the measuring station lies at a site on the Earth's surface (see the point "O" in Fig. 1.29, or Fig. 1.30) close to the upper end of the conductive path (and preferably close to a high-resistivity anomaly, see fig. 25 of Ref. [131]). One can show that the measured electric field is then appreciably stronger than in the case of a homoge-

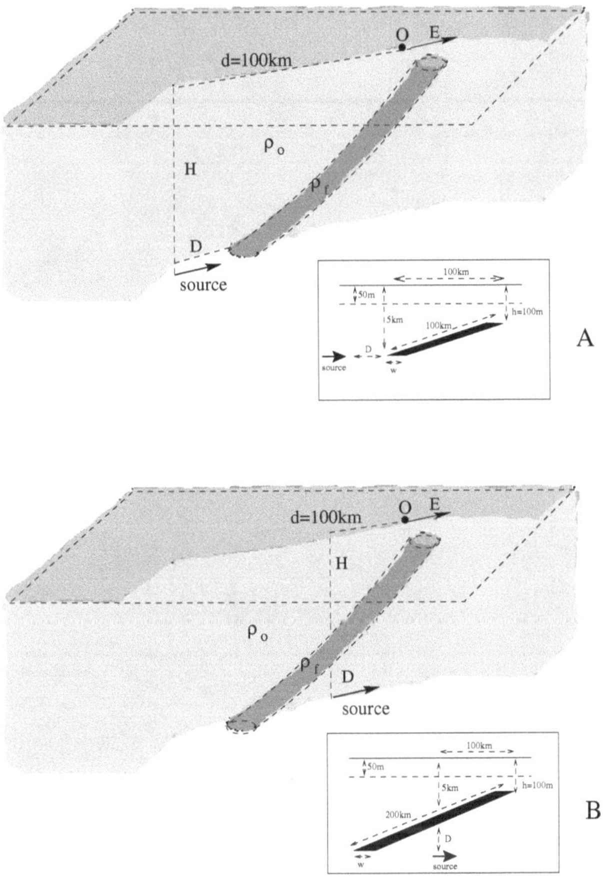

Fig. 1.29 Schematic representation of the SES transmission model suggested by Varotsos and Alexopoulos [129] (see also Fig. 1.30). In general, the dipole source may be parallel (B) or perpendicular to the neighboring conductive path; the case of perpendicular orientation (see Fig. 1.32) is practically equivalent (as far as the amplitude of the electric field above the upper end of the channel is concerned) with the case A. Varotsos and Alexopoulos [129] (see also fig. 25 of Ref. [131]) suggested that the case A is more probable than B; this seems to coincide with aspects that there is always a significant component of the emitting dipole perpendicular to the conductive path (§ 1.7.3). The symbol ρ_0 stands for ρ_{host} used in the text. Taken from Ref. [92]

neous or horizontally layered Earth. In particular, the consequences of the high conductivity path terminating inside the host medium discussed later (e.g. see § 1.7.2.3) conclude that, for large conductivity ratios, the "*edge effects*" play a major role, leading to strong electric field in the host medium close to the termination of the conductive path. In addition, we draw attention to the following case discussed in detail in sections 6.5 and 6.6 of Ref. [125]:

The increase of the electric field value may reach a factor even larger than the conductivity ratio σ/σ'. This is termed *over-amplification* (see § 1.7.2.3).

The two configurations A and B of the electric dipole (with respect to the neighboring path) depicted in Fig. 1.29 can be in principle envisaged. However, Varotsos and Alexopoulos [129] suggested a current source close to the bottom end of the path (e.g., see fig. 25 of Ref. [131], i.e., the case of Fig. 1.29(A) which can be better visualized in

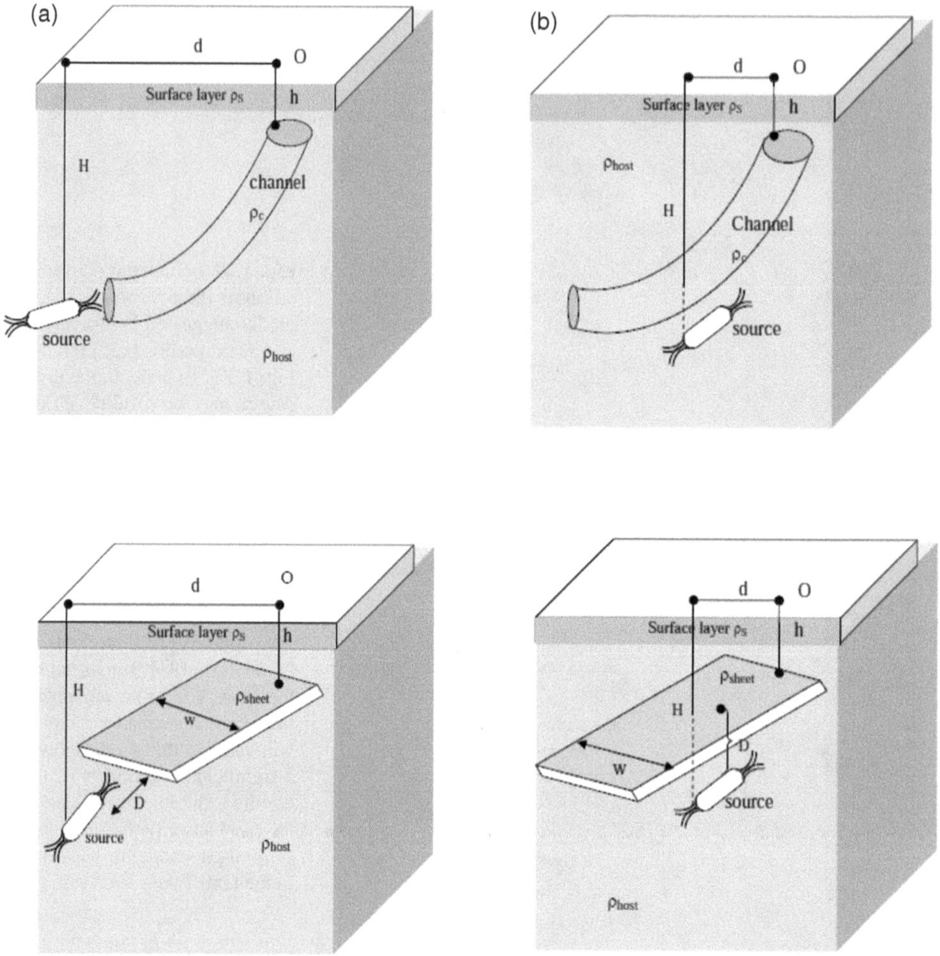

Fig. 1.30 Schematic model for the explanation of the selectivity effect suggested by Varotsos and Alexopoulos [129]. This is an enlarged version of Fig. 1.29, the caption of which applies also here, for better visualization of the parameters. (a) and (b) correspond to the cases of Figs. 1.29(A) and 1.29(B) respectively. Reprinted from Ref. [125], Copyright (2005), with permission of TerraPub.

Fig. 1.30(a)). This is the case, as explained in § 1.7.2.3, which is practically equivalent – as far as the amplitude of the electric field values above the upper end of the channel is concerned – with an electric dipole source being perpendicular to the neighboring conductive path. We note that:

Only when there is a significant component of the emitting dipole source perpendicular to the conductive path, the phenomenon of the "*over-amplification*" may become of paramount importance (§ 1.7.2.3 and § 1.7.3).

1.7.2 Analytical studies related to the explanation of the SES properties

All the analytical studies mentioned in this section have been made in the static approximation, but we draw attention to the point that, as summarized in Section 1.8 (and explained in detail in chapter 8 of Ref. [125]), approximately the *same* conclusions hold for low frequencies which are of interest for the case of SES (i.e., $\lesssim 1$ Hz).

The electric field due to a dipole source being oriented either parallel or perpendicular to a neighboring high conductivity path of conductivity σ embedded in a less conductive medium of conductivity σ' (occasionally also labeled σ_{host}), was investigated in chapter 6 of Ref. [125] (see also Refs. [147, 144]). Two ideal paths were considered: a cylindrical channel (radius R) of infinite length, and a conductive layer (width w) of infinite extent, see Figs. 1.31(a) and (b), respectively when the dipole source is parallel to the conductive path, and Figs. 1.32(a) and (b), respectively when the dipole source is perpendicular. When

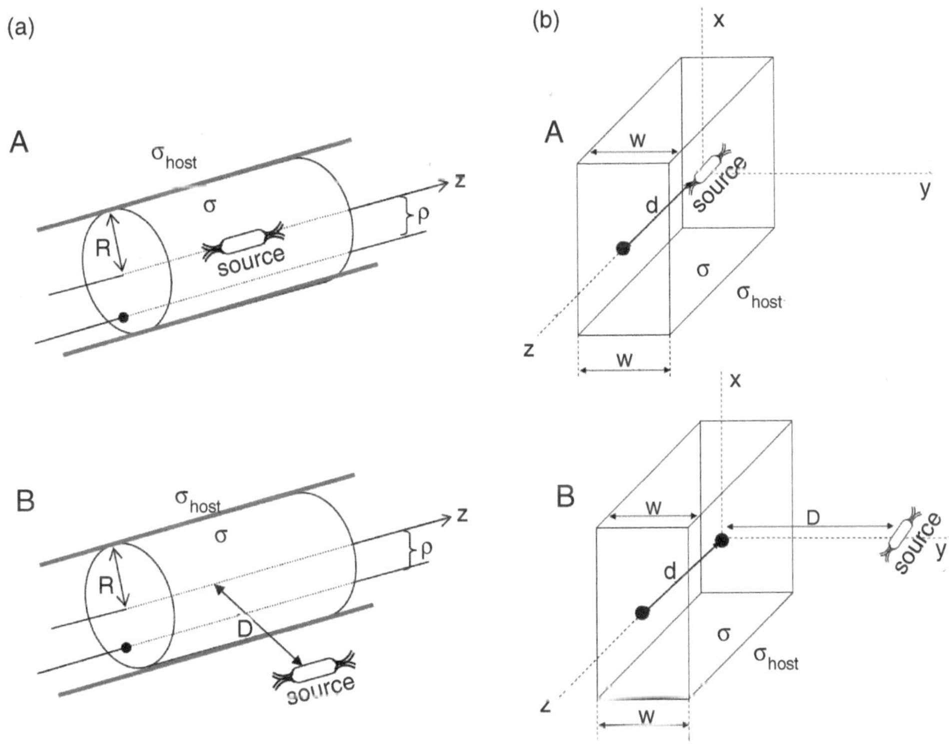

Fig. 1.31 (a) Current dipole lying inside (case A) or outside (case B) but parallel to the main axis of a conductive cylinder (radius R and conductivity σ) of infinite length embedded in a host medium with conductivity $\sigma'(\sigma > \sigma')$. (b) Current dipole lying inside (case A) or outside (case B) but parallel to the surface of a conductive layer of infinite extent (width w, conductivity σ), which is embedded in a host medium with conductivity $\sigma'(\sigma > \sigma')$. Note that the host medium conductivity σ' is labeled σ_{host} in the figure (and occasionally used in the text). Reprinted from Ref. [125], Copyright (2005), with permission from TerraPub.

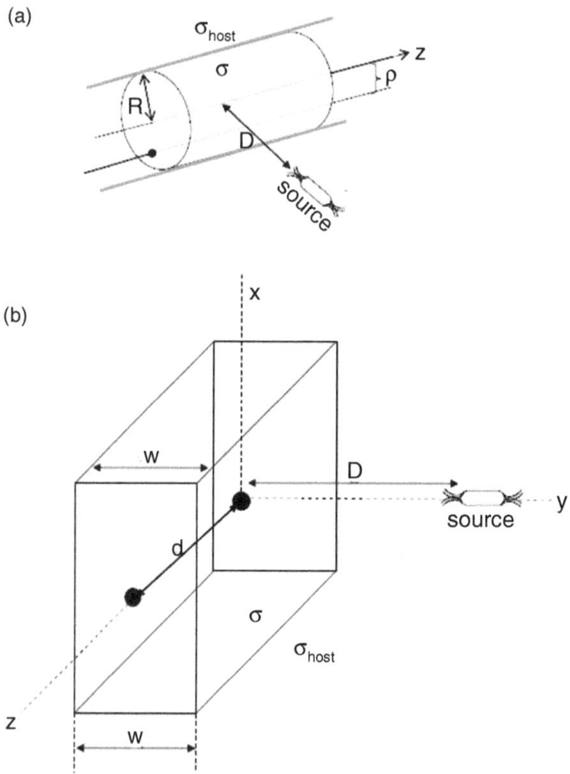

Fig. 1.32 A current dipole source perpendicular to a conductive path. The emitting dipole lies at a distance D from a conductive cylinder of infinite length and radius R (a) or from a conductive layer of infinite extent and width w (b). The host medium conductivity, labeled σ_{host} in this figure, is alternatively label σ' elsewhere (while the conductivity of the path is always labeled σ). Reprinted from Ref. [125], Copyright (2005), with permission from TerraPub.

the source lies outside the conductive path, the distance D between the source and the path is also shown in Figs. 1.29 to 1.32.

The amplitude of the component of the electric field E_z at points lying on the z-axis, shown in Figs. 1.31 and 1.32, and thus inside the path is hereafter labeled E_{inside}.

As an example, we consider the case A of Fig. 1.31(a) and present in Fig. 1.33(a) the amplitude E_{inside} of the electric field component E_z along the axis of the cylinder at $\rho = 0$ versus the distance $d = z$ from the dipole. The ratio of E_{inside} over the corresponding amplitude of the electric field for a full space of conductivity σ' (labeled E_{host}) is shown in Fig. 1.33(b). For a given conductivity ratio, the ratio E_{inside}/E_{host} reaches a maximum value, larger than unity, at a certain (critical) reduced distance – labeled $(d/R)_{crit.\|}$ – and then decreases approaching unity at larger distances. This reflects, if we recall that E_{host} varies with distance as $1/d^3$, the following conclusion.

When studying at (reduced) distances smaller than $(d/R)_{crit.\|}$, the value of E_{inside} decreases (versus distance) at a rate slower than $1/d^3$. In particular for distances smaller than the inflection point $(d/R)_{infl}$, an almost parabolic increase of the ratio E_{inside}/E_{host} is no-

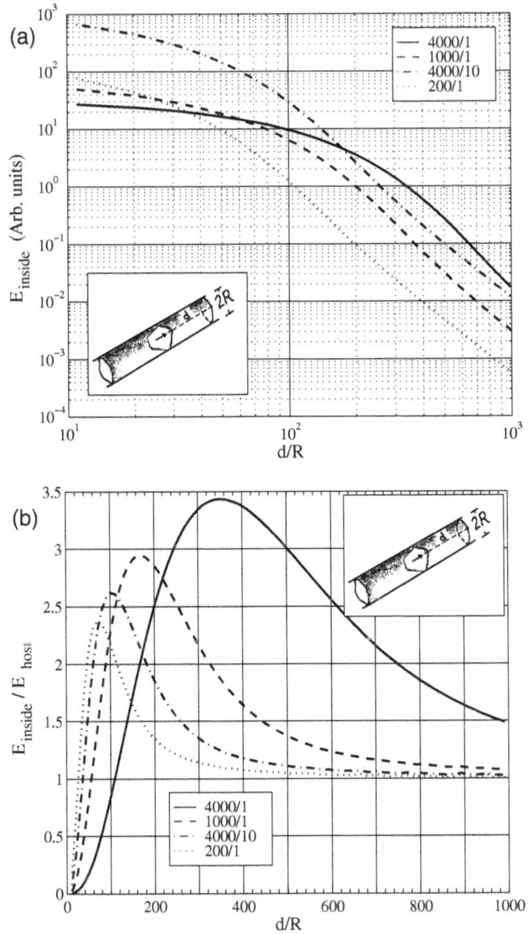

Fig. 1.33 The amplitude E_{inside} of the electric field (a) and the ratio E_{inside}/E_{host} (b), versus the reduced distance d/R from a current dipole lying inside a conductive cylinder (of radius R and conductivity σ) of infinite length embedded in a medium (host) with conductivity $\sigma'(\sigma > \sigma')$. The curves correspond to the conductivity ratios: $\sigma/\sigma' = 4000/1, 1000/1, 4000/10$ and $200/1$, respectively. Reprinted with permission from Ref. [147]. Copyright (1998), American Institute of Physics.

ticed in Fig. 1.33(b), and hence $E_{inside} \propto 1/d$. Estimating the inflection point $(d/R)_{infl}$ of the curves in Fig. 1.33(b), we find that $(d/R)_{crit.\|}$ exceeds $(d/R)_{infl}$ by a factor of around 2.5. At distances much larger than $(d/R)_{crit.\|}$, the value of E_{inside} varies as $1/d^3$ approaching E_{host}.

We now summarize below the main results deduced [125] for large conductivity ratios σ/σ' and for distances (d), between the source and the measuring site, of practical interest.

1.7.2.1 Dipole source parallel to the path of infinite length

This is the case depicted in Fig. 1.31. The main conclusions are [125]:

(a) The amplitude E_{inside} of the electric field inside a cylindrical channel is found to vary as $1/d$, compared to an $1/d^3$ behavior in an isotropic and homogeneous medium; this holds, as mentioned in the example of Fig. 1.33, almost up to a certain (reduced) distance $(d/R)_{infl}$, which increases with the conductivity ratio σ/σ'.

(b) The value of E_{inside}, inside a layer of width w, varies as $1/d^2$; this holds up to a certain (reduced) distance, d/w, which is around $(1/e)(\sigma/\sigma')$.

(c) For conductive paths (cylinder or layer), and at distances d appreciably longer than those mentioned above in (a) and (b), the amplitude of the electric field E_{inside} becomes comparable to E_{host}, i.e., to the value that would be measured for a full volume of conductivity σ'.

This holds irrespective of the fact that the source lies inside the path or outside of it (under the condition that the distance D of the dipole source from the path is much smaller than d). It implies a high value of the current density inside the path, j_{inside}, because $j_{inside}/j_{host} \approx \sigma/\sigma'$.

1.7.2.2 Dipole source perpendicular to the path of infinite length

This is the case depicted in Fig. 1.32. The main conclusions are [125]:

(a) The amplitude E_{inside} of the electric field inside a cylindrical channel increases with the distance d up to a certain distance from the source, and then, at larger distances, decreases. Furthermore, for long (reduced) distances d/R, the value of E_{inside}, decreases upon increasing the distance D of the source from the path, in a way approximately equal to $1/D$; this holds only up to a value of D which exceeds R by a factor of around 10. In general, for long distances d/R, and for small values of D/R, the value of E_{inside} varies as $1/(D/R)$.

The value of E_{inside} at long (reduced) distances d/R becomes comparable to E_{host} for a dipole source relatively close to the path, e.g., $D/R \approx 5$–10 (which implies a high value of the current density inside the path, because $j_{inside}/j_{host} \approx \sigma/\sigma'$). At even smaller distances, e.g., $D/R \sim 2$, E_{inside} may exceed E_{host} by a significant factor, e.g., by a factor of 10, thus implying that j_{inside}/j_{host} is appreciably larger than σ/σ'.

The ratio E_{inside}/E_{host} versus d/R reaches a maximum at a certain value $(d/R)_{crit.\perp}$, which is approximately *equal* to that found $(d/R)_{crit.\parallel}$ for a dipole parallel to the path. This maximum, which significantly exceeds unity when the dipole lies at a small distance $D/R = 2$ from the path, e.g., see Fig. 1.34, has a special importance for the effect of *over-amplification*, see below.

(b) The amplitude E_{inside} of the electric field inside the conductive layer increases with the distance d/w, reaches a maximum when d starts to become comparable with D, and at larger distances (d/w), decreases reaching finally the $1/d^3$ behavior.

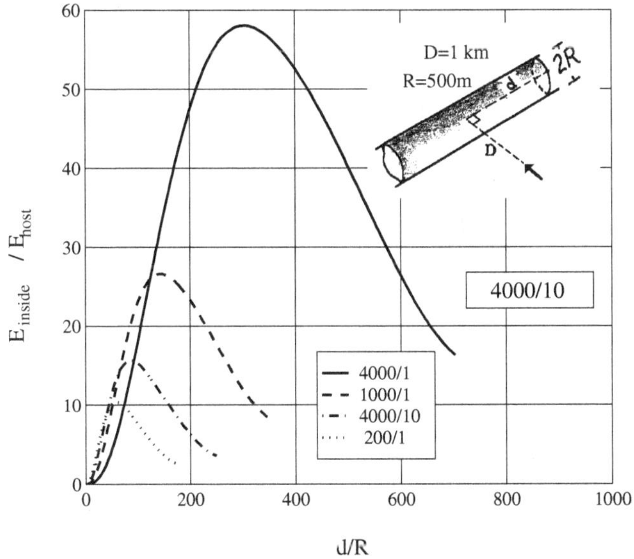

Fig. 1.34 The ratio E_{inside}/E_{host}, versus the reduced distance d/R for a given conductive cylinder ($R = 500$ m) and for the same D value, i.e., $D = 1$ km; the various curves correspond to the conductivity ratios $4000/1, 1000/1, 4000/10$ and $200/1$. The current source dipole is assumed perpendicular to this conductive path. Taken from Ref. [144].

At very long (reduced) distances (e.g., $d/w = 10^2$), the value of E_{inside} does not practically depend either on D or w (for a dipole source relatively close to the path) and becomes comparable to E_{host}, thus implying a high value of the current density inside the path.

1.7.2.3 Dipole source close to a path terminating inside the host (more resistive) medium. The case of over-amplification

If a conductive path terminates inside the host (more resistive) medium, then at some points of the host medium close to the termination of the path the amplitude of the electric field, labeled $E_{outside}$, exceeds E_{host} by a factor $\sim \sigma/\sigma'$ due to current conservation, e.g. see Fig. 1.44. We focus, however, below on certain cases where the ratio $E_{outside}/E_{host}$ becomes even larger than σ/σ' which, as mentioned (§ 1.7.1), is called *"over-amplification"*.

In general, in order to have *over-amplification*, i.e., $E_{outside}/E_{host} > \sigma/\sigma'$, the condition required is $E_{inside}/E_{host} > 1$ along the perpendicular to the interface at the termination point. Therefore, cases of conductive paths with infinite length that have been found to fulfill, in a certain region of d/R values, the condition $E_{inside}/E_{host} > 1$, are candidates for *over-amplification* when they happen to terminate (inside the more resistive medium) approximately in this region. Such regions of d/R values have been noted above in both cases, i.e., when the current dipole source is either parallel or perpendicular to a highly conductive cylinder of infinite length, see Figs. 1.33(b) and 1.34, respectively. It was mentioned that, for a given conductivity ratio, the ratio E_{inside}/E_{host} versus d/R shows a maximum at a distance $(d/R)_{crit}$. Comparing the corresponding values from these two figures and considering that the curves E_{inside}/E_{host} versus d/R have a considerable full-width half maximum (see Figs. 1.33(b) and 1.34), we may conclude that, for the same conductivity ratio, they maximize almost at the same region of d/R values. This fact, i.e., $(d/R)_{crit.\perp}$ is almost equal to $(d/R)_{crit.\parallel}$ (already mentioned above), implies that the corresponding critical value $(d/R)_{crit}$ when the dipole source forms *any* angle between 0 and 90° is almost the same, i.e.,

$$(d/R)_{crit} \approx (d/R)_{crit.\parallel} \approx (d/R)_{crit.\perp} \tag{1.71}$$

It should be emphasized, however, that, when comparing the two orientations of the dipole source, the corresponding critical values of the ratio E_{inside}/E_{host}, which will be labeled $(E_{inside}/E_{host})_{crit}$, are quite different.

In particular, a current dipole source perpendicular to a conductive cylinder is in general expected to show, for small values of D/R, a stronger over-amplification than that of the parallel one when the termination occurs in the range of $(d/R)_{crit}$. In other words, the value of $(E_{inside}/E_{host})_{crit.\perp}$ is markedly larger than $(E_{inside}/E_{host})_{crit.\parallel}$.

Concerning the electric field $\mathbf{E}_{outside}$ when a dipole source forms a certain angle with a neighboring highly conductive path terminating *inside* the host medium, the following points emerge [144] (see also section 6.4 of Ref. [125]).

(1) The direction of $\mathbf{E}_{outside}$ close to an edge is regulated by the angle between the emitting dipole and the conductive path *as well as* by the distance from the source.

(2) The amplitude $E_{outside}$ is usually larger than that of E_{host} by a factor of around σ/σ', but there are also some cases of *over-amplification*, i.e., the value of $E_{outside}/E_{host}$ exceeds the conductivity ratio σ/σ'. Such an *over-amplification* may also occur in the cases of conductive paths that are not connected.

(3) The amplitude $E_{outside}$ versus the distance r from the edge varies only slowly, i.e., $E_{outside} \propto 1/r^{\theta}$ where θ is around (but smaller than) unity.

Concerning an elongated conductive spheroid, which may approximate an elongated conductive body, the following conclusions hold as far as the over-amplification is concerned [144] (see also pp. 149–153 of Ref. [125]):

(i) A dipole source produces *over-amplification* when lying either perpendicular to a neighboring conductive path (terminating at a critical reduced length) or close to (and lying on the major axis of) an elongated conductive spheroid (body).

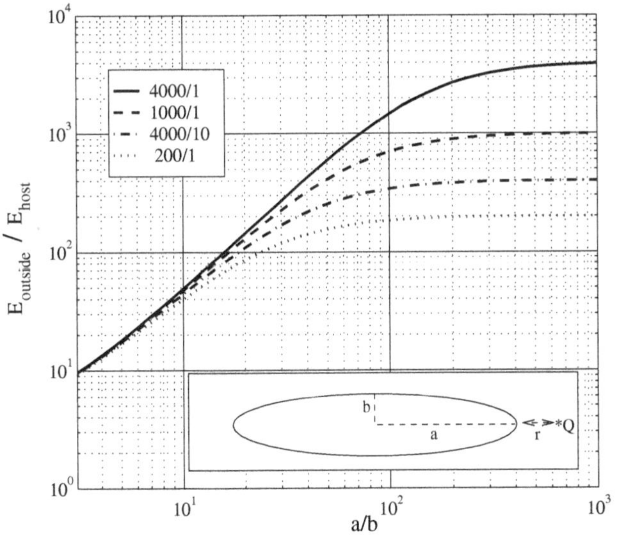

Fig. 1.35 The ratio $E_{outside}/E_{host}$ for a conductive spheroid versus a/b embedded in a more resistive medium (for various conductivity ratios σ/σ'). The emitting current dipole source lies (along the major axis) at a remote distance from the one end (this is almost the so-called "plane wave" response), while the measurements are made at a point Q, lying very close to the other end. Taken from Ref. [144].

(ii) Elongated conductive spheroids "amplify" the remote electric fields by a factor of around σ/σ' at the most (see Fig. 1.35), while with a nearby emitting electric dipole source (lying on the major axis) they may lead to *over-amplification without requiring so much elongation.*

(iii) An electric dipole of finite length that approaches *perpendicularly* an elongated conductive body, may give rise to *over-amplification* provided that one pole of the source is very close to the body.

As for the amplitude of the electric field, the following two properties have been proven (see pp.150–151 of Ref. [125]).

When changing the distance D of a dipole source that lies on the major axis and approaches the one end of an elongated conductive spheroid, the amplitude of the electric field surrounding the other end varies as $\sim 1/D$.

A dipole source at a small distance D from one end of an elongated conductive spheroid, produces electric field values which vary with the distance r from the other edge as $1/r^{\theta}$, where $\theta < 1$, in accordance to the point (3) mentioned above.

1.7.3 Direction of the maximum principal stress with respect to the neighboring earthquake fault

In the late 1980s it was reported that the direction of the maximum horizontal principal stress near the San Andreas fault is in general nearly perpendicular to the fault (see Ref. [166] and references therein). However, Scholz [96] pointed out that the directions reported to be nearly fault normal in central California do not correspond to regional stresses but are a result of active folding within folds that have been rotated 20°–30° clockwise from their original orientations. A debate followed on this issue (e.g. see Scholz [97]), which could be alternatively seen as diversity of views on whether the San Andreas fault is weak (or strong) relative to the surrounding crust. Some details can be found in pp. 157–158 of Ref. [125], but here we keep as a conclusion that one could compromise the current aspects as follows:

The maximum horizontal principal stress near the fault, is normally oriented at an angle of 30°–60° with respect to the normal vector to the fault, but may be perpendicular to the fault if the fault has a low coefficient of friction.

This, under the simplifying assumption that the emitting current dipole source is oriented parallel to the maximum principal stress along a fault (which seems to be true, at least, in the frame of either the pressure stimulated currents SES generation model (§ 1.6.2) or the deformation-induced charge flow phenomenon (§ 1.6.4), implies the following: The existence of a significant component perpendicular to the fault cannot be anyhow ignored, because as explained above (§ 1.7.2.3) a dipole source perpendicular to a nearby conductive path may result in over-amplification.

Thus, we conclude that it is very likely that there exists a considerable current dipole moment component perpendicular to the fault. This may be of major importance when calculating the electric and the magnetic field at sites close to the upper end of the channel, because it may give rise to *over-amplification*.

1.7.4 Explanation of the SES properties based on analytical studies

Earthquakes (EQ) occur by slip on faults which may have lengths of several tens of kilometers and widths of a few to several hundred meters (for example, in the San Andreas fault, the cataclasite fault *core* has a width of the order of 10–100 m, e.g., see Ref. [14]). The resistivities of faults have been found to be around a few Ω m to 10 Ω m, thus being 10^2 to 10^3 times more conductive than the surrounding medium, which, at a usual depth of 5–30 km, has a resistivity of 10^3 to 10^4 Ω m. Thus, in the case of SES, the emitting source (i.e., the EQ preparation zone, where stresses are "accumulating" before rupture) *should* lie at a small distance D from a neighboring conductive path with a large conductivity ratio, e.g., $\sigma/\sigma' = 10^2$ to 10^3 or so. The length of a fault, which is of the order of several tens of km, is drastically larger than its other two dimensions. Therefore, the situation, although in reality lies between the two ideal conductive paths discussed in § 1.7.2, may be better approximated by a conductive cylinder (radius R) rather than a conductive layer (width w).

1.7.4.1 Explanation of the selectivity effect

We show below that, if the electric field measurements are carried out at a distance d (e.g., $d \approx 100$ km) much longer than the distance D between the emitting source and the neighboring conductive path, the SES detectability does not depend on the exact values of the distance D or the width of the path.

Since $d \gg D, w, R$, the *emitting source* is usually assumed to be a point electric dipole. A value of dipole moment $p = 8 \times 10^{-4}$ Cb·m was estimated by Slifkin [100, 101], see § 1.6.3.2, assuming a *modest* value for the density of dislocations and considering a slab with a length of $L \approx 1$ km and a cross-sectional area of 1,000 m × 100 m = 0.1 km^2. This equivalently results in a current dipole moment $Il = 22.6$ A km when assuming $\rho_{host} = 4{,}000\Omega$ m and considering $Il = \frac{p}{\varepsilon_0 \rho_{host}}$ giving rise, as mentioned (§ 1.6.3.2), to an electric field of 7 mV/km at distances of 10 km. A typical seismic source of an EQ with (magnitude) M $\approx$ 5, however, has a length of around 5 km [13] and a cross-sectional area of around 1 km^2.

Therefore, when disregarding – for the purpose of our calculation – factors of around 2, or so, we may estimate that the relevant emitting source for a M $\approx$ 5 EQ should have [147] a dipole moment two orders of magnitude larger than that estimated by Slifkin [101], hence $Il \approx 22.6 \times 10^2$ A km.

This value will be used below (see also § 1.8.4), although we could have accepted larger values due to a probably larger density of dislocations.

Estimation of the electric field at remote distances. For long distances of practical interest, e.g., $d/R \approx 200$ or $d/w \approx 200$, if the dipole source is parallel to the path for any value of D lying in the range from a few to 10 km, the amplitude of E_{inside} is of the order of E_{host}, see point (c) in § 1.7.2.1. This is of the order of 1 mV/km ($\approx E_{host}$) at distances $d \approx 100$ km, as can be seen for example in Fig. 1.36, see the asterisks. The same is true for a dipole with an orientation perpendicular to the path; see § 1.7.2.2 (in cases where there is no overamplification).

The electric field E_{host} ($\approx$1 mV/km) will be enhanced, as mentioned in § 1.7.2.3, by a factor almost equal to the conductivity ratio σ/σ' (or even larger due to *overamplification*; see below) if the measurements are carried out in the vicinity of the termination of the conductive path. Thus, at certain sites lying at distances $d \approx 100$ km, electric field values of the order of at least around 10 mV/km are expected which are well above typical noise levels.

Hence, if we make reasonable assumptions, the SES detectability at certain remote sites (sensitive sites) becomes clear as long as $d \gg R$ (or $d \gg w$), and $d \gg D$.

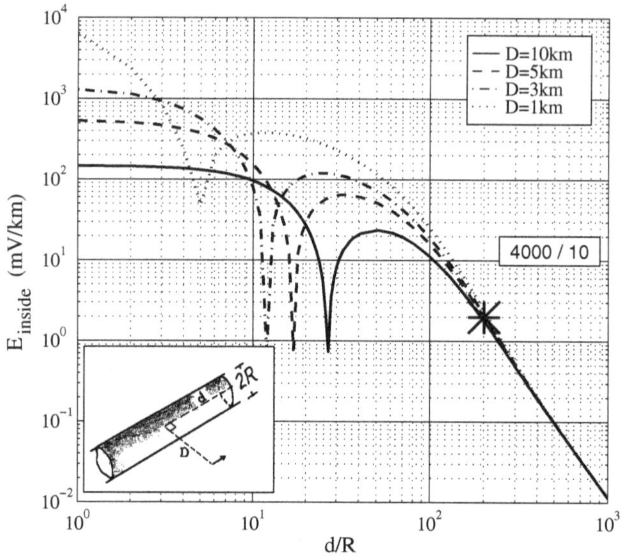

Fig. 1.36 The amplitude E_{inside} versus d/R, for various values of the distance D (= 1, 3, 5, 10 km, see the inset) of the emitting dipole from a given conductive cylinder ($R = 500$ m). For values $d/R \approx 2 \times 10^2$ or larger, all curves practically coincide. The points with asterisks correspond to $d \approx 100$ km. The dipole moment is 10^2 times that estimated by Slifkin [100, 101]. Conductivity ratio $\sigma/\sigma' = 4000/10$. Reprinted with permission from Ref. [147]. Copyright (1998), American Institute of Physics.

In other words, an explanation of the *selectivity effect* of SES becomes evident, if we just consider the fact that the emitting current dipole source lies in the vicinity of a path appreciably more conductive than the surrounding medium and that the measurements are carried out in the more resistive medium close to the termination of the conductive path.

The orientation of the emitting dipole source is likely to be around the normal to the neighboring fault (see § 1.7.3) which may result in the *over-amplification* phenomenon, thus strengthening the importance of the "edge effects" concerning the SES detectability at remote sites.

Therefore, detectable electric field values are found in two regions ("sensitive sites") only: first, at sites in the region just above the source (e.g., $d \approx 10$ km, due to the small attenuation *if* the source depth is not too large) and second, at those remote sites (e.g., $d \approx 100$ km) of the Earth's surface lying close to the upper end of a path (if it terminates near the Earth's surface).

We clarify, however, that when the depth of the source increases, the electric field in the first region becomes smaller and hence (for appreciably large depths) may lie below the detectability limit (see the electric field numerical calculations below in § 1.7.5.2); on the other hand, the values in the second region remain practically unchanged provided that the termination of the conductive path is close to the Earth's surface.

Hence, the electric field *may* reach measurable values at remote sites, but may be *not* at shorter. This behavior has been experimentally observed (§ 1.3.4).

1.7.4.2 Explanation of the use of the ratio of the two SES components to determine the EQ epicenter

As already mentioned in § 1.7.2.3, the direction of the electric field $\mathbf{E}_{outside}$ close to the termination of a conductive path (at sites lying in the more resistive medium) is regulated by the angle between the emitting dipole source and the conductive path as well as by the distance d of the source from the measuring site. Therefore, if we assume that, for a given seismic area, the mechanism of the build up of stresses remains the same (which of course does not always hold, e.g., see Ref. [125]), thus reflecting almost the same direction of the emitting dipole source, the relevant earthquakes should give, at the same measuring (remote) site, SES with the same direction, and hence the same ratio of the two SES components. On the other hand, two different epicentral areas, which are likely to have in general different directions of accumulating stresses as well as different epicentral distances, should correspond to different ratios of the SES components (measured at the same remote site).

This provides an explanation of the fact that the ratio of the two SES components has been used (see § 1.3.5) for the estimation of the epicentral area of an impending EQ.

1.7.5 Electric field numerical calculations explaining the selectivity effect

1.7.5.1 The procedure for the numerical simulation of the selectivity model

We now explain how we simulate the model of Fig. 1.29 studied in Ref. [90]. Since we are interested in the case where the focal depth is within the range of 5–50 km, we can safely assume that the "host rock" has a host resistivity between 10^3 Ω m and 10^4 Ω m [147]. As a first approximation, we select the value $\rho_{host} = 4 \times 10^3 \Omega$ m. The surface layer with depth 50 m has a typical resistivity value $\rho_S = 200$ Ω m. The resistivity ρ_f of a fault is known to be around 10 Ω m (or smaller see § 1.7.4), giving the conductivity σ_c of the channel, $\sigma_c = (1/\rho_f) = 0.1$ S/m (e.g., Ref. [145]). Concerning the width w of the channel, we may assume values of the order of 100 to 1000 m (§ 1.7.4). Calculations have been made with various w values, e.g., $w = 500$ m in Refs. [145, 90] or $w = 800$ m in Ref. [91], and it was found that the essence of the main conclusions is not affected by the exact w value chosen. A value $\Delta z \approx 500$ m was assumed for the thickness Δz and hence the channel conductance $\tau = \sigma_c \Delta z$ is 50 S. The same value of τ results, if we alternatively consider for example $\rho_f = 2$ Ω m, $\Delta z = 100$ m or $\rho_f = 1$ Ω m, $\Delta z = 50$ m, etc, thus the resistivity contrast ρ_{host}/ρ_f is of the order of 10^3.

For reasons explained in § 1.7.4.1, the dipole source associated with a $M \approx 5.0$ EQ should be around $Il = 22.6 \times 10^2$ A km and *hereafter* the calculations will be carried out by using this value except otherwise stated. The current dipole is assumed to be oriented along either the x-axis or the y-axis and its projection on the Earth's surface lies at a distance of 100 km from the point "O" (see Fig. 1.29) with coordinates (0,0,0). The latter point represents the projection of the channel's upper end on the Earth's surface.

In summary, the conductivity structure involves a two layered Earth (with a 50 m surface layer with resistivity $\rho_S = 200\ \Omega$ m, and a host with resistivity $\rho_{host} = 4000\ \Omega$ m) and a conductive channel with resistivity $\rho_c = 10\ \Omega$ m. The dimensions of the channel were taken either 500 m $\times$ 500 m $\times$ 100 km or 500 m $\times$ 500 m $\times$ 200 km for the cases A and B of Fig. 1.29, respectively. The conductive channel was modeled by a thin sheet (depicted in the insets of Fig. 1.29(A) and 1.29(B)) of conductance $\tau = 50$ S. In other words, the conductive channel was modeled by a slightly dipping conductive thin sheet. The results were obtained in a frequency range around 10^{-2} Hz by running the EM1DSH program (Hoversten and Becker [37]), with two different rectangular grids of 6 $\times$ 150 and 5 $\times$ 200 cells, on HP 735 or a Sun Ultra Enterprise 450 digital computer (note that a larger number of cells have been also used; see § 1.7.5.3). Only the electric field values that have discrepancy less than 20% between the two discretizations were used for further analysis; the error bars in the diagrams depict the extent of this discrepancy. The real problem was studied by a scaled model $L_m = 10^{-7} L_w$, following the "similitude relationship" (see Refs. [145, 165]) that relates the corresponding frequency ω, magnetic permeability μ, conductivity σ, and length scale L of a real world (w) problem to a model (m) problem:

$$\omega_m \mu_m \sigma_m L_m^2 = \omega_w \mu_w \sigma_w L_w^2. \tag{1.72}$$

Of course, the current source was scaled by the corresponding factor 10^{-7} mentioned above.

1.7.5.2 Numerical results when assuming point dipole current source

Let us consider the case when the dipole source (oriented along the x-axis) is almost parallel with a neighboring conductive sheet. Following § 1.7.5.1, Fig. 1.37 depicts the results of the calculation for the case of a model shown in Fig. 1.29(B) (with $D = 5$ km and $w = 500$ m, see Fig. 1.37(D)). All the values have been reduced by the magnitude E_{host} (-100,0,0) of the electric field ($\approx$700 mV/km) that would be measured just above the source on the Earth's surface in *absence* of the channel. The distribution of the absolute values of the horizontal component E_x of the electric field on the XZ plane, $Y = 0$ and on the XY plane, $Z = 0$, i.e., at the Earth's surface, is shown in Figs. 1.37(A) and 1.37(B) respectively, while Fig. 1.37(C) depicts E_y at the Earth's surface. Since "artificial" noise is usually of the order of 1 mV/km, Fig. 1.37(B) indicates that:

For a $M \approx 5$ earthquake, there are two regions on the Earth's surface in which E_x is detectable, i.e., larger than 10 mV/km: one region lies close to the channel's upper end and another one just above the source.

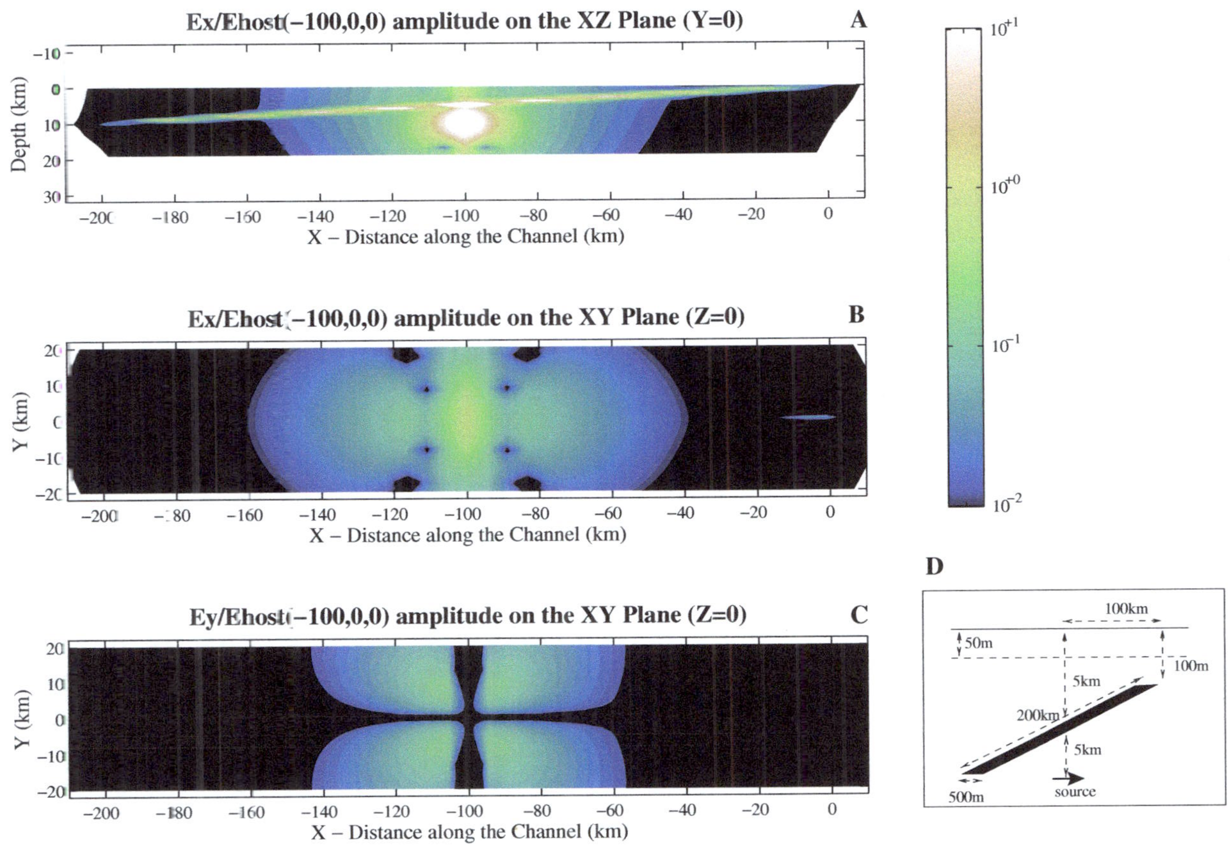

Fig. 1.37 Calculated values of the amplitude of various components of the electric field (reduced by the value E_{host} (−100,0,0) above the source) and the schematic diagram (D) of the model of Fig. 1.29(B) used in the calculation. Taken from Ref. [90]. Copyright (1999), American Geophysical Union. Reproduced by permission of American Geophysical Union.

Repeating the calculation for larger depths of the source and comparing the values in these two regions, we find the following [90]: the electric signals may become stronger close to the channel's upper end than directly above the source at a depth $z = 50$ km. In such cases a signal may be detected at larger epicentral distances, but not at shorter.

As for the case of Fig. 1.29(A), i.e., a dipole source close to the bottom of an almost horizontal conductive sheet, the results can be found in Ref. [145].

1.7.5.3 Numerical results when assuming a current dipole source of finite length

In order to better approximate [91] the real situation, the emitting source is now assumed to be an electric dipole of finite length, i.e., $l = 2$ km, centered at $x = -100$ km, $y = -1.5$ km at a depth $z = 5$ km with $Il = 22.6 \times 10^2$ A km. The dipole is oriented either perpendicular (i.e, along the y-axis) or almost parallel (i.e, along the x-axis) with respect to a 200 km long conductive sheet of width $w = 800$ m. As in § 1.7.5.1, the sheet was buried in a two layer Earth, with a surface layer having resistivity 200 Ω m and thickness 50 m and a host medium with $\rho_{host} = 1/\sigma' = 4000$ Ω m; the conductance of the sheet $\tau = \sigma\Delta_z$ was again chosen to be 50 S. Discretizations of the order of 2,000 to 3,000 cells were used. The error bars in Fig. 1.38(a) depict the standard deviation on the average result obtained for each case after running several discretizations on a SUN Enterprise 450 digital computer.

Figure 1.38 depicts the results when the current electric dipole is perpendicular to the conductive sheet. In Fig. 1.38(a), the ratio E/E_{host} is shown at sites along the projection of the channel on the Earth's surface. At sites above the upper end of the sheet, the ratio reaches appreciably large values, i.e., of the order of 10^4. This is probably due to "over-amplification" (see §1.7.2.3), if we recall that the electric field may reach values even one order of magnitude higher than $E_{host}(\sigma/\sigma')(= 400E_{host}$, if $\rho = 1/\sigma = 10$ Ω m). Figure 1.38(b) shows the electric field lines on the Earth's surface in the region above the source. In Fig. 1.38(c), we depict the logarithm of their ratio E/E_{host} in the region around the upper end of the channel; this figure (when also considering that $E_{host} \approx 1$ mV/km, see Fig. 1.36 and § 1.7.4.1) reveals the existence of a SES-sensitive region ($E \approx 10$ mV/km) with dimensions of the order 20 km $\times$ 4 km elongated along the projection of the channel on the Earth's surface. This compares favorably with experimental results at IOA (§ 1.3.4).

Thus, the numerical calculations reveal that the extent of a SES sensitive site is of the order: (a few tens of km) $\times$ (a few km), as the experiments show.

Note that the corresponding results when the extended dipole is oriented parallel with the sheet show [91] that the values of the ratio E/E_{host} are roughly one order of magnitude less than those in Fig. 1.38(a), thus indicating that no "over-amplification" occurs in this case.

The case of an "artificial" noise. The aforementioned numerical results reveal that actually a dipole source perpendicular (and very close) to the path produces appreciably larger electric fields compared to those of the parallel orientation. The situation changes (see Fig. 1.39), however, when considering a source lying *on* the surface of the Earth,

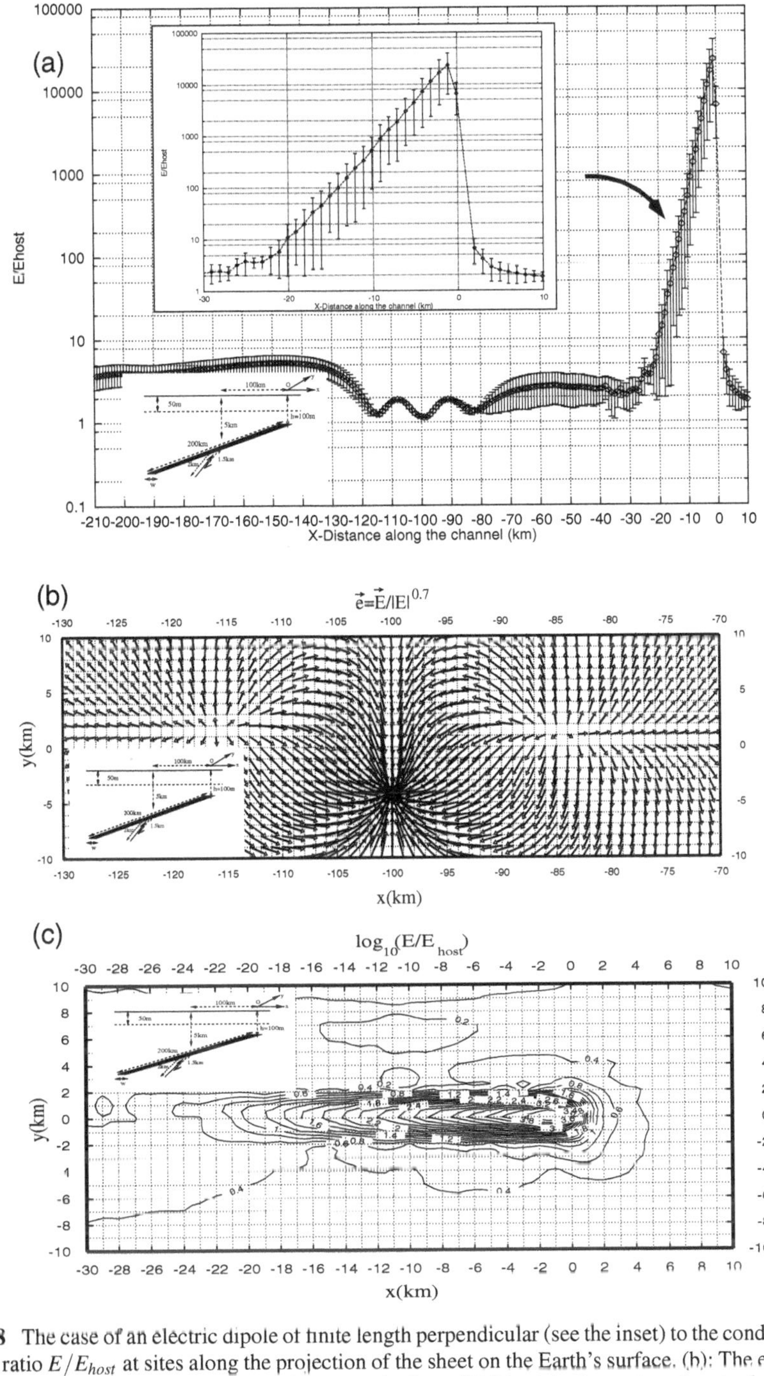

Fig. 1.38 The case of an electric dipole of finite length perpendicular (see the inset) to the conductive path. (a): The ratio E/E_{host} at sites along the projection of the sheet on the Earth's surface. (b): The electric field lines in the region above the source. (c): The quantity $\log_{10}(E/E_{host})$ in the region close to the projection of the upper top of the channel on the Earth's surface. Reprinted from Ref. [125], Copyright (2005), with permission from TerraPub.

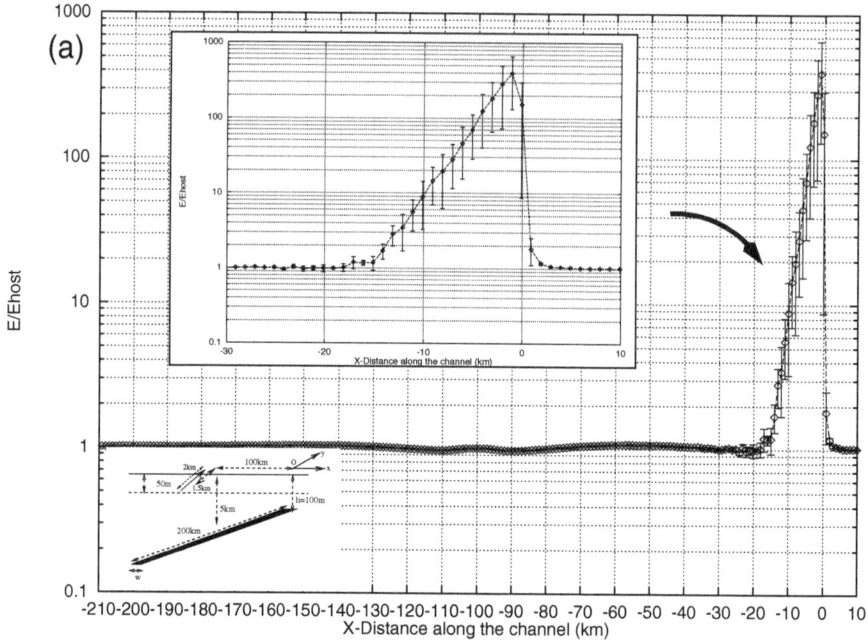

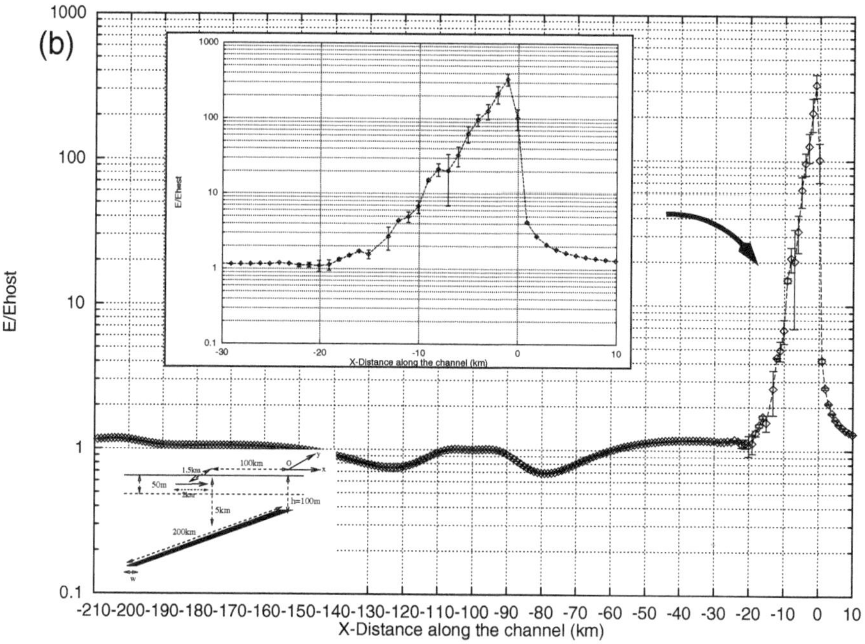

Fig. 1.39 The case of the emission of an "artificial" noise, e.g., an extended electric dipole lying on the Earth's surface: the ratio E/E_{host} at sites along the projection of the sheet on the Earth's surface, when the dipole is either perpendicular (a) or parallel (b) to this projection. Taken from Ref. [91].

as in the case of an "artificial" (man-made) noise. In this case the results are depicted in Figs. 1.39(a) and 1.39(b), which correspond to the same dipole source located either perpendicular (a) or parallel (b) to the projection of the sheet on the Earth's surface. They show that the values of the ratio E/E_{host} are now almost independent of the orientation of the source.

More importantly, a comparison of Fig. 1.39(a), or 1.39(b), with Fig. 1.38(a), reveals that an "artificial" noise source produces appreciably smaller ($\approx$2%) electric field values than the SES-emitting dipole source.

1.7.5.4 Main conclusions of the electric field numerical calculations

The electric field numerical calculations reveal the following:

(1) The simple model of Fig. 1.29 (or Fig. 1.30) assumes that a dipole current source lies in the vicinity of a conductive channel, which terminates below the Earth's surface. This results in electric field E that is significantly intensified (compared to the case with no channel) in the region *above the upper end of the channel (hereafter called AEC)*. Assuming that AEC lies at epicentral distances of $d \approx 100$ km, this model can explain that: (a) the electric field values (hereafter called E values) at AEC are larger than those measured at points on the Earth's surface that may lie at shorter epicentral distances (note that for source depths larger than a certain value, the E values at AEC may even become larger than those measured at the Earth's surface just above the source); (b) assuming current source dipole moments consistent with M $\approx$ 5 EQ, the E values at AEC are detectable (>10 mV/km).

(2) Comparing the numerical results when a dipole of finite length is oriented either perpendicular to or parallel with a neighboring conductive path, we find that, in the former case, the E values above the AEC are appreciably larger than in the latter.

(3) Current aspects (see § 1.7.3) reveal that a SES-emitting dipole source has probably a significant component perpendicular to the neighboring fault. Therefore, in view of the previous conclusion, the intensification of the E values above the upper end of the conductive path is expected to be large (since *over-amplification* then occurs). In this case, the numerical calculations show that the SES-sensitive region is an elongated one, having dimensions of the order a few tens of km $\times$ a few km (e.g., see Fig. 1.38(c)) in agreement with the experimental findings.

1.7.6 Magnetic field calculations

Analytical studies of the magnetic field in the presence of highly conductive bodies, e.g., paraboloid, spheroid and cylinder, embedded in a less conductive medium can be found in chapter 10 of Ref. [125] (see also Ref. [146]). Here, we solely focus on numerical simulations of the magnetic field related with the model described in § 1.7.1, that has been suggested for the explanation of the SES selectivity effect. In particular, since the presence of a conductive sheet may change both the horizontal and the vertical component of the magnetic field, we will summarize below whether our modeling reveals that generally the following two inequalities hold:

$$B_h/B_{h,layer} < B_z/B_{z,layer} \tag{1.73}$$

and

$$B_h > B_z \tag{1.74}$$

where the symbols B_h and B_z denote the horizontal (h) and the vertical (z) components of the magnetic field in the presence of the conductive sheet, while the subscript "layer" refer to the cases in the absence of the sheet. The inequality (1.73) is reminiscent of the well known result [18] that near the edges of conductivity contrasts the vertical magnetic field is enhanced for surface (or MT) sources.

We now summarize the main features of the results for the magnetic field near the upper end of an almost horizontal conductive sheet:

(1) When the dipole source lies in the same vertical plane (xz plane) with the main axis x of the elongated sheet as in Fig. 1.29(B), the presence of the sheet results in: (a) larger (relative) increase of the vertical component of the magnetic field compared to that of the horizontal one, i.e., the inequality (1.73) holds; (b) the B_h value still remains larger than B_z, i.e., the relation (1.74) holds ($B_h > B_z$).

(2) When the dipole source lies almost in the same horizontal plane with the sheet (e.g., if the dipole depth in Fig. 1.29(B) is decreased to 5 km): (a) If the dipole is parallel to the main axis of the sheet, the magnetic field is mainly horizontal as in the case of no sheet. (b) With a dipole perpendicular to the main axis of the sheet, the magnetic field is tilted leading to a dominant vertical component. Studying various orientations (see Fig. 1.40) of the dipole with respect to the main axis of the sheet, we find that, when the dipole is oriented around the normal to the sheet, the magnetic field near the upper end of the sheet has a dominant vertical component.

(3) When the electric dipole is located close to the center of the conductive sheet (for example, if in Fig. 1.29(B) the dipole lies 5 km deeper than the center of the channel, i.e., at a total depth of 10 km), and is oriented $-52°$ to $+53°$ with respect to the normal to the sheet, the magnetic field has a dominant vertical component near the upper end of the sheet.

In view of the aforementioned features, the magnetic field variations accompanying the SES are likely to have a dominant vertical component, since the SES current dipole source is expected to have a considerable component directed around the normal to the fault (see § 1.7.3).

In order to get an insight into the amplitude of the magnetic field variations that accompany SES for shallow (of $\approx$5 km depth) earthquakes, we proceed below to the calculation of the magnetic field for the current dipole source of finite length considered in § 1.7.5.3.

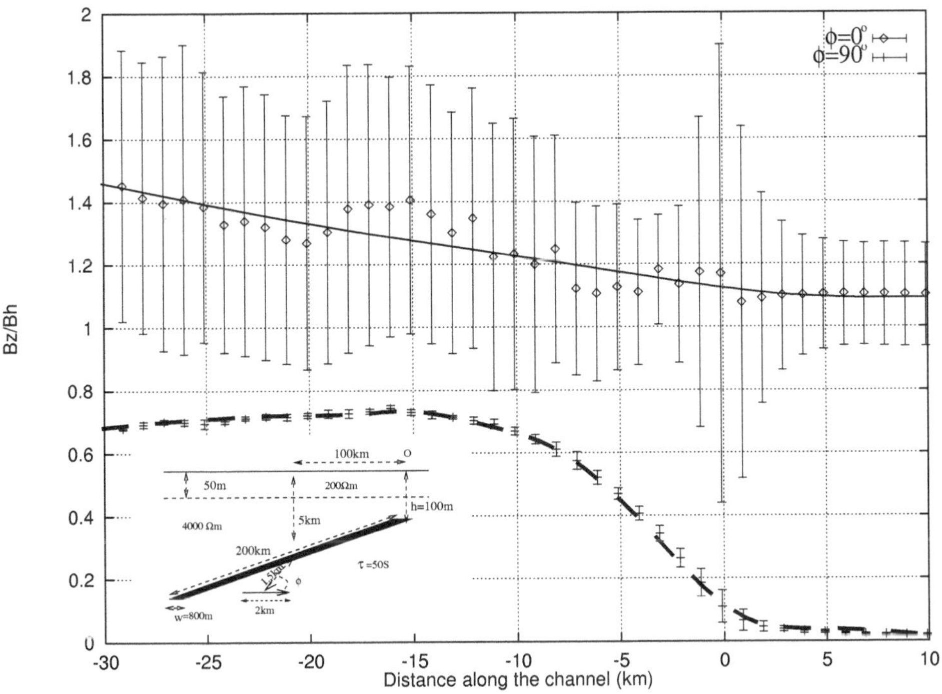

Fig. 1.40 Calculated values of the ratio B_z/B_h along the axis $y=2$ km, $z=0$ for the dipole $l=2$ km, located at $D_y=1.5$ km, which is also considered in Fig. 1.41. The results for both orientations of the dipole, i.e., almost parallel ($\phi=90°$, broken lines) or normal ($\phi=0°$, solid lines) to the sheet ($w=800$ m), are depicted. The spline curves have been drawn as a guide to the eye. Taken from Ref. [92].

Hence, the emitting source is assumed at a depth $z=5$ km with $Il=22.6\times10^2$ A km which corresponds to a magnitude M $\approx$ 5 earthquake (§ 1.7.4.1). Figure 1.41 depicts the results when the current dipole source is perpendicular (i.e., along the y-axis) to the conductive sheet, thus probably causing "over-amplification", see also Fig. 1.38. Figures 1.41(a) and 1.41(b) show the amplitude of the total magnetic field B on the surface of the Earth, in the region above the source (epicentral area) and above the upper end of the sheet, respectively. The feature of the contours for the logarithm of the amplitude of the ratio B/B_{host}, in both regions, can be seen in Fig. 1.41(c). For both orientations of the emitting dipole source, the calculated values of the ratio B_z/B_h on the Earth's surface, along the axis $y=2$ km, $z=0$, are plotted in Fig. 1.40.

In view of these results, we now comment on the amplitude of the magnetic field variations accompanying SES. We clarify two cases:

(1) *Epicentral area.* A few km from the epicenter, the calculated B values are found detectable, since they exceed 1 nT (see Fig. 1.41(a))

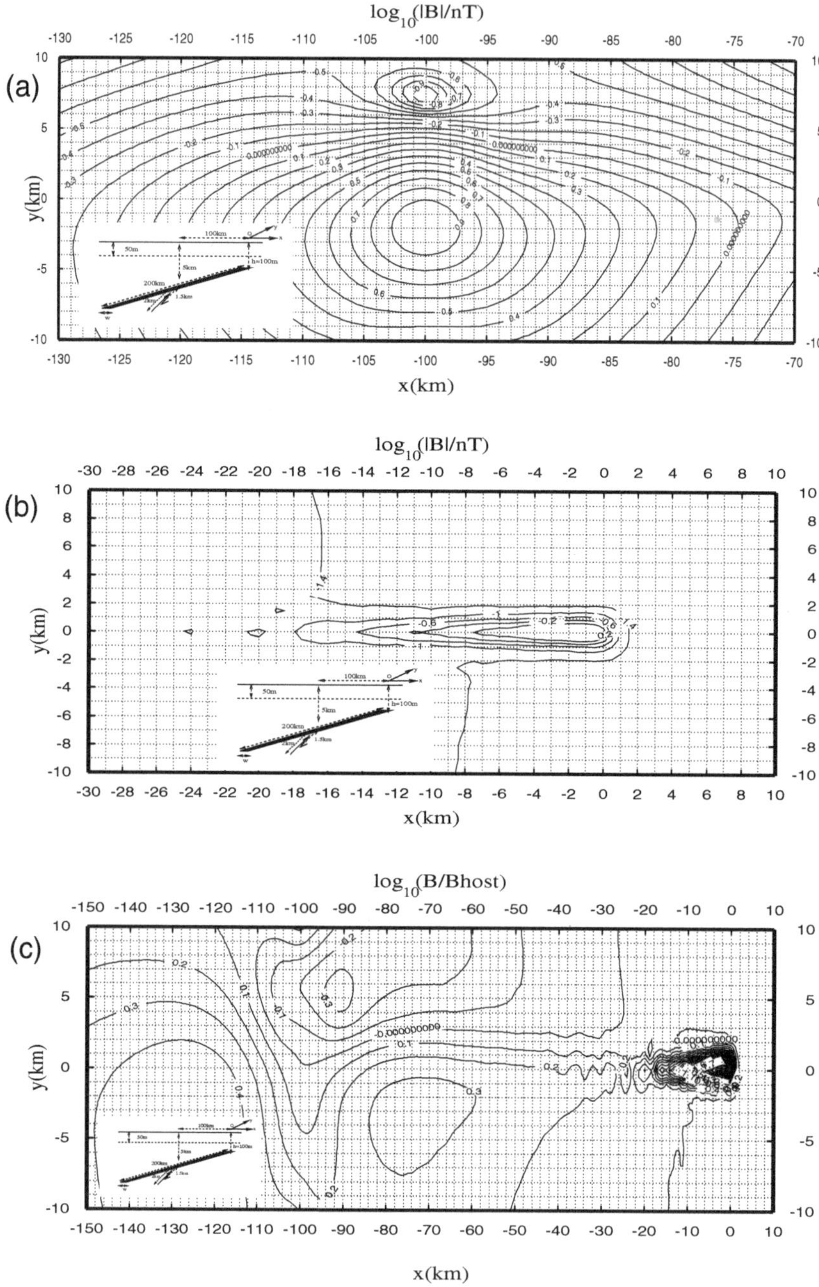

Fig. 1.41 Contours of the total magnetic field on the Earth's surface, in $\log_{10}(B/nT)$, for the case of a current dipole source with finite length perpendicular (see the inset) to the conductive path: (a) in the region above the source, (b) close to the projection of the upper top of the channel. The contours of $\log_{10}(B/B_{host})$ that correspond to the whole region are depicted in (c). A current dipole source $Il = 22.6 \times 10^2$ A km, which corresponds to M $\approx$ 5 EQ, is considered. Reprinted from Ref. [125], Copyright (2005), with permission from TerraPub.

In addition we find that for current dipole moments related to stronger EQs (i.e., M $\gtrsim$ 6.5), the B values at distances of the order of 10 km are well above 1 nT, thus being clearly detectable.

This expectation, compares favorably with the experimental observations, e.g., those by Fraser-Smith et al. [25].

(2) *Upper end of the conductive path.* When the measuring site is located at a reasonable distance (e.g., about 1 km) from the projection of the conductive path on the Earth's surface: (a) for EQs of magnitude M $\approx$ 5, the B values are of the order of 10^{-1}–10^{-2} nT (Fig. 1.41(b)) which are *not* readily detectable (because they are usually smaller than the cultural noise); (b) if we repeat the calculation for larger current dipole moments, e.g., for M $\approx$ 6.5 (by applying Eq. (1.1)), we find B values, with a prominent B_z component, of the order of 1 nT.

It is therefore expected that at epicentral distances of $\approx$100 km the B-variations are detectable *only* for earthquakes with M $\approx$ 6.5 or larger. This has been actually experimentally observed (see § 1.3.6).

1.7.7 The physical background of the $\Delta V/L$ criterion to distinguish SES from noise

Here, we review the physical basis of the $\Delta V/L$ criterion (see Section 1.2 and § 1.2.3) on the basis of which, if long dipoles and short dipoles are simultaneously operating, a SES identification can be achieved. In § 1.7.7.1, the simple case of a homogeneous half-space or a horizontally layered Earth is presented. The study is extended in § 1.7.7.2 to the case of a conductive edge embedded in a homogeneous medium by making use of the analytical solutions for the electric field in the following cases: a paraboloidal edge and a spheroid. Numerical investigations are also presented in § 1.7.7.3 when studying the SES transmission model discussed in § 1.7.1. Throughout this subsection, the measuring long- and short-dipoles are assumed to be horizontal.

1.7.7.1 Homogeneous half-space or horizontally layered Earth

We first discuss below the case where a single long (measuring) dipole is operating, which is collinear with the noise source and a short (measuring) dipole. This is the case (2) mentioned in § 1.2.3.

Let us consider a short (measuring) dipole AB (e.g., $L = 50$ m) and a long (measuring) dipole $A\Gamma$ (e.g., $L = 5$ km) which lie on a straight line (Fig. 1.42). Although we assume here, that these dipoles have a common electrode A, this never occurs in practice (Section 1.1), because we use independent neighboring electrodes in order to avoid the electrochemical noise, e.g., due to rain (see § 1.2.2). We designate with ρ the distance of a current (point) dipole noise source N (grounded at the Earth's surface) from the measuring

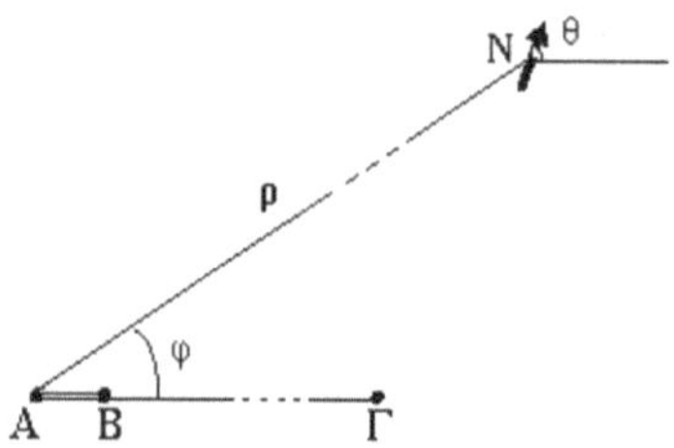

Fig. 1.42 Definition of the angles ϕ and θ (and the distance ρ of the noise-emitting dipole N from the measuring site) with two measuring dipoles: a long dipole $A\Gamma$ and a short dipole AB. Taken from Ref. [145].

site (which, by definition, is the site of the short dipole's deployment A–B), while ϕ denotes the angle between the position vector of the noise and the direction of the measuring dipoles and θ the angle between noise source dipole moment and the measuring dipoles. We assume, as mentioned, that the two measuring dipoles and the noise source lie on a straight line, i.e., $\phi = 0$ or $\phi = 180°$.

The calculations (static, since we are interested for frequencies $f \approx 10^{-2}$–10^{-3} Hz) were performed by the method of images [165]. The series of image charges contributions was summed up numerically so as to obtain an accuracy of 1 ppm. The results were compared to those obtained by the EM1DSH program (Hoversten and Becker [37]).

The $\Delta V/L$ value for the long measuring dipole is just $[\phi(A) - \phi(\Gamma)]/L$, where $\phi(A) - \phi(\Gamma)$ is the (electric) potential difference between A and Γ and L is the length $(A\Gamma)$. The calculation was made in each case: (i) either by representing the Earth with a half-space, having resistivity $\rho_0 = 4 \times 10^3$ Ω m, or (ii) with a two (*horizontal*) layer Earth having a surface layer of thickness 50 m with resistivity $\rho_s = 200$ Ω m and a basement with $\rho_0 = 4 \times 10^3$ Ω m. The detailed results can be found in Refs. [145, 143]. Here, we shall present only some typical examples, in which we assume a current dipole noise source with $Il = 1$ Am.

The following three quantities are plotted in Fig. 1.43: the $\Delta V/L$ values recorded by the short dipole ($L = 50$ m) and the long dipole ($L = 5$ km), as well as their ratio, i.e., "$\Delta V/L$ of the long dipole" / "$\Delta V/L$ of the short dipole". The latter quantity labeled "*Ratio (Long/Short)*" is depicted with a solid line (it corresponds to the left vertical scale). The former two quantities, i.e., the $\Delta V/L$ values, are depicted with dotted and broken lines, respectively (they correspond to the right vertical scale).

Case A ($\phi = 0$). *The noise source and the remote electrode of the long dipole lie on the same side with respect to the measuring site.*

Figure 1.43(a) depicts the results for the two-layer model, which show that at long distances, the "Ratio (Long/Short)" is around unity. As we move from $+\infty$ to the site of the remote electrode Γ, the "Ratio (Long/Short)" increases (note that this is one of the cases recommended by Varotsos and Lazaridou [140] *for an easy recognition* of noise; see Fig. 1.7(c)).

At shorter distances from the measuring site, i.e., $\rho < (A\Gamma)$, Fig. 1.43(a) shows that the "Ratio (Long/Short)" goes down to negative values, thus leading to an even easier recognition of noise.

As mentioned, Fig. 1.43(a) shows the results when the dipole moment is parallel to the measuring (long and short) dipoles, i.e., $\theta = 0$. The study was extended to various values of θ, i.e., from 0 to 180°, and the results show [145, 143] that although the $\Delta V/L$

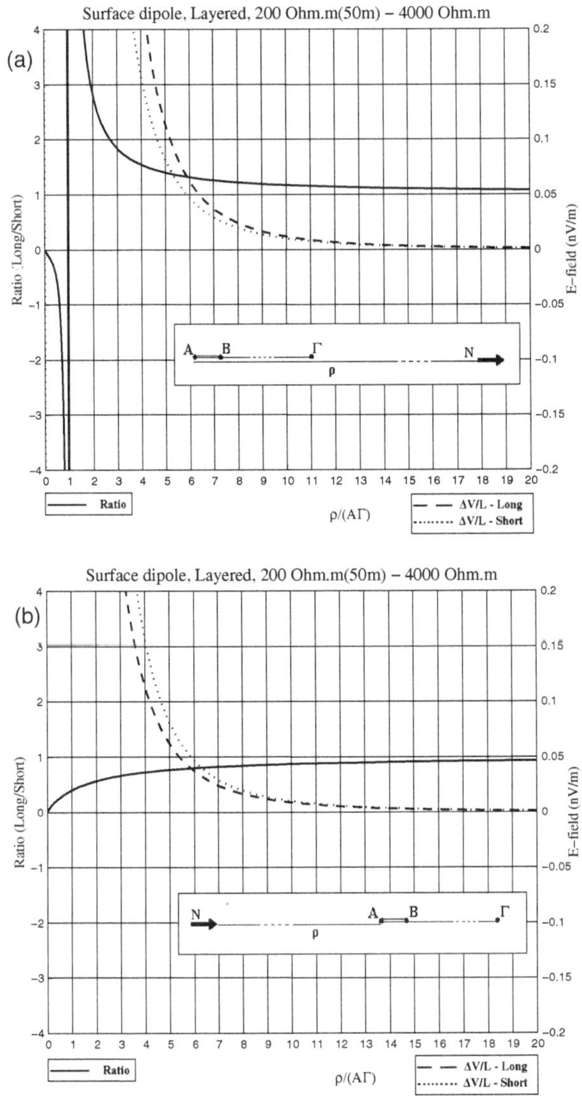

Fig. 1.43 The "Ratio (Long/Short)" and the $\Delta V/L$ values for a short dipole $L = 50$ m and a long dipole $L = 5$ km (*asymmetric* dipoles) at various (reduced) distances from an emitting surface source. The calculation was made for a dipole source located on a two layer Earth ($\rho_S = 200\ \Omega$ m, $\rho_0 = 4 \times 10^3\ \Omega$ m). (a): $\phi = 0$; (b): $\phi = 180°$. Taken from Ref. [145].

values vary from case to case, the "Ratio (Long/Short)" retains the *same* behavior as in Fig. 1.43(a) (cf. both $\Delta V/L$ values, for the short and long dipole, are positive for $\theta < 90°$ and negative for $\theta > 90°$).

Case B ($\phi = 180°$). The noise source N lies at the other side, with respect to the measuring site, of the remote electrode Γ of the long dipole.

Fig. 1.43(b) shows that the "Ratio (Long/Short)" approaches unity at large distances [i.e., $\rho \gg (A\Gamma)$]. This ratio never becomes negative (and this why such an electrode deployment was not recommended in Refs. [140, 131]), but it becomes markedly smaller than unity when the emitting source approaches the measuring site, e.g., $\rho \sim 2(A\Gamma)$, and hence the noise becomes recognizable.

The results in Fig. 1.43(b) hold when the dipole moment of the emitting source is parallel to the measuring (long and short) dipoles, i.e., $\theta = 0$. The study was extended to various values of θ and the results (for a half-space with $\rho_0 = 4 \times 10^3\ \Omega$ m) are shown in figs. 7.1.6 and 7.1.7 of Ref. [125]. These figures indicate again that, although the $\Delta V/L$ values change from case to case, the "ratio (Long/Short)" retains the *same* behavior as in Fig. 1.43(b).

Recall that in all cases treated above, the middle point of the short dipole AB does not coincide with the middle point of the long dipole $A\Gamma$. *We shall call such cases asymmetric.* We emphasize that the asymmetric configuration plays a decisive role in recognizing easily the noise emitted from nearby "artificial" sources.

In short, the configuration recommended for an easy recognition of noise is the following: the noise source and the remote electrode lie on the same side of the measuring site, or simply when the remote electrode is installed very close to the noise source.

The study of the case (3) of § 1.2.3 which applies to the general case where noise source is not necessarily collinear with the measuring dipoles, can be found in Refs. [125, 145, 143].

1.7.7.2 Conductive edges embedded in a more resistive homogeneous medium

Detailed analytical results for a paraboloidal edge have been reported in Refs. [147, 144, 90] (see also subsections 6.4.1, 6.5.1 and 7.2.1 of Ref. [125]) and for an elongated conductive spheroid in Ref. [144] (see also subsections 6.5.3, 6.6.1 and 6.6.2 of Ref. [125]).

Let us now summarize these results by starting from the case of a paraboloidal edge (see Fig. 1.44). Consider a paraboloidal region of conductivity σ embedded in a host medium of conductivity σ'. We assume the paraboloidal coordinates [69]

$$x = \lambda\mu\cos(\phi), \quad y = \lambda\mu\sin(\phi), \quad z = \frac{1}{2}(\lambda^2 - \mu^2), \tag{1.75}$$

where $\phi \in [0, 2\pi]$, $\mu, \lambda \in [0, \infty)$. The surfaces of equal λ and μ are parabolic surfaces of revolution, and the equal ϕ surfaces are planes that intersect each other along the z-axis. The results show that just outside the edge, the amplitude of the electric field $E_{outside}$ (along $\hat{\mathbf{e}}_\mu$, see Fig. 1.44(d)) close to the vertex (but not very close to it) varies as $\propto 1/\delta$, where δ denotes the distance of the measuring site from the surface of the paraboloid (but along $\hat{\mathbf{e}}_\mu$). This $1/\delta$ behavior, which seems to be a general one for "*edge effects*" (e.g., Ref. [144]), could be explained, in the frame of the accumulation of charges close to the interface between two media with different conductivities. At distances δ of the order of the (linear) dimension of the edge, the electric field decreases with δ slower than the field of a point charge (which varies as $1/\delta^2$), giving an almost $1/\delta$ dependence. As the distance becomes appreciably larger than the (linear) dimension of the edge, the field dependence turns first to $1/\delta^2$, and finally, at even longer distances, to $1/\delta^3$, see Fig. 1.44(b). Furthermore, a

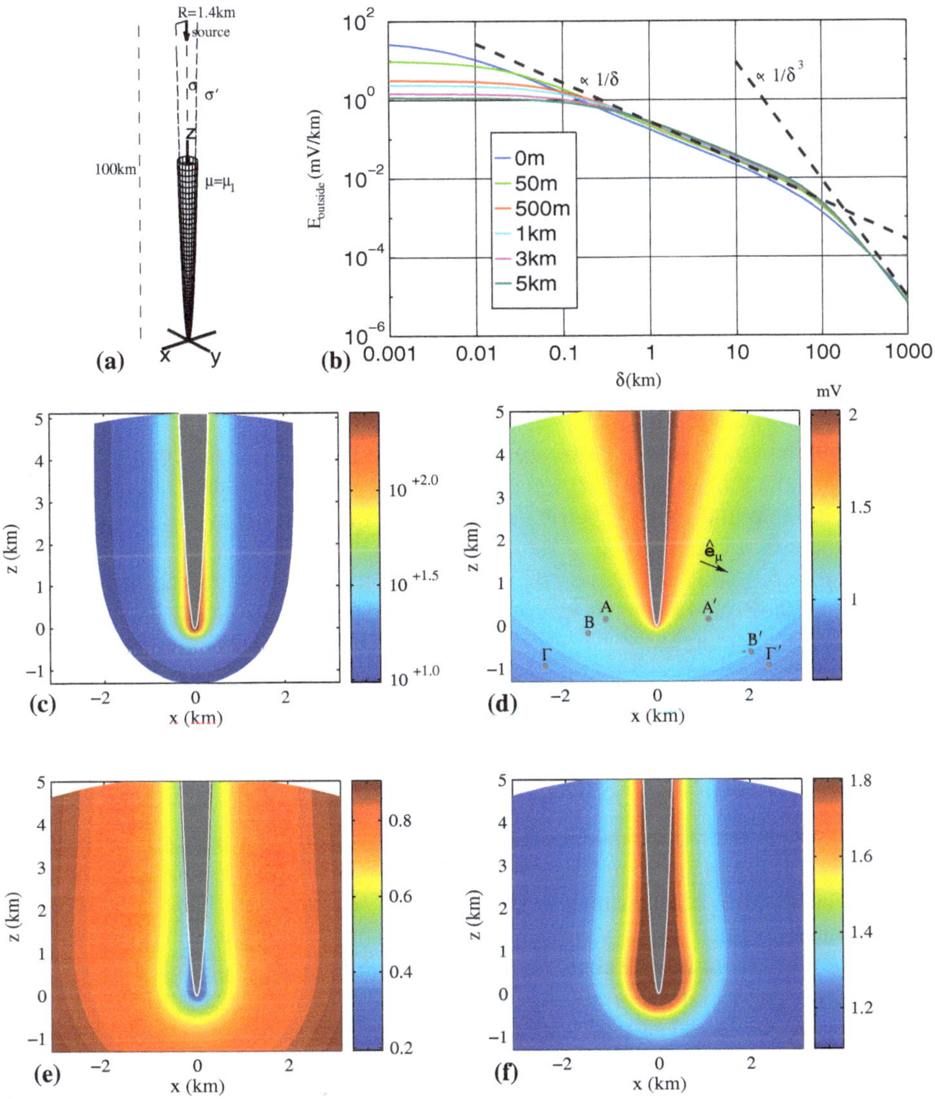

Fig. 1.44 The case of a paraboloidal edge for a conductivity ratio $\sigma/\sigma' = 1000/1$. The colors denote the values of the scale on the right. (a) Schematic diagram of the surface $\mu = \mu_1 = 0.1\sqrt{km}$ separating the regions with conductivities σ and σ'. (b) $E_{outside}$ as a function of δ (along $\hat{\mathbf{e}}_\mu$) measured from various points on its surface (their ordinate z are given in the inset). (c) The ratio $E_{outside}/E_{host}$ on the xz plane. (d) The potential (due to a dipole source: 4×22.6 A km lying inside the conductive medium at a distance from the vertex $z_0 = 100$ km), along with the two short (100 m) and long (1 km) dipole configurations (not to scale) AB, $A\Gamma$ or $A'\Gamma'$, $B'\Gamma'$. (e) and (f): The "Ratio (Long/Short)" for the two dipole configurations AB, $A\Gamma$ or $A'\Gamma'$, $B'\Gamma'$ as a function of the position of A or A', respectively. Taken from Ref. [90]. Copyright (1999), American Geophysical Union. Reproduced by permission of American Geophysical Union.

study of the equipotential surfaces surrounding a (paraboloidal) edge shows that, in order to find significant (and hence measurable) values of the electric field, we should measure along $\hat{\mathbf{e}}_\mu$. Thus, in order to detect measurable values of the electric field, according to the usual formula $\Delta V/L$, the line connecting the two electrodes should be perpendicular to the surface of the edge(i.e., oriented along $\hat{\mathbf{e}}_\mu$). Since the electric fields show a "slow" variation like $\propto 1/\delta$ – we conclude that they should show a small variation when the measurements are carried out at neighboring sites such that $L \ll \delta$. This is the basis of the $\Delta V/L$ criterion. Furthermore, the calculations show that, if two measuring dipoles have lengths around a few tens of meters and a few km respectively, they both give comparable $\Delta V/L$ values, if they are both directed perpendicularly to the paraboloidal edge; see Fig. 1.44(e) and (f).

The same conclusion is reached for other cases of conductive edges, such as the electric field $E_{outside}$ at one end of the major axis of an elongated spheroid; see point (3) in § 1.7.2.3.

By summarizing, the physical basis for the validity of the $\Delta V/L$ criterion for the SES stems from the following fact: at long distances from a dipole source but close to an "edge", the amplitude of the electric field exhibits a *slow* $1/\delta$-behavior (or, in general, $1/r^\theta$, where $\theta \approx 1$; see §1.7.2.3). Hence, the main characteristic of the distant source assumption (see Section 1.2), i.e., the *slow* variation of the amplitude of the electric field versus the distance when measuring along a certain direction at sites close to each other, practically remains.

1.7.7.3 Numerical investigation of the $\Delta V/L$ criterion in the model proposed for the explanation of the SES selectivity

In § 1.7.1, we considered the model of a conductive path terminating close to the surface of a two-layer medium and found that the SES are detectable in the region above the upper end of the conductive path. One may wonder, if the $\Delta V/L$ criterion still holds in this region. Thus, we now comment on its validity in the model shown in the insets of Figs. 1.29(A) and 1.29(B) by examining whether the "Ratio (Long/Short)", i.e., the ratio of the $\Delta V/L$ values for the long and the short dipoles, markedly differs from unity.

The case of the model of Fig. 1.29(A) has been studied in Ref. [145] by considering the "Ratio (Long/Short)" (see § 1.7.7.1) for the asymmetric configuration of a set of a long dipole (5 km) and a short dipole (50 m) placed at various distances from the projection "O" on the Earth's surface of the top of the conductive path. These (collinear) dipoles were assumed to be placed parallel to the projection of the conductive path on the Earth's surface (which is the direction of the dominant electric field). The results show that *only* at short distances from the top of the conductive path (i.e., a few times its width) the "ratio (Long/Short)" markedly deviates from unity, as expected. Otherwise, this ratio *approaches unity* (it only differs from unity by 20%, or so).

We now proceed to the case of the model of Fig. 1.29(B): Figures 1.45(A),(B) show the "Ratio (Long/Short)" for the *asymmetric* dipole configurations of a set of a long dipole and a short dipole placed at various distances x as shown in the top part of each figure, x being measured from the projection on the Earth's surface of the upper end of the conductive

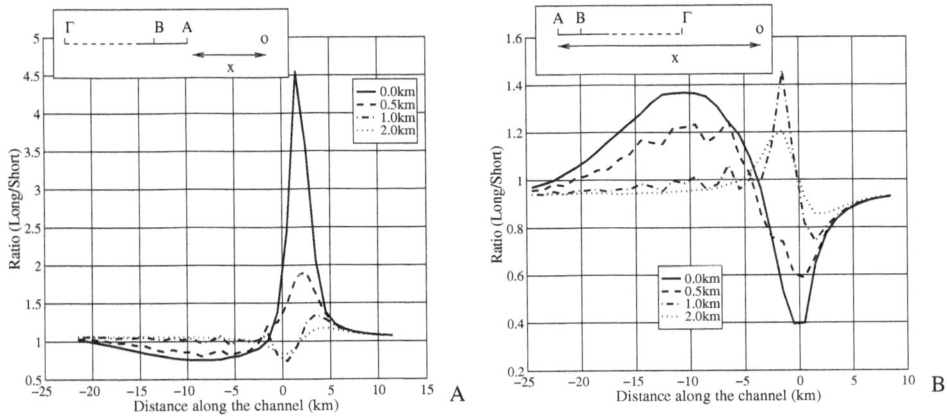

Fig. 1.45 The "Ratio (Long/Short)" for the model, depicted in Fig. 1.29(B). The collinear dipoles (short: AB, long: $A\Gamma = 3$ km) are parallel to the projection of the conductive path on the Earth's surface, at various y-values given in the inset. Taken from Ref. [90]. Copyright (1999), American Geophysical Union. Reproduced by permission of American Geophysical Union.

path. These figures again show that *only* at very short distances from the upper end of the conductive path, and at small y-values (i.e., $y \approx 0$, 0.5 km), the "Ratio (Long/Short)" markedly differs from unity, but it otherwise remains close to unity.

1.7.8 Explanation of the difference between SES polarization and MT polarization

We explain below that when measuring the SES, arising from an emitting dipole source that forms a certain angle with the neighboring conductive path, close to an "edge" the ratio of the two SES components is *different* from that of the MT variations at the same frequency range.

Consider the case of Fig. 1.29(B) or Fig. 1.30(b), i.e., a current dipole source lying at a distance D from a channel of high conductivity whose projection on the Earth's surface is assumed along x-axis. We use the same parameters (as well as the procedure of calculation) as in § 1.7.5.2. The electric current dipole source (ECD) is assumed to be oriented along either the x-axis or the y-axis. For the sake of comparison, the calculation is repeated for a source, which is considered as an incident plane electromagnetic wave (PEM) circularly polarized on the x–y plane. The results of the calculations (as far as only the direction of the field is concerned) are given in Figs. 1.46(a) and 1.46(b), which correspond to the cases when the ECD is assumed either parallel or perpendicular to the projection of the conductive channel on the Earth's surface, respectively. The polarization of the measured electrical variations due to PEM is depicted, at each site, by a red circle or ellipse, and the direction of the signal due to ECD by a short blue line.

We first discuss the results of the variations due to PEM. An inspection of Figs. 1.46(a),(b) shows that at those sites which are relatively far from the projection of the channel on the Earth's surface, there is no preferred direction and hence the results are represented by circles. As we approach the projection of the channel, the circles transform

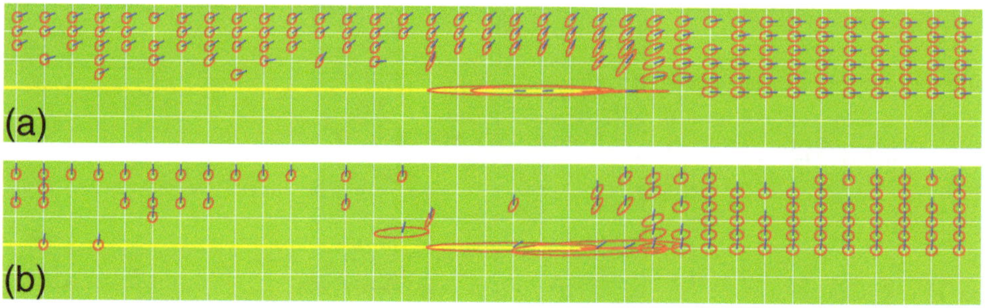

Fig. 1.46 Calculated directions (red circle or ellipse for PEM and blue lines for ECD) of the electric field at various points close to the projection on the Earth's surface (thick yellow line) of the conductive channel shown in Fig. 1.29B with $w = 500$ m. Panels (a) and (b) correspond to different orientations of the ECD, i.e., along x-axis or y-axis, respectively. Each square corresponds to 1 km × 1 km. The source is located at the point $x = -100$ km, $y = 0.5$ km, $z = 0.5$ km. For the resistivity parameters used see § 1.7.5.1. Taken from Ref. [144].

gradually to ellipses. These results are compatible with those intuitively expected: for a homogeneous half-space or for a horizontally layered medium, at large distances from the channel, the electrical measurements should not show any preferred (horizontal) direction; on the other hand, at short distances from the conductive channel, the direction along the (projection of the) channel should correspond to larger electric fields than that perpendicular to the channel.

We now turn to the discussion of the results concerning the direction of the measured electric field due to ECD. This is mainly regulated from the orientation of the emitting dipole, as can be seen from a comparison of Figs. 1.46(a) and 1.46(b). Since we are mainly interested for sites close to the edge, Fig. 1.46(b) shows that when ECD is along the y-axis, the electric field measured at sites close to the edge is in general almost perpendicular to the projection of the channel on the Earth's surface. On the other hand, in Fig. 1.46(a), where ECD is parallel to the x-axis, the electric field measured at sites close to the edge, forms with the projection of the channel an angle appreciably smaller than 90°. In other words, the "polarization" of the signal due to ECD is drastically different in Figs. 1.46(a) and 1.46(b), thus indicating that:

The information related with the directionality of the current dipole source is not "lost" when the transmission of the current is taking place mainly through the conductive channel.

It is evident that when ECD forms any angle $\varphi(0 < \varphi < 90°)$ with the conductive channel, the direction of the signals measured at sites close to the edge can be found from a combination of Figs. 1.46(a) and 1.46(b). By studying the difference in the direction of the electric field, measured at sites close to the edge between the cases ECD and PEM, we find that it minimizes when the emitting dipole is almost parallel to the conductive path and maximizes when the emitting dipole is almost perpendicular to the path.

Thus, a considerable difference between SES polarization and MT polarization seems likely to happen in most cases, since in the case of ECD there exists a significant current dipole component perpendicular to the path as explained in § 1.7.3.

In other words, the SES direction is governed by the angle of the dipole source with the neighboring conductive path. Thus, it does not have to coincide with the MT polarization coming from *circularly* polarized EM waves "traveling" vertically.

1.8 Transmission of electric signals in dielectric media: time- and frequency-dependence

1.8.1 The propagation regime and the diffusion regime of electromagnetic fields. Isotropic and homogeneous medium

Following subsection 8.1.2 of Ref. [125], we restrict ourselves below to the simple case of a plane wave of angular frequency ω propagating along the z-direction in an isotropic and homogeneous medium of conductivity σ. The electric field **E** is assumed polarized in the x-direction:

$$E_x = E_0\, \mathrm{e}^{i(\omega t - kz)} \tag{1.76}$$

and the magnetic field **H** in the y-direction:

$$H_y = \frac{E_0 k}{\mu_0 \omega}\, \mathrm{e}^{i(\omega t - kz)} \tag{1.77}$$

The complex wave number k here is determined by the condition that the fields **E**, **H** should satisfy Maxwell's equations; by inserting Eqs. (1.76) and (1.77) into the equation for the propagation of an electromagnetic field (which results from Maxwell's equations):

$$\nabla^2 E_x = \sigma\mu_0 \frac{\partial E_x}{\partial t} + \varepsilon\mu_0 \frac{\partial^2 E_x}{\partial t^2}\,, \tag{1.78}$$

we find the following relation between the wavenumber, angular frequency and the material properties σ, $\varepsilon = \varepsilon' - i\varepsilon''$:

$$k^2 = \left(\varepsilon - i\frac{\sigma}{\omega}\right)\mu_0\omega^2 = (1 - i\tan\delta)\,\mu_0\varepsilon'\omega^2, \tag{1.79}$$

for $\tan\delta$ see Eq. (1.86), below. This relation contains *all* the essential information concerning the transmission of signals and leads to:

$$E_x = E_0 e^{\;k'' z}\, \mathrm{e}^{i(\omega t - k' z)} \tag{1.80}$$

The *imaginary* part k'' of k describes how the field strength decreases with distance, $E_0\, \mathrm{e}^{-k'' z}$. It is associated with energy dissipation. This quantity k'' can be determined by measuring the amplitude at two locations z_1, z_2 and applying the relation:

$$\text{amplitude ratio} = \exp\left[-k''(z_2 - z_1)\right] \tag{1.81}$$

The *induction skin depth* is equal to $1/k''$ and simply denotes the distance over which the field strength attenuates to $1/e$ of its original value.

The *real part* k' describes the propagation of the electric field and its oscillations through the factor $e^{i(\omega t - k'z)}$, where $k' = 2\pi/\lambda$ and λ is the wavelength. The quantity k' is experimentally determined through the relation:

$$\text{phase shift} = (z_2 - z_1)k' \tag{1.82}$$

The quantities k', k'' are explicitly connected to the material properties through:

$$k' = \omega\sqrt{\mu_0\varepsilon'}\sqrt{\frac{1+\cos\delta}{2\cos\delta}} \quad \text{and} \quad k'' = \omega\sqrt{\mu_0\varepsilon'}\sqrt{\frac{1-\cos\delta}{2\cos\delta}} \tag{1.83}$$

Measurements of the amplitude ratio and the phase shift, lead as mentioned to the determination of k', k''. The real ε' and imaginary ε'' parts of ε are obtained from:

$$\varepsilon' = \frac{k'^2 - k''^2}{\mu_0\omega^2} \tag{1.84}$$

$$\varepsilon'' = \frac{\sigma}{\omega} + 2\frac{(k'^2 - k''^2)}{\mu_0\omega^2} \tag{1.85}$$

and hence the *loss tangent* is given by [142]:

$$\tan\delta = \frac{\sigma}{\omega\varepsilon'} + \frac{\varepsilon''}{\varepsilon'} = 2\frac{k'k''}{(k'^2 - k''^2)} \tag{1.86}$$

Low frequencies, i.e., $\omega\varepsilon' < \sigma$. At low frequencies the displacement currents are negligible in comparison to the conductivity currents. In this case Eqs. (1.78) and (1.79) take the form:

$$\frac{\partial^2 E_x}{\partial z^2} = \sigma\mu_0\frac{\partial E_x}{\partial t} \tag{1.87}$$

$$k = (1-i)\left(\frac{\sigma\omega\mu_0}{2}\right)^{1/2} \tag{1.88}$$

Equation (1.87) shows that the electric field obeys the *diffusion* equation (and therefore this frequency range is called the *diffusion regime*). The typical $t^{1/2}$ dependence, which characterizes a diffusion process, can now be presented as follows.

For an electromagnetic pulse emitted at $z = 0$ at $t = 0$: the electric field amplitude measured at a fixed time t versus the distance z, reaches a maximum at a distance $z = z_m$:

$$z_m = \left(\frac{2t}{\mu_0\sigma}\right)^{1/2}. \tag{1.89}$$

This maximum of the pulse travels with a velocity:

$$\mathrm{v}_m \equiv \frac{\mathrm{d}z_m}{\mathrm{d}t} = (2\mu_0\sigma t)^{-1/2} \tag{1.90}$$

and its amplitude decreases versus the time as $1/t$. The phase velocity v_{ph} can be found by inserting Eq. (1.83) into the relation $\mathrm{v}_{ph} = \omega/k'$:

$$\mathrm{v}_{ph} = \frac{1}{\sqrt{\mu_0\varepsilon'}}\sqrt{\frac{2\cos\delta}{1+\cos\delta}} \tag{1.91}$$

The electric field measured at a given distance z from the source varies with time, so that the signal is detected over a long time interval; the maximum occurs after a time Δt:

$$\Delta t = \frac{\mu_0\sigma z^2}{6} \tag{1.92}$$

This time Δt is appreciably larger than the time $z/\sqrt{\kappa'}$ (where $\kappa' = \varepsilon'/\varepsilon_0$) corresponding to the transmission of either a high-frequency signal in a conductive medium or of (any frequency) signal in an insulating medium.

High frequencies, i.e., $\omega\varepsilon' > \sigma$. The conductivity currents are negligible in comparison to the displacement currents. In this case Eq. (1.86) shows that $\tan\delta = \varepsilon''/\varepsilon'$ (and considering that $\varepsilon''/\varepsilon' \ll 1$), Eqs. (1.78) and (1.79) take the form:

$$\frac{\partial^2 E_x}{\partial z^2} = \varepsilon\mu_0\frac{\partial^2 E_x}{\partial t^2} \tag{1.93}$$

$$k = \omega(\mu_0\varepsilon')^{1/2}\left(1 - i\frac{\varepsilon''}{\varepsilon'}\right)^{1/2} \approx \omega(\mu_0\varepsilon')^{1/2}\left(1 - \frac{i}{2}\frac{\varepsilon''}{\varepsilon'}\right) \tag{1.94}$$

Equation (1.93) shows that the electromagnetic field satisfies the wave equation so that this high-frequency regime is known as the *propagation regime*. In this range (because $c' \gg c''$), the waves propagate without significant attenuation and dispersion. In absence of all dissipation, i.e., $\delta = 0$, Eq. (1.91) gives:

$$\mathrm{v}_{ph} = \frac{1}{\sqrt{\mu_0\varepsilon_0\kappa'}} = \frac{c}{\sqrt{\kappa'}} \tag{1.95}$$

We see that, for very high frequencies, there is no dispersion, i.e., all velocities are essentially constant and they have a value lower than c by a factor $1/\sqrt{\kappa'}$.

Thus, in short, in a conductive medium at low frequencies, the electric field obeys a diffusion type equation (*diffusion regime*), while at high frequencies, the electromagnetic field satisfies the wave equation (*propagation regime*).

As a result, when considering *low-frequency* electric signals transmission in a conductive medium as in the case of SES: (1) The velocity v is appreciably smaller than the value $c/\sqrt{\kappa'}$ (c the speed of light in vacuum), which only holds in absence of dissipation. (2) The time Δt, that is necessary for the transmission to a distance z, is appreciably larger than that expected from the usual rough guess $\Delta t \approx z/(c/\sqrt{\kappa'})$. (3) The duration recorded at a remote site is appreciably longer than the emitted true duration of a short-duration pulse.

1.8.2 Electric field from a dipole current source lying close to a conductive path. Frequency dependence

The frequency dependence of the electric field produced by a current dipole source lying inside or very close to a cylinder of infinite length (Fig. 1.31(a)) and high conductivity (σ), embedded in a significantly less conductive medium (σ'), was investigated in detail in Ref. [148]. The study was also extended to the case when the dipole is located inside or very close to a highly conductive layer (σ), of infinite extent, embedded in a significantly less conductive medium (σ') (Fig. 1.31(b)).

The main conclusions were the following. At large distances d from the source and for appreciably low frequencies, i.e., smaller than around $f_c = 1/(2\pi\mu\sigma R^2)$ for the case of the cylinder or some $f_0 = 1/(\mu\sigma w^2)$ for the case of the layer, the electric field follows mainly the properties of the outer (host) medium and hence its attenuation is governed by a skin depth δ_{OUT} of the outer medium, i.e., that for a full space of conductivity σ'. For higher frequencies, i.e., $f > f_c$ or $f > f_0$ for the case of the cylinder or layer, respectively, the electric field is attenuated with a skin depth significantly smaller than δ_{OUT}, with a lower limit δ_{IN}, the skin depth in a full space of conductivity σ. The static dependence is valid as long as the distance d is appreciably smaller than the "wavelength" in the host medium, but larger than the "wavelength" in the internal highly conductive medium.

An example of practical interest is given for a highly conductive layer in Fig. 1.47. It depicts the electric field E_z measured at the points in the middle of the conductive layer versus d/w, for various distances D of the source from the layer (see the inset of Fig. 1.47). The calculation of Fig. 1.47 was made for the frequency 0.1 Hz, for various D values and for a certain width $w = 500$ m (the calculation was repeated at the same frequency, for a certain D value, i.e., $D = 5$ km, but for various widths $w = 100$ to 1000 m and the results are given in Ref. [148]). These results show that:

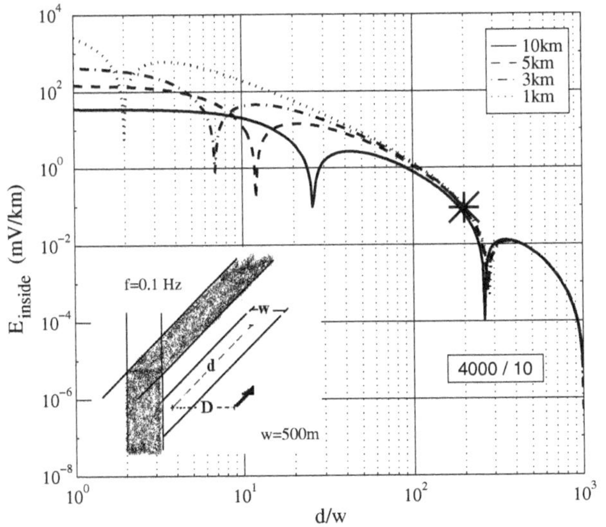

Fig. 1.47 The amplitude of the electric field E_{inside} versus d/w, for various values of the distance D (= 1, 3, 5, 10 km, see the inset) of the emitting dipole from a given conductive layer ($w = 500$ m). For values $d/w = 10^2$ or larger, all curves practically coincide. Conductivity ratio $\sigma/\sigma' = 4000/10$. Source 22.6×10^2 A km. Frequency $f = 0.1$ Hz. The point with asterisk corresponds to $d = 100$ km. Taken from Ref. [148].

At long distances, i.e. $d \gg D$, $d \gg w$, the amplitude of the electric field practically does *not* depend either on the exact distance D from the conductive layer (or from the exact value of its width w) and almost follows the properties of the external more resistive medium.

1.8.3 The electric signal recorded at a remote site. Time domain

The characteristics (amplitude, duration) of the signal at remote distances d from a current dipole source emitting with a time dependence of the form $f(t) = \Theta(t)\exp(-t/\tau)$ were studied in detail in Refs. [160, 148]. The dipole source is located either inside a full volume of conductivity σ' or inside a conductive half-space at a depth $h(\ll d)$:

(1) In the case of a homogeneous medium of conductivity σ', a certain time scale $\tau_0 = \mu\sigma' d^2/4$ exists (governed solely by the properties of the medium and the distance d from the source), so that the following main conclusions hold: (a) If $\tau \ll \tau_0$, the recorded signal has an enlarged duration of around τ_0 and an amplitude orders of magnitude smaller than that corresponding to the static case. (b) If $\tau > \tau_0$, the recorded signal has duration almost equal to that of the emitted signal, it arrives after a time of around τ_0 (due to the "diffusion regime"), and its amplitude approaches that of the static case. An example is given in Fig. 1.48. This figure depicts the amplitude of the electric field measured along the direction of the dipole for various values of the "relaxation time" τ of an emitted signal. An inspection of this figure shows that, for long "relaxation times", i.e., $\tau \gg \tau_0$ (τ_0 is of the order of 1 sec if $\sigma' \approx 10^{-3}$ S/m and $d = 100$ km), the maximum amplitude of the recorded signal approaches the value expected from the static (i.e., if $f(t) = 1$) calcula-

tion; on the other hand, for short "relaxation times", i.e., $\tau \ll \tau_0$, the maximum amplitude is smaller by order(s) of magnitude. In the latter case the amplitude is almost proportional to the value of τ.

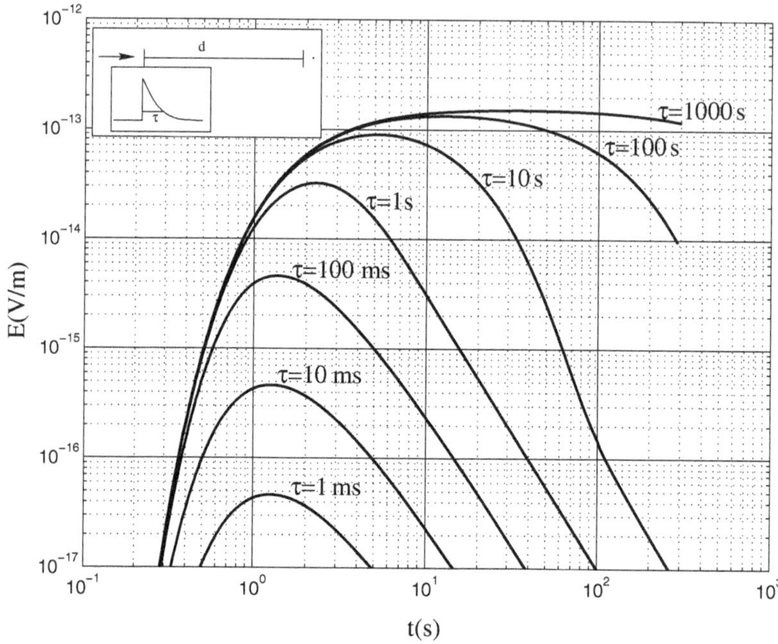

Fig. 1.48 Amplitude of the electric field versus time of a signal recorded at a distance $d = 100$ km from a current dipole source $Il\Theta(t)\exp(-t/\tau)$, where $Il = 1$ A m, located in a homogeneous conductive medium with resistivity $\rho'(=1/\sigma') = 1000\ \Omega$ m. The curves correspond to various τ values lying between $\tau = 1$ ms and 1000 s, as calculated in Ref. [160]. Note that for short values of τ, the maximum value of the signal is almost proportional to τ. Taken from Ref. [148].

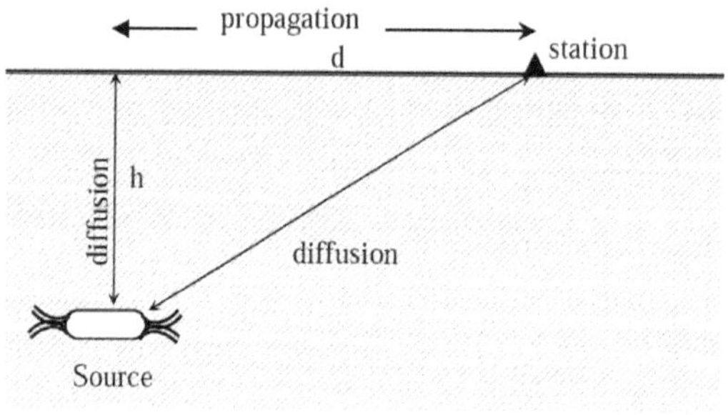

Fig. 1.49 Schematic diagram that shows how the signal (emitted from a horizontal current dipole source at a depth h and recorded at an observation point located on the surface of a half-space) arrives in two parts: "semi-diffusing" path h–d (that arrives first), and "solely-diffusing" direct path (that arrives second). Reprinted from Ref. [125], Copyright (2005), with permission of TerraPub.

(2) In the case of the current source located within a conductive half-space, the results could be interpreted in simple words as follows. The signal recorded at a remote observation site on the interface, arrives in two parts (Fig. 1.49): the first part "diffuses" vertically from the source to the interface and then propagates horizontally to the measuring site. It has a duration significantly smaller than the second part, which "diffuses" directly from the source to the easuring site through the conductive medium. The role of the second part is more important for larger conductivity σ' and becomes dominant for cases of practical interest like SES transmission in the Earth.

For example, let us consider the case of a current dipole source of the form $Il(t) = Il\Theta(t)\exp(-t/\tau)$ with $Il = 1$ A m located inside a conductive half-space with resistivity $\rho'(=1/\sigma') = 1{,}000\ \Omega$ m. The results of the calculation for short "relaxation times", e.g., $\tau = 10$, 100 and 200 ms given in Ref. [148] show that the *first* part (arriving after almost 0.1 s in all cases) has an amplitude which is larger than the *second* part. On the other hand, the second part becomes dominant at significantly longer "relaxation times", e.g. $\tau = 10$ and 100 s, see Figs. 1.50(A),(B) and the amplitude of this solely diffusing part reaches the static value. For example, the amplitude becomes 1.6×10^{-13} V/m and 2.4×10^{-13} V/m for $\tau = 10$ and 100 s respectively, which is comparable with the value of around 3×10^{-13} V/m in the static case.

1.8.4 Discussion on the explanation of the SES detectability and selectivity

We focus on the detectability at long distances of the low frequencies only. The results mentioned in § 1.8.2 confirm the analytical results summarized in § 1.7.2 and § 1.7.4 in the following sense: for appreciably low frequencies (e.g., smaller than around 1 Hz or 0.1 Hz as in the case of SES) and for distances $d \approx 100$ km in media with σ' in the range 10^{-4}–10^{-3} S/m, the electric field values approach those of the static case. Thus, we can follow the same arguments as in § 1.7.4.1 and explain the SES selectivity.

We now turn to § 1.8.3, which concludes that for an emitted signal with $\tau \geq 10$ s, the amplitude of the second solely diffusing part (which dominates, see Fig. 1.50) becomes almost comparable to that of the static case. The overlapping of a considerable number of such signals (in view of the results of the static calculation in § 1.7.2.3) may reach detectable values at remote distances, a point to which we now turn. Let us assume a SES generation mechanism like that of charged dislocations discussed in § 1.6.3. This mechanism has a relaxation time [160] of $\tau = 10$–100 s. We consider a sequence of neighboring blocks, each one emitting a signal with $\tau \geq 10$ s. Each block may have [101] a width, say, a few hundreds of meters (§ 1.6.3) and the mechanical relaxation needs a time $\tau_m \lesssim 0.1$ s to "travel" through each block (i.e., with a velocity of a few to several kilometers per second). Hence, we may assume that every tenth of a second or so an electric signal is emitted and, after taking into account the dimensions of the source (which, of course, for EQs depends on the magnitude, and may reach several kilometers), we may consider a sequence of the order of 10^7 such signals. So the electric field value at remote distances ($d \approx 100$ km) will be equal to the number of blocks multiplied by the value emitted from each block. This

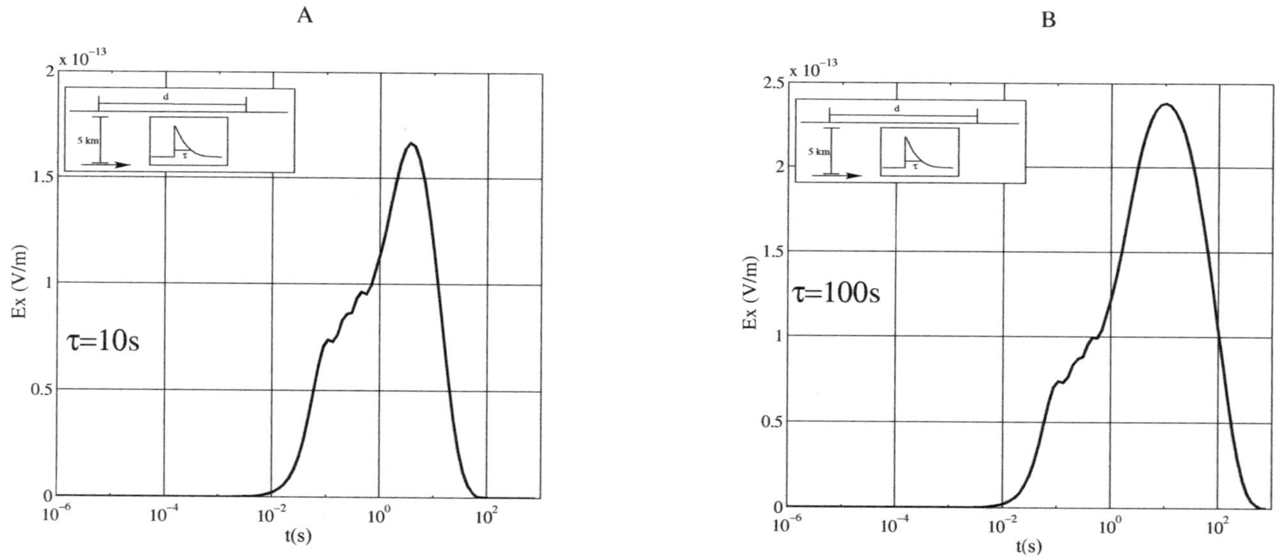

Fig. 1.50 Amplitude of the electric field versus time of a signal recorded at a distance $d = 100$ km from a current dipole source $Il(t) = Il\Theta(t)\exp(-t/\tau)$, with $Il = 1$ A m, located at a depth of $h = 5$ km from the surface of a half-space with resistivity $\rho' = 1{,}000\ \Omega$ m. The observation site lies on the surface of the half-space while the source lies inside the conductive half-space. (A) and (B) correspond to $\tau = 10$ s and 100 s, respectively. Taken from Ref. [148].

alternatively justifies why in the static case of § 1.7.4.1 we used a value 10^2 larger than that corresponding to each of the blocks considered by Slifkin [101], on the basis of which we deduced significant electric field values, i.e., larger than 10 mV/km, at distances of $d \approx 100$ km.

In other words, in the Earth, a sequence of signals emitted every ≈0.1 s, with long duration each (e.g., with $\tau \approx 10$–100 s), may lead to significant electric field values at remote distances ($d \approx 100$ km) due to the overlapping of their "solely diffusing" parts.

1.8.5 Discussion on the time-difference between the SES electric field variation and the associated magnetic field recordings

We first recall (see § 1.3.6.1) that the SES electric field variations before the M_w6.6 Grevena-Kozani EQ precede the associated magnetic (DMM) recordings (note that the latter correspond to dB/dt, *not* B) by a time of 1–2 sec. We also recall that (see § 1.8.3), the time scale $\tau_0 = \mu\sigma' d^2/4$ calculated, for example, for a medium with resistivity 2000 Ω m at a distance 80 km (which is the distance between IOA and the Kozani-Grevena EQ epicenter) is around 1 second.

If a dipole current source is located within a conductive half-space (with the aforementioned resistivity) at a depth (say 5 to 10 km) appreciably smaller than the epicentral distance, the signal arrives in two parts (see § 1.8.3 and Fig. 1.49). It is the second solely diffusing part that dominates the recordings at epicentral distances of the order of 100 km for the reasons discussed above in § 1.8.3 and § 1.8.4. Thus, we focus below our discussion on this "solely diffusing" part. In view of the SES transmission model (i.e., a current dipole source lying close to a highly conductive path, see § 1.7.1), we study below, in the time domain, the case of a conductive cylinder embedded in a more resistive medium.

We solve the problem in the frequency domain and finally calculate the result in the time domain by inverse Fourier transform. The detailed procedure for the latter calculation can be found in Ref. [161]. As an example, we present in Fig. 1.51 the results for a cylinder of infinite length with radius $R = 500$ m (having a small resistivity 2 Ω m, typical of a fault) embedded in a more resistive medium (with resistivity 2000 Ω m, typical of the Earth's upper crust). We assume that the main axis of the cylinder lies along the z-axis of a cylindrical system of coordinates (ρ, ϕ, z). The fields are studied at a distance $z =$ 80 km far from a (point) current dipole source, which, for simplicity, is oriented along the z-axis (a more realistic case should consider of course a current dipole source having a considerable component perpendicular to the conductive path for reasons discussed in § 1.7.3). The observation point is taken at a distance $\rho = 1$ km from the cylinder axis, i.e., at $\rho = 2R = 1$ km. The current emitted $i(z,t)[= \delta(z)\Theta(t)\Theta(T_d - t)]$, is a boxcar pulse with duration $T_d = 11$ s comparable to the typical duration (see table I of Ref. [153]) of the pulses in the corresponding SES activities.

Two main results emerge from this calculation. (1) Concerning the electric field components: the component E_ρ (i.e., perpendicular to the surface of the cylinder, associated with

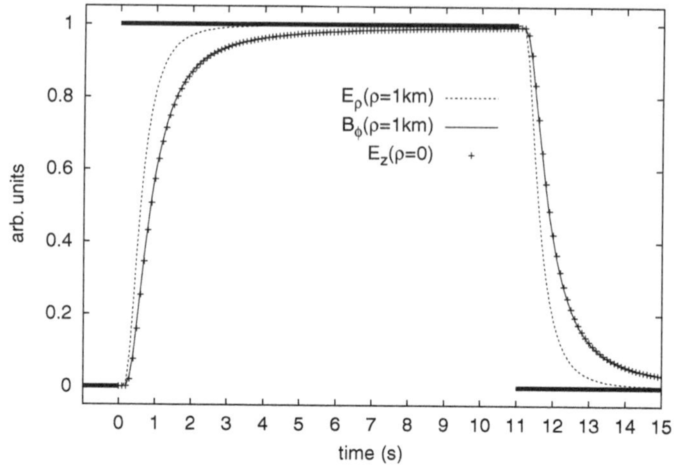

Fig. 1.51 Results for a conductive cylinder of infinite extent (with its axis along the z-axis) embedded in a less conductive medium. A point current dipole source (oriented along the z-axis) is located at the origin (0,0,0) of cylindrical coordinates (ρ, ϕ, z). The fields E_ρ and B_ϕ at an observation point with $\rho = 1$ km, $z = 80$ km together with E_z ($\rho = 0$, $z = 80$ km) are depicted versus time. The thick line depicts the emitted pulse. Taken from Ref. [161].

the accumulation of charges on the cylinder's surface) reaches detectable values *earlier than* the component E_z (which accompanies the high current density inside the cylinder). (2) The magnetic field B_ϕ (which is measured again at $\rho = 2R = 1$ km, $z = 80$ km) appears practically simultaneously with E_z (see Fig. 1.51) and their forms seem to coincide, as expected.

In other words, the magnetic field B_ϕ, which accompanies the high current density flowing inside the cylinder, is *"delayed"* compared to the field E_ρ that signifies the accumulation of charges at the interface.

In the simple model of Fig. 1.51, no frequency dispersion of the dielectric constant (as well as of the conductivities, inside and outside the cylindrical path) was assumed. The results do show that the E_ρ-field *"precedes"* the B-field. However, a careful inspection of Fig. 1.51 reveals that the model cannot explain that the E-field is recorded even before $\mathrm{d}B/\mathrm{d}t$ (note that the latter is not plotted), as the field experiments show (§ 1.3.6.1). At this point, we recall the limitations of the Poynting theorem [161]: when considering non-linear relations between the four vectors **D**, **E**, **B**, **H** and wave-vector dispersive interactions discussed in section 8.5 of Ref. [125], the Poynting theorem does not apply. Maybe, such nonlinear relations as well as wave-vector dispersive interactions should also be considered together with criticality, in order to achieve a satisfactory explanation of the experimental fact that the E-field markedly precedes $\mathrm{d}B/\mathrm{d}t$ [161].

Appendix

The instrumentation for the magnetic field measurements. The permanent recordings of the magnetic field variations are carried out by three DANSK coil magnetometers (DMM) directed along the directions EW, NS and Z. Furthermore, since 1996 the portable MT-1 system of Electro-Magnetic Instruments (EMI) has been also used for temporary measurements. Concerning the calibration of DMM (details are given in Ref. [149]), beyond a laboratory calibration, an *in situ* (i.e., at IOA station) calibration was performed by comparing the DMM recordings $V_m(t)$ to those of the EMI-magnetometers and relying on the accurate laboratory calibration of the latter which has been reported by the manufacturer.

The DMM calibration led to a Heaviside (or unit step) response function $H(t)$:

$$V_m(t) = \int_{-\infty}^{\infty} H(\xi) \frac{dB}{dt}(t-\xi)\, d\xi , \tag{1.96}$$

that vanishes for $t < 0$, while for $t \geq 0$

$$\begin{aligned} H(t) = & A \frac{\tau_d^4}{(\tau_d - \tau_r)^4} \left[\exp\left(-\frac{t}{\tau_d}\right) - \exp\left(-\frac{t}{\tau_r}\right) \right] \\ & - A \frac{\tau_d^3}{(\tau_d - \tau_r)^3} \frac{t}{\tau_r} \exp\left(-\frac{t}{\tau_r}\right) \\ & - \frac{A}{2} \frac{\tau_d^2}{(\tau_d - \tau_r)^2} \left(\frac{t}{\tau_r}\right)^2 \exp\left(-\frac{t}{\tau_r}\right) \\ & - \frac{A}{6} \frac{\tau_d}{(\tau_d - \tau_r)} \left(\frac{t}{\tau_r}\right)^3 \exp\left(-\frac{t}{\tau_r}\right) \end{aligned} \tag{1.97}$$

with $\tau_d \approx 0.025$ s and $\tau_r \approx 0.007$ s (see Ref. [149]).

In summary, this calibration showed that, for magnetic field variations with periods larger than around half a second, DMM magnetometers act as dB/dt detectors. Furthermore, upon the "arrival" of a Heaviside unit step magnetic variation, i.e, $B(t) = B_0\Theta(t)$, their output is "neutralized" after 200 ms.

The instrumentation for the electric field measurements. Low pass "1 Hz" or "10 Hz" filters have been used. They are fourth-order active low pass filters, having two symmetric second-order poles in the complex f-plane with a frequency response:

$$R(f) = \frac{A \exp(-2\pi i f \tau_d)}{[1-(f/f_p)^2 + i\sqrt{2}(f/f_p)]^2}, \tag{1.98}$$

where A is the amplification, f_p is the half-response frequency corner of the filter and τ_d is the time delay of the filter. The expression (1.98) was applied to the laboratory measured data for both amplitude and phase, and a nonlinear least-squares fitting was performed (see

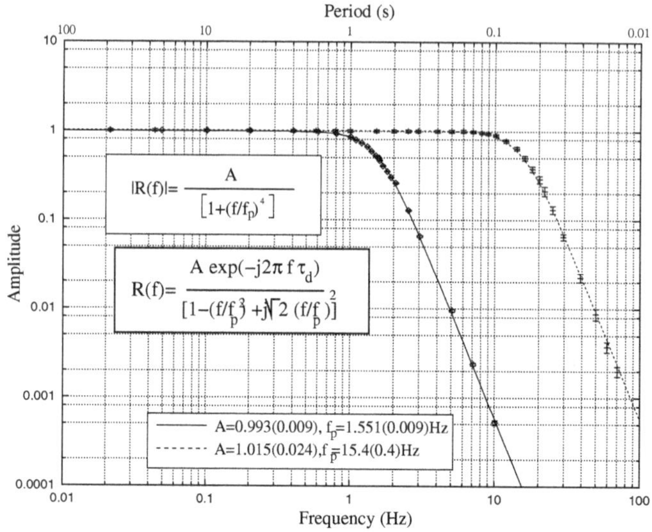

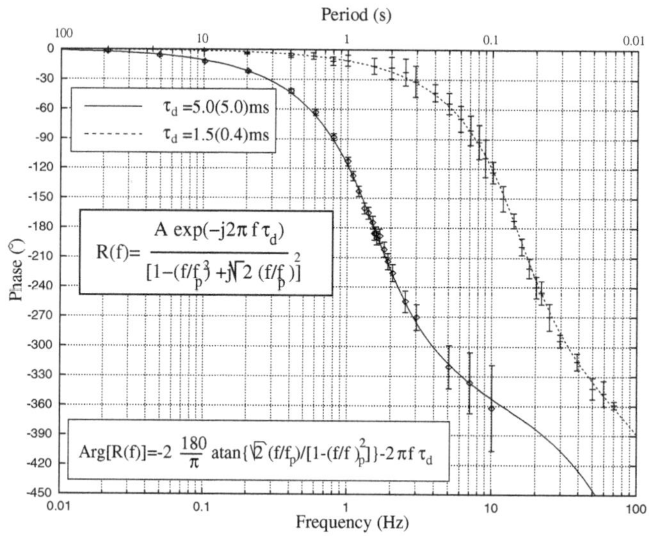

Fig. 1.52 The frequency response of the "1 Hz" (solid lines) and "10 Hz" (broken lines) low-pass filters. Reprinted from Ref. [125], Copyright (2005), with permission from TerraPub.

Fig. 1.52) using the constant chi-square ($p = 95\%$) boundaries for the determination of the errors in the fitting parameters which are also shown in Fig. 1.52.

Finally, Eq. (1.98) leads to an impulse response:

$$I(t) = 2\frac{A}{T_p} \exp\left[-\frac{(t-\tau_d)}{T_p}\right]\left[\sin\left(\frac{t-\tau_d}{T_p}\right) - \left(\frac{t-\tau_d}{T_p}\right)\cos\left(\frac{t-\tau_d}{T_p}\right)\right]\Theta(t-\tau_d). \quad (1.99)$$

In the case of "10 Hz" filters, for example, $T_p = 14.6 \pm 0.4$ ms and $\tau_d = 1.5 \pm 0.4$ ms, determined by the aforementioned laboratory calibration; see Fig. 1.52.

References

1. Antonopoulos, G., Kopanas, J., Eftaxias, K., Hadjicontis, V.: On the experimental evidence of SES vertical component. Tectonophysics **224**, 47–49 (1993)
2. Ashkenazy, Y., Hausdorff, J.M., Ivanov, P.C., Stanley, H.E.: A stochastic model of human gait dynamics. Physica A **316**, 662–670 (2002)
3. Ashkenazy, Y., Ivanov, P.C., Havlin, S., Peng, C.K., Yamamoto, Y., Goldberger, A.L., Stanley, H.E.: Decomposition of heartbeat time series: Scaling analysis of the sign sequence. Comput. Cardiol. **27**, 139–143 (2000)
4. Ausloos, M., Vandewalle, N., Boveroux, P., Minguet, A., Ivanova, K.: Applications of statistical physics to economic and financial topics. Physica A **274**, 229–240 (1999)
5. Bartsch, R., Plotnik, M., Kantelhardt, J.W., Havlin, S., Giladi, N., Hausdorff, J.M.: Fluctuation and synchronization of gait intervals and gait force profiles distinguish stages of Parkinson's disease. Physica A **383**, 455–465 (2007)
6. Bashan, A., Bartsch, R., Kantelhardt, J.W., Havlin, S.: Comparison of detrending methods for fluctuation analysis. Physica A **387**, 5080–5090 (2008)
7. Bernard, P.: Plausibility of long distance electrotelluric precursors to earthquakes. J. Geophys. Res. **97**, 17,531–17,546 (1992)
8. Box, G.E.P., Jenkins, G.M., Reinsel, G.C.: Time Series Analysis (3rd ed.). Prentice-Hall, Inc., Upper Saddle River, NJ (1994)
9. Brace, W.F., Byerlee, J.D.: Stick slip as a mechanism for earthquakes. Science **153**, 990–992 (1966)
10. Bray, A.J.: Theory of phase-ordering kinetics. Adv. Phys. **43**, 357–459 (1994)
11. Bray, A.J., Briant, A.J., Jervis, D.K.: Breakdown of scaling in the nonequilibrium critical dynamics of the two-dimensional XY model. Phys. Rev. Lett. **84**, 1503–1506 (2000)
12. Bunde, A., Havlin, S., Kantelhardt, J.W., Penzel, T., Peter, J.H., Voigt, K.: Correlated and uncorrelated regions in heart-rate fluctuations during sleep. Phys. Rev. Lett. **85**, 3736–3739 (2000)
13. Burton, P.W.: Dicing with earthquakes. Geophys. Res. Lett. **23**, 1379–1382 (1996)
14. Chester, F.M., Evans, J.P., Biegel, R.L.: Internal structure and weakening mechanisms of the San Andreas fault. J. Geophys. Res. **98**, 771–786 (1993)
15. Cremer, D.: General and theoretical aspects of the peroxide group, in The Chemistry of Functional Groups, Peroxides, ed. S. Patai. John Wiley, New York (1983)
16. Dobrovolsky, I.P., Gershenzon, N.L., Gokhberg, M.B.: Theory of electrokinetic effect occurring at the final state in the preparation of a tectonic earthquake. Phys. Earth Planet. Inter. **57**, 144–156 (1989)
17. Dukhin, S.S., Derjaguin, B.V.: Electrokinetic Phenomena. John Wiley & Sons, New York (1974)
18. Edwards, R.N., Nabighian, M.N.: The magnetometric resistivity method. In: Nambighian (ed.) Electromagnetic methods in Applied Geophysics, Vol. 2, pp. 43–104. SEG, Tulsa, Oklahoma (1996)
19. Ernst, R., Head, J.W., Parfitt, E., Grosfils, E., Wilson, L.: Giant radiating dyke swarms on Earth and Venus. Earth Science Reviews **39**, 1–58 (1995)
20. Fenoglio, M.A., Johnston, M.J.S., Byerlee, J.: Magnetic and electric fields associated with changes in high pore pressure in fault zones, application to the Loma Prieta ULF emissions. J. Geophys. Res. **100**, 12,951–12,958 (1995)
21. Fischbach, D., Nowick, A.: Some transient electrical effects of plastic deformation in NaCl crystals. J. Phys. Chem. Solids **5**, 302–315 (1958)
22. Fischbach, D.B., Nowick, A.S.: Deformation-induced charge flow in NaCl crystals. Phys. Rev. **99**, 1333–1334 (1955)
23. Fitterman, D.V.: Theory of electrokinetic and magnetic anomalies associated with dilatant regions in a layered Earth. J. Geophys. Res. **83**, 5923–5928 (1978)
24. Fitterman, D.V.: Calculation of self-potential anomalies near vertical contacts. Geophys. **44**, 195–205 (1979)
25. Fraser-Smith, A.C., Bernardi, A., McGill, P.R., Ladd, M.E., Helliwell, R.A., Villard, O.G.: Low-frequency magnetic-field measurements near the epicenter of the Ms-7.1 Loma Prieta earthquake. Geophys. Res. Lett. **17**, 1465–1468 (1990)

26. Freund, F.: Hydrogen and carbon in solid solution in oxides and silicates. Phys. Chem. Minerals **15**, 1–18 (1987)
27. Freund, F.: Earthquake prediction is worthy of study. EOS Trans. AGU **80**, 230–232 (1999)
28. Freund, F.: Charge generation and propagation in igneous rocks. J. Geodynamics **33**, 543–570 (2002)
29. Freund, F., Freund, M.M., Battlo, F.: Critical review of electrical conductivity measurements and charge distribution analysis of magnesium oxide. J. Geophys. Res. **98(B12)**, 22,209–22,229 (1993)
30. Freund, F., Whang, E.J., Jonathan, L.: Highly mobile hole charge carries in minerals: Key to the enigmatic electrical earthquake phenomena? In: M. Hayakawa, Y. Fujinawa (eds.) Electromagnetic Phenomena Related to Earthquake Prediction, pp. 271–291. TerraPub, Tokyo (1994)
31. Freund, M.M., Freund, F., Battlo, F.: Highly mobile oxygen holes in magnesium oxide. Phys. Rev. Lett. **63**, 2096–2099 (1989)
32. Gershenzon, N., Gokhberg, M.: On the origin of electrotelluric disturbances prior to an earthquake in Kalamata, Greece. Tectonophysics **224**, 169–174 (1993)
33. Gueguen, Y., Palciauskas, V.: Introduction to the Physics of Rocks. Princeton University Press, New Jersey (1994)
34. Guterman, V.G., Khazan, Y.M.: A possible mechanism of strong shallow earthquake preparation. Geophys. J. **16**, 22–29 (1994)
35. Hanks, T.C., Kanamori, H.: Moment magnitude scale. J. Geophys. Res. **84(B5)**, 2348 (1979)
36. Hernádez-Pérez, R., Guzmán-Vargas, L., Ramírez-Rojas, A., Angulo-Brown, F.: Pattern synchrony in electrical signals related to earthquake activity. Physica A **389**, 1239–1252 (2010)
37. Hoversten, G.M., Becker, A.: EM1DSH with EMMODEL a Motif GUI, Numerical Modeling of multiple thin 3D sheets in a layered earth, University of California at Berkeley, Engineering Geoscience Department, June 12, 1995.
38. Hu, K., Ivanov, P.C., Chen, Z., Carpena, P., Stanley, H.E.: Effect of trends on detrended fluctuation analysis. Phys. Rev. E **64**, 011114 (2001)
39. Hu, K., Ivanov, P.C., Chen, Z., Hilton, M.F., Stanley, H.E., Shea, S.A.: Non-random fluctuations and multi-scale dynamics regulation of human activity. Physica A **337**, 307–318 (2004)
40. Ikeya, M., Kinoshita, Y., Matsumoto, H., Takaki, S., Yamanaka, C.: A model experiment of electromagnetic wave propagation over long distances using waveguide terminology. Jpn. J. Appl. Phys. **36**, 1558–1561 (1997)
41. Ikeya, M., Takaki, S., Matsumoto, H., Tani, A., Komatsu, T.: Pulsed charge model of fault behavior producing Seismic Electric Signals (SES). J. Circuit, Systems and Computers **7**, 153–164 (1997)
42. Ishido, T., Mizutani, H.: Experimental and theoretical basis of electrokinetic phenomena in rock-water systems and its applications to Geophysics. J. Geophys. Res. **86**, 1763–1775 (1981)
43. Ishido, T., Pritchett, J.W.: Numerical simulation of electrokinetic potentials associated with subsurface fluid flow. J. Geophys. Res. **104(B7)**, 15,247–15,260 (1999)
44. Ivanov, P.C.: Scale-invariant aspects of cardiac dynamics – observing sleep stages and circadian phases. IEEE Eng. Med. Biol. **26**, 33–37 (2007)
45. Ivanov, P.C., Ma, Q.D.Y., Bartsch, R.P., Hausdorff, J.M., Nunes Amaral, L.A., Schulte-Frohlinde, V., Stanley, H.E., Yoneyama, M.: Levels of complexity in scale-invariant neural signals. Phys. Rev. E **79**, 041920 (2009)
46. Ivanov, P.C., Yuen, A., Podobnik, B., Lee, Y.: Common scaling patterns in intertrade times of U. S. stocks. Phys. Rev. E **69**, 056107 (2004)
47. Ivanova, K., Ausloos, M.: Application of the detrended fluctuation analysis (DFA) method for describing cloud breaking. Physica A **274**, 349–354 (1999)
48. Jouniaux, L., Pozzi, J.P.: Streaming potential and permeability of saturated sandstones under triaxial stress: Consequences for electrotelluric anomalies prior to earthquakes. J. Geophys. Res. **100(B6)**, 10,197–10,209 (1995)
49. Jouniaux, L., Pozzi, J.P.: Laboratory measurements anomalous 0.1–0.5 Hz streaming potential under geochemical changes: Implications for electrotelluric precursors to earhquakes. J. Geophys. Res. **102(B7)**, 15,335–15,343 (1997)
50. Kanamori, H., Anderson, D.L.: Theoretical basis of some empirical relations in seismology. Bull. Seism. Soc. Am. **65**, 1073–1096 (1975)
51. Kantelhardt, J.W., Koscielny-Bunde, E., Rego, H.H.A., Havlin, S., Bunde, A.: Detecting long-range correlations with detrended fluctuation analysis. Physica A **295**, 441–454 (2001)

52. Karakelian, D., Beroza, G.C., Klemperer, S.L., Fraser-Smith, A.C.: Analysis of ultralow-frequency electromagnetic field measurements associated with the 1999 M 7.1 Hector Mine, California, earthquake sequence. Bull. Seism. Soc. Am. **92**, 1513–1524 (2002)
53. Kondo, S., Uyeda, S., Nagao, T.: The selectivity of the Ioannina VAN station. J. Geodynamics **33**, 433–461 (2002)
54. Kou, S.C., Xie, X.S.: Generalized langevin equation with fractional Gaussian noise: Subdiffusion within a single protein molecule. Phys. Rev. Lett. **93**, 180603 (2004)
55. Lazarus, D.: Short term earthquake prediction in Greece by Seismic Electric Signals. In: Sir J. Lighthill (ed.) The Critical Review of VAN: Earthquake Prediction from Seismic Electric Signals, pp. 91–96. World Scientific, Singapore (1996)
56. Li, M., Lim, S.C.: A rigorous derivation of power spectrum of fractional Gaussian noise. Fluctuation and Noise Letters (FNL) **6**, C33–C36 (2006)
57. Li, M., Lim, S.C.: Fractional Gaussian noise and network traffic modeling. In: Chen, S. Y. and Li, Q. (ed.) The 8th WSEAS International Conference on Applied Computer & Applied Computational Science, pp. 34–39 (2009)
58. Liu, Y.H., Cizeau, P., Meyer, M., Peng, C.K., Stanley, H.E.: Correlations in economic time series. Physica A **245**, 437–440 (1997)
59. Lockner, D.A., Johnston, M.J.S., Byerlee, J.D.: A mechanism to explain the generation of earthquake lights. Nature **302**, 28–33 (1983)
60. Mandelbrot, B.B., van Ness, J.W.: Fractional noises and applications. SIAM Rev. **10**, 422–437 (1968)
61. Mizutani, H., Ishido, T.: A new interpretation of magnetic field variation associated with Matsushiro earthquakes. J. Geomagn. Geoelectr. **28**, 179–188 (1976)
62. Mizutani, H., Ishido, T., Yokokura, T., Ohnishi, S.: Electrokinetic phenomena associated with earthquakes. Geophys. Res. Lett. **3**, 365–368 (1976)
63. Molchanov, O.A.: Fracturing as underlying mechanisms of seismo-electric signals. In: Atmospheric and Ionospheric Electromagnetic Phenomena Associated with Earthquakes, ed. M. Hayakawa, pp. 349–356. TERRAPUB, Tokyo (1999)
64. Molchanov, O.A., Kopytenko, Y.A., Voronov, P.M., Kopytenko, E.A., Matiashvili, T.G., Fraser-Smith, A.C., Bernardi, A.: Results of ULF magnetic-field measurements near the epicenters of the Spitak (Ms = 6.9) and Loma Prieta (Ms = 7.1) earthquakes – comparative-analysis. Geophys. Res. Lett. **19**, 1495–1498 (1992)
65. Morgan, D.: A model for the explanation of SES-generation based on electrokinetic effect, paper presented In the International Conference on Measurements and Theoretical Models of the Earth's Field Variations Related to Earthquakes. Univ. Athens, Athens. (1990)
66. Morgan, F.D.: Electrodynamics of the earthquake source. EOS Trans. AGU **F173** (1995)
67. Morgan, F.D., Nur, A.: Enhanced streaming potentials with two-phase flow (abstract). EOS Trans. AGU **67(44)**, 1203 (1986)
68. Morgan, F.D., Williams, E.R., Madden, T.R.: Streaming potential properties of westerly granite with applications. J. Geophys. Res. **94**, 12,449–12,461 (1989)
69. Morse, P., Feshbach, H.: Methods of Theoretical Physics. McGraw-Hill, New York (1954)
70. Multhaupt, R.G., Gross, B., Sessler, G.M.: Recent Progress in Electret Research. In Electrets, ed. G. M. Sessler, 2nd edition. Springer-Verlag, Heidelberg (1987)
71. Nagahama, H., Teisseyre, R.: Thermodynamics of line defects and transient electric current: electromagnetic field generation in earthquake preparation zone. Acta Geophys. Pol. **46**, 35–54 (1998)
72. Nagao, T., Orihara, Y., Yamaguchi, T., Takahashi, I., Hattori, K., Noda, Y., Sayanagi, K., , Uyeda, S.: Coseismic geoelectric potential changes observed in Japan. Geophys. Res. Lett. **27**, 1535–1538 (2000)
73. Nagao, T., Uyeshima, M., Uyeda, S.: An independent check of VAN's criteria for signal recognition. Geophys. Res. Lett. **23**, 1441–1444 (1996)
74. Nomicos, K., Chatzidiakos, P.: A telemetric system for measuring electrotelluric variations in Greece and its application to earthquake prediction. Tectonophysics **224**, 39–46 (1993)
75. Nomicos, K., Makris, J., Kefalas, M.: The telemetric system of VAN group In: Sir J. Lighthill (ed.) The Critical Review of VAN: Earthquake Prediction from Seismic Electric Signals, p. 77. World Scientific, Singapore (1996)

76. Ohnaka, M.: A constitutive scaling law and a unified comprehension for frictional slip failure, shear fracture of intact rock, and earthquake rupture. J. Geophys. Res. **108(B2)**, 2080 (2003)
77. Papathanassiou, A.N., Sakellis, I., Grammatikakis, J.: Negative activation volume for dielectric relaxation in hydrated rocks. Tectonophysics **490**, 307–309 (2010)
78. Park, S., Johnston, M., Madden, T., Morgan, D., Morrison, F.: Electromagnetic precursors to earthquakes in the VLF band: A review of observations and mechanisms. Rev. Geophys. **31**, 117–132 (1993)
79. Peng, C.K., Buldyrev, S.V., Goldberger, A.L., Havlin, S., Simons, M., Stanley, H.E.: Finite-size effects on long-range correlations: Implications for analyzing DNA sequences. Phys. Rev. E **47**, 3730–3733 (1993)
80. Peng, C.K., Buldyrev, S.V., Havlin, S., Simons, M., Stanley, H.E., Goldberger, A.L.: Mosaic organization of DNA nucleotides. Phys. Rev. E **49**, 1685–1689 (1994)
81. Peng, C.K., Havlin, S., Stanley, H.E., Goldberger, A.L.: Quantification of scaling exponents and crossover phenomena in nonstationary heartbeat time series. CHAOS **5**, 82 (1995)
82. Pride, S.R.: Governing equations for the coupled electromagnetics and acoustics of porous media. Phys. Rev. B. **50**, 15,678–15,696 (1994)
83. Ramírez-Rojas, A., Flores-Márquez, E. L., Guzmán-Vargas, L., Gálvez-Coyt, G., Telesca, L., Angulo-Brown, F.: Statistical features of seismoelectric signals prior to M7.4 Guerrero-Oaxaca earthquake (México). Nat. Hazards and Earth Syst. Sci. **8**, 1001–1007 (2008)
84. Ramírez-Rojas, A., Telesca, L., Angulo-Brown, F.: Entropy of geoelectrical time series in the natural time domain. Nat. Hazards and Earth Syst. Sci. **11**, 219–225 (2011)
85. Reppert, P.M.: Electrokinetics in the Earth. MIT, Cambridge, Ph.D. Thesis (2000)
86. Reppert, P.M., Morgan, F.D.: Streaming potential collection and data processing techniques. J. Colloid Interface Sci. **233**, 348–355 (2001)
87. Reppert, P.M., Morgan, F.D., Lesmes, D., Jouniaux, L.: Frequency dependent streaming potentials. J. Colloid Interface Sci. **234**, 194–203 (2001)
88. Rokityansky, I.I.: Anomalous phenomena preceding earthquakes (one more approach to discussion on SES nature and VAN-prediction reliability). Phys. Chem. of the Earth (A) **24**, 731–734 (1999)
89. Rundle, J.B., Turcotte, D.L., Klein, W.: (eds.) GeoComplexity and the Physics of Earthquakes. AGU, Washington, DC (2000)
90. Sarlis, N., Lazaridou, M., Kapiris, P., Varotsos, P.: Numerical model of the selectivity effect and $\Delta V/L$ criterion. Geophys. Res. Lett. **26**, 3245–3248 (1999)
91. Sarlis, N., Varotsos, P.: Comments on the transmission of electric signals in dielectric media. Acta Geophys. Pol. **49**, 277–285 (2001)
92. Sarlis, N., Varotsos, P.: Magnetic field near the outcrop of an almost horizontal conductive sheet. J. Geodynamics **33**, 463–476 (2002)
93. Sarlis, N.V., Skordas, E.S., Lazaridou, M.S., Varotsos, P.A.: Investigation of seismicity after the initiation of a Seismic Electric Signal activity until the main shock. Proc. Japan Acad., Ser. B **84**, 331–343 (2008)
94. Sata, N., Eberman, K., Eberl, K., Maier, J.: Mesoscopic fast ion conduction in nanometre-scale. Nature (London) **408**, 946–949 (2000)
95. Scholz, C.H.: Earthquakes and friction laws. Nature (London) **391**, 37–42 (1998)
96. Scholz, C.H.: Evidence for a strong San Andreas fault. Geology **28**, 163–166 (2000)
97. Scholz, C.H.: A fault in the 'weak San Andreas' theory. Nature **406**, 234 (2000)
98. Skordas, E., Kapiris, P., Bogris, N., Varotsos, P.: Field experimentation on the detectability of co-seismic electric signals. Proc. Japan Acad., Ser. B **76**, 51–56 (2000)
99. Slifkin, L.: Surface and dislocation effects on diffusion in ionic crystals, in Diffusion in Materials, eds. A.L. Laskar, J.L. Bocquet, G. Brebec and C. Monty, vol. 179, pp. 471–483. Kluwer, Dordrecht (1990)
100. Slifkin, L.: Seismic Electric Signals from displacement of charged dislocations. Tectonophysics **224**, 149–152 (1993)
101. Slifkin, L.: A dislocation model for Seismic Electric Signals. In: Sir J. Lighthill (ed.) The Critical Review of VAN: Earthquake Prediction from Seismic Electric Signals, pp. 97–104. World Scientific, Singapore (1996)

102. Sornette, A., Sornette, D.: Earthquake rupture as a critical point: Consequences for telluric precursor. Tectonophysics **179**, 327–334 (1990)
103. Sornette, D.: Earthquakes: from chemical alteration to mechanical rupture. Phys. Reports **313**, 237–292 (1999)
104. Sornette, D.: Mechanochemistry: An hypothesis for shallow earthquakes. In: R. Teisseyre, J. Mazewski (eds.) Earthquake Thermodynamics and Phase Transformations in the Earth's Interior, pp. 329–366. Academic Press, San Diego (2001)
105. Sornette, D., Lagier, M., Roux, S., Hansen, A.: Critical piezoelectricity in percolation. J. Phys. France **50**, 2201–2216 (1989)
106. Stanley, H.E.: Scaling, universality, and renormalization: Three pillars of modern critical phenomena. Rev. Mod. Phys. **71**, S358–S366 (1999)
107. Stanley, H.E., Buldyrev, S.V., Goldberger, A.L., Havlin, S., Peng, C.K., Simon, M.: Scaling features of noncoding DNA. Physica A **273**, 1–18 (1999)
108. Stauffer, D.: Scaling theory of percolation clusters. Phys. Reports **54**, 1–74 (1979)
109. Surkov, V.V., Uyeda, S., Tanaka, H., Hayakawa, M.: Fractal properties of medium and seismoelectric phenomena. J. Geodynamics **33**, 477–487 (2002)
110. Talkner, P., Weber, R.O.: Power spectrum and detrended fluctuation analysis: Application to daily temperatures. Phys. Rev. E **62**, 150–160 (2000)
111. Taqqu, M.S., Teverovsky, V., Willinger, W.: Estimators for long-range dependence: An empirical study. Fractals **3**, 785–798 (1995)
112. Teisseyre, R.: Earthquake premonitory processes: Evolution of stresses and current generation. Terra Nova **4**, 509–513 (1992)
113. Teisseyre, R.: Electric field generation in earthquake premonitory processes. In: R. Teisseyre (ed.) Theory of Earthquake Premonitory and Fracture Processes, pp. 282–292. PWN, Warsaw (1995)
114. Teisseyre, R.: Motion and flow equation of stresses. Acta Geophys. Pol. **44**, 19–29 (1996)
115. Teisseyre, R.: Dislocation-stress relations and evolution of dislocation fields. Acta Geophys. Pol. **45**, 205–214 (1997)
116. Teisseyre, R.: Dislocation dynamics and related electromagnetic excitation. Acta Geophys. Pol. **49**, 55–73 (2001)
117. Teisseyre, R., Nagahama, H.: Dislocation field evolution and dislocation source/sink function. Acta Geophys. Pol. **46**, 13–33 (1998)
118. Turnhout, J.: Thermally Stimulated Discharge of Polymer Electrets. Elsevier, New York (1975)
119. Uyeda, S., Hayakawa, M., Nagao, T., Molchanov, O., Hattori, K., Orihara, Y., Gotoh, K., Akinaga, Y., Tanaka, H.: Electric and magnetic phenomena observed before the volcano-seismic activity in 2000 in the Izu Island region, Japan. Proc. Natl. Acad. Sci. USA **99**, 7352–7355 (2002)
120. Uyeda, S., Kamogawa, M., Tanaka, H.: Analysis of electrical activity and seismicity in the natural time domain for the volcanic-seismic swarm activity in 2000 in the Izu Island Region, Japan. J. Geophys. Res. **114**, B02310 (2009)
121. Vandewalle, N., Ausloos, M.: Crossing of two mobile averages: A method for measuring the roughness exponent. Phys. Rev. E **58**, 6832–6834 (1998)
122. Varotsos, P.: The dielectric loss of X-irradiated $LiD+Mg^{+2}$. J. Phys. France **41**, 377–379 (1980)
123. Varotsos, P.: Earthquake prediction in Greece based on Seismic Electric Signals: Period January 1, 1984 to March 18, 1986. Bollettino di Geodesia e Scienze Affini XLV **2**, 191–202 (1986)
124. Varotsos, P.: Fracture and friction: A review. Acta Geophys. Pol. **52**, 105–142 (2004)
125. Varotsos, P.: The Physics of Seismic Electric Signals. TERRAPUB, Tokyo (2005)
126. Varotsos, P., Alexopoulos, K.: Calculation of the migration volume of vacancies in ionic solids from macroscopic parameters. Phys. Status Solidi A **47**, K133–136 (1978)
127. Varotsos, P., Alexopoulos, K.: Physical properties of the variations of the electric field of the Earth preceding earthquakes, I. Tectonophysics **110**, 73–98 (1984)
128. Varotsos, P., Alexopoulos, K.: Physical properties of the variations of the electric field of the Earth preceding earthquakes, II. Tectonophysics **110**, 99–125 (1984)
129. Varotsos, P., Alexopoulos, K.: Thermodynamics of Point Defects and their Relation with Bulk Properties. North Holland, Amsterdam (1986)
130. Varotsos, P., Alexopoulos, K.: Physical properties of the variations of the electric field of the Earth preceding earthquakes, III. Tectonophysics **136**, 335–339 (1987)

131. Varotsos, P., Alexopoulos, K., Lazaridou, M.: Latest aspects of earthquake prediction in Greece based on Seismic Electric Signals, II. Tectonophysics **224**, 1–37 (1993)
132. Varotsos, P., Alexopoulos, K., Nomicos, K.: Seven-hour precursors to earthqaukes determined from telluric currents. Practica of Athens Academy **56**, 417–433 (1981)
133. Varotsos, P., Alexopoulos, K., Nomicos, K.: Comments on the pressure variation of the Gibbs energy for bound and unbound defects. Phys. Status Solidi B **111**, 581–590 (1982)
134. Varotsos, P., Bogris, N., Kyritsis, A.: Comments on the depolarization currents stimulated by variations of temperature and pressure. J. Phys. Chem. Solids **53**, 1007–1011 (1992)
135. Varotsos, P., Eftaxias, K., Lazaridou, M., Antonopoulos, G., Makris, J.: Recent VAN results on earthquake prediction in Greece (in Japanese). Zisin (Journal of the Seismological Society of Japan) **17**, 18–26 (1994)
136. Varotsos, P., Eftaxias, K., Lazaridou, M., Antonopoulos, G., Makris, J., Poliyiannakis, J.: Summary of the five principles suggested by Varotsos et al. [1996] and the additional questions raised in this debate. Geophys. Res. Lett. **23**, 1449–1452 (1996)
137. Varotsos, P., Eftaxias, K., Lazaridou, M., Bogris, N., Makris, J.: Note on the extension of the SES sensitive area at Ioannina station, Greece. Acta Geophys. Pol. **46**, 55–60 (1998)
138. Varotsos, P., Eftaxias, K., Lazaridou, M., Nomicos, K., Sarlis, N., Bogris, N., Makris, J., Antonopoulos, G., Kopanas, J.: Recent earthquake prediction results in Greece based on the observation of Seismic Electric Signals. Acta Geophys. Pol. **44**, 301–327 (1996)
139. Varotsos, P., Hadjicontis, V., A.S. Nowick, A.S.: The physical mechanism of Seismic Electric Signals. Acta Geophys. Pol. **49**, 415–421 (2001)
140. Varotsos, P., Lazaridou, M.: Latest aspects of earthquake prediction in Greece based on Seismic Electric Signals. Tectonophysics **188**, 321–347 (1991)
141. Varotsos, P., Lazaridou, M., Eftaxias, K., Antonopoulos, G., Makris, J., Kopanas, J.: Short term earthquake prediction in Greece by Seismic Electric Signals. In: Sir J. Lighthill (ed.) The Critical Review of VAN: Earthquake Prediction from Seismic Electric Signals, pp. 29–76. World Scientific, Singapore (1996)
142. Varotsos, P., Mourikis, S.: Difference in conductivity between LiD and LiH crystals. Phys. Rev. B **10**, 5220–5224 (1974)
143. Varotsos, P., Sarlis, N., Bogris, N., Makris, J., Kapiris, P., Abdulla, A.: A comment on the $\Delta V/L$-criterion for the identification of seismic electric signals. In: Atmospheric and Ionospheric Electromagnetic Phenomena Associated with Earthquakes, ed. M. Hayakawa, pp. 1–45. TERRAPUB, Tokyo (1999)
144. Varotsos, P., Sarlis, N., Lazaridou, M.: Transmission of stress induced electric signals in dielectric media. Part II. Acta Geophys. Pol. **48**, 141–177 (2000)
145. Varotsos, P., Sarlis, N., Lazaridou, M., Kapiris, P.: A plausible model for the explanation of the selectivity effect of Seismic Electric Signals. Practica of Athens Academy **71**, 283–354 (1996)
146. Varotsos, P., Sarlis, N., Lazaridou, M., Kapiris, P.: Transmission mechanism of Seismic Electric Signals, II. Practica of Athens Academy **72**, 270–302 (1997)
147. Varotsos, P., Sarlis, N., Lazaridou, M., Kapiris, P.: Transmission of stress induced electric signals in dielectric media. J. Appl. Phys. **83**, 60–70 (1998)
148. Varotsos, P., Sarlis, N., Skordas, E.: Transmission of stress induced electric signals in dielectric media. Part III. Acta Geophys. Pol. **48**, 263–297 (2000)
149. Varotsos, P., Sarlis, N., Skordas, E.: Magnetic field variations associated with SES. the instrumentation used for investigating their detectability. Proc. Jpn. Acad., Ser. B **77**, 87–92 (2001)
150. Varotsos, P., Sarlis, N., Skordas, E.: Magnetic field variations associated with the SES before the 6.6 Grevena-Kozani earthquake. Proc. Jpn. Acad., Ser. B **77**, 93–97 (2001)
151. Varotsos, P., Sarlis, N., Skordas, E.: A note on the spatial extent of the Volos SES sensitive site. Acta Geophys. Pol. **49**, 425–435 (2001)
152. Varotsos, P., Sarlis, N., Skordas, E.: On the difference in the rise times of the two SES electric field components. Proc. Jpn. Acad., Ser. B **80**, 276–282 (2004)
153. Varotsos, P.A., Sarlis, N.V., Skordas, E.S.: See (the freely available) EPAPS Document No. E-PLEEE8-68-116309 originally from P.A. Varotsos, N.V. Sarlis and E.S. Skordas, *Phys. Rev. E* **68**, 031106 (2003). For more information on EPAPS, see http://www.aip.org/pubservs/epaps.html.

154. Varotsos, P.A., Sarlis, N.V., Skordas, E.S.: Long-range correlations in the electric signals that precede rupture. Phys. Rev. E **66**, 011902 (2002)
155. Varotsos, P.A., Sarlis, N.V., Skordas, E.S.: Attempt to distinguish electric signals of a dichotomous nature. Phys. Rev. E **68**, 031106 (2003)
156. Varotsos, P.A., Sarlis, N.V., Skordas, E.S.: Long-range correlations in the electric signals that precede rupture: Further investigations. Phys. Rev. E **67**, 021109 (2003)
157. Varotsos, P.A., Sarlis, N.V., Skordas, E.S.: Time-difference between the electric field components of signals prior to major earthquakes. Appl. Phys. Lett. **86**, 194101 (2005)
158. Varotsos, P.A., Sarlis, N.V., Skordas, E.S.: Detrended fluctuation analysis of the magnetic and electric field variations that precede rupture. CHAOS **19**, 023114 (2009)
159. Varotsos, P.A., Sarlis, N.V., Skordas, E.S., Lazaridou, M.S.: Electric pulses some minutes before earthquake occurrences. Appl. Phys. Lett. **90**, 064104 (2007)
160. Varotsos, P.A., Sarlis, N.V., Lazaridou, M.S.: Interconnection of defect parameters and stress-induced electric signals in ionic crystals. Phys. Rev. B **59**, 24–27 (1999)
161. Varotsos, P.A., Sarlis, N.V., Skordas, E.S.: Electric fields that "arrive" before the time derivative of the magnetic field prior to major earthquakes. Phys. Rev. Lett. **91**, 148501 (2003)
162. Vicsek, T.: Fractal growth phenomena. World Scientific, Singapore (1989)
163. Weron, A., Burnecki, K., Mercik, S., Weron, K.: Complete description of all self-similar models driven by Lévy stable noise. Phys. Rev. E **71**, 016113 (2005)
164. Yoshii, H.: Social impacts of earthquake prediction in Greece. Tectonophysics **224**, 251–255 (1993)
165. Zhdanov, M.S., Keller, G.V.: The Geoelectrical Methods in Geophysical Exploration. Elsevier, Amsterdam (1994)
166. Zoback, M.D.: Strength of the San Andreas. Nature **405**, 31–32 (2000)

Part II
Natural Time Foundations

2. Natural Time. Background

Abstract. Time and not space poses the greatest challenge to science. Conventional time is modeled as the one-dimensional continuum $\mathscr{R}$ of real numbers. This continuity, however, does *not* stem from *any* fundamental principle. On the other hand, natural time χ, which is a new time domain introduced by the authors in 2001, is *not* continuous and its values as well as those of the energy form *countable* sets. Novel dynamical features hidden behind time series in complex systems can emerge upon analyzing them in natural time, which conforms to the desire to reduce uncertainty and extract signal information as much as possible. The fluctuations, under time reversal, of the natural time can serve in time series for the quantification of the long-range dependence. Natural time analysis also enables the study of the dynamical evolution of a complex system and identifies *when the system enters a critical state*. In particular, the normalized power spectrum $\Pi(\omega)$ is introduced in natural time, and its Taylor expansion leads, at low natural (cyclic) frequencies ω ($\omega \to 0$), to the expression $\Pi(\omega) \approx 1 - \kappa_1 \omega^2$. The values of the coefficient κ_1, which is just the variance of natural time, i.e., $\kappa_1 = \langle \chi^2 \rangle - \langle \chi \rangle^2$, are useful in identifying the approach to a critical point such as SES whose κ_1 value is shown to be 0.070. In addition, natural time analysis enables the distinction between the two origins of self-similarity, i.e., whether self-similarity solely results from long-range temporal correlations (the process's memory only) or solely from the process's increments' infinite variance (heavy tails in their distribution). In general, however, the self-similarity may result from both these origins, a case that can be also identified by natural time.

2.1 Introduction to natural time

In this Section, we follow Ref. [50]. In reviewing the state of physics today, a consensus seems to emerge that we are missing something absolutely fundamental, e.g., Refs. [2, 17]. Furthermore, there is a widespread belief that, it is not space but time that in the end poses the greatest challenge to science (e.g., p. 18 of Ref. [71]) as it will be further discussed in the next subsection.

2.1.1 Time and not space poses the greatest challenge to science

Time, according to Weyl (see p. 5 of Ref. [67]) for example, is "the primitive form of the stream of consciousness. It is a fact, however, obscure and perplexing to our minds, that ... one does not say this *is* but this is *now*, yet no more" or according to Gödel "that mysterious and seemingly self-contradictory being which, on the other hand, seems to form the basis of the world's and our own existence" (p. 111 of Ref. [71]). The challenge seems to stem from the fact that special relativity and quantum mechanics, which are the two great (and successful) theories of twentieth-century physics, are based on entirely different ideas, which are not easy to reconcile. (In general, the former theory, according to Einstein [10], is an example of "principled theory" in the sense that you start with the principles that underlie the theory and then work down to deduce the facts, while the latter is a "constructive theory" meaning that it describes phenomena based on some known facts but an underlying principle to explain the strangeness of the quantum world has not yet been found.) In particular, special relativity puts space and time on the *same* footing, but quantum mechanics treats them very differently, e.g., see p. 858 of Ref. [69]. (In quantum gravity, space is fluctuating and time is hard to define, e.g., Ref. [70].) More precisely, as far as the theory of special relativity is concerned, let us recall the following wording of Einstein [11]:

"Later, H. Minkowski found a particularly elegant and suggestive expression ..., which reveals a formal relationship between Euclidean geometry of three dimensions and the space time *continuum* of physics ... . From this it follows that, in respect to its *rôle* in the equations of physics, though not with regard to its physical significance, time is equivalent to the space co-ordinates (apart from the relations of reality). From this point of view, physics is, as it were, Euclidean geometry of four dimensions, or, more correctly, a statics in a four-dimensional Euclidean *continuum*." – whereas in quantum mechanics, Von Neumann complains [28]:

"First of all we must admit that this objection points at an essential weakness which is, in fact, the *chief weakness* of quantum mechanics: its non-relativistic character, which distinguishes the time t from the three space coordinates x, y, z, and presupposes an objective simultaneity concept. In fact, while all other quantities (especially those x, y, z, closely connected with t by the Lorentz transformation) are represented by operators, there corresponds to the time an ordinary number-parameter t, just as in classical mechanics."

Note also that Pauli [33] has earlier shown that there is no operator canonically conjugate to the Hamiltonian, if the latter is bounded from below. This means that for many systems a time operator does *not* exist. In other words, the introduction of an operator t is basically forbidden and the time must necessarily be considered as an ordinary number (but recall the long-standing question that Schrödinger's equation, as well as Einstein's general theory of relativity, is symmetric under time reversal in contrast to the fact that our world is *not*, e.g., Ref. [35]). These observations have led to a quite extensive literature mainly focused on time-energy (as well as on "phase-action") uncertainty relation, proposing a variety of attempts to overcome these obstacles. The discussion of this literature, however, lies beyond the scope of the present monograph. We just summarize here that the (conventional) time t is *currently* modeled as the one-dimensional *continuum* $\mathscr{R}$ of the real numbers, e.g., p. 10 of Ref. [70] (or p. 12 of Ref. [67] in which it is stated that

"... the straight line ... is homogeneous and a linear *continuum* just like time"). It is the consequences of this *continuity* that will be compared to the newly introduced concept of natural time, in a sense that will be discussed later in Section 2.7.

2.1.2 Definition of natural time

In a time series comprising N events, the *natural time*

$$\chi_k = k/N \qquad (2.1)$$

serves as an index for the *occurrence* of the k-th event [51, 52], and it is smaller than, or equal to, unity (note that the symbol χ originates from the ancient Greek word $\chi\rho\acute{o}\nu o\varsigma$ (chronos), see the cover page, which means "time").

In natural time analysis the evolution of the pair of two quantities (χ_k, Q_k) is considered, where $\chi_k = k/N$, N being the total number of events, and Q_k denotes in general a quantity proportional to the energy of the individual (k-th) event [51, 52]. Equivalently with Q_k, one can consider the quantity

$$p_k = \frac{Q_k}{\sum_{n=1}^{N} Q_n}, \qquad (2.2)$$

$$\sum_{k=1}^{N} p_k = 1, \qquad (2.3)$$

where p_k is the *normalized* energy emitted during the k-th event. In other words, the evolution of the pair either (χ_k, Q_k) or (χ_k, p_k) is considered in natural time analysis.

For example, to perform the analysis of dichotomous electric signals (Fig. 2.1(a)), which is frequently the case of a SES activity (see Chapter 1), we consider Q_k as being proportional to the duration of the k-th pulse [51, 52, 55, 54]. As another example, we refer to the analysis of seismic events (Fig. 2.1(b)): we then consider the evolution of the pair (χ_k, M_{0_k}) where M_{0_k} stands for the seismic moment of the k-th earthquake [51, 53, 61, 60], since M_{0_k} is proportional to the energy emitted in that earthquake (note that M_{0_k} differs essentially from the magnitude M, but they are interconnected [21] $\mathrm{M}_{0_k} \propto 10^{c\mathrm{M}}$ where $c \approx 1.5$, see also Section 6.1). Other examples elaborated in this monograph are: first, the analysis of electrocardiograms (see Fig. 2.2) which will be discussed in detail in Chapter 9. Second, the case of long-duration SES activities of non-obvious dichotomous nature, which is treated in Section 4.11. Third, the analysis of various dynamical models (among which a case of quasi-periodic Q_k, see Fig.(8.4)) in natural time which is discussed in detail in Chapter 8.

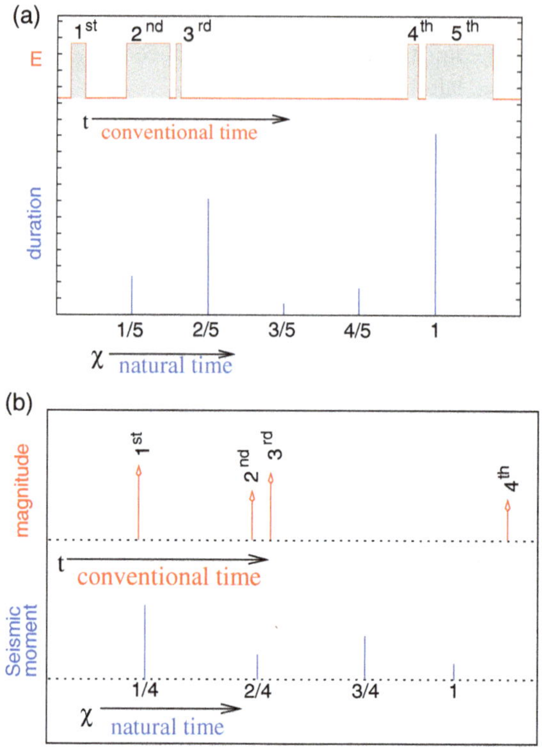

Fig. 2.1 (a) How a dichotomous series of electric pulses in conventional time t (upper panel, red) can be read in natural time χ (lower panel, blue). The symbol E stands for the electric field. (b) The same as in (a), but for a series of seismic events.

2.1.3 The "uniform" distribution

Among the various applications of natural time that will be discussed throughout this monograph, there is the fundamental paradigm of the "uniform" distribution that corresponds for example to the case when the system under study is in a stationary state emitting uncorrelated bursts of energy:

As a "uniform" distribution we consider the case when Q_k are *positive independent and identically distributed* (p.i.i.d.) random variables.

In this case, the expectation value $\mathscr{E}(p_k)$ of the point probabilities p_k is $\mathscr{E}(p_k) = 1/N$ by virtue of Eq. (2.3).

Let us now consider the distribution

$$p(\chi) = \sum_{k=1}^{N} p_k \delta(\chi - \chi_k) = \sum_{k=1}^{N} p_k \delta\left(\chi - \frac{k}{N}\right) \tag{2.4}$$

that corresponds to the point probabilities p_k. (Note that, $\delta(x)$ stands for the usual Dirac delta function.)

As $N \to \infty$, $p(\chi)$ for a "uniform" distribution tends to

$$p(\chi) = 1, \tag{2.5}$$

leading to an average value of natural time

$$\langle \chi \rangle = \int_0^1 \chi p(\chi)\, \mathrm{d}\chi = \frac{1}{2}. \tag{2.6}$$

2.2 Time reversal and natural time

In a time series comprising N events, the effect of the time-reversal operator $\hat{T}$ on Q_k is given by

$$\hat{T} Q_k = Q_{N-k+1}, \tag{2.7}$$

so that the first pulse ($k = 1$) is positioned last in the time reversed time-series, the second becomes last but one, etc.

Thus, the time reversal operator $\hat{T}$ in natural time acting on p_k results in

$$\hat{T} p_k = p_{N-k+1} \tag{2.8}$$

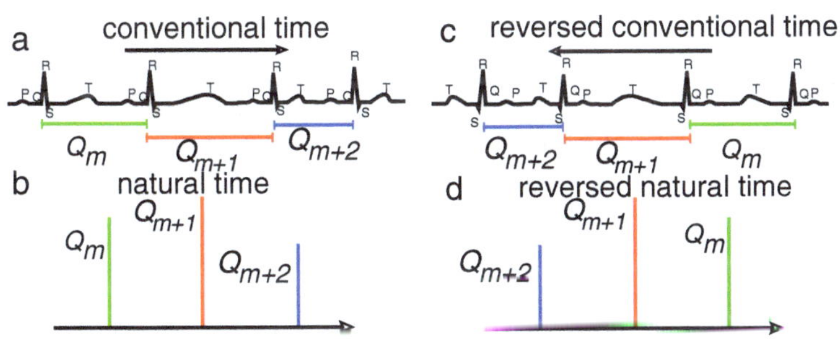

Fig. 2.2 (a) Schematic diagram (not to scale) of a four heartbeat excerpt of an ECG (for the notation of the inflection points see § 9.1.1) in the usual (conventional) time domain. The durations Q_m, Q_{m+1}, Q_{m+2} of the three RR intervals are shown. (b) The RR interval time series of (a) read in natural time; the vertical bars are equally spaced, but the length of each bar denotes the duration of the corresponding RR interval marked in (a). In (c) and (d) we depict (a) and (b), respectively, but under time reversal. Reprinted with permission from Ref. [57]. Copyright (2007), American Institute of Physics.

Let us consider an example from the case of electrocardiogram (ECG) analysis discussed in detail in Chapter 9. Figure 2.2(a) provides a schematic diagram of a four-heartbeat excerpt of an ECG in the conventional time domain. The durations Q_m, Q_{m+1} and Q_{m+2} of the three RR (beat to beat) intervals are marked in green, red and blue, respectively. In Fig. 2.2(b), we show how the RR interval time series of Fig. 2.2(a) is read in natural time: the vertical bars are *equally* spaced and the length of each bar denotes the duration of the corresponding RR interval marked in Fig. 2.2(a). We now turn to the effect of the time reversal: Fig. 2.2(c) depicts how the four heartbeat excerpt of Fig. 2.2(a) becomes upon reversing the conventional time (thus the sequential order of colors–durations in Fig. 2.2(a) has been reversed) and Fig. 2.2(b) turns to Fig. 2.2(d). Time reversal may reveal important elements of the dynamics of the system as will become clear, for example, in identifying the occurrence time of an impending cardiac arrest; see § 9.4.1.

2.2.1 Interconnection of the average value of natural time with the effect of a small linear trend on a "uniform" distribution

The way through which natural time captures the influence of the effect of a small linear trend on a "uniform" distribution is studied on the basis [60, 58] of the parametric family of probability density functions (cf. Eq. (2.5)):

$$p(\chi;\varepsilon) = 1 + \varepsilon(\chi - 1/2), \tag{2.9}$$

where the parameter ε quantifies the linear trend. Such a family of pdfs shares the interesting property

$$\hat{T}p(\chi;\varepsilon) = p(\chi;-\varepsilon), \tag{2.10}$$

i.c, the action of the time reversal is obtained by simply changing the sign of ε. A *linear* measure of ε in natural time is [58] the average of the natural time itself since:

$$\langle\chi\rangle = \int_0^1 \chi p(\chi;\varepsilon)\,\mathrm{d}\chi = \frac{1}{2} + \frac{\varepsilon}{12}. \tag{2.11}$$

In the following subsection, we shall show that if we consider the fluctuations of this simple measure upon time reversal, we can obtain information on the long-range dependence of Q_k.

2.2.2 Quantification of the long-range dependence from the fluctuations of the average value of natural time under time reversal

As discussed in § 1.4.1, in order to study the long-range dependence in a time series, e.g., Q_k, we have to define a scale-dependent measure (for example, $F(s)$ of Eq. (1.12) constitutes such a measure in DFA; see § 1.4.2).

We shall show that such a scale-dependent measure is the one that quantifies how the average value of natural time fluctuates upon time reversal when considering a window of length l (= number of) consecutive events sliding through the time series Q_k.

In a window of length l starting from Q_{m_0} (thus ending at Q_{m_0+l-1}), the values of natural time are $\chi_k = k/l$ for $k = 1,2,\ldots,l$ and correspond to the point probabilities $p_k = Q_{m_0+k-1}/\sum_{i=1}^{l} Q_{m_0+i-1}$. Since under time reversal we have $\hat{T}p_k = p_{l-k+1}$, the fluctuations of the average value of natural time under time reversal could be quantified by

$$\Delta\chi_l^2 \equiv \mathscr{E}[(\langle\chi\rangle - \langle\hat{T}\chi\rangle)^2] = \mathscr{E}\left\{\left[\sum_{k=1}^{l}\frac{k}{l}(p_k - p_{l-k+1})\right]^2\right\}, \tag{2.12}$$

where the symbol $\mathscr{E}[\ldots]$ denotes the expectation value obtained when a window of length l is sliding through the time series Q_k. The evaluation of $\mathscr{E}[\ldots]$ can be carried out either by full computation or by Monte Carlo; the full computation refers to the case when the window is sliding consecutively event by event, i.e., m_0 takes all the $N-l+1$ ($m_0 = 1,2,\ldots N-l+1$) possible values, whereas in Monte Carlo evaluation m_0 is selected randomly. The argument of $\mathscr{E}[\ldots]$ is computed by substituting $p_k = Q_{m_0+k-1}/\sum_{i=1}^{l} Q_{m_0+i-1}$ and $p_{l-k+1} = Q_{m_0+l-k}/\sum_{i=1}^{l} Q_{m_0+i-1}$. The sum of the resulting values over the number of the selected segments (different m_0) is assigned to $\mathscr{E}[\ldots]$.

By expanding the square in the last part of Eq. (2.12), we obtain

$$\Delta\chi_l^2 = \sum_{k=1}^{l}\left(\frac{k}{l}\right)^2 \mathscr{E}[(p_k - p_{l-k+1})^2] + \sum_{k\neq m}\frac{km}{l^2}\mathscr{E}[(p_k - p_{l-k+1})(p_m - p_{l-m+1})]. \tag{2.13}$$

Equation (2.3) constitutes the basic relation that interrelates p_k, i.e., $\sum_{k=1}^{l} p_k = 1$ or equivalently $p_k = 1 - \sum_{m\neq k} p_m$. By subtracting from the last expression its value for $k = l-k+1$, we obtain $p_k - p_{l-k+1} = -\sum_{m\neq k}(p_m - p_{l-m+1})$, and hence

$$(p_k - p_{l-k+1})^2 = -\sum_{m\neq k}(p_k - p_{l-k+1})(p_m - p_{l-m+1}). \tag{2.14}$$

By substituting Eq. (2.14) into Eq. (2.13), we obtain

$$\begin{aligned}\Delta\chi_l^2 = &-\sum_{k=1}^{l}\left(\frac{k}{l}\right)^2 \sum_{m\neq k}\mathscr{E}[(p_k - p_{l-k+1})(p_m - p_{l-m+1})] \\ &+ \sum_{k\neq m}\frac{km}{l^2}\mathscr{E}[(p_k - p_{l-k+1})(p_m - p_{l-m+1})]\end{aligned} \tag{2.15}$$

which simplifies to

$$\Delta\chi_l^2 = -\sum_{k,m}\frac{(k-m)^2}{l^2}\mathscr{E}[(p_k - p_{l-k+1})(p_m - p_{l-m+1})]. \tag{2.16}$$

The negative sign appears because $(p_k - p_{l-k+1})$ and $(p_m - p_{l-m+1})$ are in general anti-correlated in view of Eq. (2.14). We notice that the quantity $-\mathscr{E}[(p_k - p_{l-k+1})(p_m - p_{l-m+1})]$ in Eq. (2.16) is similar to the covariance $\mathrm{Cov}(p_k, p_m) \equiv \mathscr{E}\{[p_k - \mathscr{E}(p_k)][p_m - \mathscr{E}(p_m)]\}$, thus capturing the correlations between p_k and p_m as they appear within the window length l under time reversal. Hence, $\Delta\chi_l^2$ due to Eq. (2.16) may reveal non-trivial correlations between the elements of the time series Q_k.

Let us now assume that Q_k are long-range correlated, thus it may be justified to use the ansatz (see § 1.5.1.1):

$$-\mathscr{E}[(p_k - p_{l-k+1})(p_m - p_{l-m+1})] \propto \frac{(k-m)^{2\chi_H}}{l^2}, \tag{2.17}$$

where χ_H is a scaling exponent and we divided by l^2 because the probability p_k is expected to scale with $1/l$ in view of $\sum_{k=1}^{l} p_k = 1$. Substituting Eq. (2.17) into Eq. (2.16), we have

$$\Delta\chi_l^2 \propto l^{4+2\chi_H}/l^4 \tag{2.18}$$

so that

$$\Delta\chi_l \left(\equiv \sqrt{\Delta\chi_l^2}\right) \propto l^{\chi_H}. \tag{2.19}$$

Equation (2.19) reveals that the scaling exponent χ_H can be determined from the slope of the $\log\Delta\chi_l$ versus $\log l$ plot.

2.2.2.1 An example from fractional Brownian motion and fractional Gaussian noise time series

In order to examine the validity of the above result of Eq. (2.19) when the quantities Q_k come from fractional Brownian motion (fBm) or fractional Gaussian noise (fGn) (see § 1.5.1.1), we employ the following procedure. First, we generate fBm (or fGn) time series X_k (consisting of 2×10^4 points) for a given value of the self-similarity index H using the Mandelbrot–Weierstrass function [25, 44, 13] of Eq. (3.37) described in detailed later in § 3.4.3; see also Ref. [60]. Second, since Q_k should be positive, we normalize the resulting X_k time series to zero mean and unit standard deviation and then add to the normalized time series N_k a constant factor c to ensure the positivity of $Q_k = N_k + c$ (for the purpose of the present study we used $c = 10$). The resulting Q_k time series is then used for the calculation of the fluctuations of $\Delta\chi_l$ versus the scale l which are shown in Figs. 2.3(a) and 2.3(d) for fGn and fBm, respectively. The upper three panels of Fig. 2.3 correspond to fGn and the lower three to fBm. We observe that:

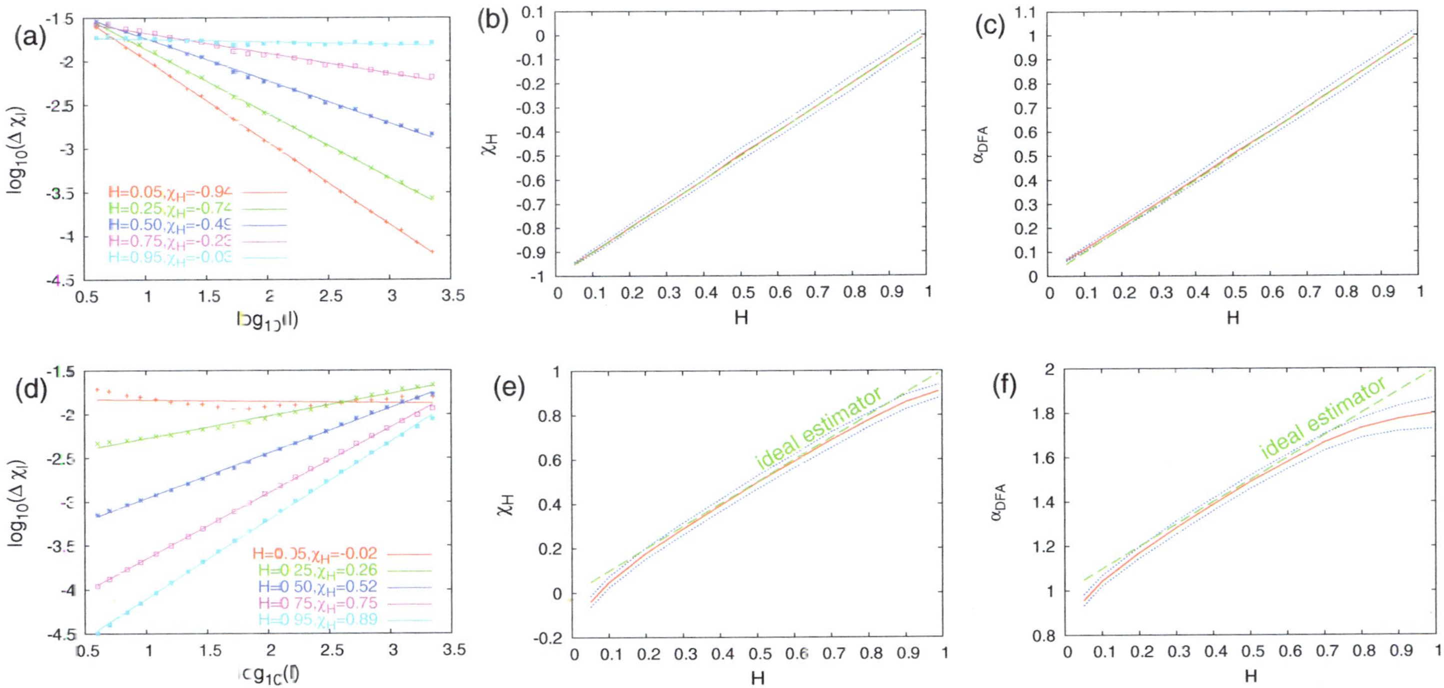

Fig. 2.3 Examples of log-log plot of the fluctuations $\Delta\chi_l$ of the natural time under time reversal versus the scale l for fGn (a) and fBm (d). (b) and (e) depict the values of the scaling exponent χ_H versus the self-similarity index H for fGn and fBm, respectively. For the sake of comparison, (c) and (f) are similar to (b) and (e), respectively, but for the DFA exponent α_{DFA}. The (blue) dotted curves show the $\pm\sigma$ deviation from the average value (obtained after 10^2 runs) depicted by the (red) solid curves. The (green) dashed straight lines correspond to the ideal behavior of each exponent and have been drawn as a guide to the eye. Reprinted with permission from Ref. [58]. Copyright (2008), American Institute of Physics.

For fGn we have the interconnection (see Fig. 2.3(b)) $\chi_H \approx H - 1$ corresponding to *descending* curves(see Fig. 2.3(a)).

For fBm the interconnection turns (see Fig. 2.3(e)) to $\chi_H \approx H$ corresponding to *ascending* curves (see Fig. 2.3(d)).

In order to judge the merits or demerits of the procedure proposed here for the determination of the scaling exponent, we compare Figs. 2.3(b) and 2.3(e) with Figs. 2.3(c) and 2.3(f), respectively, that have been obtained by DFA (§ 1.4.2). This comparison reveals that the results are more or less comparable for fGn, while for fBm the exponent χ_H deviates less from the behavior of an ideal estimator of the true scaling exponent (drawn in dashed green) compared to the exponent α_{DFA} obtained from the DFA method, especially for the largest H values.

2.3 Characteristic function. Mathematical background

Here, we recapitulate some useful properties related to the notion of the characteristic function in Probability Theory. These are given here without proofs, which can be found in Ref. [12]. For further studies see Ref. [7].

2.3.1 Definition of the characteristic function

Definition 2.1. Let X be a random variable with probability distribution F. The characteristic function of F (or of X) is the function φ defined for real ζ by

$$\varphi(\zeta) = \int_{-\infty}^{+\infty} e^{i\zeta X} F\{dX\} = u(\zeta) + iv(\zeta), \tag{2.20}$$

where $u(\zeta) = \Re[\varphi(\zeta)]$ and $v(\zeta) = \Im[\varphi(\zeta)]$.

For distributions F with a probability distribution function f

$$\varphi(\zeta) = \int_{-\infty}^{+\infty} e^{i\zeta X} f(X)\, dX. \tag{2.21}$$

According to Ref. [12], we make the following terminological note. In the accepted terminology of Fourier analysis φ is the *Fourier–Stieltjes transform of* F. Such transforms are defined for all bounded measures and the term "characteristic function" emphasizes that the measure has unit mass. (No other measures have characteristic functions.) On the other hand, integrals of the form (2.21) occur in many connections and one can say that Eq. (2.21) defines the *ordinary Fourier transform of* f. The characteristic function of F is the ordinary Fourier transform of the pdf f (when the latter exists), but the term Fourier transform applies also to other functions.

We now note that the function $\Phi(\omega)$, defined as

$$\Phi(\omega) = \frac{\sum_{k=1}^{N} Q_k \exp\left(i\omega\frac{k}{N}\right)}{\sum_{n=1}^{N} Q_n} = \sum_{k=1}^{N} p_k \exp\left(i\omega\frac{k}{N}\right), \tag{2.22}$$

is a characteristic function of p_k for *all* $\omega \in \mathscr{R}$.

2.3.2 Properties of the characteristic function

Definition 2.2. The moments m_n and the absolute moments M_n of X are given by

$$m_n = \int_{-\infty}^{+\infty} X^n F\{\mathrm{d}X\}, \tag{2.23}$$

and

$$M_n = \int_{-\infty}^{+\infty} |X|^n F\{\mathrm{d}X\}. \tag{2.24}$$

The following important theorem holds [12]:

Theorem 2.1. *If $M_n < \infty$, the n-th derivative of φ exists and is a continuous function given by*

$$\varphi^{(n)}(\zeta) = i^n \int_{-\infty}^{+\infty} \mathrm{e}^{i\zeta X} X^n F\{\mathrm{d}X\} \tag{2.25}$$

leading to

$$\varphi'(0) = im_1, \tag{2.26}$$

$$\varphi''(0) = -m_2, \tag{2.27}$$

$$\varphi'''(0) = -im_3, \text{ etc.} \tag{2.28}$$

It is important to note that the *converse* in Eq. (2.27) is also true: If $\varphi''(0)$ exists, then $m_2 < \infty$. For example, the function $\varphi_\alpha(\zeta) = \exp(-|\zeta|^\alpha)$ is *not* acceptable as a characteristic function *when* $\alpha > 2$, because the second moment of a distribution should be non-vanishing (note that this fact is important for understanding the applications of Lévy α-stable distributions in physics, e.g., see Refs. [27, 46, 47])

Thus, the moments m_n of the distribution are calculated from the behavior of the characteristic function as $\zeta \to 0$.

There exists [12] another important theorem which describes the behavior of the characteristic function for large values of ζ: if F has a pdf f, then $\varphi(\zeta) \to 0$ as $\zeta \to \pm\infty$. If f has integrable derivatives $f', f'', \ldots, f^{(n)}$, then $|\varphi(\zeta)| = o(|\zeta|^{-n})$ as $|\zeta| \to \infty$.

2.4 The normalized power spectrum $\Pi(\omega)$ or $\Pi(\phi)$ and the variance κ_1 of natural time

For the purpose of natural time analysis, the following *continuous* function $\Phi(\omega)$, recall Eq. (2.22), was introduced [51, 52]:

$$\Phi(\omega) = \frac{\sum_{k=1}^{N} Q_k \exp\left(i\omega\frac{k}{N}\right)}{\sum_{n=1}^{N} Q_n} = \sum_{k=1}^{N} p_k \exp\left(i\omega\frac{k}{N}\right) = \sum_{k=1}^{N} p_k \, \mathrm{e}^{i\omega\chi_k} \tag{2.29}$$

where

$$\omega = 2\pi\phi, \tag{2.30}$$

ϕ standing for the frequency in natural time, termed *natural* frequency.

We then compute the normalized power spectrum $\Pi(\omega)$ as

$$\Pi(\omega) = |\Phi(\omega)|^2 = \left|\sum_{k=1}^{N} p_k \, \mathrm{e}^{i\omega\frac{k}{N}}\right|^2 \tag{2.31}$$

which does not change of course under time reversal. The function $\Phi(\omega)$ should not be confused with the *discrete* Fourier transform because ω is here a *continuous* variable.

Using Eq. (2.4), we have

$$\int_0^1 \mathrm{e}^{i\omega\chi} p(\chi)\,\mathrm{d}\chi = \sum_{k=1}^{N} \left[\int_0^1 p_k \delta(\chi-\chi_k)\,\mathrm{e}^{i\omega\chi}\,\mathrm{d}\chi\right] = \sum_{k-1}^{N} p_k \, \mathrm{e}^{i\omega\chi_k}, \tag{2.32}$$

thus $\Phi(\omega)$ can be written as

$$\Phi(\omega) = \int_0^1 \mathrm{e}^{i\omega\chi} p(\chi)\,\mathrm{d}\chi = \sum_{k=1}^{N} p_k \, \mathrm{e}^{i\omega\chi_k}. \tag{2.33}$$

If we regard $p(\chi)$ in Eq. (2.33) as the probability density function of χ, in analogy with probability theory, its Fourier transform $\Phi(\omega)$ may be regarded as the characteristic function of χ, representing the expectation value of $\mathrm{e}^{i\omega\chi}$ (see Eq. (2.21) in § 2.3.1).

By differentiations at the origin, i.e., as $\omega \to 0$, $\Phi(\omega)$ gives (see Theorem 2.1) the statistical properties of $p(\chi)$, such as the mean, variance etc. In view of Eq. (2.31), we now focus on the small ω values of $\Pi(\omega)$ by considering [51] its Taylor expansion, around $\omega = 0$,

$$\Pi(\omega) = 1 - \kappa_1\omega^2 + \kappa_2\omega^4 + \kappa_3\omega^6 + \kappa_4\omega^8 + \ldots \tag{2.34}$$

where

$$\kappa_1 = -\frac{1}{2}\frac{\mathrm{d}^2\Pi(\omega)}{\mathrm{d}\omega^2}\bigg|_{\omega=0}. \tag{2.35}$$

We now consider

$$\frac{\mathrm{d}^2\Pi(\omega)}{\mathrm{d}\omega^2} = \Phi^*(\omega)\frac{\mathrm{d}^2\Phi(\omega)}{\mathrm{d}\omega^2} + \Phi(\omega)\frac{\mathrm{d}^2\Phi^*(\omega)}{\mathrm{d}\omega^2} + 2\frac{\mathrm{d}\Phi(\omega)}{\mathrm{d}\omega}\frac{\mathrm{d}\Phi^*(\omega)}{\mathrm{d}\omega} \tag{2.36}$$

and taking into account Eq. (2.29) along with the fact that $\Phi(0) = 1$, we find:

$$\begin{aligned}\kappa_1 &= -\frac{1}{2}\left[-\sum_k p_k\chi_k^2 - \sum_k p_k\chi_k^2 + 2\left(\sum_k p_k\chi_k\right)^2\right] \\ &= \langle\chi^2\rangle - \langle\chi\rangle^2,\end{aligned} \tag{2.37}$$

where

$$\langle\chi^n\rangle = \sum_{k=1}^{N} p_k\chi_k^n \tag{2.38}$$

denote the moments of the natural time χ 'weighted' by p_k.

Thus, the quantity κ_1 corresponds to the variance of natural time:

$$\kappa_1 = \langle\chi^2\rangle - \langle\chi\rangle^2 = \sum_{k=1}^{N} p_k\left(\frac{k}{N}\right)^2 - \left(\sum_{k=1}^{N}\frac{k}{N}p_k\right)^2. \tag{2.39}$$

Since the normalized power spectrum $\Pi(\omega)$ does not change under time reversal, the same holds for κ_1.

The remaining terms of Eq. (2.34) can be shown [51] to be equal to

$$\kappa_2 = \frac{\langle\chi^2\rangle^2}{4} + \frac{\langle\chi^4\rangle}{12} - \frac{\langle\chi\rangle\langle\chi^3\rangle}{3} \tag{2.40}$$

$$\kappa_3 = \frac{\langle\chi^3\rangle^2}{36} + \frac{\langle\chi\rangle\langle\chi^5\rangle}{60} - \frac{\langle\chi^6\rangle}{360} - \frac{\langle\chi^2\rangle\langle\chi^4\rangle}{24} \tag{2.41}$$

$$\kappa_4 = \frac{\langle\chi^8\rangle}{20160} + \frac{\langle\chi^2\rangle\langle\chi^6\rangle}{720} + \frac{\langle\chi^4\rangle^2}{576} - \frac{\langle\chi^3\rangle\langle\chi^5\rangle}{360} - \frac{\langle\chi\rangle\langle\chi^7\rangle}{2520} \tag{2.42}$$

When considering the symmetric expansion of $p(\chi)$ in the region $[-1,1]$ which is obtained by selecting $p(0) \equiv \lim_{\chi\to 0} p(\chi)$ and $p(-\chi) \equiv p(\chi)$, we obtain that $p(\chi)$ can be expanded [51, 55] in a cosine Fourier series for $\chi \subset (0,1]$:

$$p(\chi) = 1 + \sum_{n=1}^{\infty} \mathrm{p}_n \cos(n\pi\chi) \tag{2.43}$$

where

$$\mathrm{p}_n = 2\int_0^1 p(\chi)\cos(n\pi\chi)d\chi\,, \tag{2.44}$$

are the cosine Fourier series expansion coefficients. Equation (2.43) could give insight into what one should expect for the normalized power spectra $\Pi(\omega)$.

We recall that the lowest frequency included in this expansion, in addition to $\phi = 0$, is $\phi = 0.5$ corresponding to $\omega = \pi$.

Furthermore, $\Pi(\omega)$ for $\omega \ll \pi$, or $\phi \ll 0.5$, by virtue of the Taylor expansion (2.34) and Eqs. (2.39), (2.40), (2.41) and (2.42) resembles the properties of the characteristic function $\Phi(\omega)$ for $p(\chi)$ since its Taylor expansion coefficients are explicitly related to the moments of natural time χ. Of course, these moments do not appear in such a simple way as they appear in Theorem 2.1.

The detailed study of the quantity κ_1 shows that it exhibits (see Section 3.3) positivity, concavity, experimental stability and reveals that it has interesting physical properties; see Chapters 4 to 8.

By combining Eqs. (2.33), (2.35), (2.43) and (2.44), the following interrelation between κ_1 and the Fourier coefficients of $p(\chi)$ can be found [51]

$$\kappa_1 = \langle\chi^2\rangle - \langle\chi\rangle^2 = \frac{1}{12} + \frac{1}{2\pi^2}\sum_{n=1}^{\infty}\frac{\mathrm{p}_{2n}}{n^2} - \left[\frac{1}{2\pi^2}\sum_{k=0}^{\infty}\frac{\mathrm{p}_{2n+1}}{(n+1/2)^2}\right]^2 . \tag{2.45}$$

We now calculate the limit for the variance κ_1 in the case of a *"uniform" distribution*, see § 2.1.3, for which $p(\chi) = 1$ and $\mathrm{p}_n = 0$. Thus, Eq. (2.45) leads to $\kappa_1 = 1/12$. This will be hereafter labeled κ_u, i.e.,

$$\kappa_u = \frac{1}{12} = 0.0833\ldots \tag{2.46}$$

The κ_1 value has been calculated in a variety of cases discussed in the present monograph. In particular, for SES activities it is theoretically obtained in § 2.4.2 and given in Table 4.6 for several experimental cases. The latter table also includes the κ_1 value for various "artificial" noises, and Table 4.4 the ionic current fluctuations in membrane channels. The κ_1 value for the case when the increments of the time series of Q_k are p.i.i.d. random variables of finite variance is calculated in § 2.5.3 and for power law distributed (uncorrelated) energy bursts in § 2.5.4. For the case of fBm time series the κ_1 value will be discussed later in § 3.4.3 and for short-range correlated time series in § 3.4.5. As for dichotomous Markovian time series the κ_1 value will be treated in Chapter 4; see Fig. 4.22. Moreover, the κ_1 value for long-term seismicity will be discussed in Chapter 6, while for the seismicity that evolves after the initiation of SES activities and before the mainshock occurrence will be treated in Chapter 7 for several cases. Finally, for various dynamical models discussed in Chapter 8, the results for the κ_1 value when the critical point is approached are compiled in Table 8.1. Note also that the κ_1 values for a case when Q_k are *quasi-periodic* are depicted in Fig. 8.4.

The largest κ_1 value obtained either from experimental data or from theoretical models is 0.25. A theoretical explanation of this fact is given in § 3.3.2.1.

2.4.1 The normalized power spectrum for the "uniform" distribution

Using Eqs. (2.31) and (2.33), we obtain

$$\Pi(\omega) = \left| \int_0^1 e^{i\omega\chi} p(\chi)\, d\chi \right|^2 = \int_0^1 \int_0^1 p(\chi)p(\psi)\, e^{i\omega(\chi-\psi)}\, d\chi\, d\psi \tag{2.47}$$

After the transformation of variables: $X = (\chi + \psi)/2$ and $\delta = (\chi - \psi)$, the double integral in Eq. (2.47) becomes

$$\Pi(\omega) = 2\int_0^1 \cos(\omega\delta) \int_{\frac{\delta}{2}}^{1-\frac{\delta}{2}} p\left(X - \frac{\delta}{2}\right) p\left(X + \frac{\delta}{2}\right) dX\, d\delta \tag{2.48}$$

Equation (2.48) can be also written as

$$\Pi(\omega) = 2\int_0^1 \cos(\omega\delta) G(\delta)\, d\delta \tag{2.49}$$

with

$$G(\delta) = \int_{\frac{\delta}{2}}^{1-\frac{\delta}{2}} p\left(X - \frac{\delta}{2}\right) p\left(X + \frac{\delta}{2}\right) dX \tag{2.50}$$

We can now estimate the normalized power spectrum $\Pi_u(\omega)$ for the *"uniform"* distribution. As already mentioned this is the case when Q_k are p.i.i.d. random variables. Thus, the pdf $p(\chi)$ becomes $p(\chi) = 1$ for all $\chi \in (0,1]$; Eq. (2.50) simply results in $G(\delta) = \int_{\frac{\delta}{2}}^{1-\frac{\delta}{2}} dX = 1 - \delta$ leading, see Eq. (2.49), to the normalized power spectrum

$$\Pi_u(\omega) = 2\int_0^1 (1-\delta)\cos(\omega\delta)\, d\delta = \frac{\sin^2(\omega/2)}{(\omega/2)^2} \tag{2.51}$$

When expanding $\Pi_u(\omega)$ of Eq. (2.51) around $\omega \to 0$, we obtain

$$\Pi_u(\omega) \approx \left[1 - \frac{1}{3!}\left(\frac{\omega}{2}\right)^2\right]^2 \approx 1 - \frac{2}{3!}\left(\frac{\omega}{2}\right)^2 = 1 - \frac{1}{12}\omega^2 \tag{2.52}$$

When considering the expansion of Eq. (2.34), we observe that Eq. (2.52) results to $\kappa_1 = \kappa_u = 1/12$ in accordance with Eq. (2.46).

2.4.2 The normalized power spectrum of seismic electric signals

Here, we focus on the normalized power spectrum of SES activities which are emitted when criticality is approached [51, 52]. Thus, we rely on the physics behind their generation discussed in Section 1.6. We first consider the following two laboratory measurements. (i) Indentation experiments even in simple ionic crystals showed that transient electric signals are emitted, without the action of any external electric field, due to (formation and motion of) point and linear defects, e.g., see Ref. [62]. (ii) Independent measurements [37] revealed that, as the glass transition is approached, a polarization time series is emitted which probably arises from the reorientation process of electric dipoles; this process includes a large number of atoms (*cooperativity*). The feature of this polarization time series is strikingly similar [48] to the measured SES activities. This similarity is reminiscent of the pressure stimulated currents model [49] discussed in § 1.6.2, which suggests that upon a gradual variation of the pressure (stress) P on a solid, when approaching the *critical* pressure (stress) P_{cr}, transient electric signals are emitted arising from the (re)orientation of electric dipoles (formed due to defects). This emission occurs when the following condition is obeyed (which is just Eq. (1.48) of § 1.6.2):

$$\left.\frac{\mathrm{d}P}{\mathrm{d}t}\right|_T \frac{v^{m,b}}{kT} = -\frac{1}{\tau(P_{cr})}, \tag{2.53}$$

where $\left.\frac{\mathrm{d}P}{\mathrm{d}t}\right|_T$ is the pressure rate and $\tau(P_{cr})$ is the relaxation time of the dipoles at the critical pressure. It has been argued, see p. 404 of Ref. [49], that the values of the migration volume $v^{m,b}$ associated with SES generation should exceed the mean atomic volume by orders of magnitude, and this entails that the relevant (re)orientation process should involve the motion of a large number of "atoms" . Thus, the laboratory measurements fortify the suggestion [48] that the emission of the SES activities could be discussed in the frame of the theory of *dynamic phase transitions* (critical phenomena). The very stochastic nature of the relaxation process has been repeatedly discussed in the literature (see p. 350 of Ref. [19] and references therein; other suggestions have been reviewed in Ref. [31], while illuminating aspects have been forwarded in Ref. [66]). A stochastic analysis was based on the concept of clusters, the structural rearrangement of which develops in time [19]. According to this analysis the exponential relaxation of the polarization is arrested at a random time variable η_i and the instantaneous orientation reached at this instant is "frozen" at a value $\exp(-\beta_i\eta_i)$ where $\beta_i = b =$ constant (see fig. 11.19 of Ref. [19]). Assuming that η_i itself follows an exponential distribution, with a time constant $\tau_0 \ll \tau(P_{cr})$, an almost constant current would be expected for as long as this unit "lives" (i.e., for a duration η_i).

As a result of *cooperativity*, the duration Q_k of a SES activity pulse is envisaged as the sum of n_k such identical units, thus $Q_k = \sum_{i=1}^{n_k} \eta_i$. Under this assumption, the duration Q_k of the k-th pulse in a SES activity follows the gamma distribution with a mean lifetime $n_k\tau_0$ and variance $n_k\tau_0^2$ (e.g., see lemma 8.1.6.5. of Ref. [30]), i.e., the average duration is given by:

$$\mathscr{E}(Q_k) = n_k\tau_0 \tag{2.54}$$

and its variance by:

$$\mathscr{E}(Q_k^2) - n_k^2\tau_0^2 = n_k\tau_0^2. \tag{2.55}$$

As already mentioned (§ 1.6.2), the SES activity is emitted when the focal area enters into the critical regime. The approach of a system to a critical point can be characterized by a feature that events begin to be temporally correlated, which is equivalent to a persistent avalanching. The condition for the persistent avalanching can be expressed as

$$\mathscr{E}(Q_{k+1}) = Q_k \tag{2.56}$$

which means that the average Q_{k+1} value of the $k+1$-th event is maintained at the level already reached by the previous one. This is reminiscent of the aspect that the reorientation of a spin in the random-field Ising Hamiltonian, will cause on average just one more spin to flip at the critical point [23]. Since Q_{k+1} is assumed to be distributed according to the gamma distribution, we also have:

$$\begin{aligned}\mathscr{E}\left\{[Q_{k+1} - \mathscr{E}(Q_{k+1})]^2\right\} &= \mathscr{E}(Q_{k+1})\tau_0 \Rightarrow \\ \mathscr{E}(Q_{k+1}^2) &= Q_k\tau_0 + Q_k^2\end{aligned} \tag{2.57}$$

We now turn to the evaluation of the normalized power spectrum $\Pi(\omega)$, see Eqs. (2.49) and Eq. (2.50), for the SES activities. We will first attempt to evaluate the average value $\tilde{G}(\delta)$

$$\tilde{G}(\delta) = \int_{\frac{\delta}{2}}^{1-\frac{\delta}{2}} \mathscr{E}\left[Q_{X-\frac{\delta}{2}} Q_{X+\frac{\delta}{2}}\right] \mathrm{d}X \tag{2.58}$$

as it results from all SES activities comprising N pulses. Note that $\tilde{G}(\delta)$ is similar to $G(\delta)$ of Eq. (2.50) apart from the fact that it does *not* involve the normalized pdfs $p(X - \frac{\delta}{2})p(X + \frac{\delta}{2})$. When for example $X - \frac{\delta}{2} = k/N$ and $X + \frac{\delta}{2} = l/N$, we have

$$\mathscr{E}[Q_{X-\frac{\delta}{2}} Q_{X+\frac{\delta}{2}}] = \mathscr{E}[Q_k Q_l] = \underbrace{\int \cdots \int}_{N} Q_k Q_l \, \mathrm{d}\mathscr{P}_1 \, \mathrm{d}\mathscr{P}_2 \ldots \, \mathrm{d}\mathscr{P}_k \ldots \, \mathrm{d}\mathscr{P}_l \ldots \, \mathrm{d}\mathscr{P}_N \tag{2.59}$$

where $\mathscr{P}_1, \mathscr{P}_2 \ldots \mathscr{P}_N$ are the pdfs for the durations $Q_1, Q_2, \ldots Q_N$, respectively. Using the normalization condition of the pdfs, we can eliminate the integrals over $\mathscr{P}_{l+1}$ to $\mathscr{P}_N$

$$\mathscr{E}[Q_k Q_l] = \underbrace{\int \dots \int}_{l} Q_k Q_l \, \mathrm{d}\mathscr{P}_1 \, \mathrm{d}\mathscr{P}_2 \dots \mathrm{d}\mathscr{P}_k \dots \mathrm{d}\mathscr{P}_l \tag{2.60}$$

and using Eq. (2.56) we can integrate over $\mathscr{P}_l$ down to $\mathscr{P}_k$

$$\mathscr{E}[Q_k Q_l] = \underbrace{\int \dots \int}_{k} Q_k^2 \, \mathrm{d}\mathscr{P}_1 \, \mathrm{d}\mathscr{P}_2 \dots \mathrm{d}\mathscr{P}_k \tag{2.61}$$

Performing now the integration over $\mathscr{P}_k$ by using the recursive relation of Eq. (2.57) for $k = k-1$, we obtain

$$\mathscr{E}[Q_k Q_l] = \underbrace{\int \dots \int}_{k-1} (Q_{k-1}\tau_0 + Q_{k-1}^2) \, \mathrm{d}\mathscr{P}_1 \, \mathrm{d}\mathscr{P}_2 \dots \mathrm{d}\mathscr{P}_{k-1} \tag{2.62}$$

whereas a second application of the recursive relations of Eqs. (2.56) and (2.57) into Eq. (2.62) results in

$$\mathscr{E}[Q_k Q_l] = \underbrace{\int \dots \int}_{k-2} (2Q_{k-2}\tau_0 + Q_{k-2}^2) \, \mathrm{d}\mathscr{P}_1 \, \mathrm{d}\mathscr{P}_2 \dots \mathrm{d}\mathscr{P}_{k-2} , \tag{2.63}$$

a third one to

$$\mathscr{E}[Q_k Q_l] = \underbrace{\int \dots \int}_{k\ 3} (3Q_{k-3}\tau_0 + Q_{k-3}^2) \, \mathrm{d}\mathscr{P}_1 \, \mathrm{d}\mathscr{P}_2 \dots \mathrm{d}\mathscr{P}_{k-3} , \tag{2.64}$$

and so on. Finally, we obtain

$$\mathscr{E}[Q_{X-\frac{\delta}{2}} Q_{X+\frac{\delta}{2}}] = \mathscr{E}[Q_k Q_l] = \int \left[(k-1) Q_1 \tau_0 + Q_1^2 \right] \, \mathrm{d}\mathscr{P}_1 . \tag{2.65}$$

Restoring $k = (X - \frac{\delta}{2})N$ into Eq. (2.65), we obtain

$$\mathscr{E}[Q_{X-\frac{\delta}{2}} Q_{X+\frac{\delta}{2}}] = \alpha \left(X - \frac{\delta}{2} \right) + \beta , \tag{2.66}$$

where $\alpha = \int N\tau_0 Q_1 \, \mathrm{d}\mathscr{P}_1$ and $\beta = \int \left(Q_1^2 - \tau_0 Q_1 \right) \mathrm{d}\mathscr{P}_1 = (\int Q_1 \, \mathrm{d}\mathscr{P}_1)^2$. Substituting Eq. (2.66) into Eq. (2.58), we obtain

$$\tilde{G}(\delta) = \int_{\frac{\delta}{2}}^{1-\frac{\delta}{2}} \left[\alpha \left(X - \frac{\delta}{2} \right) + \beta \right] \mathrm{d}X = \alpha \frac{(1-\delta)^2}{2} + \beta (1-\delta) . \tag{2.67}$$

Equation (2.67) provides $\tilde{G}(\delta)$ for the SES activities comprising N pulses. We note the existence of two terms in the right-hand side of Eq. (2.67): The last term, which is simply proportional to $(1-\delta)$, originates from the positivity of Q_k and is also present in the case of the "uniform" distribution, see Eq. (2.51). On the other hand, the first term $\frac{(1-\delta)^2}{2}$ comes from the *memory* of the critical process as reflected in Eq. (2.66), which states that the expectation $\mathscr{E}[Q_{X-\frac{\delta}{2}}Q_{X+\frac{\delta}{2}}]$ depends solely on $X-\frac{\delta}{2}$, i.e., the natural time elapsed since the initiation of the process.

In order to determine the normalized power spectrum for SES activities through a formula similar to Eq. (2.49), e.g.,

$$\Pi(\omega) = 2\int_0^1 \cos(\omega\delta)\mathscr{G}(\delta)\,\mathrm{d}\delta \tag{2.68}$$

we need also to average over all possible values of N to obtain an appropriate $\mathscr{G}(\delta)$. The quantity of $G(\delta)$ in Eq. (2.49), as well as $\mathscr{G}(\delta)$ in Eq. (2.68), is dimensionless since it results from the pdf $p(\chi)$ in Eq. (2.49). Equation (2.67) was obtained, however, without normalizing $Q_{X-\frac{\delta}{2}}$ and $Q_{X+\frac{\delta}{2}}$ by the appropriate factor $\left(\int Q_\chi\,\mathrm{d}\chi\right)^2$ because the inclusion of such a factor in the denominator would hinder the integration procedure followed. As a first approximation, we construct a dimensionless quantity from Eq. (2.67), thus for example we divide $\tilde{G}(\delta)$ by α:

$$\frac{\tilde{G}(\delta)}{\alpha} = \frac{(1-\delta)^2}{2} + \frac{\beta}{\alpha}(1-\delta). \tag{2.69}$$

The quantity $\mathscr{G}(\delta)$ is expected to be a weighted sum of the right-hand side of Eq. (2.69) for various values of N, thus it will be of the form

$$\mathscr{G}(\delta) \propto \frac{(1-\delta)^2}{2} + \overline{\left(\frac{\beta}{\alpha}\right)}(1-\delta), \tag{2.70}$$

where $\overline{\left(\frac{\beta}{\alpha}\right)}$ stands for the corresponding average – renormalized – value of the ratio

$$\frac{\beta}{\alpha} = \frac{\left(\int Q_1\,\mathrm{d}\mathscr{P}_1\right)^2}{\int N\tau_0 Q_1\,\mathrm{d}\mathscr{P}_1} = \frac{n_1}{N}. \tag{2.71}$$

Let us now impose (natural time) scale invariance which should hold for *criticality*. This means that the result should be independent of N. Hence, the time scale τ_0, so far arbitrary, should be such that the results obtained from Eq. (2.71) for various N lead to a value (labeled $\overline{\left(\frac{\beta}{\alpha}\right)}$ in Eq. (2.70)) *independent* of N. This is satisfied when $\tau_0 = \text{const.} \times Q_1/N$ as it is evident from Eq. (2.71). Since when a single SES pulse is emitted the only reasonable time scale to assume is that of the duration of the single pulse, we should impose $\tau_0 = Q_1/N$. Thus, we may write

$$\overline{\left(\frac{\beta}{\alpha}\right)} = \frac{\left(\int Q_1\,\mathrm{d}\mathscr{P}_1\right)^2}{\int Q_1^2\mathrm{d}\mathscr{P}_1}. \tag{2.72}$$

Assuming that at the initiation of the SES activity, only one unit is available, i.e, $n_1 = 1$, the duration Q_1 in Eq. (2.72) is exponentially distributed (see Eqs. (2.54) and (2.55)) leading to

$$\overline{\left(\frac{\beta}{\alpha}\right)} = \frac{1}{2}. \tag{2.73}$$

Equation (2.70) then reads

$$\mathscr{G}(\delta) \propto \frac{(1-\delta)^2}{2} + \frac{(1-\delta)}{2}. \tag{2.74}$$

Inserting Eq. (2.74) into Eq. (2.68), we obtain that for the SES activities (*critical dynamics*) the normalized power spectrum equals to [51]

$$\Pi(\omega) = \frac{18}{5\omega^2} - \frac{6\cos\omega}{5\omega^2} - \frac{12\sin\omega}{5\omega^3}. \tag{2.75}$$

Expanding Eq. (2.75) around $\omega = 0$ (see Eqs. (2.34) and (2.35)), we get

$$\Pi(\omega) \approx 1 - \kappa_1 \omega^2, \tag{2.76}$$

where

$$\kappa_1 = 0.070. \tag{2.77}$$

An inspection of Fig. 4.7 shows that for the region of natural frequencies $0 \leq \phi < 0.5$ (recall the shaded remark after Eq. (2.44)) the experimental results for the SES activities agree favorably with Eq. (2.75). In addition, for the SES activities observed to date, see Table 4.6, the validity of Eq. (2.77) has been confirmed.

An alternative derivation that $\kappa_1 \approx 0.070$ for SES activities, can be given on the basis of the Ising model if we also consider its qualitative similarity under certain conditions with the pressure-stimulated currents model (§ 1.6.2) for the SES generation, as will be explained in § 8.4.1.

Note that the relation $\kappa_1 = 0.070$, i.e., Eq. (2.77), emerges for several dynamical models approaching criticality which are compiled in Table 8.1.

2.5 Distinction of the origins of self-similarity

A large variety of natural systems exhibit irregular and complex behavior which at first looks erratic, but in fact possesses scale-invariant structure (e.g., see Refs. [34, 20]). As explained in § 1.5.1, a process $\{X(t)\}_{t \geq 0}$ is called self-similar [24] with index $H > 0$, if it has the property

$$X(\lambda t) \stackrel{d}{=} \lambda^H X(t) \qquad \forall \quad \lambda > 0. \tag{2.78}$$

Equation (2.78) means a "scale invariance" of the finite-dimensional distributions of $X(t)$, which does *not* imply, in stochastic processes, the same for the sample paths (e.g., see Ref. [65]). In this Section, following Ref. [59], we will explain how natural time enables the distinction of the two origins of self-similarity.

2.5.1 The two origins of self-similarity. Background

Examples of self-similar processes are Brownian, fractional Brownian (fBm), Lévy stable and fractional Lévy stable motion (fLsm). Lévy stable distributions (which are followed by many natural processes, e.g., see Refs. [46, 47]) differ greatly from the Gaussian ones because they have heavy tails and their variance is infinite (e.g., see Refs. [65, 38]).

An important point in analyzing data from natural systems that exhibit scale-invariant structure is the following. In several systems this nontrivial structure points to long-range *temporal* correlations; in other words, the self-similarity results from the process's memory *only* (e.g., the case of fBm discussed in § 1.5.1.1). Alternatively, the self-similarity may solely result from the process's increments' infinite variance, e.g., Lévy stable motion. (Note that in distributions that are applicable to a large variety of problems, extreme events have to be *truncated* for physical reasons, e.g., finite size effects – when there is no infinity [6] – and this is why we write hereafter "infinite".) In general, however, the self-similarity may result from both these origins (e.g., fLsm). It is the main aim of this Section to discuss how a distinction of the two origins of self-similarity (i.e., process's memory and process's increments' "infinite" variance) can be in principle achieved by employing natural time analysis.

Before proceeding, the following clarifications are necessary as far as the aforementioned two sources of self-similarity are concerned. Long-range temporal correlations, which are quoted above as a first origin of self-similarity, are an immediate consequence of Eq. (2.78) with $H > \frac{1}{2}$ defining a self-similar process. We stress, however, that long-range correlations do not automatically imply self-similarity of a process. Multifractal processes provide a large class of counter-examples (note that the natural time analysis of multiplicative cascades is discussed in § 6.2.5). The second origin of self-similarity comes from the statistical properties of the increments of the process. We emphasize, however, that the statistics of these increments does not automatically lead to nontrivial self-similarity of the process. Specifically, a process which is invariant under shuffling of the increments has independent increments and is characterized by the self-similarity index $\frac{1}{2}$.

2.5.2 The expectation value of κ_1 when a (natural) time window of length l is sliding through a time series

Here, we focus on the expectation value $\mathscr{E}(\kappa_1)$ of the variance (κ_1) of natural time when sliding a (time) window of length l through a time series of $Q_k > 0, k = 1,2,\dots N$ (while in § 2.2.2 we calculated the fluctuations of the average value of the natural time itself under time reversal). In a window of length l starting at $k = k_0$, the quantities $p_j = Q_{k_0+j-1}/\sum_{m=1}^{l} Q_{k_0+m-1}$, $j = 1,2,\dots,l$ are obtained, which satisfy the necessary conditions

$$p_j > 0, \tag{2.79}$$

$$\sum_{j=1}^{l} p_j = 1 \tag{2.80}$$

to be considered as point probabilities. We then define [51, 55] the moments of the natural time $\chi_j = j/l$ as $\langle \chi^q \rangle = \sum_{j=1}^{l} (j/l)^q p_j$ and hence

$$\kappa_1 = \sum_{j=1}^{l} \left(\frac{j}{l}\right)^2 p_j - \left[\sum_{j=1}^{l} \frac{j}{l} p_j\right]^2. \tag{2.81}$$

Note that κ_1 is a nonlinear functional of $\{p_j\}$. Let us consider the expectation value $\mu_j \equiv \mathscr{E}(p_j)$ of p_j. For the purpose of our calculation the relation between the variance of p_j, $\mathrm{Var}(p_j) \equiv \mathscr{E}[(p_j - \mu_j)^2]$, and the covariance of p_j and p_m, $\mathrm{Cov}(p_j, p_m) \equiv \mathscr{E}[(p_j - \mu_j)(p_m - \mu_m)]$, is important. In view of Eqs. (2.79) and (2.80), the quantities μ_j, $\mathrm{Var}(p_j)$ and $\mathrm{Cov}(p_j, p_m)$ are always finite independent of the presence of heavy tails in Q_k. Using the constraint of Eq. (2.80), leading to $p_j - \mu_j = \sum_{m \neq j} (\mu_m - p_m)$, we obtain

$$\mathrm{Var}(p_j) = -\sum_{m \neq j} \mathrm{Cov}(p_j, p_m). \tag{2.82}$$

We now turn to the evaluation of $\mathscr{E}(\kappa_1)$, and study its difference from the one that corresponds to the average time series $\mathscr{M} = \{\mu_k\}$ which is labeled $\kappa_{1,\mathscr{M}}$,

$$\kappa_{1,\mathscr{M}} = \sum_{j=1}^{l} \left(\frac{j}{l}\right)^2 \mu_j - \left[\sum_{j=1}^{l} \frac{j}{l} \mu_j\right]^2. \tag{2.83}$$

Hence,

$$\mathscr{E}(\kappa_1) - \kappa_{1,\mathscr{M}} = \mathscr{E}\left[\sum_{m=1}^{l} \frac{m^2}{l^2} (p_m - \mu_m) - \left(\sum_{m=1}^{l} \frac{m}{l} p_m\right)^2 + \left(\sum_{m=1}^{l} \frac{m}{l} \mu_m\right)^2\right]. \tag{2.84}$$

In view of the definition of μ_m, the first term in the right-hand side of Eq. (2.84) vanishes, whereas the latter two terms reduce to the variance of $\langle \chi \rangle$:

$$\mathscr{E}(\kappa_1) - \kappa_{1,\mathscr{M}} = -\mathscr{E}\left\{\left[\sum_{m=1}^{l} \frac{m}{l} (p_m - \mu_m)\right]^2\right\}. \tag{2.85}$$

Expanding this variance, we get

$$\kappa_{1,\mathscr{M}} - \mathscr{E}(\kappa_1) = \sum_{m=1}^{l} \frac{m^2}{l^2} \mathrm{Var}(p_m) + 2\sum_{j=1}^{l-1} \sum_{m=j+1}^{l} \frac{jm}{l^2} \mathrm{Cov}(p_j, p_m). \tag{2.86}$$

which, upon using Eq. (2.82), leads to

$$\mathscr{E}(\kappa_1) - \kappa_{1,\mathscr{M}} = \sum_{j=1}^{l-1} \sum_{m=j+1}^{l} \frac{(j-m)^2}{l^2} \mathrm{Cov}(p_j, p_m) = \frac{1}{2} \sum_{j=1}^{l} \sum_{m=1}^{l} \frac{(j-m)^2}{l^2} \mathrm{Cov}(p_j, p_m). \tag{2.87}$$

This relation turns to

$$\mathscr{E}(\kappa_1) = \kappa_{1,\mathscr{M}} + \sum_{\text{all pairs}} \frac{(j-m)^2}{l^2} \mathrm{Cov}(p_j, p_m), \tag{2.88}$$

where $\sum_{\text{all pairs}} \equiv \sum_{j=1}^{l-1} \sum_{m=j+1}^{l}$ (compare Eq. (2.88) with Eq. (2.16) in which a term similar to the covariance $\mathrm{Cov}(p_j, p_m)$ has been discussed).

The case when Q_k do not exhibit temporal correlations: This is the case for example of randomly *shuffled* data. As the window is sliding through the whole time series, Q_k takes of course every position j within the window of length l. Then, Eq. (2.80) leads to

$$\mathscr{E}(p_j) = \frac{1}{l}, \tag{2.89}$$

and $\mathrm{Cov}(p_j, p_m)$ becomes independent of j and m, thus Eq. (2.82) becomes

$$\mathrm{Cov}(p_j, p_m) = -\frac{\mathrm{Var}(p)}{(l-1)}. \tag{2.90}$$

Since $\mathrm{Var}(p_j)$ is also independent of j, $\mathrm{Var}(p_j)$ was merely substituted by $\mathrm{Var}(p)$. Moreover, $\kappa_{1,\mathscr{M}}$ reduces to $\kappa_{1,c}$, where $\kappa_{1,c}$ corresponds to the constant time series $\mathscr{K} = \{x_k\} : x_k = 1/l, k = 1, 2, \ldots l$, which is given by

$$\kappa_{1,c} = \sum_{m=1}^{l} \frac{m^2}{l^3} - \left(\sum_{m=1}^{l} \frac{m}{l^2} \right)^2 = \kappa_u \left(1 - \frac{1}{l^2} \right), \tag{2.91}$$

where $\kappa_u = 1/12 \approx 0.0833$. Turning now to Eq. (2.86) and by adding and subtracting $\frac{\mathrm{Var}(p)}{l-1} \sum \frac{m^2}{l^2}$, we obtain that:

For *shuffled* data

$$\mathscr{E}(\kappa_1) = \kappa_u \left(1 - \frac{1}{l^2} \right) - \kappa_u (l+1)\, \mathrm{Var}(p). \tag{2.92}$$

In view of Eqs. (2.79) and (2.80), $\mathrm{Var}(p) < \mathscr{E}(p^2) \leq \mathscr{E}(p) = 1/l$, and thus the second term in Eq. (2.92) remains finite for $l \to \infty$.

The l-dependence of $\mathrm{Var}(p)$ when Q_k have a finite second moment is obtained from

$$\mathrm{Var}(p_k) = \frac{1}{l^2} \mathscr{E} \left[\left(\frac{l Q_k}{\sum_{n=1}^{l} Q_n} - 1 \right)^2 \right], \tag{2.93}$$

where the quantity $\mathscr{E}[(lQ_k/\sum_{n=1}^{l} Q_n - 1)^2]$ is *asymptotically* l-independent. The latter arises as follows: if $\mathscr{E}(Q_k) = \mu$ and $\mathrm{Var}(Q_k) = \sigma^2 (< \infty)$, as a result of the central limit theorem [12], we have $\mathscr{E}(\sum_{k=1}^{l} Q_k/l) = \mu$ and $\mathrm{Var}(\sum_{k=1}^{l} Q_k/l) = \sigma^2/l$. The latter two equations, for large enough l imply that $\mathscr{E}[(lQ_k/\sum_{n=1}^{l} Q_n - 1)^2] \approx \mathscr{E}[(Q_k/\mu - 1)^2] = \sigma^2/\mu^2$. Thus, Eq. (2.93) becomes (note that $\mathrm{Var}(p_k)$ is independent of k)

$$\mathrm{Var}(p) = \frac{\sigma^2}{l^2\mu^2}. \tag{2.94}$$

For Q_k which do not exhibit time correlations, e.g., randomly shuffled data:

If Q_k do not exhibit heavy tails and have finite variance, $\mathrm{Var}(p)$ scales (Eq. (2.94)) as $1/l^2$, thus $\mathscr{E}(\kappa_1)$, as l increases in Eq. (2.92), converges to κ_u. The same holds for the most probable value $\kappa_{1,p}$ of κ_1.

Otherwise, the expectation value $\mathscr{E}(\kappa_1)$ differs from κ_u -pointing that $\kappa_{1,p}$ also differs from κ_u, i.e., $\kappa_{1,p} \neq \kappa_u$ – thus identifying the presence of heavy tails in the examined time series.

2.5.2.1 Comments on the expectation value of κ_1 for a given window length l

Let us now comment on the expectation value $\mathscr{E}(\kappa_1)$ of κ_1 when a (natural) time window of length l is sliding through a time series of $Q_k > 0$, which as mentioned (see Eq. (2.88)) is given by

$$\mathscr{E}(\kappa_1) = \kappa_{1,\mathscr{M}} + \sum_{\text{all pairs}} \frac{(j-m)^2}{l^2} \mathrm{Cov}(p_j, p_m), \tag{2.95}$$

Let us first discuss the case when Q_k are shuffled randomly. Equation (2.95) then turns to (see Eq. (2.92))

$$\mathscr{E}(\kappa_{1,shuf}) = \kappa_u \left(1 - \frac{1}{l^2}\right) - \kappa_u (l+1)\, \mathrm{Var}(p). \tag{2.96}$$

If Q_k do not exhibit heavy tails and have finite variance, Eq. (2.96) reveals (see the discussion above, § 2.5.2) that $\mathscr{E}(\kappa_{1,shuf})$ rapidly converges to κ_u. For example, this is the case of the SES activities [60] discussed in Chapter 4, e.g., see § 4.7.1. Otherwise, $\mathscr{E}(\kappa_{1,shuf})$ differs from κ_u, thus pointing to $\kappa_{1,p} \neq \kappa_u$. This is the case, for example, of the earthquakes discussed in Chapter 6.

Second, if Q_k do exhibit time correlations, the difference between the $\kappa_{1,p}$ for the original and the shuffled time series most likely originates from the difference of Eqs. (2.95) and (2.96), respectively. The extent to which the latter difference is nonzero accounts for the time correlations irrespective if Q_k exhibit heavy tails. For example, this is clearly the case of aftershocks and the case of earthquake catalogs in general (both of which exhibit heavy tails) discussed in detail in Section 6.3 (e.g., see Figs. 6.14 and 6.13, respectively).

The application of the above results to two important examples are given in the next two subsections.

2.5.3 The case when the increments of the time series of Q_k are positive i.i.d. random variables of finite variance

We first discuss the case when the increments of the time series of Q_k are p.i.i.d. random variables r_n of *finite* variance. In this case $Q_k = \sum_{n=1}^{k} r_n$, and hence Q_k is on average linearly related to k. Thus, it is expected that the continuous distribution $p(\chi)$, that corresponds to p_k see Eq. (2.4), is $p(\chi) = 2\chi$. Using

$$\kappa_1 = \int_0^1 p(\chi)\chi^2 \, \mathrm{d}\chi - \left(\int_0^1 p(\chi)\chi \, \mathrm{d}\chi \right)^2 , \tag{2.97}$$

a direct calculation leads to the value $\kappa_1 = \frac{1}{18} \approx 0.056$ which significantly differs from that $\kappa_u \approx 0.083$ of the "uniform" distribution (see Eq. (2.46)). In view of the fact that the increments have finite variance, the distribution of Q_k for a given N has also finite variance. Hence, as shown in the previous subsection, we expect that when Q_k are shuffled randomly the resulting κ_1 values should scatter around κ_u. A numerical example for exponentially distributed increments is shown in Fig. 2.4.

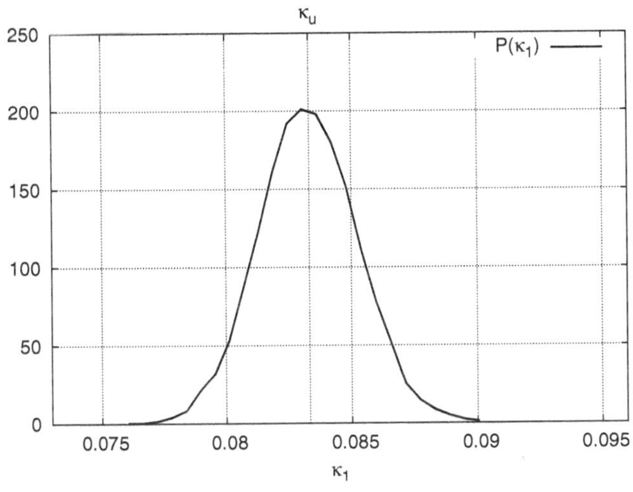

Fig. 2.4 The pdf of κ_1 that has been obtained by shuffling the Q_k *randomly* in the case of exponential increments, i.e., r_n are randomly drawn from an exponential distribution. Here, $N = 500$ and the original time series results in $\kappa_1 = 0.055$. See also Fig. 3.3.

2.5.4 The value of κ_1 when a (natural) time window is sliding through power law distributed energy bursts

We now study a case of self-similarity resulting from the process's increments' "infinite" variance. Here, we restrict ourselves to (slowly driven) systems that emit energy bursts obeying a power law distribution

$$P(E) \sim E^{-\gamma} \tag{2.98}$$

where γ is constant. In a large variety of such systems in diverse fields, an inspection of the experimental data reveals that the γ exponent lies in a narrow range $1.5 \leq \gamma \leq 2.1$ and mostly within even narrower bounds, i.e., $\gamma = 1.5$ to 1.8. To realize the diversity of the phenomena that exhibit the aforementioned property, we compile some indicative examples in Table 2.1, which are the following.

Table 2.1 Compilation of the experimental values of the power law exponent γ determined in different physical processes. Taken from Ref. [59].

Process / type of measurement	γ	References
Dislocation glide in hexagonal ice single crystals (acoustic emission)	1.6	[26]
Intermittent plastic flow in nickel microcrystals	1.6	[9]
Solar flares	1.5–2.1	[5, 32, 18, 29]
Microfractures before the breakup of wood (acoustic emission)	1.5	[14, 1]
Microfractures before the breakup of fiberglass (acoustic emission)	2.0	[14, 1]
Earthquakes	1.5–1.8	See Ref. [36] and references therein
Icequakes	$\approx$1.8	See p.212 of Ref. [64] and references therein

First, crystalline materials subjected to an external stress, display bursts of activity owing to the nucleation and motion of dislocations. These sudden local changes produce acoustic emission waves which reveal that a large number of dislocations move cooperatively in an intermittent fashion (e.g., see Ref. [22] and references therein). As a precise example, we include in Table 2.1 the results of acoustic emission experiments on stressed single crystals of ice under viscoelastic deformation (creep), which show that the probability distribution of energy bursts intensities obey a power law distribution with $\gamma = 1.6$ spanning many decades (see fig. 1 of Ref. [26]). Second, the same exponent is found [9] (i.e., $\gamma = 1.60 \pm 0.02$) in the analysis of intermittent plastic flow observations (i.e., measurements of discrete slip events for loadings above the elastic–plastic transition) on nickel microcrystals (see fig. 2 of Ref. [9]). Third, we consider the case of solar flares that represent impulsive energy releases in the solar corona (e.g. see Ref. [29] and references therein; see also Ref. [4] in which it is concluded that earthquakes and solar flares exhibit the same distributions of sizes, inter-occurrence times, and the same temporal clustering). This energy release is observed in various forms: thermal, soft and hard X-ray emissions, accelerated particles etc. The statistical analysis of these impulsive events show that the energy distribution exhibits, over several orders of magnitude, a power law with exponents

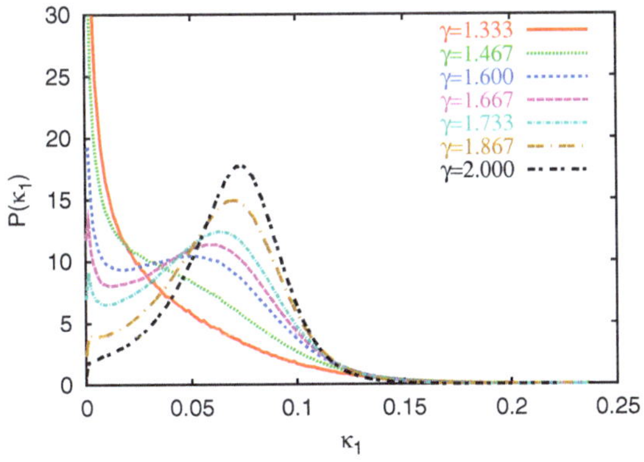

Fig. 2.5 The probability density function $P(\kappa_1)$ versus κ_1 for several values of γ. Taken from Ref. [59].

γ ranging from 1.5 to approximately 2.1 (depending on the experimental procedure and the geometrical assumptions adopted in the analysis). Other examples are: acoustic emission from microfractures before the breakup of heterogeneous materials (wood, fiberglass), ice-quakes and earthquakes.

The following procedure is now applied. We generate a large amount (500,000) of artificial data obeying Eq. (2.98) for a certain γ value with energy $E \geq 1$ and *randomly* shuffle them. This was repeated for various γ values by keeping the total number of events *constant* (which implies that when changing γ, the maximum energy involved in the calculation also changes). These *randomized* ("shuffled" [63, 56]) data are subsequently analyzed [61] in the natural time domain: the calculation of the variance κ_1 is made for an event taking time windows for $l = 6$ to 40 consecutive events (i.e., while in § 2.5.2 the value of l was kept constant, here l varies within certain limits and no κ_1 averaging is made). The choice of the precise value of the upper limit of l is not found decisive, since practically the same results are obtained even if the number of consecutive events was changed from 6–40 to 6–100. And second, this process was performed for all the events (for all the l

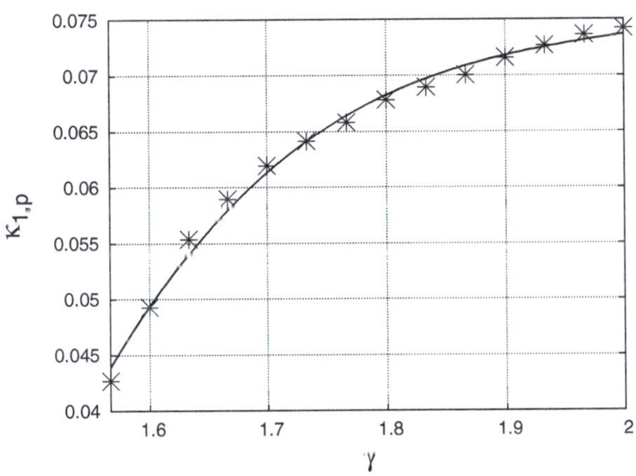

Fig. 2.6 The values of $\kappa_{1,p}$ as a function of γ for power law distributed data. The continuous line has been drawn as a guide to the eye. Note that $\kappa_{1,p} \approx 0.070$ for $\gamma \approx 1.87$, see also Fig. 2.5. Taken from Ref. [59].

values, e.g. between $l = 6$ to $l = 40$) by scanning the whole dataset. In Fig. 2.5, we plot the pdf $P(\kappa_1)$ versus κ_1 for several γ values. The most probable value $\kappa_{1,p}$ (for γ = constant) is also plotted in Fig. 2.6 versus the corresponding γ value.

This curve interrelates κ_1 and γ for the shuffled data (thus an eventual process's memory is destroyed) and hence the plotted $\kappa_{1,p}$ values (which differ markedly from κ_u) correspond to the self-similarity resulting *solely* from the heavy-tailed distribution.

Note that the study of the origin of the self-similarity in real earthquake data will be elaborated in Chapter 6.

2.5.5 Conclusions

In summary, the origin of self-similarity may be distinguished as follows:

If self-similarity exclusively results from the process's memory, the κ_1 value should *change* to $\kappa_u = 1/12$ for the (randomly) shuffled data. This is the case of the SES activities, e.g. see § 4.7.1.

On the other hand, if the self-similarity results from process's increments' "infinite" variance *only*, the most probable value $\kappa_{1,p}$ should be the same (but differing from κ_u) for the original and the (randomly) shuffled data.

When both origins of self-similarity are present, the relative strength of the contribution of the one origin with respect to that of the other can be quantified on the basis of Eqs. (2.95) and (2.96), e.g., see § 6.3.2.

2.6 Origin of the optimality of the natural time representation

Here we address the problem [3] of optimality of the natural time representation of time series resulting from complex systems. For this purpose, we first study the structures of the time-frequency representations [7] of the signals by employing the Wigner function [68] to compare the natural time representation with the ones, either in conventional time or in other possible reparametrizations. We shall see that significant enhancement of the signal is observed in the time-frequency space if natural time is used, in marked contrast to other time domains. To quantify this localization property, we examine the generalized entropic measure proposed by Tsallis [45], which has been widely discussed in the studies of complex dynamical systems (see also Section 6.5).

In time series analysis, it is desired to reduce uncertainty and extract signal information as much as possible. Consequently, the most useful time domain should maximize the information measure, and hence minimize the entropy. We find that this can statistically be ascertained in natural time, by investigating a multitude of different time domains.

Consider a signal $\{x(t)\}$ represented in conventional time, t. The normalized time-frequency Wigner function associated with it is defined by

$$W(t,\omega) = A \int \mathrm{d}\tau\, \mathrm{e}^{-i\omega\tau} x(t-\tau/2)x(t+\tau/2), \tag{2.99}$$

where $A = \left[\pi \int \mathrm{d}t x^2(t)\right]^{-1}$ is the normalization constant and ω is the frequency. Numerically, it is necessary to discretize and make finite both time and frequency, and the integral has to be replaced by a sum. To make comparison of the natural time analysis with Eq. (2.99), it is convenient to rescale χ_k by $N\chi_k$, which is precisely the pulse number, $k \equiv t_k$. The quantity, Q_k , has a clear meaning for dichotomous time series (Fig. 2.1(a)), whereas for nondichotomous time series, threshold should be appropriately applied (e.g., the mean value plus half of the standard deviation) to transform it to a dichotomous one. The normalized Wigner function associated with Q_k is now given as follows:

$$W(k,\tilde{\omega}) = B \sum_{i=0}^{N-1} Q_{k-i} Q_{k+i} \cos\left[\tilde{\omega}(t_{k+i} - t_{k-i})\right], \tag{2.100}$$

where $B = \left[\pi \sum_{k=1}^{N} Q_k^2\right]^{-1}$ stands for the normalization constant and $\tilde{\omega}$ is the dimensionless "frequency". In the sum, Q_k with $k \leq 0$ and $k > N$ should be set equal to zero. Note that Eq. (2.100) is a discrete version of the continuous Wigner function in Eq. (2.99) and unlike the ordinary definition, the transformation in Eq. (2.100) is not orthogonal in general.

Figure 2.7 depicts the Wigner functions in the time-frequency spaces for the conventional time (a) and the natural time (b). Remarkably, significant enhancement of the signal is observed in the latter case, with the scale of enhancement being about 10 times. A localized structure emerges in natural time, in contrast to a moderate profile in the conventional time representation.

In the natural time domain, the time difference between two consecutive pulses (i.e., inter-occurrence time) is equally spaced and dimensionless, and is here taken to be unity: $t_{k+1} - t_k = 1$. However, for the sake of comparison, we will later consider various time domains in which the occurrence time $t_k = Nu_k$ in Eq. (2.100) is made random. The conventional time representation is characterized by a constant time increment Δt (e.g., 1 sec), and the occurrence of the i-th event is at $t_i = i\Delta t$. To generate the random time domains artificially, we consider uniformly distributed u_k so that the average inter-occurrence time is again unity. Performing Monte-Carlo simulation, we have constructed more than 1,000 different time domains and integrated over ω ($\tilde{\omega}$) over 0 to π [rad/sec] ([rad]), which can cover the regimes of interest (recall that when $t_k = k$, $W(k,\omega+\pi) = W(k,\omega)$).

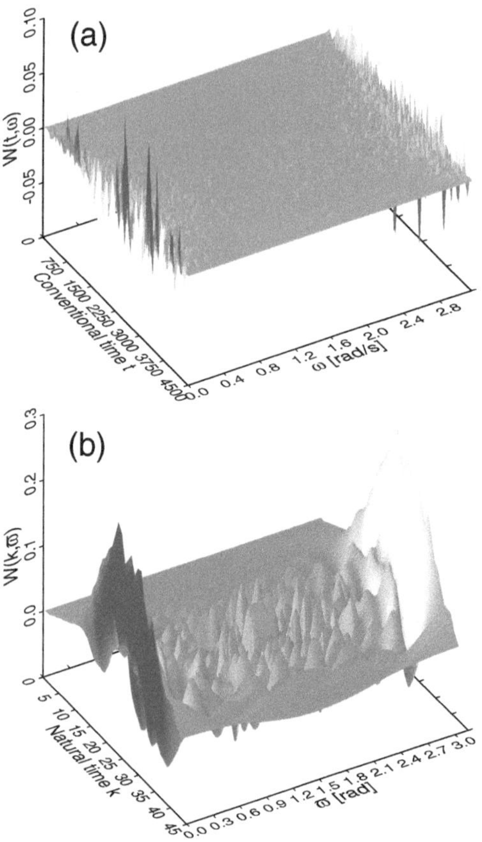

Fig. 2.7 The plots of the Wigner functions of the SES activity A of Fig. 2.8 given below in (a) the conventional time domain and (b) the natural time domain. Significant enhancement of the signal is recognized in the natural time domain at both edges but mainly in the localized structures in the intermediate region. Note that, instead of χ_k, $N\chi_k = k$ is used (see the text). ω has the unit [rad/sec], whereas has $\tilde{\omega}$ has [rad]. Taken from Ref. [3].

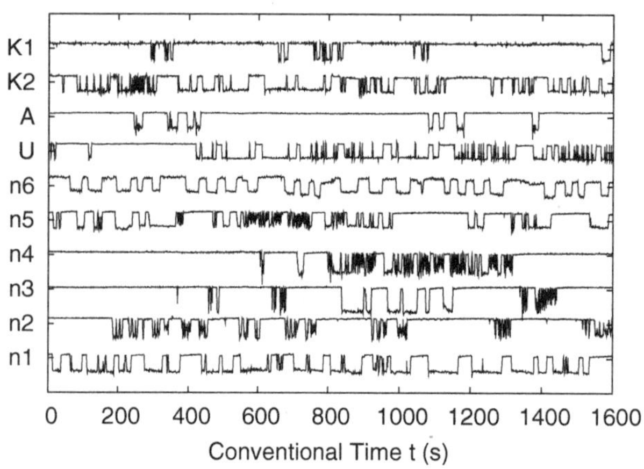

Fig. 2.8 Excerpts of 4 SES activities, labeled K1, K2, A, U and 6 "artificial" noises, labeled n1–n6, in normalized units, see the caption of Fig. 4.2. Taken from Ref. [3].

To quantify the degrees of disorder in the time-frequency spaces with various time domains, we employ as mentioned the Tsallis entropy [45] defined by

$$S_q = \frac{1}{1-q}\left(\int \mathrm{d}\mu W^q - 1\right), \tag{2.101}$$

where $\int \mathrm{d}\mu$ is the collective notation for integral and sum over the time-frequency space and q is the positive entropic index. In the limit $q \to 1$, this quantity tends to the form of the Boltzmann–Gibbs–Shannon entropy $S = -\int \mathrm{d}\mu W \ln W$. This limit cannot however be taken, since the Wigner function is a pseudo-distribution and takes negative values, in general. The quantity S_q is, however, well defined if q is even. Thus, we propose to use the value

$$q = 2, \tag{2.102}$$

which, by considering Eqs. (2.100) and (2.101), results in:

$$S_2 = 1 - \frac{1}{2\pi} \times \left\{ \frac{\sum_{k=1}^{N}\sum_{l=0}^{N-1}\sum_{l'=0}^{N-1} Q_{k-l}Q_{k+l}Q_{k-l'}Q_{k+l'} \frac{\sin[\pi(t_{k+l}-t_{k-l}+t_{k+l'}-t_{k-l'})]}{\pi(t_{k+l}-t_{k-l}+t_{k+l'}-t_{k-l'})}}{\left[\sum_{k=1}^{N}\left(Q_k^2 + \sum_{l=1}^{N-1} Q_{k-l}Q_{k+l}\frac{\sin[\pi(t_{k+l}-t_{k-l})]}{\pi(t_{k+l}-t_{k-l})}\right)\right]^2} + \frac{\sum_{k=1}^{N}\sum_{l=0}^{N-1}\sum_{l'=0}^{N-1} Q_{k-l}Q_{k+l}Q_{k-l'}Q_{k+l'}\left(\delta_{l+l',0} + \delta_{l,l'} + \frac{\sin[\pi(t_{k+l}-t_{k-l}-t_{k+l'}+t_{k-l'})]}{\pi(t_{k+l}-t_{k-l}-t_{k+l'}+t_{k-l'})}\right)}{\left[\sum_{k=1}^{N}\left(Q_k^2 + \sum_{l=1}^{N-1} Q_{k-l}Q_{k+l}\frac{\sin[\pi(t_{k+l}-t_{k-l})]}{\pi(t_{k+l}-t_{k-l})}\right)\right]^2} \right\} \tag{2.103}$$

Table 2.2 The values of $\mathrm{Prob}(S_2 < S_2^{nat})$ together with the number of pulses N for the electric signals of Fig. 2.8 with $N > 50$. The estimation error is at the most 1.6%.

Signal	N	$\mathrm{Prob}(S_2 < S_2^{nat})(\%)$
K1	312	3.7
K2	141	6.9
U	80	8.1
n1	216	5.7
n2	1080	<0.1
n3	259	2.7
n4	396	1.6
n5	432	2.8

To examine how the natural time representation is superior to other ones, in Ref. [3] we made comparison of the values of S_2 for 10 different time series [54] of electric signals (see Fig. 2.8, whereas Fig. 4.9 depicts their natural time representation): 4 SES activities and 6 "artificial" noises. The results of 8 (out of the 10) signals comprising more than 50 pulses are compiled in Table 2.2 in which we give the values of $\mathrm{Prob}(S_2 < S_2^{nat})$, i.e., the probability that S_2 calculated for a time domain different than the natural time domain to be smaller than the value S_2^{nat} calculated for natural time (note that this value comes from Eq. (2.103) and should not be confused with the entropy S in natural time defined by Eq. (3.1), see Chapter 3). This probability Prob($S_2 < S_2^{nat}$) was estimated as follows. For each time domain produced by Monte–Carlo the corresponding S_2 value was calculated through Eq. (2.103) and compared to S_2^{nat}. For signals with a reasonable number of pulses, e.g., larger than 50, Table 2.2 reveals that the quantity S_2^{nat}, in fact, tends to be minimum compared to those of other representations attempted. In addition, it is mentioned that S_2^{nat} is also appreciably smaller than S_2 in conventional time (see Fig. 2.7).

In conclusion, we investigated if natural time yields an optimal representation for enhancing the signals in the time-frequency space by employing the Wigner function and measuring its localization property by means of the Tsallis entropy. For this purpose, we compared the values of the entropy for various time series (being either SES activities or "artificial" noises) represented in a multitude of different time domains. We find that the entropy is highly likely to be minimum for natural time, implying the least uncertainty in the time-frequency space. This explains why dynamical evolution of diverse systems can be better described in natural time.

2.7 Is time continuous?

Natural time χ, from its definition, is *not continuous* and takes values which are *rational* numbers in the range (0,1]. (In these numbers, as the complex system evolves, the numerators are just the natural numbers (except 0), which denote the *order* of appearance of the consecutive events.) Hence, one of the fundamental differences between (conventional) time and natural time refers to the fact that the former is based on the idea of *continuum*, while the latter is *not*. Following Ref. [50], here we aim at raising some consequences of this difference, and in particular those that stem from the *set theory* developed by Cantor, having in mind the following *crucial* remark made by Schrödinger (see pp. 62–63 of Ref. [40]):

"We are familiar with the idea of the *continuum*, or we believe ourselves to be. We are *not* familiar with the enormous difficulty this concept presents to the mind, unless we have studied very modern mathematics (Dirichlet, Dedekind,Cantor)."

2.7.1 Differences between natural time and conventional time on the basis of set theory

We clarify in advance that we do not tackle here the case (since it is inapplicable to our universe [16]) raised by Gödel in 1949 who discovered [15] unexpected solutions to the equations of general relativity corresponding to universes in which *no universal temporal ordering is possible* (see also Refs. [8, 71] and references therein). This solution acquires its simplest form (see p.86 in Ref. [39]) "with *two* of the coordinate-line-elements time-like (the other two space-like)". Interestingly, Schrödinger in an early version of Ref. [39], which was published almost simultaneously with Gödel's work, had also emphasized that "there is no necessity for just three of the four line-elements being space-like, one time-like ...".

We now recapitulate some points of the Cantor set theory that are relevant to our present discussion.

A *transfinite number or transfinite cardinal* is the cardinality of some *infinite* set, where the term *cardinality* of a set stands for the number of members it contains, e.g., see Ref. [43].

The set of natural numbers is labeled by $\mathscr{N}$, while the number of natural numbers is designated by $\aleph_0$, i.e., $\aleph_0 = |\mathscr{N}|$ (note that the cardinality of a set S is labeled $|S|$). In this transfinite number, the zero subscript is justified by the fact that, as proved by Cantor, no infinite set has a smaller cardinality than the set of natural numbers.

It can be shown that the set of rational numbers designated by $\mathscr{Q}$ has the same cardinality as the set of natural numbers, or $|\mathscr{N}| = |\mathscr{Q}|$ (e.g., Theorem 2 in Ref. [43]). In other words, the rationals are *exactly* as numerous as the naturals.

Note that a set is *countable* iff its cardinality is either finite or equal to $\aleph_0$ and in particular is termed *denumerable* iff its cardinality is exactly $\aleph_0$ (note that as usually, for "if and only if" we write simply "iff"). A set is *uncountable* iff its cardinality is greater than $\aleph_0$; see also below.

Hence, natural time takes values (which, as mentioned, are rational numbers) that form in general a *countable* set; this becomes a *denumerable* set in the limit of infinitely large number of events (see § 2.7.2).

Further, since in natural time analysis we consider the pairs (χ_k, Q_k), the values of the quantity Q_k *should* form a set with cardinality smaller than (or equal to) $\aleph_0$. In other words, the values of the energy also form a *countable* set, which reflects of course that the energy is *not continuous*, thus the *quantization of energy* seems to emerge.

The fact that $|\mathscr{N}| = |\mathscr{Q}|$ is an astounding result in view of the following. The rational numbers are *dense* in the real numbers, which means that between any two rational numbers on the real number line we can find *infinitely more* rational numbers. In other words, although the set of rational numbers seems to contain infinities within infinities, there are just as many natural numbers as there are rational numbers. This reflects the following point.

Let us assume that we follow the evolution of a system with some (experimental) accuracy, in which, as mentioned, in the limit of infinitely large number of events the cardinality of the set of the values of natural time is $\aleph_0$. Let us assume that we now repeat the measurement with more sensitive instrumentation, i.e., counting events above an appreciably smaller energy threshold (which should be constrained by the uncertainty principle, but a further discussion on this point lies beyond the scope of the present monograph, as already mentioned in § 2.1.1); hence between two consecutive events of the former measurement a considerable number of appreciably smaller events may be monitored. The corresponding cardinality, in contrast to our intuition, is again $\aleph_0$. In other words, when considering the limit of infinitely large number of consecutive events, the natural time takes values that form a *denumerable* set and this remains so even upon increasing the accuracy (and hence lowering the uncertainty) of our measurement. The inverse, i.e., when the instrumentation becomes less sensitive, may correspond to a "coarse graining" procedure.

We now turn to the aspects of Cantor set theory related to the real numbers, which as mentioned are associated with the conventional time. It is shown that the number of points on a finite line segment is the same as the number of points on an infinite line (e.g., Theorem 13 in Ref. [43]). Considering the definition: the number of real numbers is the same as the number of points on an infinite line (or in the jargon, the *numerical continuum* has the same cardinality as the *linear continuum*), let "c" designate the cardinality of the continuum – or equivalently the cardinality of the set of real numbers. (Hence $c = |\mathscr{R}|$ by definition.) It is proven (e.g., Theorem 16 in Ref. [43]) that the set of real numbers is uncountable, or $|\mathscr{R}| > \aleph_0$. (Equivalently, this theorem asserts that $c > \aleph_0$.)

Hence, the values of conventional time form an *uncountable* set, in contrast to that of natural time which in general as mentioned is countable.

In order to further inspect this fundamental difference, we resort to the continuum hypothesis (CH) which was formulated (but not proved) by Cantor.

Continuum hypothesis, after Euclid's parallel postulate, was the first major conjecture to be proved *undecidable* by standard mathematics [43].

We first clarify that the power set $^*S$ of a set S, which is the set of all subsets of S, has a cardinality $|^*S| = 2^{|S|}$ when S is finite. According to Cantor's Theorem the cardinality of the power set of an *arbitrary* set has a greater cardinality than the original arbitrary set, i.e., $|^*S| > |S|$ (e.g., Theorem 4 in Ref. [43]). This theorem is trivial for finite sets, but fundamental for infinite sets. Hence, for any infinite cardinality, there is a larger infinite cardinality, namely, the cardinality of its power set.

The continuum hypothesis asserts that there is *no* cardinal number α such that $\aleph_0 < \alpha < c$.

Then it follows that the next largest transfinite cardinal after $\aleph_0$ (labeled $\aleph_1$) is c, thus $c = \aleph_1$. Since Cantor proved (e.g., Theorem 17 in Ref. [43]) that $\aleph_1 = 2^{\aleph_0}$, CH leads to: $c = 2^{\aleph_0}$ (thus, this is the number of points on an infinite line).

Hence, if we assume CH, the cardinality of the set of the values of natural time – in the limit of infinitely large number of events – corresponds to $\aleph_0$, while that of the conventional time is $2^{\aleph_0}$.

The values of the former, as mentioned, are rational numbers, while almost all the values of the latter are *irrational*, because, since $2^{\aleph_0} \gg \aleph_0$, almost all reals are irrational numbers. (On the other hand, without assuming CH we have essentially *no idea* which transfinite number corresponds to c, and we would know the cardinality of the naturals, integers, and rationals, but *not* the cardinality of the reals, e.g., see Ref. [43].) As for the values of Q_k, they are *not* necessarily rational, because in general when taking $\aleph_0$ (at the most) out of $2^{\aleph_0}$ values they may all be irrational.

Hence, in the limit of infinitely large number of events, even upon gradually improving the accuracy of our measurements, both sets $\{\chi_k\}$ and $\{Q_k\}$ remain denumerable, the former consisting of rational numbers only.

2.7.2 Proof of the cardinality of the set of the values of natural time

We now indicate how in the limit of infinitely large number of events we conclude that the cardinality of the set of the values of natural time equals to $\aleph_0$. Let us tabulate the values of natural time upon the occurrence of each event:

after the first event	1					
after the second event	$\frac{1}{2}$	$\frac{2}{2}$				
after the third event	$\frac{1}{3}$	$\frac{2}{3}$	$\frac{3}{3}$			
after the fourth event	$\frac{1}{4}$	$\frac{2}{4}$	$\frac{3}{4}$	$\frac{4}{4}$		
...						
after the Nth event	$\frac{1}{N}$	$\frac{2}{N}$	$\frac{3}{N}$	$\frac{4}{N}$	...	$\frac{N}{N}$

This indicates that the cardinality of the set of the values of natural time $|\{\chi_k\}|$ should be greater than (or equal to) N(number of entries in the first column) and smaller than (or equal to) N^2 (number of entries in the square $N \times N$ matrix), i.e.,

$$N \leq |\{\chi_k\}| \leq N^2 \tag{2.104}$$

Thus, for $N \to \infty$ we have $\aleph_0 \leq |\{\chi_k\}| \leq \aleph_0^2$ and since $\aleph_0^2 = \aleph_0$ (see Theorem 22 of Ref. [43]), we find that $|\{\chi_k\}| = \aleph_0$.

2.7.3 Is natural time compatible with Schrödinger's point of view?

Schrödinger, in order to point out "the intricacy of the continuum", used the following example (see pp. 138–143 of Ref. [41]): Let us consider the interval [0,1], you first take away the whole middle third including its left border point, thus the points from 1/3 to 2/3 (but you leave 2/3). Of the remaining two-thirds you again take away "the middle thirds", including their left border points, but leaving their right border points. With the remaining "four ninths" you proceed in the same way and so on. The cardinality of the set that remains *ad infinitum* is no less than that of [0,1] because it can be shown [41] that there is a one-to-one correspondence between their elements. Moreover, since it is a subset of [0,1], its cardinality is also no greater, so it must in fact be equal. In particular, Schrödinger concludes [41] as follows: "The remarkable fact about our remaining set is that, although it covers no measurable interval, yet it still has the vast extension of any continuous range. This astonishing combination of properties is, in mathematical language, expressed by saying that our set has still the 'potency' of the continuum, although it is 'of measure zero'." In other words, the cardinality of the aforementioned remaining set considered by Schrödinger exceeds drastically that of the set of the values of natural time.

Let us now comment on the common view that (conventional) time is continuous, keeping in the frame that, as pointed out by Schrödinger (p. 145 of Ref. [42]) "our sense perceptions constitute our sole knowledge about things". In short, it seems that the continuity of time does not stem from *any* fundamental principle, but probably originates from the following demand on continuity discussed by Schrödinger (see p. 130 of [41]):

"From our experiences on a large scale ... physicists had distilled the one clear-cut demand that a truly clear and complete description of any physical happening has to fulfill: it ought to inform you precisely of what happens at any point in space at any moment of time .... We may call this demand the *postulate of continuity of the description*."

Schrödinger, however, subsequently commented on this demand as follows (see p. 131 of Ref. [41]): "It is this postulate of continuity that appears to be unfulfillable!..." and furthermore added: "We *must not admit the possibility of continuous observation*." Considering these important remarks, we may say that the concept of natural time is not inconsistent with Schrödinger's point of view.

2.7.4 Conclusions

Conventional time is currently assumed continuous, but this does not necessarily result from *any* fundamental principle. Its values form an uncountable set, almost all of which may be *irrational* numbers. On the other hand, natural time is not continuous, and its values form a countable set consisting of rational numbers only; further, the values of the energy also form a countable set but they are not necessarily rational. In the limit of infinitely large number of events, the cardinality of the set of the values of natural time is $\aleph_0$ (irrespective of whether we increase the accuracy of the measurement), thus being drastically smaller than that of conventional time, which equals to $2^{\aleph_0}$ if we accept the validity of the continuum hypothesis.

References

1. A. Guarino, S. Ciliberto, A. Garcimartín, M. Zei, R. Scorretti: Failure time and critical behaviour of fracture precursors in heterogeneous materials. Eur. Phys. J. B **26**(2), 141–151 (2002)
2. Concluding the 23rd Solvey Conference (Dec. 2005), David Gross, compared the state of Physics today to that during the first Solvay conference in 1911 and said: "They were missing something absolutely fundamental" he said. "We are missing perhaps something as profound as they were back then", see New Scientist, December 10, pp.6,7 (2005).
3. Abe, S., Sarlis, N.V., Skordas, E.S., Tanaka, H.K., Varotsos, P.A.: Origin of the usefulness of the natural-time representation of complex time series. Phys. Rev. Lett. **94**, 170601 (2005)
4. de Arcangelis, L., Godano, C., Lippiello, E., Nicodemi, M.: Universality in solar flare and earthquake occurrence. Phys. Rev. Lett. **96**, 051102 (2006)
5. Aschwanden, M.J., Nightingale, R.W., Tarbell, T.D., Wolfson, C.J.: Time variability of the "quiet" Sun observed with TRACE, II. Physical parameters, Temperature Evolution, and Energetics of Extreme-Ultraviolet-Nanoflares. Astrophys. J. **535**, 1047–1065 (2000)
6. Ausloos, M., Lambiotte, R.: Brownian particle having a fluctuating mass. Phys. Rev. E **73**, 011105 (2006)
7. Cohen, L.: Time-Frequency Analysis. Theory and Applications. Prentice-Hall, Upper Saddle River, NJ (1994)
8. Davis, M.: Gödel's universe. Nature **435**, 19–20 (2005)
9. Dimiduk, D.M., Woodward, C., LeSar, R., Uchic, M.D.: Scale-free intermittent flow in crystal plasticity. Science **312**, 1188–1190 (2006)
10. Einstein, A.: *Ideas and Opinions* (Crown Publishers, 1954; new edition by Souvenir Press, 2005); see also p. 54 in A. Gefter in New Scientist, 10 December, 2005
11. Einstein, A.: A brief outline of the developement of the theory of relativity. Nature **106**, 782–784 (1921)
12. Feller, W.: An Introduction to Probability Theory and Its Applications, Vol. II. Wiley, New York (1971)
13. Frame, M., Mandelbrot, B., Neger, N.: Fractal Geometry, Yale University, available from http://classes.yale.edu/fractals/, see http://classes.yale.edu/Fractals/RandFrac/fBm/fBm4.html
14. Garcimartín, A., Guarino, A., Bellon, L., Ciliberto, S.: Statistical properties of fracture precursors. Phys. Rev. Lett. **79**, 3202–3205 (1997)
15. Gödel, K.: An example of a new type of cosmological solutions of Einstein's field equations of gravitation. Rev. Mod. Phys. **21**, 447–450 (1949)
16. Hentschel, K.: Review of: A world without time: The forgotten legacy of Gödel and Einstein. Physics Today **58(12)**, 60–61 (2005)
17. Hooft, G.T.: In *Does God play dice*? Physics World, December 2005, (http://physicsweb.org/articles/world)
18. Hughes, D., Paczuski, M., Dendy, R.O., Helander, P., McClements, K.G.: Solar flares as cascades of reconnecting magnetic loops. Phys. Rev. Lett. **90**, 131101 (2003)
19. Jonscher, A.K.: Universal Relaxation Law. Chelsea Dielectric Press, London (1996)
20. Kalisky, T., Ashkenazy, Y., Havlin, S.: Volatility of linear and nonlinear time series. Phys. Rev. E **72**, 011913 (2005)
21. Kanamori, H.: Quantification of earthquakes. Nature **271**, 411–414 (1978)
22. Koslowski, M., LeSar, R., Thomson, R.: Avalanches and scaling in plastic deformation. Phys. Rev. Lett. **93**, 125502 (2004)
23. Kuntz, M.C., Sethna, J.P.: Noise in disordered systems: The power spectrum and dynamic exponents in avalanche modelss. Phys. Rev. B **62**, 011699 (2000)
24. Lamperti, J.W.: Semi-stable stochastic processes. Trans. Am. Math. Soc. **104**, 62–78 (1962)
25. Mandelbrot, B.B., Wallis, J.R.: Some long run properties of geophysical records. Water Resources Research **5**, 321–340 (1969)
26. Miguel, M.C., Vespignani, A., Zapperi, S., Weiss, J., Grasso, J.R.: Intermittent dislocation flow in viscoplastic deformation. Nature **410**, 667–671 (2001)

27. Montroll, E.W., Bendler, J.T.: On Lévy (or stable) distributions and the Williams-Watts model of dielectric relaxation. J. Stat. Phys. **34**, 129–162 (1984)
28. von Neumann, J.: Mathematical Foundations of Quantum Mechanics. Princeton University Press, Princeton N.J. (1955)
29. Nigro, G., Malara, F., Carbone, V., Veltri, P.: Nanoflares and mhd turbulence in coronal loops: A hybrid shell model. Phys. Rev. Lett. **92**, 194501 (2004)
30. NIST, SEMATECH: (2000). *NIST/SEMATECH e-Handbook of Statistical Methods*, www.itl.nist.gov/div898/handbook
31. Nowick, A.S., Vaysleyb, A.V., Kuskovsky, I.: Universal dielectric response of variously doped CeO_2 ionically conducting ceramics. Phys. Rev. B **58**, 8398–8406 (1998)
32. Parnell, C.E., Jupp, P.E.: Statistical analysis of the energy distribution of nanoflares in the quiet Sun. Astrophys. J. **529**, 554–569 (2000)
33. Pauli, W.: *Die allgemeinen Prinzipien der Wellenmechanik.* in K. Geiger and H. Scheel (eds), p.140 Handbuck der Physik 2nd Edition, Vol. 245 (Berlin Springer, 1933)
34. Peng, C.K., Buldyrev, S.V., Goldberger, A.L., Havlin, S., Mantegna, R.N., Simon, M., Stanley, H.E.: Statistical properties of DNA sequences. Physica A **221**, 180–192 (1995)
35. Penrose, O.: An asymmetric world. Nature **438**, 919 (2005)
36. Rundle, J.B., Turcotte, D.L., Shcherbakov, R., Klein, W., Sammis, C.: Statistical physics approach to understanding the multiscale dynamics of earthquake fault systems. Rev. Geophys. **41**, 1019 (2003)
37. Russell, E.V., Israeloff, N.E.: Direct observation of molecular cooperativity near the glass transition. Nature **408**, 695–698 (2000)
38. Scafetta, N., West, B.J.: Multiscaling comparative analysis of time series and geophysical phenomena. Complexity **10(4)**, 51–56 (2005)
39. Schrödinger, E.: Space-Time Structure. Cambridge Univ. Press, Cambridge (1985)
40. Schrödinger, E.: Nature and the Greeks. Cambridge Univ. Press, Cambridge 1954; Canto Edition with *Science and Humanism* (1996)
41. Schrödinger, E.: Science and Humanism. Cambridge Univ. Press, Cambridge 1951; Canto Edition with *Nature and the Greeks* (1996)
42. Schrödinger, E.: Mind and Matter. Cambridge Univ. Press, Cambridge 1958; Canto Edition with *Autobiographical Sketches* and *What is Life*, Tenth printing (2003)
43. Suber, P.: A crash course in the mathematics of infinite sets, http://www.earlham.edu/~peters/writing/infapp.htm, published in the *St. John's Review*, **XLIV**, 2, 35–59 (1998)
44. Szulga, J., Molz, F.: The Weierstrass Mandelbrot process revisited. J. Stat. Phys. **104**, 1317 (2001)
45. Tsallis, C.: Possible generalization of Boltzmann-Gibbs statistics. J. Stat. Phys. **52**, 479–487 (1988)
46. Tsallis, C., Levy, S.V.F., Souza, A.M.C., Maynard, R.: Statistical-mechanical foundation of the ubiquity of Lévy distributions in nature. Phys. Rev. Lett. **75**, 3589–3593 (1995)
47. Tsallis, C., Levy, S.V.F., Souza, A.M.C., Maynard, R.: Statistical-mechanical foundation of the ubiquity of the Lévy distributions in nature. Phys. Rev. Lett. **77**, 5442 (1996)
48. Varotsos, P.: A review and analysis of electromagnetic precursory phenomena. Acta Geophys. Pol. **49**, 1–42 (2001)
49. Varotsos, P., Alexopoulos, K.: Thermodynamics of Point Defects and their Relation with Bulk Properties. North Holland, Amsterdam (1986)
50. Varotsos, P.A.: Is time continuous? arXiv:cond-mat/0605456v1 [cond-mat.other] (18 May 2006)
51. Varotsos, P.A., Sarlis, N.V., Skordas, E.S.: Spatio-temporal complexity aspects on the interrelation between Seismic Electric Signals and seismicity. Practica of Athens Academy **76**, 294–321 (2001)
52. Varotsos, P.A., Sarlis, N.V., Skordas, E.S.: Long-range correlations in the electric signals that precede rupture. Phys. Rev. E **66**, 011902 (2002)
53. Varotsos, P.A., Sarlis, N.V., Skordas, E.S.: Seismic Electric Signals and seismicity: On a tentative interrelation between their spectral content. Acta Geophys. Pol. **50**, 337–354 (2002)
54. Varotsos, P.A., Sarlis, N.V., Skordas, E.S.: Attempt to distinguish electric signals of a dichotomous nature. Phys. Rev. E **68**, 031106 (2003)
55. Varotsos, P.A., Sarlis, N.V., Skordas, E.S.: Long-range correlations in the electric signals that precede rupture: Further investigations. Phys. Rev. E **67**, 021109 (2003)
56. Varotsos, P.A., Sarlis, N.V., Skordas, E.S., Lazaridou, M.S.: Entropy in natural time domain. Phys. Rev. E **70**, 011106 (2004)

57. Varotsos, P.A., Sarlis, N.V., Skordas, E.S., Lazaridou, M.S.: Identifying sudden cardiac death risk and specifying its occurrence time by analyzing electrocardiograms in natural time. Appl. Phys. Lett. **91**, 064106 (2007)
58. Varotsos, P.A., Sarlis, N.V., Skordas, E.S., Lazaridou, M.S.: Fluctuations, under time reversal, of the natural time and the entropy distinguish similar looking electric signals of different dynamics. J. Appl. Phys. **103**, 014906 (2008)
59. Varotsos, P.A., Sarlis, N.V., Skordas, E.S., Tanaka, H.K., Lazaridou, M.S.: Attempt to distinguish long-range temporal correlations from the statistics of the increments by natural time analysis. Phys. Rev. E **74**, 021123 (2006)
60. Varotsos, P.A., Sarlis, N.V., Skordas, E.S., Tanaka, H.K., Lazaridou, M.S.: Entropy of seismic electric signals: Analysis in the natural time under time reversal. Phys. Rev. E **73**, 031114 (2006)
61. Varotsos, P.A., Sarlis, N.V., Tanaka, H.K., Skordas, E.S.: Similarity of fluctuations in correlated systems: The case of seismicity. Phys. Rev. E **72**, 041103 (2005)
62. Varotsos, P.A., Sarlis, N.V., Lazaridou, M.S.: Interconnection of defect parameters and stress-induced electric signals in ionic crystals. Phys. Rev. B **59**, 24–27 (1999)
63. Varotsos, P.A., Sarlis, N.V., Skordas, E.S., Tanaka, H.K.: A plausible explanation of the b-value in the Gutenberg-Richter law from first principles. Proc. Japan Acad., Ser. B **80**, 429–434 (2004)
64. Weiss, J.: Scaling of fracture and faulting of ice on Earth. Surveys in Geophysics **24**, 185–227 (2003)
65. Weron, A., Burnecki, K., Mercik, S., Weron, K.: Complete description of all self-similar models driven by Lévy stable noise. Phys. Rev. E **71**, 016113 (2005)
66. Weron, K., Jurlewicz, A., Jonscher, A.K.: Energy criterion in interacting cluster systems. IEEE Trans. on Dielectrics and Electrical Insulators **8**, 352–358 (2001)
67. Weyl, H.: Space-Time-Matter. Dover, New York (1952)
68. Wigner, E.: On the quantum correction for thermodynamic equilibrium. Phys. Rev. **40**, 749–759 (1932)
69. Wilczek, F.: Nobel lecture: Asymptotic freedom: From paradox to paradigm. Rev. Mod. Phys. **77**, 857–870 (2005)
70. Wilczek, F.: Whence the Force F=ma? III: Cultural Diversity. Physics Today **58(7)**, 10–11 (2005)
71. Yourgrau, P.: A World Without Time:The forgotten legacy of Gödel and Einstein. Basic Books, Cambridge MA (2005)

3. Entropy in Natural Time

Abstract. Entropy is a concept equally applicable to deterministic as well as stochastic processes. An entropy S is defined in natural time, which exhibits positivity, concavity and Lesche's (experimental) stability. The entropy S_- deduced from analyzing in natural time the time series obtained upon time reversal, is in general different from S, thus the entropy in natural time does satisfy the condition to be "causal" (while the variance $\kappa_1 = \langle\chi^2\rangle - \langle\chi\rangle^2$ does not). The physical meaning of the change $\Delta S \equiv S - S_-$ of the entropy in natural time under time reversal, which is of profound importance for the study of the dynamical evolution of a complex system, is discussed. For a fractional Brownian motion time series with self-similarity exponent H close to unity, as well as for an on–off intermittency model when the *critical* value is approached from below, both values of S and S_- are smaller than the entropy $S_u \approx 0.0966$ of a "uniform" distribution. When a (natural) time window of length l is sliding through a time series, the entropy S exhibits fluctuations, a measure of which is the standard deviation δS. Complexity measures are introduced that quantify the δS variability upon changing the length scale l as well as the extent to which δS is affected when shuffling the consecutive events randomly (for $l =$ const.). In a similar fashion, complexity measures can be defined for the fluctuations of the quantity ΔS whose standard deviation is designated $\sigma[\Delta S]$. For the case that Q_k are independent and identically distributed positive random variables, as in the case of data shuffled randomly, their σ/μ value is interrelated with δS and $\sigma[\Delta S]$.

3.1 The entropy in dynamical systems and the advantages of its use

Before defining the entropy S in natural time time (see Section 3.2), we recapitulate some background knowledge on the entropy in dynamical systems (see chapter 13 of Ref. [36]).

Following Eckmann and Ruelle [8] (see pp. 637, 638), we note that a system with sensitive dependence on initial conditions produces information. This is because two initial conditions that are different but indistinguishable (at a certain experimental precision) will evolve into distinguishable states after a finite time. If ρ is an ergodic probability measure for a dynamical system, the concept of the mean rate of creation of information, $h(\rho)$, also known as the Kolmogorov–Sinai (KS) invariant or entropy, was introduced. This, which (is, in fact, an entropy per unit time and) will be hereafter labeled h_{KS}, is not the same physical quantity as the thermodynamic entropy when studying the dynamics of dissipative physicochemical systems (see also below).

Grassberger and Procaccia [12] proposed a method to estimate a very good lower bound for h_{KS} directly from a time signal (see also Refs. [6] and [7] for other estimations and/or relevant discussion). The value of h_{KS} is zero in an ordered system, and a constant different than zero (but positive, see p. 649 of Ref. [8]) in a chaotic (deterministic) system.

The greater the h_{KS}-value, the stronger are the chaotic features of the system in question. In a random system h_{KS} is infinite (see p. 38 of Ref. [21]).

An interconnection between h_{KS} (which is a single number characteristic of the chaotic dynamical system under consideration) and the time evolution of the entropy of the second law of thermodynamics, is not yet well established. The latter is a function of time; this function depends on both (i) the particular dynamical system considered and (ii) the choice of an initial probability distribution for the state of the system.

We now summarize [43, 36] the advantages when using the concept of the entropy for the study of a dynamical system for which an agreement whether the system dynamics (e.g., normal heart dynamics) are chaotic or not is lacking. The most commonly used nonlinear complexity measures are *fractal dimensions* of various kinds (e.g., correlation dimension, Rényi dimensions). We emphasize, however, that each of them measures different aspects of the *statistics* on the attractor. On the other hand, Liapunov exponents and KS entropy and entropy rates are measures of the *dynamics* on an attractor.

Except for the KS entropy and the entropy rates, the other categories of complexity measures assume a purely deterministic system. On the other hand, entropy is a concept equally applicable to deterministic as well as stochastic processes.

Thus, in a time series that may be due to a mixed process, stochastic and deterministic, the use of fractal dimensions can be criticized. Furthermore, the following point should be stressed: complexity measures based on *static* entropy (e.g., Shannon entropy) quantify *statistical* properties in the time series. The underlying key property of these complexity measures is the probability distribution of the (dwell times in the) data analyzed; thus, the result of such computations should be *independent* of permutations performed on the (sequence of the dwell times in the) time series as in surrogate (randomized) dataset obtained by *shuffling*. On the other hand, the entropy S in natural time that will now be defined (and the relevant complexity measures, see Section 3.6) considers the *sequential* order of pulses (events); in other words, S is a *dynamic* entropy, i.e., it captures characteristics of the dynamics of the system. We emphasize, however, that an interrelation between h_{KS} and S still remains to be explored.

3.2 Entropy in natural time. Definition

The derivative of the fluctuation function $\langle\chi^q\rangle - \langle\chi\rangle^q$, with respect to q, i.e., $\langle\chi^q \ln\chi\rangle - \langle\chi\rangle^q \ln\langle\chi\rangle$ in the region around $q=1$ leads [40] to the quantity $\langle\chi \ln\chi\rangle - \langle\chi\rangle \ln\langle\chi\rangle$. The latter is reminiscent of an excessive "entropy" (see pp. 26–28 of Ref. [37], but recall that the usual expressions of the thermodynamic potentials, in terms of macroscopic variables, break down [37] far from equilibrium and the behavior of entropy is still a matter of intensive investigation). Thus,

Definition 3.1. The entropy in natural time is [38, 40]

$$S \equiv \langle\chi \ln\chi\rangle - \langle\chi\rangle \ln\langle\chi\rangle \tag{3.1}$$

or equivalently

$$S = \sum_{k=1}^{N} \frac{k}{N} \ln\left(\frac{k}{N}\right) p_k - \left(\sum_{l=1}^{N} \frac{l}{N} p_l\right) \ln\left[\sum_{m=1}^{N} \frac{m}{N} p_m\right] \tag{3.2}$$

Note that S should not be confused with $\mathrm{Cov}(\chi, \ln\chi) \equiv \langle\chi \ln\chi\rangle - \langle\chi\rangle\langle\ln\chi\rangle$ since in general $\langle\ln\chi\rangle \neq \ln\langle\chi\rangle$.

The entropy S consists of two terms: $\mathrm{S}_\chi \equiv -\langle\chi \ln\chi\rangle$ and $\mathrm{S}_{\langle\chi\rangle} \equiv -\langle\chi\rangle \ln\langle\chi\rangle$ and hence $S = -\mathrm{S}_\chi + \mathrm{S}_{\langle\chi\rangle}$. Upon employing $p(\chi)$, Eq. (3.1) can be written as

$$S = \int_0^1 p(\chi)\chi \ln\chi \, d\chi - \left(\int_0^1 p(\chi)\chi \, d\chi\right) \ln\left[\int_0^1 p(\chi')\chi' \, d\chi'\right]. \tag{3.3}$$

Recalling that for the "uniform" (u) distribution $p(\chi) = 1$ and $\langle\chi\rangle = \frac{1}{2}$ (see § 2.1.3), we find that, since $\frac{d}{dx}[x^2/4 - (x^2/2)\ln x] = -x \ln x$, the entropy S_u of the "uniform" distribution has the value

$$S_u = \frac{\ln 2}{2} - \frac{1}{4} \approx 0.0966 \tag{3.4}$$

3.3 Properties of the entropy in natural time

An entropic functional $\Sigma[p]$, where $\{p_i\}_{i=1,2,\ldots N}$ is a probability distribution for a given system (beyond *positivity*, e.g., see Ref. [16]) should be *concave* (e.g., pp. 52–53 of Ref. [5]). For the central importance of this concavity on both the 0th and the 2nd principle of thermodynamics; see for example Ref. [33]. Another important issue which has attracted a strong interest is the *stability* or experimental robustness of the entropies, e.g., see Refs. [31, 2, 16, 17, 15, 24, 33, 19]. In particular, this investigation is usually made in terms of an early suggestion by Lesche [18] (Lesche stability criterion), which states that an entropic measure is stable if its change upon an arbitrarily small deformation of the distribution (representing fluctuations of experimental data [31]) remains small. By means of this stability criterion, Lesche [18] showed that the well known Boltzmann–Gibbs–Shannon (BGS) entropy $S_{BGS} = -k_B \sum_{i=1}^{N} p_i \ln p_i$ is stable, while the Rényi-entropy [28] is unstable. Abe later proved [2] that Tsallis entropy is also stable, while the escort-entropy is not. Finally, the stability was also shown for the κ-entropy [17, 31], while it became clear [34] that the Landsberg–Vedral.-entropy does not obey this criterion.

To sum up, an entropic functional should exhibit positivity, concavity and Lesche stability.

The mathematical proofs concerning the positivity, concavity and uniform continuity (or as usually called Lesche stability) for both the variance κ_1 and entropy S in natural time are now given below. In particular, we first present in § 3.3.1 some background material, while §§ 3.3.2, 3.3.3 and 3.3.4 provide in detail the proof of the positivity, concavity and Lesche stability of *both* the variance κ_1 and the entropy S in natural time.

Despite this similarity of the properties of κ_1 and S, however, we note that upon time reversal they exhibit an essential difference, i.e., the former (κ_1) always remains the same while the latter (S) does not, see § 3.4.1.

Finally, § 3.3.5 is reserved for the presentation of a more general theorem.

3.3.1 Background material

Here, we review some of the basic properties of the real functions $g(x) = x^2$ and $f(x) = x \ln x$ defined on the closed interval [0,1] (more accurately we consider $f(x) = \{x \ln x : \forall x \in (0,1],\ 0 : \text{if } x = 0\}$). These are depicted in Fig. 3.1, and note that the following two inequalities hold:

$$0 \le g(x) \le 1, \tag{3.5}$$

$$0 \ge f(x) \ge -\frac{1}{e}. \tag{3.6}$$

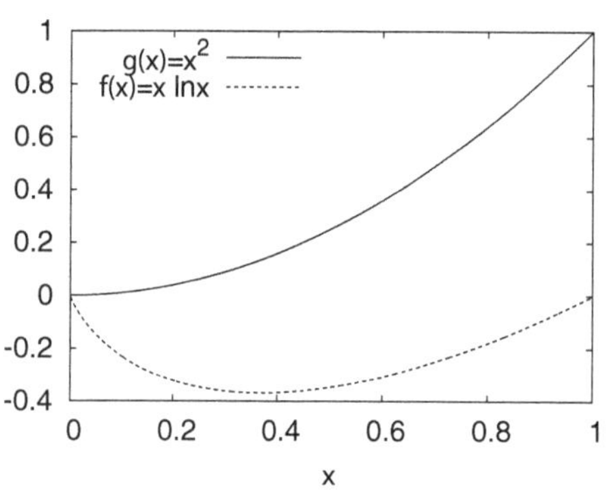

Fig. 3.1 The functions $g(x) = x^2$ and $f(x) = x \ln x$ in the closed interval [0,1].

Equation (3.6) clearly means that

$$|f(x)| \leq \frac{1}{e}. \tag{3.7}$$

We now proceed to two very simple Lemmas:

Lemma 1 *Both $g(x)$ and $f(x)$ are continuous in the interval [0,1].*

Proof. For $g(x)$ this is trivial; for $f(x)$ it is also trivial for $x \in (0,1]$ and since $\lim_{x\to 0} f(x) = 0$, $f(x)$ is also continuous at $x = 0$. □

Lemma 2 *Both $g(x)$ and $f(x)$ are convex in the interval (0,1].*

Proof. It is sufficient to show that the second derivatives of these twice differentiable functions are positive. Indeed $g''(x) = 2$ and $f''(x) = 1/x$ which are both positive for $x > 0$. □

3.3.2 The positivity of κ_1 and S

We first recall Eqs. (2.39) and (3.1) and that the symbol $\langle F(\chi)\rangle$ stands for

$$\langle F(\chi)\rangle = \sum_{k=1}^{N} p_k F\left(\frac{k}{N}\right). \tag{3.8}$$

Second, in order to prove the positivity of κ_1 and S, we shall make use of the following well known theorem [1] (see also **12.411** at page 1101 of Ref. [11]):

Theorem 3.1. *(Jensen's inequality) If F is a convex function on the interval $[a,b]$, then*

$$F\left(\sum_{k=1}^{n} \lambda_k x_k\right) \leq \sum_{k=1}^{n} \lambda_k F(x_k)$$

where $0 \leq \lambda_k \leq 1$, $\lambda_1 + \lambda_2 + \cdots + \lambda_n = 1$ and each $x_k \in [a,b]$.

Due to Lemma 2 both $g(x) = x^2$ and $f(x) = x \ln x$ are convex in (0,1]. Using in Jensen's inequality $\lambda_k = p_k$, $x_k = k/N(> 0)$ and Eq. (3.8), we obtain:

$$\langle\chi\rangle^2 \leq \langle\chi^2\rangle \tag{3.9}$$

and

$$\langle\chi\rangle \ln\langle\chi\rangle \leq \langle\chi \ln \chi\rangle, \tag{3.10}$$

respectively. Obviously, Eqs. (3.9) and (3.10) imply the positivity of both κ_1 and S.

3.3.2.1 Upper bounds of κ_1 and S

Another important property of κ_1 and S is that they are not only bounded from below by zero, but also bounded from above by N-independent bounds. For κ_1, in view of Eq. (3.5), we have:

$$0 \leq \kappa_1 = \langle \chi^2 \rangle - \langle \chi \rangle^2 < \langle \chi^2 \rangle + \langle \chi \rangle^2 \leq \sum_{k=1}^{N} p_k \left(\frac{k}{N} \right)^2 + 1 \leq 2. \tag{3.11}$$

As for S, in view of Eq. (3.7), we get:

$$0 \leq S = \langle \chi \ln \chi \rangle - \langle \chi \rangle \ln \langle \chi \rangle \leq |\langle \chi \ln \chi \rangle| + |\langle \chi \rangle \ln \langle \chi \rangle| \leq \sum_{k=1}^{N} p_k \left| \frac{k}{N} \ln \frac{k}{N} \right| + \frac{1}{e} < \frac{2}{e}. \tag{3.12}$$

Thus, in summary

$$0 \leq \kappa_1 < 2 \tag{3.13}$$

$$0 \leq S < \frac{2}{e} \tag{3.14}$$

Note that the usual values of κ_1 and S are much smaller than these upper bounds. For example, since κ_1 is the variance of the natural time χ that varies from 0 to 1, it is expected to be maximum when $p(\chi)$ has support only at the extreme points 0 and 1. In such a case, $\kappa_1 = p - p^2$ and $S = -p \ln p$, where $p \equiv p(\chi = 1)$. Since $0 < p \leq 1$, the *maximum* value of κ_1 is obtained when $p = \frac{1}{2}$ and then κ_1 equals $\frac{1}{4}$. Moreover, $S = -p \ln p$ maximizes for $p = 1/e$ taking the value $1/e \approx 0.3679$. We note that the latter value is the *maximum* value of S since, when using Eq. (3.6) we have $S = \langle \chi \ln \chi \rangle - \langle \chi \rangle \ln \langle \chi \rangle \leq -\langle \chi \rangle \ln \langle \chi \rangle \leq 1/e$.

3.3.3 The concavity of κ_1 and S

The concavity of κ_1 and S with respect to p_k is based on the fact that they both have negative second derivatives:

$$\frac{\partial^2 \kappa_1}{\partial p_k \partial p_l} = -\frac{kl}{N^2}, \tag{3.15}$$

$$\frac{\partial^2 S}{\partial p_k \partial p_l} = -\frac{kl}{N^2} \left(\sum_{m=1}^{N} p_m \frac{m}{N} \right)^{-1}. \tag{3.16}$$

Thus, the matrix elements of the corresponding Hessians ($H_{k,l} = \frac{\partial^2}{\partial p_k \partial p_l} \Sigma[p]$) have the form $H_{k,l} = -\varepsilon V_k V_l$, where $V = (1/N, 2/N, \ldots, 1)$ and $\varepsilon = 1 (> 0)$ when $\Sigma[p] = \kappa_1$ or $\varepsilon = \left(\sum_{m=1}^{N} p_m \frac{m}{N} \right)^{-1} (> 0)$ when $\Sigma[p] = S$. Such Hessians cannot have a positive eigenvalue λ, because $H_{k,l} e_\lambda = \lambda e_\lambda \Rightarrow e_\lambda^T H_{k,l} e_\lambda = \lambda |e_\lambda| = \lambda = -\varepsilon e_\lambda^T V^T V e_\lambda = -\varepsilon |V e_\lambda| \leq 0$, where $e_\lambda (\in \mathscr{R}^N)$ is any normalized eigenvector of the symmetric real matrix $H_{k,l}$. Since they cannot have a positive eigenvalue these Hessians are negative semi-definite and hence the corresponding entropic measures κ_1 and S are concave [11].

3.3.4 Lesche stability (or experimental robustness) of κ_1 and S

Lesche stability [18], as mentioned, is considered [3, 24, 15, 16] as an important property to be satisfied by an entropic measure $\Sigma[p]$.

Lesche stability implies [3] that for two slightly different distributions $\{p_i\}_{i=1,2,...N}$ and $\{p'_i\}_{i=1,2,...N}$, the corresponding entropic measures $\Sigma[p]$ and $\Sigma[p']$ do not change drastically (and also in a uniform way, see below).

Mathematically

$$\forall \varepsilon > 0 \, \exists \delta \, : \|p - p'\| < \delta \Rightarrow \left| \frac{\Sigma[p] - \Sigma[p']}{\Sigma_{max}} \right| < \varepsilon \tag{3.17}$$

for any value of N, with the metric $\|p\| = \sum_{i=1}^{N} |p_i|$ and Σ_{max} is the maximum value of $\Sigma[p]$.

We note [24] that, for a fixed value of N, Lesche stability implies uniform continuity which is a rather trivial statement, because a continuous function on a compact set is automatically uniformly continuous (see Theorem 3.3 below).

It was pointed out [16] that Lesche condition is a definition of natural uniform metric continuity.

The power of Lesche stability condition arises from the fact that uniform continuity may not survive in the $N \to \infty$ limit [15]. Thus, to avoid confusion, one should consider [15] that the mapping $\Sigma[p]$, where $p \in (\mathscr{R}^+)^N$, taken as a function of N, converges to a uniformly continuous function in a uniform manner, i.e., $\forall \varepsilon > 0$ there exists δ_ε (which depends *only* on ε) such that $\forall \, p, p' \in (\mathscr{R}^+)^N$ and for every $N \in \mathscr{Z}^+$

$$\|p - p'\| < \delta_\varepsilon \Rightarrow \left| \frac{\Sigma[p] - \Sigma[p']}{\Sigma_{max}} \right| < \varepsilon. \tag{3.18}$$

In our case of κ_1 and S, there is at least one distribution $\{p_i\}_{i=1,2,...N}$, the constant one with all $p_i = 1/N$, for which for all N the corresponding values $\kappa_{1,c}$ and S_c:

$$\kappa_{1,c}(N) = \sum_{k=1}^{N} \frac{k^2}{N^3} - \left(\sum_{k=1}^{N} \frac{k}{N^2} \right)^2, \tag{3.19}$$

$$S_c(N) = \sum_{k=1}^{N} \frac{k}{N^2} \ln\left(\frac{k}{N}\right) - \sum_{k=1}^{N} \frac{k}{N^2} \ln\left(\sum_{l=1}^{N} \frac{l}{N^2} \right), \tag{3.20}$$

as well as those in the limit $N \to \infty$:

$$\lim_{N\to\infty} \kappa_{1,c}(N) = \kappa_u = \frac{1}{12}, \tag{3.21}$$

$$\lim_{N\to\infty} S_c(N) = S_u = \frac{\ln 2}{2} - \frac{1}{4}. \tag{3.22}$$

reach well-defined finite and positive values. We note that both $\kappa_{1,c}(N)$ and $S_c(N)$ are monotonically increasing with respect to N and hence:

$$\frac{1}{16} = \kappa_{1,c}(2) \leq \kappa_{1,c}(N), \tag{3.23}$$

$$\frac{5\ln 2 - 3\ln 3}{4} = S_c(2) \leq S_c(N). \tag{3.24}$$

Since Σ_{max} should be by definition greater or equal than each of these values for all N, we can replace Σ_{max} in the definition of Lesche stability by either $\frac{1}{16}$ or $\frac{5\ln 2-3\ln 3}{4}$, respectively. Then, these positive numbers can be absorbed in ε and thus we retain the usual definition of uniform metric continuity in a uniform manner (independent of N). This is what we shall prove:

$$\forall \varepsilon > 0, N \in \mathscr{Z}^+ \exists \delta(\varepsilon) : \|p - p'\| < \delta(\varepsilon) \Rightarrow |\Sigma[p] - \Sigma[p']| < \varepsilon. \tag{3.25}$$

Theorem 3.2. *(Stability of κ_1) The variance κ_1 in natural time:*

$$\kappa_1[p] = \sum_{k=1}^{N} p_k \left(\frac{k}{N}\right)^2 - \left(\sum_{k=1}^{N} \frac{k}{N} p_k\right)^2 \tag{3.26}$$

satisfies the condition (3.25) and hence is Lesche stable.

Proof. For every $\varepsilon > 0$, we can consider $\delta(\varepsilon) = \varepsilon/3$ so that if $\|p - p'\| < \delta(\varepsilon)$ we have:

$$\begin{aligned}
|\kappa_1[p] - \kappa_1[p']| &= \left| \sum_{k=1}^{N} \left(\frac{k}{N}\right)^2 (p_k - p'_k) - \left(\sum_{k=1}^{N} \frac{k}{N} p_k\right)^2 + \left(\sum_{k=1}^{N} \frac{k}{N} p'_k\right)^2 \right| = \\
&= \left| \sum_{k=1}^{N} \left(\frac{k}{N}\right)^2 (p_k - p'_k) + \left(\sum_{k=1}^{N} \frac{k}{N} p_k\right) \sum_{k=1}^{N} \frac{k}{N}(p'_k - p_k) \right. \\
&\quad \left. + \left(\sum_{k=1}^{N} \frac{k}{N} p'_k\right) \sum_{k=1}^{N} \frac{k}{N}(p'_k - p_k) \right| \leq \\
&\leq \left| \sum_{k=1}^{N} \left(\frac{k}{N}\right)^2 (p_k - p'_k) \right| + \left| \sum_{k=1}^{N} \frac{k}{N} p_k \right| \left| \sum_{k=1}^{N} \frac{k}{N}(p'_k - p_k) \right| + \\
&\quad + \left| \sum_{k=1}^{N} \frac{k}{N} p'_k \right| \left| \sum_{k=1}^{N} \frac{k}{N}(p'_k - p_k) \right| \leq \\
&\leq \sum_{k=1}^{N} \left(\frac{k}{N}\right)^2 |p_k - p'_k| + \left| \sum_{k=1}^{N} \frac{k}{N}(p'_k - p_k) \right| + \left| \sum_{k=1}^{N} \frac{k}{N}(p'_k - p_k) \right| \leq \\
&\leq \sum_{k=1}^{N} |p_k - p'_k| + \sum_{k=1}^{N} \left|\frac{k}{N}\right| |p'_k - p_k| + \sum_{k=1}^{N} \left|\frac{k}{N}\right| |p'_k - p_k| \leq \\
&\leq 3 \sum_{k=1}^{N} |p_k - p'_k|
\end{aligned} \tag{3.27}$$

but since $\|p - p'\| = \sum_{k=1}^{N} |p_k - p'_k| < \varepsilon/3$, inequality (3.27) implies that

$$|\kappa_1[p] - \kappa_1[p']| < \varepsilon \tag{3.28}$$

This relation shows that the condition (3.25) is obeyed for κ_1. □

Thus, the quantity κ_1 is Lesche stable.

Now, before proceeding to the final proof for the stability of the entropy S, we make use of a well-known theorem [51]:

Theorem 3.3. *(Heine 1870) If a function $F(x)$ of a real variable x is continuous when $a \leq x \leq b$, then $F(x)$ is uniformly continuous throughout the range $a \leq x \leq b$.*

In Lemma 1 we proved that $f(x) = \{x \ln x : \forall x \in (0,1],\ 0 : \text{if } x = 0\}$ is continuous in the closed interval [0,1], and hence it is also uniformly continuous in the same interval. Uniform continuity implies that

$$\forall \frac{\varepsilon}{2} > 0, x, y \in [0,1]\, \exists \delta_1(\varepsilon/2) : |x - y| < \delta_1(\varepsilon/2) \Rightarrow |x \ln x - y \ln y| < \frac{\varepsilon}{2}. \tag{3.29}$$

Now, we can show that S is Lesche stable.

Theorem 3.4. *(Stability of S) The entropy S in natural time:*

$$S[p] = \sum_{k=1}^{N} p_k \frac{k}{N} \ln \frac{k}{N} - \left(\sum_{k=1}^{N} p_k \frac{k}{N} \right) \ln \sum_{k=1}^{N} p_k \frac{k}{N} \tag{3.30}$$

satisfies the condition (3.25) and hence it is Lesche stable.

Proof. For every $\varepsilon > 0$, we can consider $\delta(\varepsilon) = \min\left[\frac{e\varepsilon}{2}, \delta_1(\varepsilon/2)\right]$ so that if $\|p - p'\| < \delta(\varepsilon)$ we have:

$$\begin{aligned} |S[p] - S[p']| &= \left| \sum_{k=1}^{N} (p_k - p'_k) \frac{k}{N} \ln \frac{k}{N} - \left(\sum_{k=1}^{N} \frac{k}{N} p_k \right) \ln \sum_{k=1}^{N} \frac{k}{N} p_k + \right. \\ &\quad \left. + \left(\sum_{k=1}^{N} \frac{k}{N} p'_k \right) \ln \sum_{k=1}^{N} \frac{k}{N} p'_k \right| \leq \\ &\leq \left| \sum_{k=1}^{N} (p_k - p'_k) \frac{k}{N} \ln \frac{k}{N} \right| + |x \ln x - y \ln y|, \end{aligned} \tag{3.31}$$

where $x = \sum_{k=1}^{N} \frac{k}{N} p_k$ and $y = \sum_{k=1}^{N} \frac{k}{N} p'_k$. We now consider that

$$|x - y| = \left| \sum_{k=1}^{N} \frac{k}{N} (p_k - p'_k) \right| \leq \sum_{k=1}^{N} \left| \frac{k}{N} \right| |p_k - p'_k| \leq \sum_{k=1}^{N} |p_k - p'_k| < \delta(\varepsilon) \leq \delta_1(\varepsilon/2) \tag{3.32}$$

and hence (see condition (3.29))

$$|x \ln x - y \ln y| < \frac{\varepsilon}{2}. \tag{3.33}$$

Now, we return to inequality (3.31) to complete the proof:

$$\begin{aligned}
|S[p] - S[p']| &\leq \left| \sum_{k=1}^{N} (p_k - p'_k) \frac{k}{N} \ln \frac{k}{N} \right| + |x \ln x - y \ln y| < \\
&< \left| \sum_{k=1}^{N} (p_k - p'_k) \frac{k}{N} \ln \frac{k}{N} \right| + \frac{\varepsilon}{2} \leq \\
&\leq \sum_{k=1}^{N} |p_k - p'_k| \left| \frac{k}{N} \ln \frac{k}{N} \right| + \frac{\varepsilon}{2} \leq \\
&\leq \sum_{k=1}^{N} |p_k - p'_k| \frac{1}{e} + \frac{\varepsilon}{2},
\end{aligned} \tag{3.34}$$

since we assumed $\|p - p'\| = \sum_{k=1}^{N} |p_k - p'_k| < \delta(\varepsilon) \leq \frac{e\varepsilon}{2}$, the inequality (3.34) becomes:

$$|S[p] - S[p']| < \frac{\varepsilon}{2} + \frac{\varepsilon}{2} = \varepsilon. \tag{3.35}$$

This relation shows that the condition (3.25) is obeyed for S. □

Thus, the entropy S is Lesche stable.

3.3.5 A more general theorem for entropic functionals in natural time

The following general theorem holds

Theorem 3.5. *Let $F(x) : [0,1] \to \mathscr{R}$ which is:*

1. *uniformly continuous in [0,1]*
2. *strictly convex in (0,1]*
3. *twice differentiable in (0,1]*

then the functional:

$$\Sigma[p] = \sum_{k=1}^{N} p_k F\left(\frac{k}{N}\right) - F\left(\sum_{k=1}^{N} p_k \frac{k}{N}\right)$$

is:

1. *positive*
2. *concave*
3. *Lesche stable.*

Proof. The proof of this Theorem is given in Section V of Ref. [49]. □

3.4 Entropy under time reversal

It is believed (e.g., see Ref. [20] and references therein) that in general there is a relation between the irreversibility of *thermodynamic* processes as expressed by the breaking of time-reversal symmetry and the entropy production in such processes. An essential characteristic of these processes is that the time-reversal invariance of the microscopic dynamics is apparently broken [20]. It means that out of equilibrium a particular sequence of macrostates and its time reversal can have a very different plausibility (this, basically, must be the reason for the positivity of entropy production [20]). Since in general we are dealing with out of equilibrium processes, the above motivated us to investigate what happens when calculating the S values upon time reversal of the original time series. We find that in general S is *not* invariant under time reversal [50].

3.4.1 Definition of the entropy in natural time under time reversal

The value of the entropy deduced upon analyzing in natural time the time series obtained upon considering the time reversal $\hat{T}$ of the original time series, i.e., $\hat{T} p_k = p_{N-k+1}$ see Eq. (2.8), is designated by S_-. This differs from the S value which results from the analysis of the original time series. On the other hand κ_1 does not change upon time reversal – since it results [39] from a power spectrum (see Section 2.4) – in a similar fashion as the exponents obtained from DFA (see § 1.4.2) and Hurst [14] analysis (see Section 4.3). Since the value of S_- is in general different [50, 48, 47] from S, the entropy in natural time does satisfy the condition to be "causal" in the following sense (see Ref. [50] and references therein):

When studying a dynamical system evolving in time, a "causal" operator should be able to represent the evolution of the system according to the true time arrow, thus such an operator can reveal the differences arising upon time reversal. This holds for S but *not* for κ_1.

3.4.1.1 A simple example clarifying the physical meaning of the entropies S and S_-

Let us study the influence of a linear trend on the "uniform" distribution by considering the example of the parametric family of pdfs discussed in § 2.2.1, i.e., $p(\chi;\varepsilon) = 1+\varepsilon(\chi-1/2)$, for small $\varepsilon(\ll 1)$, see Eq. (2.9). Such a family of pdfs shares the interesting property of Eq. (2.10), i.e., the action of time reversal is obtained by simply changing the sign of ε. The calculation of the entropy $S(\varepsilon) \equiv S[p(\chi;\varepsilon)]$, as well as that of the entropy under time reversal $S_-(\varepsilon) \equiv S[\hat{T} p(\chi;\varepsilon)] = S(-\varepsilon)$, can be done analytically by using Eq. (3.3). This yields

$$S(\varepsilon) = -\frac{1}{4} + \frac{\varepsilon}{72} - \left(\frac{1}{2} + \frac{\varepsilon}{12}\right) \ln \left(\frac{1}{2} + \frac{\varepsilon}{12}\right). \tag{3.36}$$

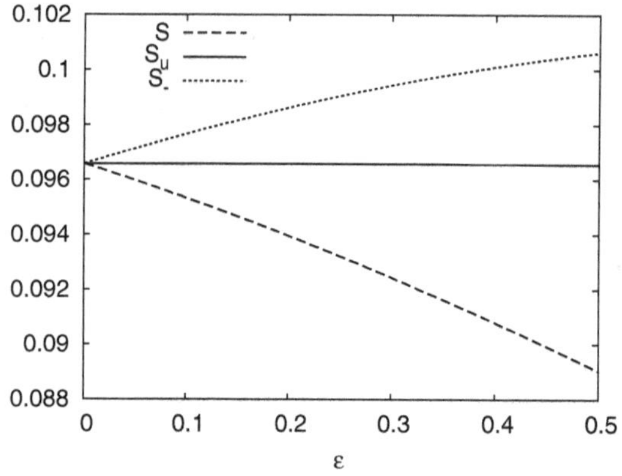

Fig. 3.2 The values of S (dashed) and S_- (dotted) as a function of the linear trend parameter ε. The solid line corresponds to S_u and is drawn for the sake of comparison. Taken from Ref. [48].

Figure 3.2 shows the values of S and S_- as a function of the linear trend parameter ε. We observe that they lie above and below S_u, respectively. In simple words:

A (small) linearly increasing (decreasing) trend superimposed on a "uniform" distribution leads to an entropy S smaller (larger) than S_u, while S_- is larger (smaller) than S_u.

3.4.2 The case when the increments of the time series of Q_k are positive i.i.d. random variables of finite variance

In this case, as discussed in § 2.5.3, $p(\chi) = 2\chi$. By using Eq. (2.97), we find $\kappa_1 = \frac{1}{18} \approx 0.056 \neq \kappa_u$ (see § 2.5.3). Moreover, Eq. (3.3) leads to the value $S = \frac{2}{3}\ln\frac{2}{3} - \frac{2}{9} \approx 0.048$ and, in addition when considering $\hat{T}p(\chi) = 2(1-\chi)$, we get $S_- = \frac{1}{3}\ln 3 - \frac{5}{18} \approx 0.088$. Both S and S_- values significantly differ from $S_u \approx 0.0966$ of the "uniform" distribution, see Eq. (3.4). On the other hand, when Q_k are shuffled randomly, in view of the fact that the increments have a finite variance, the distribution of Q_k for a given N has also finite variance. Thus, the results for the shuffled case correspond [43] to those treated later in § 3.4.6, which lead (see Eqs. (3.56) and (3.48)) to $S_{shuf} \rightarrow S_u$ and $\kappa_{1,shuf} \rightarrow \kappa_u$ as $N \rightarrow \infty$. A numerical example for exponentially distributed increments of Q_k is shown in Fig. 3.3.

3.4.3 Fractional Brownian motion time series

The fBm as mentioned in (§ 1.5.1.1) is H self-similar with stationary increments and constitutes [29] the *only* Gaussian process with such properties for $0 < H < 1$. This can be simulated [23, 25, 30]; see also pp. 321–323 of Ref. [22], by randomizing a construction due to Weierstrass, i.e., using the Weierstrass–Mandelbrot function [10]:

$$w(t) = \sum_{l=1}^{\infty} c_l \frac{\sin(b^l t + d_l)}{b^{lH}}, \tag{3.37}$$

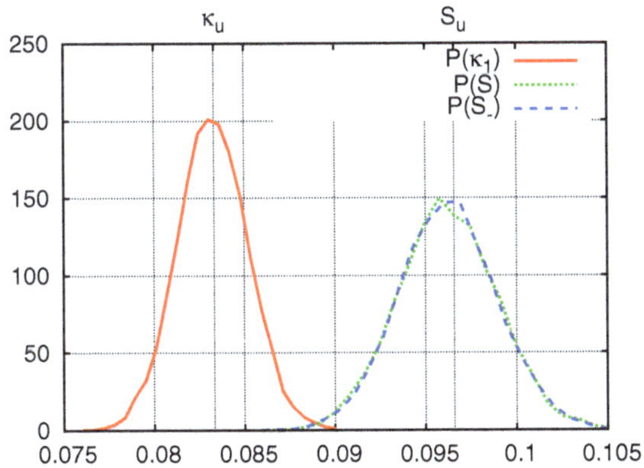

Fig. 3.3 The pdfs of κ_1, S and S_- that have been obtained by shuffling the Q_k randomly in the case of exponentially distributed increments (see also Fig. 2.4). Here, $N = 500$ and the original time series results in $\kappa_1 = 0.055$, $S = 0.048$ and $S_- = 0.088$. Taken from Ref. [47].

where $b > 1$, c_l normally distributed with mean 0 and standard deviation 1, and d_l are uniformly distributed in the interval $[0, 2\pi]$ (note that when using the *increments* of Eq. (3.37) one can also produce fractional Gaussian noise of a given H, see the note after Eq. (1.31)).

By using Eq. (3.37), fBm for various values of H were produced [48], the one-sided segments of which were analyzed in natural time (an example showing how the one-sided segments of a fractional Brownian motion are read in natural time is given in Fig. 3.4). This means that if we denote by w_i, $i = 0, 1, 2, \ldots N+1$, some $N+2$ consecutive fBm values obtained from Eq. (3.37) with $w_0 w_1 < 0$ and $w_N w_{N+1} < 0$ whereas all $w_n, n = 1, 2, \ldots N$, have the same sign – thus constituting an one-sided segment – then the p_k, $k = 1, 2, \ldots N$, used in the calculation are given by $p_k = w_k / \sum_{n=1}^{N} w_n$ and correspond to the "energies" $Q_k = w_k$ mentioned in § 2.1.2. A Monte Carlo calculation was made by analyzing a large number of such segments in natural time. The results obtained for each one-sided segment include the values of the entropies S, S_- (and the value of κ_1 see below) together with the exponent α_{DFA} of the DFA described in § 1.4.2. For segments of a small number of points N (note that only segments with $N > 40$ were considered), the values of α_{DFA} may vary significantly, but they scatter around that expected for a given value of H (see fig. 11 of Ref. [41]); in this sense, the DFA exponent α_{DFA} is consistent with the H-index used to generate fBm by means of Eq. (3.37). The method of DFA was preferred, because it is one of the few well-defined and robust estimators of the scaling properties for such segments, see § 4.6.1.1. The results are shown in Fig. 3.5, in which we plot the S and S_- values versus α_{DFA}. Since, as will be explained later in § 4.4.2, the analysis of the SES activities in natural time leads [41, 40] to DFA exponents α_{DFA} around unity, we are solely focused in Fig. 3.5 on the range $0.8 < \alpha_{DFA} < 1.5$. An inspection of this figure reveals the following conclusions. First, despite the large standard deviation, we may say that these computations do *not* result in a definite sign for $S - S_-$ and that *both* S and S_- are smaller than S_u ($\approx$0.0966) when $\alpha_{DFA} \approx 1$. Second, S and S_- are more or less comparable. Third, the computed S and S_- values are ≈ 0.08 for $\alpha_{DFA} \approx 1$. Interestingly, when plotting the most probable value $\kappa_{1,p}$ of κ_1 versus α_{DFA} we find that $\kappa_{1,p} \approx 0.070$ when $\alpha_{DFA} = 1$, see Fig. 3.6.

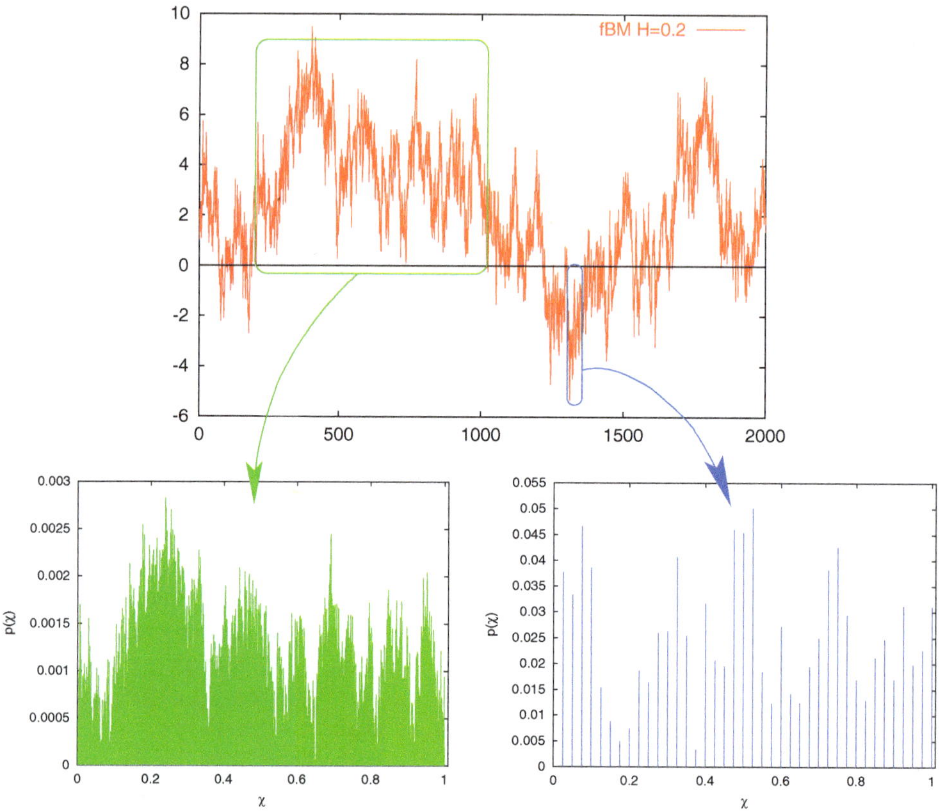

Fig. 3.4 An example showing how the one-sided segments of a fractional Brownian motion (upper panel) are defined and then read in natural time (lower panel).

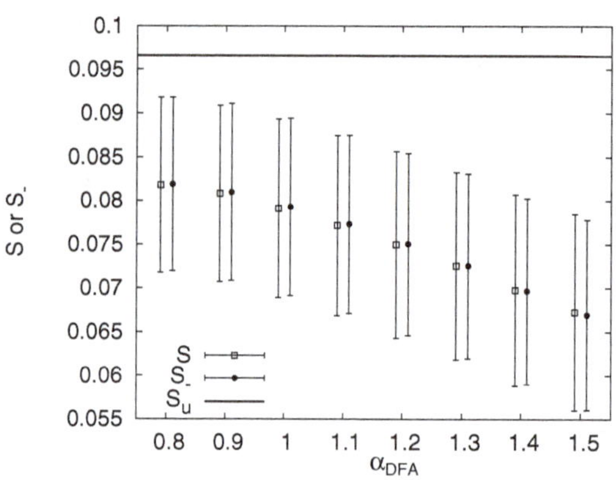

Fig. 3.5 Calculated values of S (open squares) and S_{-} (filled circles) versus the DFA exponent α_{DFA}. The error bars indicate the region of one standard deviation ($\pm\sigma$). The horizontal line corresponds to S_u. Taken from Ref. [48].

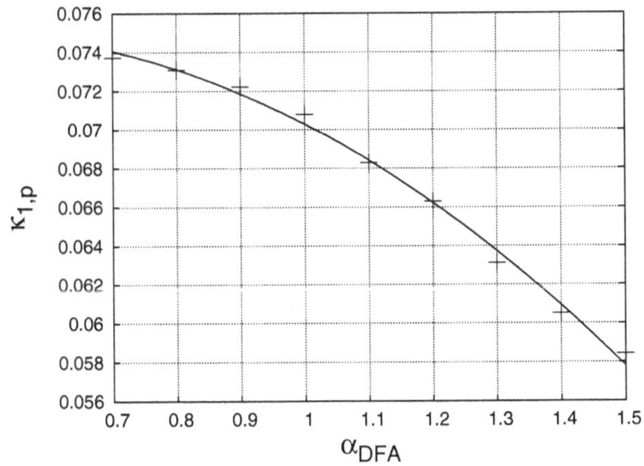

Fig. 3.6 The most probable value $\kappa_{1,p}$ (plus) of κ_1 as a function of α_{DFA} for the case of the fBm simulations of § 3.4.3. The solid curve has been drawn as a guide to the eye. Taken from Ref. [47]

Thus, the results deduced from a numerical simulation in fBm time series show that when $\alpha_{DFA} \approx 1$ the corresponding values are $\kappa_1 \approx 0.070$ and $S \approx S_- \approx 0.080$.

3.4.4 An on–off intermittency model

We clarify that on–off intermittency is a phase-space mechanism that allows dynamical systems to undergo bursting (bursting is a phenomenon in which episodes of high activity are alternated with periods of inactivity). This mechanism is different from the well-known Pomeau–Manneville scenario for the behavior of a system in the proximity of a saddle-node bifurcation [27]. Here, we use the simple model of the driven logistic map

$$X_{t+1} = A(Y_t)X_t(1 - X_t) \tag{3.38}$$

where we assume that the quantity $A(Y_t)$ is *monotonic* function of Y_t and that $0 \leq A \leq 4$ (A is further specified below). The system has the invariant manifold $X = 0$ and the level of its activity is measured by X_t; see Ref. [32]. In order to have the on–off mechanism in action, we specialize to the case of a noise-driven logistic map, with

$$A(Y_t) = A_0 + \alpha Y_t \tag{3.39}$$

where Y_t is δ-correlated noise which is uniformly distributed in the interval [0,1] and A_0 and α are parameters. In order to have $0 \leq A \leq 4$, we assume [32] $A_0 \geq 0, \alpha \geq 0$ and $A_0 + \alpha \leq 4$. The relevant parameter plane for the noise-driven system of Eqs. (3.38) and (3.39) (as well as the parameter range for which the fixed point $X = 0$ is stable) can be found in fig. 1 of Ref. [32], while the description of the intermittent dynamics is given in Refs. [26, 13, 4]. Bursting is observed in the temporal evolution of X_t as the stability of the fixed point $X = 0$ varies. Following Ref. [13], for $A_0 = 0$ there is a *critical* value

$\alpha_c > 1$, below which the system asymptotically tends to the fixed point $X = 0$, without any sustained intermittent bursting. For this case, i.e., $A_0 = 0$, the value $\alpha_c = e \equiv 2.71828\ldots$ leads to on–off intermittency [32]. In the intermittent system under discussion, both the signal amplitude and the power spectrum resulted [32] in power law distributions (with low frequencies predominating in the power spectrum).

Several time series have been produced for the above on–off intermittency model with the following procedure [48]: The system was initiated at a time ($t_{in} = -200$) with a uniformly distributed value $X_{t_{in}}$ in the region $[0,1]$, and then the mapping of Eqs. (3.38) and (3.39) was followed until N events will occur after $t = 0$. The results for X_t, $t = 1,2\ldots N$, were analyzed in natural time domain (i.e., $p_k = X_k / \sum_{t=1}^{N} X_t$, where X_k here corresponds to the "energy" Q_k mentioned in § 2.1.2) and the values of S and S_- have been determined. This was repeated 10^3 times for a given number N of events and the average values of S and S_- have been deduced. These values are plotted in Fig. 3.7(a) versus $(\alpha - e)N^{1/2}$ (the factor $N^{1/2}$ stems from finite size scaling effects, since for large values of N, e.g., $N > 15{,}000$, a scaling – reminiscent of a first-order phase transition – was observed).

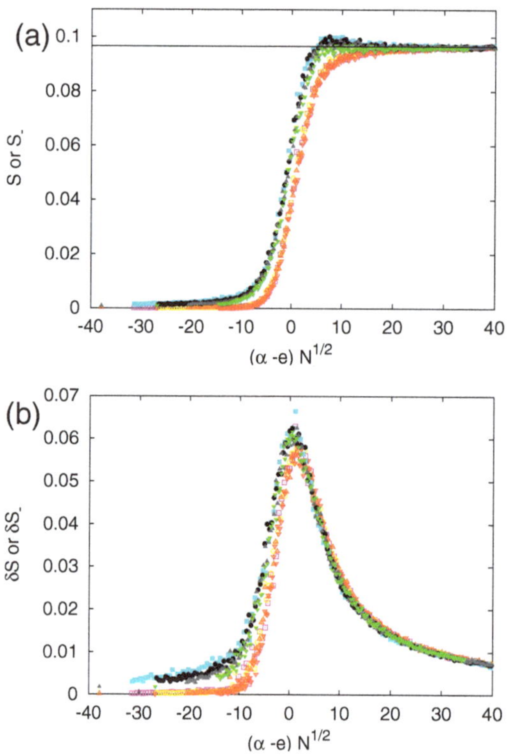

Fig. 3.7 Calculated results for the on–off intermittency model discussed in § 3.4.4: The average values of (a) S (closed symbols) and S_- (open symbols) and (b) the fluctuations δS and δS_- versus the finite size scaling variable $(\alpha - \alpha_c)N^{1/2}$. The quantity N stands for the number of the events considered in each sample time series; $N = 70000, 50000, 30000, 15000$ correspond to squares, circles, triangles and inverted triangles, respectively. The horizontal line in (a) corresponds to S_u. Taken from Ref. [48].

Figure 3.7 reveals that as the *critical* value for on–off intermittency is approached from below, i.e., $\alpha \to e_-$, the entropy S is *different* from S_- (in contrast to fBm, see § 3.4.3) and *both* S and S_- are smaller than S_u.

3.4.5 The case of signals that exhibit short-range temporal correlations

We now present results of modeling Q_k by short-ranged temporal correlated time series. Two examples are treated here by numerical simulation [46]. (i) A stationary autoregressive process $Q_k = aQ_{k-1} + g_k + c$, $|a| < 1$, where g_k are Gaussian i.i.d. random variables, and c stands for an appropriate constant to ensure positivity of Q_k. (ii) $Q_k = |aQ_{k-1} + g_k|$. Figure 3.8(a) depicts the results for S, S_- and κ_1 for the first example versus the number N of Q_k, whereas Fig. 3.8(b) refers to the second example.

In both cases S and S_- converge to S_u while κ_1 to the value $\kappa_u = 1/12$ corresponding to the "uniform" distribution.

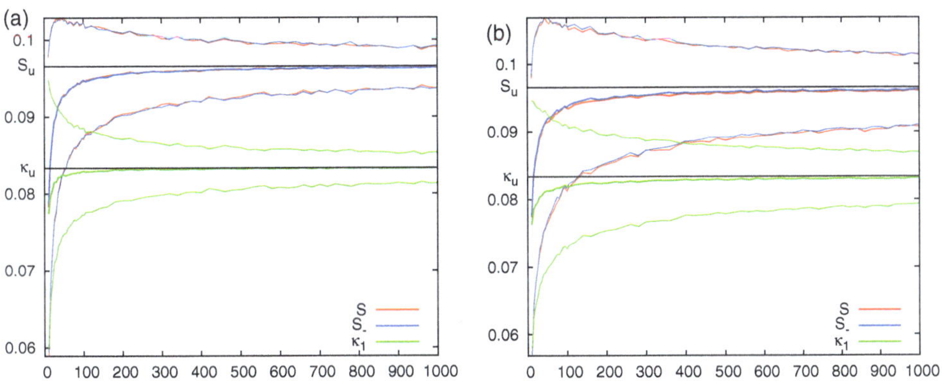

Fig. 3.8 The thick colored lines in (a) and (b) depict the average value of S (red), S_- (blue) and κ_1 (green) versus the number N of Q_k for the two examples (i) and (ii), respectively, mentioned in § 3.4.5. The thinner colored lines refer to the $\pm\sigma$ deviation from the average value. For the reader's convenience, the values of S_u and κ_u are designated by the horizontal solid black lines. Reprinted with permission from Ref. [46]. Copyright (2008), American Institute of Physics.

3.4.6 Interrelation between δS and σ/μ in the case of p.i.i.d.

This subsection is focused on the "uniform" distribution in the natural time domain (see § 2.1.3). We consider here the case when Q_k are independent and identically distributed positive random variables with *finite* second moment, i.e., $\mathrm{Var}(Q_k) = \sigma^2 (< \infty)$. This case naturally arises when an experimental time series, which does not exhibit heavy tails (§ 2.5.2), is randomly *shuffled* [42, 43]. Since Q_k are i.i.d., for $\mathscr{E}(p_k) = \mathscr{E}[Q_k/\sum_{l=1}^N Q_l]$ we have: $\mathscr{E}[\sum_{k=1}^N Q_k/\sum_{l=1}^N Q_l] = 1 = N\mathscr{E}(p_k)$, thus $\mathscr{E}(p_k) = \frac{1}{N}$ (see Eq. (2.89)). Using the constraint $\sum_{k=1}^N p_k = 1$ (see Eq. (2.3)), leading to $p_k - 1/N = \sum_{l\neq k}(1/N - p_l)$, and the fact that Q_k are i.i.d., we obtain

$$\begin{aligned} \mathrm{Var}[p_k] &= \mathscr{E}\left[\left(p_k - \frac{1}{N}\right)^2\right] = \mathscr{E}\left[\left(p_k - \frac{1}{N}\right)\sum_{l\neq k}\left(\frac{1}{N} - p_l\right)\right] \\ &= -(N-1)\mathscr{E}\left[\left(p_k - \frac{1}{N}\right)\left(p_l - \frac{1}{N}\right)\right]. \end{aligned} \tag{3.40}$$

Thus, we get

$$\mathrm{Cov}(p_k, p_l) = \mathscr{E}\left[\left(p_k - \frac{1}{N}\right)\left(p_l - \frac{1}{N}\right)\right] = -\frac{\mathrm{Var}(p_k)}{N-1} \tag{3.41}$$

which is analogous to Eq. (2.90). It then follows, when considering the central limit theorem [9], that the N-dependence of $\mathrm{Var}(p_k)$ is (see Eq. (2.94):

$$\mathrm{Var}(p_k) = \frac{\sigma^2}{N^2\mu^2}. \tag{3.42}$$

We now turn to the statistical properties of $\langle\chi^q\rangle$. Using $\mathscr{E}(p_k) = \frac{1}{N}$, we have

$$\mathscr{E}[\langle\chi^q\rangle] = \sum_{k=1}^N \left(\frac{k}{N}\right)^q \frac{1}{N}. \tag{3.43}$$

which, since [11] $\sum_{k=1}^N k^q = N^{q+1}/(q+1) + N^q/2 + o(N^q)$, reveals that $\mathscr{E}[\langle\chi^q\rangle]$ is again asymptotically N-independent because it approaches the value $1/(q+1)$ with a "small" $1/(2N)$ correction. The variance $\mathrm{Var}[\langle\chi^q\rangle][= (\delta\langle\chi^q\rangle)^2]$,

$$\mathrm{Var}[\langle\chi^q\rangle] = \mathscr{E}\left\{\left[\sum_{k=1}^N \left(\frac{k}{N}\right)^q \left(p_k - \frac{1}{N}\right)\right]^2\right\}, \tag{3.44}$$

after expanding the square and using Eqs. (3.41) and (3.42), becomes:

$$\begin{aligned} \mathrm{Var}[\,\langle\chi^q\rangle\,] &= \sum_{k=1}^N \left(\frac{k}{N}\right)^{2q} \frac{\sigma^2}{N^2\mu^2} \\ &\quad - \frac{\sigma^2}{(N-1)N^2\mu^2} \sum_{k=1}^N \left(\frac{k}{N}\right)^q \sum_{l=1,l\neq k}^N \left(\frac{l}{N}\right)^q, \end{aligned} \tag{3.45}$$

which, using Eq. (3.43), finally leads to:

$$\mathrm{Var}[\langle\chi^q\rangle] = \frac{\sigma^2}{(N-1)\mu^2}\left\{\mathscr{E}[\langle\chi^{2q}\rangle] - \mathscr{E}^2[\langle\chi^q\rangle]\right\}. \tag{3.46}$$

The expectation value of κ_1 when Q_k are p.i.i.d. can be found on the basis of Eqs. (3.43) and (3.46):

$$\begin{aligned}
\mathscr{E}[\kappa_1] &= \mathscr{E}[\langle\chi^2\rangle - \langle\chi\rangle^2] \\
&= \mathscr{E}[\langle\chi^2\rangle] - \mathscr{E}^2[\langle\chi\rangle] - \mathrm{Var}[\langle\chi\rangle] \\
&= \{\mathscr{E}[\langle\chi^2\rangle] - \mathscr{E}^2[\langle\chi\rangle]\}\left\{1 - \frac{\sigma^2}{(N-1)\mu^2}\right\} \\
&= \left\{\sum_{k=1}^{N}\frac{k^2}{N^3} - \left(\sum_{k=1}^{N}\frac{k}{N^2}\right)^2\right\}\left\{1 - \frac{\sigma^2}{(N-1)\mu^2}\right\}
\end{aligned} \tag{3.47}$$

which by considering Eq. (2.91) leads to

$$\mathscr{E}[\kappa_1] = \kappa_u\left(1 - \frac{1}{N^2}\right)\left[1 - \frac{\sigma^2}{(N-1)\mu^2}\right]. \tag{3.48}$$

Note that, as expected, Eq. (2.92) leads to the above equation when $\mathrm{Var}(Q_k) = \sigma^2(<\infty)$.

The proof of Eq. (3.46) can be generalized for all linear functionals of p_k of the form $\langle f(\chi)\rangle = \sum_{k=1}^{N} f(k/N)p_k$ and yields:

$$\mathrm{Var}[\langle f(\chi)\rangle] = \frac{\sigma^2\{\mathscr{E}[\langle f^2(\chi)\rangle] - \mathscr{E}^2[\langle f(\chi)\rangle]\}}{(N-1)\mu^2}. \tag{3.49}$$

In Fig. 3.9, we compare the theoretical result of Eq. (3.46) with synthetic (Gaussian) data which have values of μ, σ and size ($\approx$1000) similar to those in 15 min ECG (see Chapter 9). Note that when one uses the estimator $(\delta X)^2 = \sum(X-\overline{X})^2/N$, instead of the unbiased estimator $(\delta X)^2 = \sum(X-\overline{X})^2/(N-1)$, in order to find the sample variance, N should replace $N-1$ in Eq. (3.46).

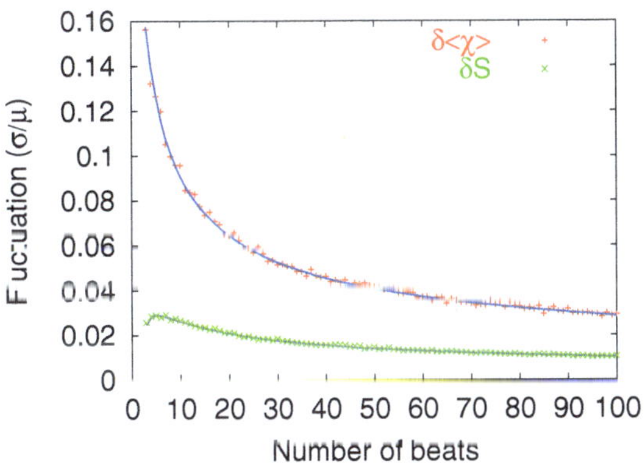

Fig. 3.9 Comparison of the theoretical estimations (solid lines) of $\delta\langle\chi\rangle$ and δS resulting from Eqs. (3.46) and (3.63), respectively, with the values obtained (plus and cross, respectively) using a Gaussian sample having values of μ, σ and size ($\approx$1000) similar to those in the case of 15 min electrocardiograms (see Chapter 9). Here the estimator $(\delta X)^2 = \sum(X-\overline{X})^2/N$ was used for the calculation of the sample variance of the synthetic data, and thus $N-1$ was replaced by N in Eqs. (3.46) and (3.63). Taken from Ref. [43].

We now proceed to the statistical properties of the entropy $S = \langle \chi \ln \chi \rangle - \langle \chi \rangle \ln \langle \chi \rangle$; see Eq. (3.1). To simplify the calculation of the expectation value $\mathscr{E}[S]$ and the variance $\mathrm{Var}(S)$ of the entropy, we define the two linear functionals

$$m[x_k] = \sum_{k=1}^{N} \frac{k}{N} x_k, \tag{3.50}$$

$$L[x_k, \xi] = \sum_{k=1}^{N} \frac{k}{N} \ln\left(\frac{k}{\xi N}\right) x_k, \tag{3.51}$$

and the constant time series $\mathscr{K} = \{x_k\} : x_k = 1/N, k = 1,2,\ldots N$ (cf. §§ 2.5.2 and 3.3.4). Note that for both functionals $m[x_k]$ and $L[x_k,\xi]$, in view of their linearity, we have

$$\mathscr{E}\left\{m\left[p_k - \frac{1}{N}\right]\right\} = \mathscr{E}\left\{L\left[p_k - \frac{1}{N}, \xi\right]\right\} = 0. \tag{3.52}$$

Moreover, we define $\overline{\chi} = m[\mathscr{K}] = \sum_{k=1}^{N} k/N^2 = (1+1/N)/2$.

The expectation value

$$\mathscr{E}(S) = \mathscr{E}\left[\sum_{k=1}^{N} \frac{k}{N} \ln\left(\frac{k}{N}\right) p_k - \sum_{k=1}^{N} \frac{k}{N} p_k \ln\left(\sum_{l=1}^{N} \frac{l}{N} p_l\right)\right] \tag{3.53}$$

of the entropy S given by Eq. (3.2) can be evaluated as follows: we add and subtract the term $\sum_{k=1}^{N} \frac{k}{N} p_k \ln \overline{\chi}$, and then expand the resulting term $\ln\left\{1 + m[p_k - \frac{1}{N}]/\overline{\chi}\right\}$ to get

$$\begin{aligned}\mathscr{E}(S) = &\sum_{k=1}^{N} \frac{k}{N^2} \ln\left(\frac{k}{N}\right) - \overline{\chi} \ln \overline{\chi} - (1 - \frac{1}{2})\mathscr{E}\left\{\frac{m^2[p_k - \frac{1}{N}]}{\overline{\chi}}\right\} \\ &+ \left(\frac{1}{2} - \frac{1}{3}\right)\mathscr{E}\left\{\frac{m^3[p_k - \frac{1}{N}]}{\overline{\chi}^2}\right\} - \left(\frac{1}{3} - \frac{1}{4}\right)\mathscr{E}\left\{\frac{m^4[p_k - \frac{1}{N}]}{\overline{\chi}^3}\right\} + \ldots\end{aligned} \tag{3.54}$$

Assuming that the contribution from the moments of $m[p_k - \frac{1}{N}]$ higher than the second are negligible, we finally obtain (see Eq. (3.46) for $q = 1$).

$$\mathscr{E}(S) = \sum_{k=1}^{N} \frac{k}{N^2} \ln\left(\frac{k}{N}\right) - \overline{\chi} \ln \overline{\chi} - \frac{\sigma^2}{2(N-1)\mu^2}\left(\frac{\overline{\chi^2}}{\overline{\chi}} - \overline{\chi}\right), \tag{3.55}$$

where $\overline{\chi^2} = \sum_{k=1}^{N} k^2/N^3 = (1+1/N)[1+1/(2N)]/3$.

Equation (3.55) reveals that $\mathscr{E}(S)$ depends slightly on σ/μ; upon increasing N the last term of Eq. (3.55) decays as $1/N$ (note that for the terms in parentheses $\lim_{N\to\infty} \overline{\chi} = \frac{1}{2}$ and $\lim_{N\to\infty} \overline{\chi^2} = \frac{1}{3}$). Equation (3.55), after considering Eqs. (2.91) and (3.20), leads to

$$\mathscr{E}(S) = \sum_{k=1}^{N} \frac{k}{N^2} \ln\left(\frac{k}{N\overline{\chi}}\right) - \kappa_u \frac{\sigma^2}{\mu^2} \frac{1}{N} = S_c(N) - \kappa_u \frac{\sigma^2}{\mu^2} \frac{1}{N}. \tag{3.56}$$

We note that Eq. (3.56) when combined with Eq. (3.22) shows that, as $N \to \infty$, $\mathscr{E}(S) \to S_u$. Using Eqs. (3.50) and (3.51) the entropy can be written, in compact form, as follows:

$$S = L[p_k, m[p_k]] = L[p_k, 1] - m[p_k] \ln m[p_k], \tag{3.57}$$

and its expectation value is written as

$$\mathscr{E}(S) = L[\mathscr{K}, 1] - m[\mathscr{K}] \ln m[\mathscr{K}] - \kappa_u \frac{\sigma^2}{\mu^2} \frac{1}{N}. \tag{3.58}$$

The variance of the entropy $\mathrm{Var}(S) = (\delta S)^2$ can then be found by adding and subtracting the term $m[p_k] \ln m[\mathscr{K}]$ and expanding the logarithms in the expression

$$\begin{aligned} m[p_k] \ln \frac{m[p_k]}{m[\mathscr{K}]} &= m[p_k - \frac{1}{N}] \ln \left\{ 1 + \frac{m[p_k - \frac{1}{N}]}{m[\mathscr{K}]} \right\} + m[\mathscr{K}] \ \ln \left\{ 1 + \frac{m[p_k - \frac{1}{N}]}{m[\mathscr{K}]} \right\} = \\ &= m \left[p_k - \frac{1}{N} \right] + \left(1 - \frac{1}{2} \right) \frac{m^2[p_k - \frac{1}{N}]}{m[\mathscr{K}]} - \left(\frac{1}{2} - \frac{1}{3} \right) \frac{m^3[p_k - \frac{1}{N}]}{m^2[\mathscr{K}]} + \dots \end{aligned} \tag{3.59}$$

This procedure gives:

$$\begin{aligned} \mathrm{Var}(S) &= \mathscr{E} \left\{ \left(L[p_k, 1] - m[p_k] \ln m[p_k] - L[\mathscr{K}, 1] + m[\mathscr{K}] \ln m[\mathscr{K}] + \kappa_u \frac{\sigma^2}{\mu^2} \frac{1}{N} \right)^2 \right\}, \\ &= \mathscr{E} \left\{ \left(L[p_k - \frac{1}{N}, 1] - m[p_k] \ln \frac{m[p_k]}{m[\mathscr{K}]} + m[\frac{1}{N} - p_k] \ln m[\mathscr{K}] + \kappa_u \frac{\sigma^2}{\mu^2} \frac{1}{N} \right)^2 \right\}, \\ &= \mathscr{E} \left\{ \left(L \left[p_k - \frac{1}{N}, m[\mathscr{K}] \right] + \kappa_u \frac{\sigma^2}{\mu^2} \frac{1}{N} - m[p_k - \frac{1}{N}] \ln e - \frac{m^2[p_k - \frac{1}{N}]}{2m[\mathscr{K}]} + \dots \right)^2 \right\}, \\ &= \mathscr{E} \left\{ \left(L \left[p_k - \frac{1}{N}, m[\mathscr{K}] e \right] + \kappa_u \frac{\sigma^2}{\mu^2} \frac{1}{N} - \frac{m^2[p_k - \frac{1}{N}]}{2m[\mathscr{K}]} + \dots \right)^2 \right\}. \end{aligned} \tag{3.60}$$

Expanding the square in Eq. (3.60), using Eq. (3.52) and keeping terms of order $1/N$, we find

$$\mathrm{Var}(S) = \mathscr{E} \left(L^2 \left[p_k - \frac{1}{N}, m[\mathscr{K}] e \right] \right), \tag{3.61}$$

which can be explicitly written as follows

$$\mathrm{Var}(S) = \mathscr{E} \left\{ \left[\sum_{k=1}^{N} \frac{k}{N} \ln \left(\frac{k}{m[\mathscr{K}] e N} \right) \left(p_k - \frac{1}{N} \right) \right]^2 \right\}. \tag{3.62}$$

The right side of Eq. (3.62) becomes similar to Eq. (3.44), if we replace χ^q by $\chi \ln(\frac{\chi}{\overline{\chi} e})$ (cf. $\overline{\chi} \equiv m[\mathscr{K}]$); thus after expanding the square and using Eqs. (3.41) and (3.42), we finally obtain

$$\mathrm{Var}(S) = \frac{\sigma^2}{(N-1)\mu^2}\left[\sum_{k=1}^{N}\left(\frac{k}{N}\ln\frac{k}{e\overline{\chi}N}\right)^2\frac{1}{N} - \left(\sum_{k=1}^{N}\frac{k}{N^2}\ln\frac{k}{e\overline{\chi}N}\right)^2\right]. \tag{3.63}$$

A comparison of Eqs. (3.61) and (3.57) reveals the following: in order to find the entropy fluctuation δS, one simply has to replace in Eq. (3.57) $m[p_k]$ with $m[\mathscr{X}]e(=\overline{\chi}e)$ and then directly take its variance according to Eq. (3.49).

3.5 The change ΔS of the entropy in natural time under time reversal

As mentioned above in § 3.4.1, in general, S_- is different from S, and hence S shows the breaking of the time reversal symmetry.

Definition 3.2. The difference $S-S_-$ will be hereafter labeled ΔS, i.e.,

$$\Delta S \equiv S - S_- \tag{3.64}$$

this may also have a subscript (ΔS_l) meaning that the calculation is made (for each S and S_-) at a scale l (= number of successive events). In such a procedure, a window of length l is sliding each time by one event through the whole time series. The entropies S and S_-, and therefrom their difference ΔS_l, are calculated within a window of length l starting from each event (see § 3.5.1 below). Thus, we form a new time series consisting of successive ΔS_l values.

3.5.1 Evaluation of ΔS_l when a (natural) time window of length l is sliding through a time series

The value of S (and S_-) calculated for a number of successive events varies within the time series, i.e., when using a time-window of length l sliding each time by one event through the whole time series (see § 2.2.2). Thus, for a window of length l when starting from the m_0-th event, we have

$$S(m_0,l) = \langle \chi \ln \chi \rangle_w - \langle \chi \rangle_w \ln \langle \chi \rangle_w \tag{3.65}$$

where

$$\langle \chi \ln \chi \rangle_w = \sum_{k=1}^{l} p_{k,w}\chi_{k,w}\ln \chi_{k,w}, \tag{3.66}$$

$$\langle \chi \rangle_w = \sum_{k=1}^{l} p_{k,w}\chi_{k,w} \tag{3.67}$$

with

$$p_{k,w} = \frac{Q_{m_0-1+k}}{\sum_{n=1}^{l} Q_{m_0-1+n}} \tag{3.68}$$

and $\chi_{k,w} = k/l$. Similarly, $S_-(m_0,l)$ is calculated by Eq. (3.65) when $p_{k,w}$ of Eq. (3.68) is substituted by

$$\hat{T} p_{k,w} = \frac{Q_{m_0+l-k}}{\sum_{n=1}^{l} Q_{m_0+l-n}}. \tag{3.69}$$

The time series of ΔS_l is obtained by the differences $\Delta S_l(m_0) \equiv S(m_0,l) - S_-(m_0,l)$, $m_0 = 1,2,\ldots,N-l$ and its variation is quantified by its standard deviation

$$\sigma[\Delta S_l] \equiv \sqrt{\mathrm{Var}[\Delta S_l]} = \sqrt{\mathscr{E}\left\{(\Delta S_l - \mathscr{E}[\Delta S_l])^2\right\}}. \tag{3.70}$$

In Eq. (3.70), the symbol $\mathscr{E}\{\ldots\}$ stands for the average obtained when all the $N-l$ values (cf. $m_0 = 1,2,\ldots,N-l$) of its argument are considered.

3.5.2 Interrelation of $\sigma[\Delta S_l]$ and σ/μ in the case of p.i.i.d.

In the previous subsection, § 3.5.1, the relevant expressions for evaluating S and S_-, when a window of length $l(=N)$ is sliding event by event through a time series were given. Let us first study $\mathscr{E}\left[S - \hat{T}S\right] = \mathscr{E}(S) - \mathscr{E}(\hat{T}S)$ for which we intuitively expect that it equals zero when Q_k are positive i.i.d. random variables. Indeed, we have (see Eq. (3.53)) that

$$\mathscr{E}(S) = \mathscr{E}\left[\sum_{k=1}^{N}\frac{k}{N}\ln\left(\frac{k}{N}\right)p_k - \sum_{k=1}^{N}\frac{k}{N}p_k\ln\left(\sum_{l=1}^{N}\frac{l}{N}p_l\right)\right] \tag{3.71}$$

and

$$\mathscr{E}(\hat{T}S) = \mathscr{E}\left[\sum_{k=1}^{N}\frac{k}{N}\ln\left(\frac{k}{N}\right)p_{N-k+1} - \sum_{k=1}^{N}\frac{k}{N}p_{N-k+1}\ln\left(\sum_{l=1}^{N}\frac{l}{N}p_{N-l+1}\right)\right]. \tag{3.72}$$

Since time reversal (see Eq. (2.8)) just "projects" p_k to p_{N-k+1}, Eq. (3.72) is just a reparametrization of Eq. (3.71) in the dummy variables where k is now $N-k+1$. Both Q_k and Q_{N-k+1} are p.i.i.d., thus the expectation values in both Eqs. (3.71) and (3.72) are equal to the result of Eq. (3.56), thus $\mathscr{E}\left[S - \hat{T}S\right] = 0$.

We now turn to the variance $\sigma^2[\Delta S] \equiv \mathscr{E}\left\{\left[S - \hat{T}S - \mathscr{E}(S - \hat{T}S)\right]^2\right\}$. For the sake of simplicity of the notation, from now on and until the end of this subsection, we use both symbols $\langle\ldots\rangle$ and $\mathscr{E}[\ldots]$ to denote the expectation value $\mathscr{E}[\ldots]$. Having this in mind and using the fact that $\delta S^2 \equiv \langle[S - \mathscr{E}(S)]^2\rangle = \mathrm{Var}[S]$ (see Eq. (3.63)) remains unchanged under time reversal, for the same reasons as $\mathscr{E}(S) = \mathscr{E}(\hat{T}S)$, we have that

$$\begin{aligned}\sigma^2[\Delta S] &= \langle\{[S - \mathscr{E}(S)] - [\hat{T}S - \mathscr{E}(\hat{T}S)]\}^2\rangle \\ &= \langle[S - \mathscr{E}(S)]^2\rangle + \langle[\hat{T}S - \mathscr{E}(\hat{T}S)]^2\rangle - 2\langle[S - \mathscr{E}(S)]\,[\hat{T}S - \mathscr{E}(\hat{T}S)]\rangle \\ &= 2\left\{\delta S^2 - \langle[S - \mathscr{E}(S)]\,[\hat{T}S - \mathscr{E}(\hat{T}S)]\rangle\right\},\end{aligned} \tag{3.73}$$

The term $\langle [S-\mathscr{E}(S)]\left[\hat{T}S-\mathscr{E}(\hat{T}S)\right]\rangle$ can be evaluated in a way similar to the one used in § 3.4.6. Namely, we add and subtract the term $\sum_{k=1}^{N}\frac{k}{N}p_k\ln\overline{\chi}$ from S and the term $\sum_{k=1}^{N}\frac{k}{N}p_{N-k+1}\ln\overline{\chi}$ from $\hat{T}S$. Then, we expand the resulting logarithmic terms $\ln[1+\sum_{l=1}^{N}\frac{l}{N}(p_l-\frac{1}{N})/\overline{\chi}]$ and $\ln[1+\sum_{l=1}^{N}\frac{l}{N}(p_{N-l+1}-\frac{1}{N})/\overline{\chi}]$ in terms of $(p_l-\frac{1}{N})$ and $(p_{N-l+1}-\frac{1}{N})$, respectively (see Eq. (3.59)). This leads to (see Eqs. (3.50), (3.51) in conjunction with the terms in parentheses in Eq. (3.60)):

$$\begin{aligned}\left\langle [S-\mathscr{E}(S)]\left[\hat{T}S-\mathscr{E}(\hat{T}S)\right]\right\rangle = \Bigg\langle \Bigg[&\sum_{k=1}^{N}\frac{k}{N}\ln\left(\frac{k}{e\overline{\chi}N}\right)\left(p_k-\frac{1}{N}\right)+\kappa_u\frac{\sigma^2}{\mu^2}\frac{1}{N} \\ &-\frac{1}{2\overline{\chi}}\sum_{k=1}^{N}\frac{k}{N}\left(p_k-\frac{1}{N}\right)\sum_{l=1}^{N}\frac{l}{N}\left(p_l-\frac{1}{N}\right)+\ldots\Bigg]\times \\ \times\Bigg[&\sum_{k'=1}^{N}\frac{k'}{N}\ln\left(\frac{k'}{e\overline{\chi}N}\right)\left(p_{N-k'+1}-\frac{1}{N}\right)+\kappa_u\frac{\sigma^2}{\mu^2}\frac{1}{N} \\ &-\frac{1}{2\overline{\chi}}\sum_{k'=1}^{N}\frac{k'}{N}\left(p_{N-k'+1}-\frac{1}{N}\right)\sum_{l'=1}^{N}\frac{l'}{N}\left(p_{N-l'+1}-\frac{1}{N}\right)+\ldots\Bigg]\Bigg\rangle . \end{aligned} \tag{3.74}$$

Restricting ourselves up to terms of order $1/N$, Eq. (3.74) simplifies to

$$\begin{aligned}\langle [S-\mathscr{E}(S)]\left[\hat{T}S-\mathscr{E}(\hat{T}S)\right]\rangle = \Bigg\langle &\left[\sum_{k=1}^{N}\frac{k}{N}\ln\left(\frac{k}{e\overline{\chi}N}\right)\left(p_k-\frac{1}{N}\right)\right]\times \\ \times &\left[\sum_{k'=1}^{N}\frac{k'}{N}\ln\left(\frac{k'}{e\overline{\chi}N}\right)\left(p_{N-k'+1}-\frac{1}{N}\right)\right]\Bigg\rangle . \end{aligned} \tag{3.75}$$

Now, using Eqs. (3.41) and (3.42), we have

$$\left\langle\left(p_k-\frac{1}{N}\right)\left(p_{N-k'+1}-\frac{1}{N}\right)\right\rangle = \frac{\sigma^2}{(N-1)N\mu^2}\delta_{k,N-k'+1}-\frac{\sigma^2}{(N-1)N^2\mu^2}, \tag{3.76}$$

where $\delta_{l,m}$ is Kronecker's delta (equal to 1 if $l=m$ and 0 otherwise). Substituting Eq. (3.76) into Eq. (3.75), we find that

$$\begin{aligned}\langle [S-\mathscr{E}(S)]\left[\hat{T}S-\mathscr{E}(\hat{T}S)\right]\rangle = \frac{\sigma^2}{(N-1)\mu^2}\Bigg\{ &\sum_{k=1}^{N}\frac{k}{N}\ln\left(\frac{k}{e\overline{\chi}N}\right)\times \\ \times &\frac{N-k+1}{N}\ln\left(\frac{N-k+1}{e\overline{\chi}N}\right)\frac{1}{N}- \\ - &\left[\sum_{k=1}^{N}\frac{k}{N^2}\ln\left(\frac{k}{e\overline{\chi}N}\right)\right]^2\Bigg\} \end{aligned} \tag{3.77}$$

Substituting Eq. (3.63) for δS^2 and Eq. (3.77) into Eq. (3.73), we finally obtain that

$$\sigma^2[\Delta S] = \frac{2\sigma^2}{(N-1)\mu^2}\left[\sum_{k=1}^{N}\left(\frac{k}{N}\ln\frac{k}{e\chi N}\right)^2\frac{1}{N} - \right.$$
$$\left. -\sum_{k=1}^{N}\frac{k}{N}\ln\left(\frac{k}{e\chi N}\right)\frac{N-k+1}{N}\ln\left(\frac{N-k+1}{e\chi N}\right)\frac{1}{N}\right]. \quad (3.78)$$

In summary, for p.i.i.d., we observe that $\sigma[\Delta S]$ is proportional to σ/μ, which is a behavior similar to the one, i.e., $\delta S \propto \sigma/\mu$, found in § 3.4.6 .

3.5.3 A simple example in which the meaning of the entropy change ΔS under time reversal seems to emerge clearly

Here, we consider, as in § 2.2.1 and § 3.4.1, the parametric family $p(\chi;\varepsilon) = 1+\varepsilon(\chi-1/2)$ for small $\varepsilon(<1)$. In this case, as mentioned, the calculation of the entropy $S(\varepsilon)$ can be done analytically and the result yields Eq. (3.36). An interrelation between $\Delta S(\varepsilon) = S(\varepsilon) - S(-\varepsilon)$ and the small linear trend parameter ε can be obtained by expanding Eq. (3.36) around $\varepsilon = 0$ which leads to

$$\Delta S(\varepsilon) = \left(\frac{6\ln 2 - 5}{36}\right)\varepsilon + O(\varepsilon^3) \quad (3.79)$$

Since $\ln 2 < 5/6$, Eq. (3.79) implies the following [45]:

From the physical point of view, a positive ε, i.e., increasing trend, corresponds to negative ΔS and vice versa.

Thus, in the ECG data for example the variation of ΔS_l may be thought as capturing the net result, at scale l, of the competing mechanisms that decrease or increase heart rate; see Section 9.4. In particular, the ΔS_l time series, at proper scales l, can serve for specifying the occurrence time of the impending sudden cardiac death, see § 9.4.1. Furthermore, ΔS_l can be used as a tool (see § 8.3.4) to investigate the predictability of a dynamical model like the case of the Olami–Feder–Christensen model for earthquakes. Additionally, the concept of ΔS_l has been used in Environmental Sciences namely in the study of the dynamical evolution of the ozone hole area over Antarctica [35].

3.6 Complexity measures using the entropy in natural time

Complexity measures have been introduced [42, 43, 44, 45] to quantify the variability of the natural entropy fluctuations upon changing either the (natural time window) length scale or shuffling the consecutive events randomly. These are classified into two categories depending on whether they make use of either the entropy S in natural time or the change ΔS of the entropy in natural time under time reversal.

3.6.1 Complexity measures that make use of the fluctuations of the entropy S in natural time

As a measure of the natural entropy fluctuations we consider the standard deviation δS (defined in § 3.4.6) when we calculate the value of S for a number of consecutive pulses (events) and study how S varies when sweeping this time window (each time by one pulse) through the whole time series. The following complexity measures for the δS variability have been suggested [42, 43, 44]:

When the natural time window length changes from a short value, e.g., 5 pulses (events), to a shorter one, e.g., 3 pulses (events), the corresponding δS value also changes. This variation in the short (s) range is quantified by the measure

$$\lambda_s = \frac{\delta S_5}{\delta S_3}, \tag{3.80}$$

where the subscript in δS denotes the time window length chosen. If a longer (L) range, e.g., 60 pulses (events), changes to a short one, e.g., 3 pulses (events), the corresponding variation is quantified by another measure

$$\lambda_L = \frac{\delta S_{60}}{\delta S_3}. \tag{3.81}$$

Thus, the values of λ_s and λ_L quantify the δS variability with the natural time window length *scale*.

Considering the surrogate data obtained by randomly *shuffling* the durations Q_k of consecutive pulses (events), we can define the measure ν as

$$\nu \equiv \frac{\delta S_{shuf}}{\delta S}, \tag{3.82}$$

where δS and δS_{shuf} denote the value of δS calculated when a time window is sweeping through the original and the *shuffled* time series, respectively. The following ratios have been investigated [43]: ν_s and ν_L for the following natural time window lengths: $s = 3$–4 pulses (events) and $L = 50$–70 pulses (events), respectively.

The quantity ν captures the extent to which the *sequential* order of pulses (events) regulates the δS value.

An alternative (but not equivalent) measure is $\lambda_{\kappa,shuf}$ (where $\kappa = s, L$ refer to the short and long-range, respectively), which is defined in a similar fashion as the measure λ mentioned above, but is calculated after shuffling the Q_k randomly.

The δS values themselves in conjuction with the aforementioned complexity measures λ_κ, $\lambda_{\kappa,shuf}$ and ν_κ (where $\kappa = s, L$), which are in fact *ratios* of the δS values, have been found of prominent importance in the analysis of electrocardiograms (ECG) and allow the distinction between healthy (H) humans and (otherwise healthy) sudden cardiac death (SD) ones; see Chapter 9.

3.6.2 Complexity measures that make use of the change ΔS of the entropy in natural time under time reversal

Complexity measures Λ and N can be defined [45], in a similar fashion with the measures λ and ν defined above in § 3.6.1, by using the quantity ΔS instead of S. For example, we can define the measure

$$N_l = \frac{\sigma[\Delta S_l^{shuf}]}{\sigma[\Delta S_l]} \tag{3.83}$$

The measure N_l quantifies the extend to which the *ordering* of the events contributes to the ΔS_l values being equal to unity for a random process.

This finds application, for example, in the identification of SD risk; see Section 9.4.

References

1. For example, see http://planetmath.org/encyclopedia/JensensInequality.html.
2. Abe, S.: Stability of Tsallis entropy and instabilities of Rényi and normalized Tsallis entropies: A basis for q-exponential distributions. Phys. Rev. E **66**, 046134 (2002)
3. Abe, S., Kaniadakis, G., Scarfone, A.M.: Stabilities of generalized entropies. J. Phys. A: Math. Gen. **37**, 10,513–10,519 (2004)
4. Balmforth, N.J., Provenzale, A., Spiegel, E.A., Martens, M., Tresser, C., Wu, C.W.: Red spectra from white and blue noise. Proc. R. Soc. London, Ser. B **266**, 311–314 (1999)
5. Beck, C., Schlögl, F.: Thermodynamics of chaotic systems, an introduction. Cambridge University Press, Cambridge, UK (1997)
6. Beck, C., Schlögl, F.: Thermodynamics of Chaotic Systems: An Introduction. Cambridge University Press, Cambridge, UK (1993)
7. Dorfman, G.R.: Introduction to Chaos in Nonequilibrium Statistical Mechanics. Cambridge University Press, Cambridge, England (1999)
8. Eckmann, J.P., Ruelle, D.: Ergodic theory of chaos and strange attractors. Rev. Mod. Phys. **57**, 617–656 (1985)

9. Feller, W.: An Introduction to Probability Theory and Its Applications, Vol. II. Wiley, New York (1971)
10. Frame, M., Mandelbrot, B., Neger, N.: Fractal Geometry, Yale University, available from http://classes.yale.edu/fractals/, see http://classes.yale.edu/Fractals/RandFrac/fBm/fBm4.html
11. Gradsteyn, I.S., Ryzhik, I.M.: Table of Integrals, Series and Products. Academic Press, San Diego (1980)
12. Grassberger, P., Procaccia, I.: Estimation of the Kolmogorov entropy from a chaotic signal. Phys. Rev. A **28**, 2591–2593 (1983)
13. Heagy, J.F., Platt, N., Hammel, S.M.: Characterization of on–off intermittency. Phys. Rev. E **49**, 1140–1150 (1994)
14. Hurst, H.E.: Long-term storage capacity of reservoirs. Trans. Am. Soc. Civ. Eng. **116**, 770–808 (1951)
15. Jizba, P., Arimatsu, T.: Observability of Rényi's entropy. Phys. Rev. E **69**, 026128 (2004)
16. Kaniadakis, G., Lissia, M., Scarfone, A.M.: Two-parameter deformations of logarithm, exponential, and entropy: A consistent framework for generalized statistical mechanics. Phys. Rev. E **71**, 046128 (2005)
17. Kaniadakis, G., Scarfone, A.M.: Lesche stability of κ-entropy. Physica A **340**, 102–109 (2004)
18. Lesche, B.: Instabilities of Rényi entropies. J. Stat. Phys. **27**, 419–422 (1982)
19. Lesche, B.: Rényi entropies and observables. Phys. Rev. E **70**, 017102 (2004)
20. Maes, C., Netocny, K.: Time-reversal and entropy. J. Stat. Phys. **110**, 269–310 (2003)
21. Majewski, E.: in Thermodynamics of chaos and fractals applied: evolution of the Earth and phase transformations, in Earthquake Thermodynamics and Phase Transformations in the Earth's Interior, R. Teisseyre and E. Majewski (eds). Academic Press, San Diego (2001)
22. Mandelbrot, B.B.: Gaussian Self-Affinity and Fractals. Springer-Verlag, New York (2002)
23. Mandelbrot, B.B., Wallis, J.R.: Some long-run properties of geophysical records. Water Resources Research **5**, 321–340 (1969)
24. Naudts, J.: Continuity of a class of entropies and relative entropies. Rev. Math. Phys. **16**, 809–822 (2004)
25. Pipiras, V., Taqqu, M.S.: Convergence of the Weierstrass-Mandelbrot process to fractional Brownian motion. Fractals **8**, 369–384 (2000)
26. Platt, N., Spiegel, E.A., Tresser, C.: On-off intermittency: A mechanism for bursting. Phys. Rev. Lett. **70**, 279–282 (1993)
27. Pomeau, Y., Manneville, P.: Intermittent transition to turbulence in dissipative dynamical-systems. Commun. Math. Phys. **74**, 189–197 (1980)
28. Rényi, A.: Probability Theory. North-Holland, Amsterdam (1970)
29. Samorodnitsky, G., Taqqu, M.S.: Stable Non-Gaussian Random Processes: Stochastic Models with Infinite Variance. Chapman & Hall/CRC, Florida (1994)
30. Szulga, J., Molz, F.: The Weierstrass Mandelbrot process revisited. J. Stat. Phys. **104**, 1317–1348 (2001)
31. Tirnakli, U., Abe, S.: Aging in coherent noise models and natural time. Phys. Rev. E **70**, 056120 (2004)
32. Toniolo, C., Provenzale, A., Spiegel, E.A.: Signature of on–off intermittency in measured signals. Phys. Rev. E **66**, 066209 (2002)
33. Tsallis, C.: What should a statistical mechanics satisfy to reflect nature? Physica D **193**, 3–34 (2004)
34. Tsallis, C., Brigatti, E.: Nonextensive statistical mechanics: A brief introduction. Continuum Mech. Thermodyn. **16**, 223–235 (2004)
35. Varotsos, C.A., Tzanis, C.: On the dynamic evolution of the ozone hole area over Antarctica. under review **1**, 123,456 (2010)
36. Varotsos, P.: The Physics of Seismic Electric Signals. TERRAPUB, Tokyo (2005)
37. Varotsos, P., Alexopoulos, K.: Thermodynamics of Point Defects and their Relation with Bulk Properties. North Holland, Amsterdam (1986)
38. Varotsos, P.A., Sarlis, N.V., Skordas, E.S.: Spatio-temporal complexity aspects on the interrelation between Seismic Electric Signals and seismicity. Practica of Athens Academy **76**, 294–321 (2001)
39. Varotsos, P.A., Sarlis, N.V., Skordas, E.S.: Long-range correlations in the electric signals that precede rupture. Phys. Rev. E **66**, 011902 (2002)

40. Varotsos, P.A., Sarlis, N.V., Skordas, E.S.: Attempt to distinguish electric signals of a dichotomous nature. Phys. Rev. E **68**, 031106 (2003)
41. Varotsos, P.A., Sarlis, N.V., Skordas, E.S.: Long-range correlations in the electric signals that precede rupture: Further investigations. Phys. Rev. E **67**, 021109 (2003)
42. Varotsos, P.A., Sarlis, N.V., Skordas, E.S., Lazaridou, M.S.: The use of the entropy in the natural time-domain to distinguish electric signals. Practica of Athens Academy **78**, 281–298 (2003)
43. Varotsos, P.A., Sarlis, N.V., Skordas, E.S., Lazaridou, M.S.: Entropy in natural time domain. Phys. Rev. E **70**, 011106 (2004)
44. Varotsos, P.A., Sarlis, N.V., Skordas, E.S., Lazaridou, M.S.: Natural entropy fluctuations discriminate similar-looking electric signals emitted from systems of different dynamics. Phys. Rev. E **71**, 011110 (2005)
45. Varotsos, P.A., Sarlis, N.V., Skordas, E.S., Lazaridou, M.S.: Identifying sudden cardiac death risk and specifying its occurrence time by analyzing electrocardiograms in natural time. Appl. Phys. Lett. **91**, 064106 (2007)
46. Varotsos, P.A., Sarlis, N.V., Skordas, E.S., Lazaridou, M.S.: Fluctuations, under time reversal, of the natural time and the entropy distinguish similar looking electric signals of different dynamics. J. Appl. Phys. **103**, 014906 (2008)
47. Varotsos, P.A., Sarlis, N.V., Skordas, E.S., Tanaka, H.K., Lazaridou, M.S.: Attempt to distinguish long-range temporal correlations from the statistics of the increments by natural time analysis. Phys. Rev. E **74**, 021123 (2006)
48. Varotsos, P.A., Sarlis, N.V., Skordas, E.S., Tanaka, H.K., Lazaridou, M.S.: Entropy of seismic electric signals: Analysis in the natural time under time reversal. Phys. Rev. E **73**, 031114 (2006)
49. Varotsos, P.A., Sarlis, N.V., Tanaka, H.K., Skordas, E.S.: See (the freely available) EPAPS Document No. E PLEEE8 71 081503 originally from P.A. Varotsos, N.V. Sarlis, H.K. Tanaka and E.S. Skordas, *Phys. Rev. E* **71**, 032102 (2005). For more information on EPAPS, see http://www.aip.org/pubservs/epaps.html.
50. Varotsos, P.A., Sarlis, N.V., Tanaka, H.K., Skordas, E.S.: Some properties of the entropy in the natural time. Phys. Rev. E **71**, 032102 (2005)
51. Whittaker, E.T., Watson, G.N.: A Course of Modern Analysis. Cambridge University Press, UK (1958)

Part III
Natural Time Applications

4. Natural Time Analysis of Seismic Electric Signals

Abstract. The natural time analysis of all the measured SES activities showed that they are characterized by very strong memory and their normalized power spectra $\Pi(\omega)$ versus ω fall on a *universal* curve having $\kappa_1 (= \langle \chi^2 \rangle - \langle \chi \rangle^2)$ value equal to 0.070. This curve coincides with the one obtained on theoretical grounds when assuming that SES are governed by *critical* dynamics. Upon shuffling the events (pulses) randomly, the memory is destroyed and the κ_1 value becomes equal to that $\kappa_u (= 1/12 \approx 0.083)$ of a "uniform" distribution. This shows that the self-similarity *solely* stems from long range temporal correlations. Concerning the distinction of SES activities from similar looking "artificial" (man-made) noises, we find the following. Modern techniques of Statistical Physics, e.g., detrended fluctuation analysis (DFA), multifractal DFA, wavelet transform, when applied to the original time series *cannot* achieve such a distinction, but when they are applied in natural time a *clear* distinction emerges. For example, for the SES activities the DFA exponent in natural time is close to unity, i.e., $\alpha \approx 1$, while for "artificial" noises it is markedly smaller, i.e., $\alpha < 0.85$. Also the entropy S in natural time can achieve such a distinction: For SES activities *both* S and S_- (where S_- stands for the entropy in natural time under time reversal) are smaller than the entropy $S_u \approx 0.0966$ of the "uniform" distribution, which is *not* the case for the "artificial" noises where S is larger than (or equal to) S_u and S_- may either be smaller or larger than S_u. Upon "shuffling" the events (pulses) randomly, both values of S and S_- in the SES activities turn out to be equal to S_u, which conforms with the aforementioned conclusion that in SES activities the self-similarity originates solely from long range temporal correlations. Finally, when investigating the dependence of the fluctuations $\Delta\chi_l$ of the average value of natural time under time reversal versus the window length l, we can also achieve a distinction between SES activities and "artificial" noises. In particular, when studying the log-log plot of $\Delta\chi_l$ versus l, the former give *ascending* curves, in contrast to the latter that result in *descending* curves.

4.1 Dichotomous time series. Markovian and non-Markovian processes

4.1.1 Difference between natural time analysis and earlier studies of dichotomous time series. The Markovian process

The following point should be stressed concerning one of the key differences of the natural time analysis compared to the earlier procedures in the study of *dichotomous* time series. For such time series, the quantity Q_k (see § 2.1.2, Fig. 2.1(a)) coincides with the so-called dwell time (for the high-level state *only*) and is one of the basic characteristics of a dichotomous (i.e., on–off) process. The standard procedure consists of the determination of the dwell times distribution $P(Q)$: for a Markovian process $P(Q)$ is *exponential*, i.e., $P(Q) = \mathrm{e}^{-Q/\overline{Q}}/\overline{Q}$ (frequently the average dwell time $\overline{Q}$ is different for the high- and the low-level states). For non-Markovian (which contain some "memory") processes $P(Q)$ is non-exponential, e.g., stretched exponential, i.e., of the form $\mathrm{e}^{-(Q/\tau)^b}$ where $0 < b < 1$, or even algebraic. On the other hand, the natural time analysis is carried out in terms of the *couple* (χ_k, Q_k), which takes into account the ordering of the pulses, and hence not *solely* based on the statistics of their durations, i.e., $P(Q)$.

We just mention here that ionic current fluctuations in membrane channels (ICFMC), the long-range correlations of which have been studied in Ref. [21], can be also approximated by dichotomous time series. Further, we clarify that (see Ref. [9] and references therein) single ionic channels in a membrane open and close spontaneously in a stochastic way, resulting in current and voltage changes which resemble the realizations of random telegraph signals (RTS, dichotomous noise). The channel's opening state can be determined [21] on the basis of the ion current: a low current corresponds to a closed channel state, while high current values indicate an open state (see Fig. 4.1). It has been shown [8] that the action of membrane-embedded enzymes depends critically on fluctuations of the membrane potential, and that the main source of these fluctuations originates in the fluctuations of ionic concentrations due to the action of ion channels. Recall that the SES activities have also an RTS feature, e.g., see Figs. 2.8 and 4.2. These figures also depict a number of "artificial" noises (see § 1.2.3) that have been intentionally selected to exhibit a RTS feature similar to that of SES activities. Note that N1–N5 and N9 of Fig. 4.2 correspond to n1–n5 and n6 of Fig. 2.8, respectively.

Hence, apart from a difference in the time-scales, the feature of *all* these electric signals is *similar* to that of the SES activities (RTS shape). This similarity instigated the simultaneous study of SES activities, "artificial" noises and ICFMC by Varotsos et al. [32, 34, 33], as will be explained below.

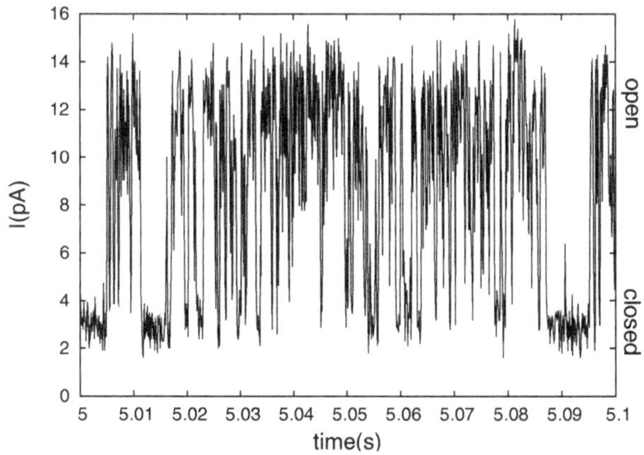

Fig. 4.1 Excerpt of the ionic current fluctuation in membrane channels (ICFMC) (f_{exp} =10 kHz) studied in Ref. [21] (see also Refs. [10, 9]).

4.1.2 Non-Markovian character of SES activities and "artificial" noises

Varotsos et al. [32] showed, by means of the Smoluchowski–Chapman–Kolmogorov functional equation (SCK equation), that the SES activities exhibit non-Markovian character (i.e., contain some "memory", see § 4.1.1). The stationarity of the signal was studied by the quantiles procedure. Subsequently, Varotsos et al. [33], in order to further investigate the non-Markovianity for both SES activities and "artificial" noises, proceeded to the study of the non-Markovian quantitative global measure G. Furthermore, they studied the coefficients of *skewness* and *kurtosis*.

The non-Markovian quantitative global measure G. Following Siwy and Fuliński [23], the definition of G can be summarized as follows: one of the properties of a Markov process is that it satisfies the SCK equation (e.g., see Ref. [32]). The deviation from this equation, i.e.,

$$D_{m,n}(t,\tau) = P(m,t|n,0) - \sum_{k=1}^{M} P(m,t|k,t-\tau)P(k,t-\tau|n,0), \tag{4.1}$$

measures the degree of non-Markovianity. In Eq. (4.1), the indices $k,m,n = 1,2,\ldots M$ number the electric field states (note that in our case we have $M = 2$ different states, labeled "high", $m = 1$, and "low", $m = 2$, respectively; we consider as *"high"-level states* those having the *largest* deflections of the electric field amplitude with respect to the background level; see the arrows in Fig. 4.2). The $P(m,t|n,s)$ stands for the field–field conditional probability that the electric field $E(t)$ is in the state number m, under the condition that at the earlier time $s < t$ the field $E(s)$ was in the state number n.

The integral measure (mean square characteristics) of the non-Markovianity is [9, 23]

$$G = G(\tau,T) = \left[\frac{1}{T}\frac{1}{M^2}\sum_{m,n}^{M}\int_{\tau}^{\tau+T} D^2_{m,n}(t,\tau)\,\mathrm{d}t\right]^{1/2} \tag{4.2}$$

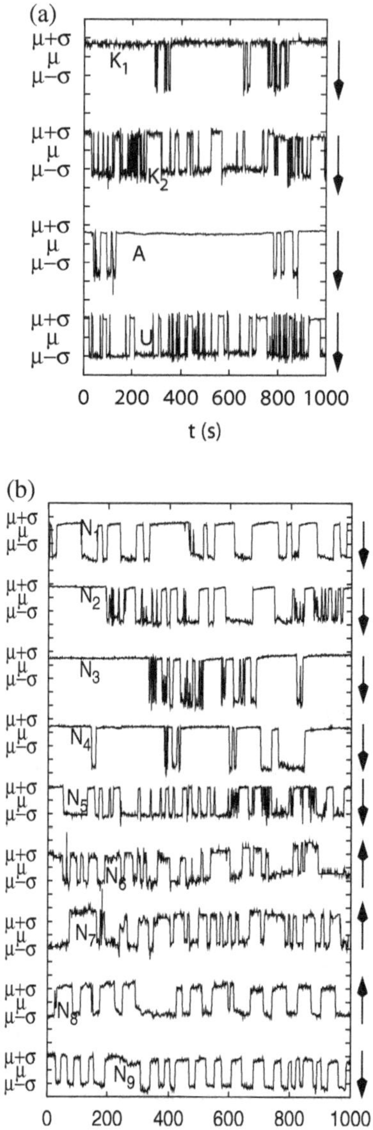

Fig. 4.2 Excerpts of: (a) four SES activities recorded on April 18, 1995 (K1), April 19, 1995 (K2), March 17, 2001 (A) and February 5, 2002 (U); (b) nine "artificial" noises recorded on November 14, 1997 (N1), November 15, 1997 (N2), November 16, 1997 (N3, N4 and N5), July 13, 2001 (N6), August 4, 2001 (N7), March 22, 2001 (N8 and N9). The SES activity U was recorded at IOA (see Ref. [34]), while for the SES activities K1, K2 and A see the caption of Fig. 4.5. The "artificial" noises were distinguished from SES activities according to the criteria discussed in Section 1.2, and collected at various stations (see the map of Fig. 1.2), i.e., N1 to N5 at VOL, N6 and N7 at IOA, N8 and N9 at LAG (this is a station lying very close to ASS). The electric field E is presented here in *normalized* units (μ and σ stand for the mean value and the standard deviation in each case, respectively). The arrows on the right indicate the polarity of the deflection from the background level to the largest (in amplitude) electric field variations (i.e., from the "low"-level to the "high"-level states). Taken from Ref. [34]. Note that N1–N5 and N9 correspond to n1–n5 and n6 of Fig. 2.8, respectively.

where T is the range of the time t and τ is the shift in the SCK equation. As an example, for the SES activity K1 and the "artificial" noise N1 mentioned in Fig. 4.2, the calculation for $T = 100$ s yields $G_{max}(= \sup_\tau G(\tau, T)) = 0.107 \pm 0.002$ and 0.135 ± 0.004, respectively. For computer-generated Markovian dichotomous series of comparable length, the corresponding G values are smaller by one order of magnitude, which also suggests the non-Markovian character of the experimental data for both cases, i.e., K1 and N1, respectively (this non-Markovianity has been also shown by employing the entropy fluctuations

δS in natural time and the relevant complexity measures; see Table 4.5 and the last paragraph of § 9.1.2).

Skewness and kurtosis. The coefficients of skewness (γ_1) and kurtosis (β_2) are (see p. 928 of Ref. [1]):

$$\gamma_1 = \mu_3/\sigma^3 \text{ and } \beta_2 = \mu_4/\sigma^4 \tag{4.3}$$

where μ_n denotes the nth central moment, i.e., $\mu_n = \sum_s (x_s - \mu)^n p_s$ for randomly distributed data x_s with point probabilities p_s. The symbol μ stands for the mean and σ for the standard deviation. For Markovian processes, the durations of the "high"- (T_h) and "low"-level states (T_l) should follow exponential distributions $p(\mathrm{T}) = \lambda \exp(-\lambda \mathrm{T})$ (see § 4.1.1), for which the values $\gamma_1 = 2$, $\beta_2 = 9$ and $\sigma^2/\mu^2 = 1$ are expected. The two coefficients γ_1 and β_2 are tabulated along with σ^2/μ^2 in table II of Ref. [28], for both series of the "high"- and "low"-level states' durations of the "artificial" noises and the SES activities depicted in Fig. 2.8. Comparing these values with those expected from an exponential distribution, we find [33] the following: None of the time series of durations, corresponding to either the SES activities or the "artificial" noises investigated, could be compatible with an exponential distribution. Moreover, the Kolmogorov-Smirnov test excludes for the SES activities the Gaussian distribution.

In short, both the SES activities and the "artificial" noises exhibit non-Markovian character.

4.1.3 Markovian dichotomous time series. Spectral analysis and detrended fluctuation analysis (DFA)

This was studied in Ref. [33]. Following Berezhkovskii and Weiss [6], in the case of a Markovian dichotomous ($M = 2$, $m = 1,2$; these are the symbols used in § 4.1.2) time series, the probability densities for the time spent in a single sojourn in the states "high" ($m = 1$) and "low" ($m = 2$) respectively are *both* exponential, i.e.,

$$p_1(T) \propto \exp(-T/\tau_{high}), \quad p_2(T) \propto \exp(-T/\tau_{low}) \tag{4.4}$$

and lead to the following expressions for the field–field conditional probabilities

$$P(1, t+\tau|1, t) = \tau_{eff}\left[\frac{1}{\tau_{low}} + \frac{\exp(-\tau/\tau_{eff})}{\tau_{high}}\right] \tag{4.5}$$

and

$$P(2, t+\tau|1, t) = \frac{\tau_{eff}}{\tau_{high}}\left[1 - \exp(-\tau/\tau_{eff})\right], \tag{4.6}$$

where $1/\tau_{eff} \equiv 1/\tau_{high} + 1/\tau_{low}$ and τ is a time lag. Note that the expressions of Eqs. (4.5) and (4.6) for the conditional probabilities satisfy the SCK functional equation (see § 4.1.2). The probability to observe the "high" state P_1 is

$$P_1 = \frac{\tau_{high}}{\tau_{low} + \tau_{high}}, \tag{4.7}$$

and the joint probability $P_{11}(\tau)$ to observe the "high" state at both the times t and $t+\tau$, due to the definition of the conditional probability, is

$$P_{11}(\tau) = P_1 P(1, t+\tau|1, t). \tag{4.8}$$

The power spectral density $S(\omega)$ is the Fourier transform of the autocovariance $\mathscr{C}(\tau) = \overline{[x(t+\tau) - \overline{x}][x(t) - \overline{x}]}$ of the stationary signal $x(t)$ [25] with average value $\overline{x}$:

$$\mathscr{C}(\tau) = \overline{x(t+\tau)x(t)} - \overline{x}^2 = \frac{1}{2\pi}\int_0^\infty S(\omega)\, \cos(\omega\tau)\, \mathrm{d}\omega. \tag{4.9}$$

If we assume that the states "low" and "high" have amplitudes 0 and ΔE, respectively, we have $\overline{x} = (\Delta E)P_1$, and $\overline{x(t+\tau)x(t)} = (\Delta E)^2 P_{11}(\tau)$, and using the expressions of Eqs. (4.5) and (4.7)–(4.9), we finally obtain

$$\mathscr{C}(\tau) = (\Delta E)^2 \frac{\tau_{eff}}{\tau_{low} + \tau_{high}} \exp\left(-\frac{\tau}{\tau_{eff}}\right) \tag{4.10}$$

Equation (4.10), using the Wiener–Khinchin theorem, leads to the power spectral density $S(\omega)$

$$S(\omega) = 4\int_0^\infty \mathscr{C}(\tau)\, \cos(\omega\tau)\, \mathrm{d}\tau = \frac{4(\Delta E)^2 \tau_{eff}^2}{(\tau_{low} + \tau_{high})(1 + \omega^2 \tau_{eff}^2)} \tag{4.11}$$

The last relation reveals that the high-frequency behavior of the spectrum becomes $S(\omega) \propto \omega^{-2}$ if $\omega \gg (1/\tau_{high} + 1/\tau_{low})$, which corresponds to *a random walk-like behavior in short time-scales*. We will come back to this point below.

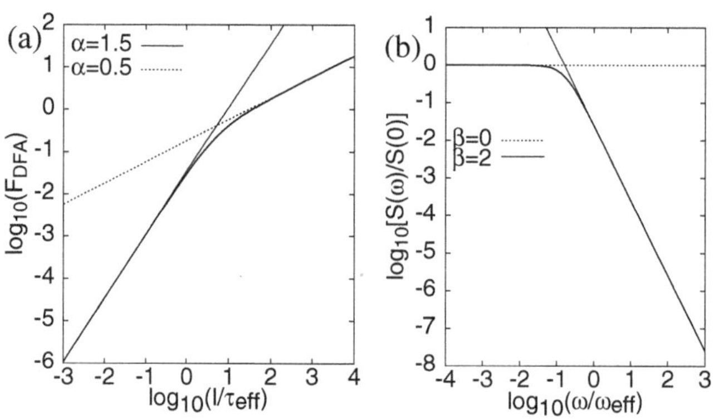

Fig. 4.3 Theoretical estimation of (a) the variability measure F_{DFA}(thick line) versus l/τ_{eff} and (b) the power spectral density $S(\omega)$ (thick line) versus ω/ω_{eff}, for a Markovian dichotomous signal (see § 4.1.3). The thin solid and dotted straight lines correspond to the short and long time ranges in each case, i.e., they are approached for $l \ll \tau_{eff}$ and $l \gg \tau_{eff}$, respectively. Taken from Ref. [33].

Following Talkner and Weber [25], the squared variability of DFA (§ 1.4.2) is given, in terms of $S(\omega)$, by:

$$F_{DFA}^2(l) = \frac{l}{2\pi}\int_0^\infty S(w/l)r_{DFA}(w)\,\mathrm{d}w \tag{4.12}$$

where w denotes the *dimensionless* frequency and $r_{DFA}(w)$ is given by the explicit form:

$$r_{DFA}(w) = [w^4 - 8w^2 - 24 - 4w^2\cos(w) + 24\cos(w) + 24w\sin(w)]/w^6. \tag{4.13}$$

In Fig. 4.3(a), the $F_{DFA}(l)$ versus l/τ_{eff} for a dichotomous Markovian process was drawn using Eqs. (4.11)–(4.13), while Fig. 4.3(b) depicts $S(\omega)$ versus ω/ω_{eff} where $\omega_{eff} \equiv 2\pi/\tau_{eff}$, using Eq. (4.11). This figure shows that [33]:

- Concerning the DFA exponent α: (i) For short time-scales (high frequencies), i.e., $\Delta t \ll \tau_{eff}$, the DFA exponent approaches the value $\alpha = 1.5$. (Note that such a behavior is expected for *any* signal with a *high* frequency spectrum as given in Eq. (4.11); see also below.) (ii) For long time-scales (low frequencies), i.e., $\Delta t \gg \tau_{eff}$, we find $\alpha = 0.5$, as expected. (iii) For intermediate scales, comparable to (or shorter than) τ_{eff}, DFA exponents exceeding unity (e.g., 1.2 or so) naturally emerge.
- Concerning the power spectrum exponent β (see § 1.4.2): it approaches the values 2 and 0 for the aforementioned short and long time-scales, respectively. For time-scales comparable to (or shorter than) τ_{eff}, values of β around unity or larger (e.g. $\beta = 1.4$) can fit the data. (In other words, data consisting, for example, of randomly distributed square pulses, if analyzed in the range $\Delta t \ll \tau_{eff}$, may approximately obey $S \propto \omega^{-\beta}$, $\beta \approx 1$.) Note that, for a given (high) frequency range, upon increasing $1/\tau_{eff}$ the calculated value of β becomes larger.

We now define for non-Markovian time series the quantity $\overline{T}$ in an analogous way with the quantity τ_{eff} introduced above for the Markovian ones, i.e., $1/\overline{T} \equiv 1/\overline{T_h}+1/\overline{T_l}$, where $\overline{T_h}$ and $\overline{T_l}$ denote the average dwell time in the "high" and the "low" state, respectively. The values of $\overline{T}$ for all SES activities and "artificial" noises mentioned in Fig. 2.8 (which are non-Markovian, e.g., see § 4.1.2) can be found in table I of Ref. [28] and vary in the range from 4 s to 20 s. In Fig. 4.4(a), we give examples of DFA plots of three Markovian time series with $\tau_{low}/2 = \tau_{high}/2 = \tau_{eff} = 4$ s, 10 s and 100 s; the first two ($\tau_{eff} = 4$ s and $\tau_{eff} = 10$ s, upper two curves) have been intentionally selected to have τ_{eff} comparable to the $\overline{T}$ of the SES activities and "artificial" noises. Comparing the DFA plots of the SES activities (that will be discussed later in § 4.4.1) with the upper two curves of Fig. 4.4(a), we find that a cross-over occurs at the *same* region $\Delta t \approx 30$ s (with almost the *same* α exponents in the short scales *only*). In other words, in short time-scales, even Markovian dichotomous time series (that have τ_{eff} values comparable to $\overline{T}$ of the SES activities and "artificial" noises) result in α values in the range $1 \leq \alpha \leq 1.5$ with a cross-over at $\Delta t \approx 30$ s. More generally, we can state [33] that not only signals of dichotomous nature, but *any signal* with a high frequency spectrum as given in Eq. (4.11) will lead to the *same* scaling behavior of $F_{DFA}(\Delta t)$ for small time lags Δt (irrespective of the particular shape of the signal; for example, a Gaussian signal with this spectrum will be much smoother and will display a continuity of values rather than only two steps).

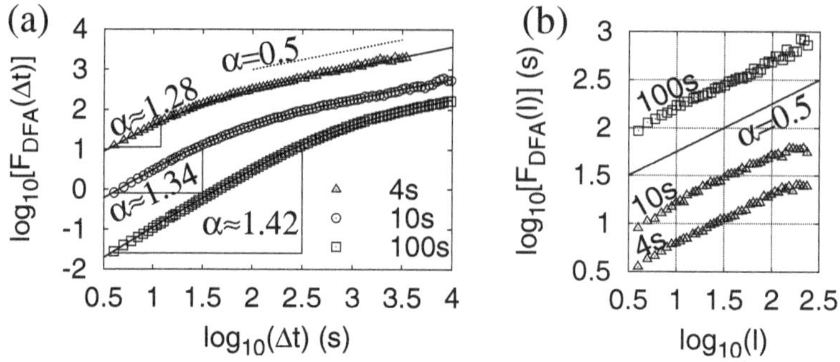

Fig. 4.4 (a) The variability measure $F_{DFA}(\Delta t)$ (in units of ΔE) for three Markovian dichotomous time series, calculated with $\tau_{eff} = 4$ (triangles), 10 (open circles) and 100 s (open squares). The solid lines, in each case, correspond to the theoretical analysis described in § 4.1.3. (b) The same as in (a), but calculated when the time series are read in natural time. The straight lines (dotted in (a), solid in (b)) correspond to $\alpha = 0.5$. The curves are shifted relative to each other by constant factors. Taken from Ref. [33].

The aforementioned points hold provided that the analysis is made in the conventional time frame. If the analysis is performed in natural time (considering as "high" either of the two states in the Markovian series), we find the following values: DFA exponent $\alpha = 0.5$ (see Fig. 4.4(b)) and power spectrum exponent $\beta = 0$. The latter values may elucidate the Markovian nature of the time series, avoiding the existence of the aforementioned characteristic intermediate scaling regions that appear in the analysis in the conventional time frame.

We now turn to the *case of spikes*. This corresponds to a very small value of τ_{eff} ($\approx \tau_{high} \ll \tau_{low}$). Recall that upon decreasing τ_{eff} (see Figs. 4.3(a) and 4.4(a)) the region described by the exponent $\alpha = 0.5$ extends to even shorter scales. This reveals that signals with superposed random spikes exhibit *uncorrelated* behavior (i.e.,$\alpha = 0.5$) at *small* scales.

By summarizing, we can state that:

For Markovian dichotomous signals, the quantity τ_{eff} – defined by $1/\tau_{eff} \equiv 1/\tau_{high} + 1/\tau_{low}$ – plays a key role. For time-scales comparable to (or shorter than) τ_{eff}, the power spectrum can be well described by a power law, $S(f) \propto 1/f^{\beta}$, with an exponent β around unity or larger, for example, $\beta = 1.0$–1.2 (note that β approaches the value 2 and 0 for short and long time-scales, i.e., in the "high"- and "low"-frequencies, respectively). In other words, this reflects that even randomly distributed square pulses could also be approximated by such a behavior. Thus, several published claims that "the appearance of a power law with an exponent β around unity constitutes a signature for critical behavior" should be examined, in each case, with extreme care.

Concerning DFA, a signal with (true) long-range correlations can be misinterpreted as having uncorrelated behavior and vice versa. Specifically: (a) truly correlated signals ($0.5 < \alpha \leq 1.5$) with superposed random spikes may show uncorrelated behavior ($\alpha = 0.5$) at short time-scales, (b) truly uncorrelated signals with superposed random square pulses, show "correlated" behavior (e.g., $\alpha \approx 1.0$–1.4) at time-scales comparable to (or shorter than) τ_{eff}. We can overcome *both* difficulties *if* the analysis is made in natural time.

4.2 Normalized power spectrum of SES activities. The universality emerged in natural time

4.2.1 Normalized power spectrum of SES activities and "artificial" noises in natural time. A universality for SES activities

Figure 4.5(a) depicts the SES activities recorded before the mainshocks labeled K, E and A of Fig. 4.5(b) (excerpts of these SES activities have been shown in Figs. 2.8 and 4.2). Once a SES activity has been recorded, we can read it in natural time and then proceed to its analysis. As an example, let us consider the SES activity K1 (see Fig. 4.5; see also Fig. 1.11(a)) recorded on April 18, 1995, that preceded the $M_w 6.6$ earthquake that occurred

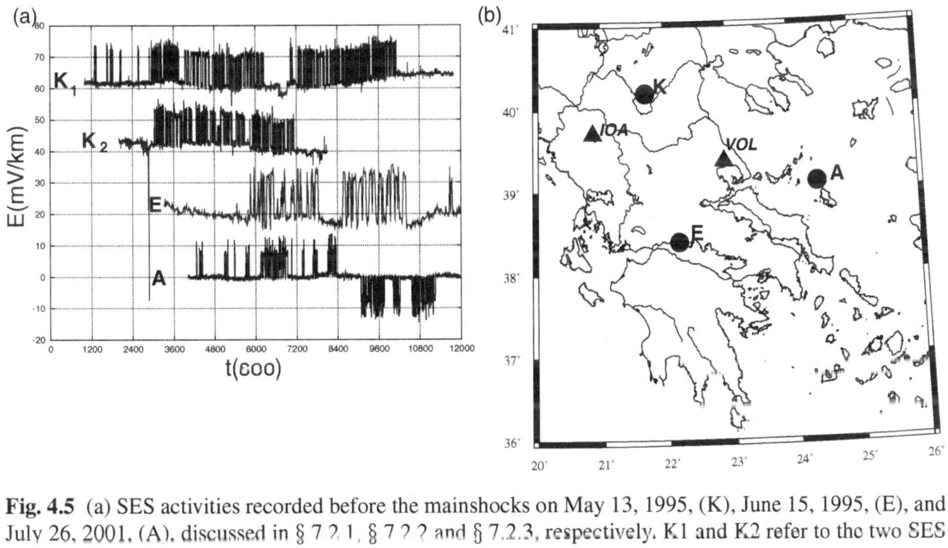

Fig. 4.5 (a) SES activities recorded before the mainshocks on May 13, 1995, (K), June 15, 1995, (E), and July 26, 2001, (A), discussed in § 7.2.1, § 7.2.2 and § 7.2.3, respectively. K1 and K2 refer to the two SES activities before the EQ labeled K (they are also depicted in Fig. 1.11(a),(b)). The upper two SES activities were recorded at IOA, while the lower two at VOL (note that the SES polarities, for drawing convenience, are *arbitrary* here; the correct polarities can be found, for example, in Fig. 4.2). (b) Map showing the EQ epicenters (circles) and the sites (triangles) of the measuring SES stations. Taken from Ref. [31].

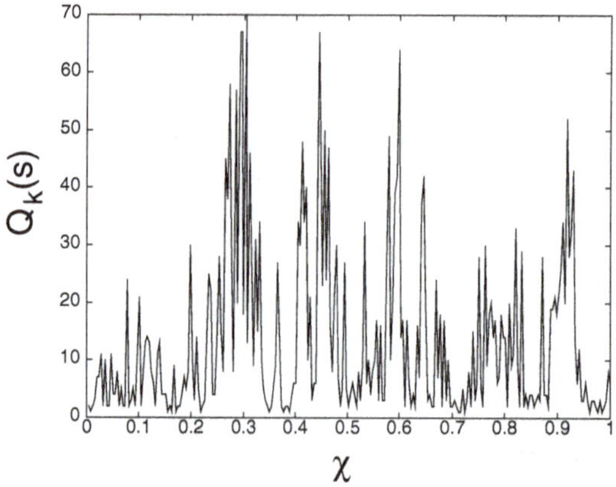

Fig. 4.6 How the SES activity K1 mentioned in Fig. 4.2 (see also Fig. 1.11(a)) is read in natural time; it depicts the durations Q_k as function of the natural time $\chi(=\chi_k)$ but drawn with continuous lines for the convenience of the reader (in reality, this should be plotted as in the lower part of Fig. 2.1(a) or as in Fig. 7.2(b)). Taken from Ref. [32].

at Grevena-Kozani on May 13, 1995 (see § 7.2.1). This lasted for around three and a half hours and was collected with a sampling rate $f_{exp} = 1$ sample/sec (thus we have $N =$ 11,900 data points). Figure 4.6 shows how the SES activity K1 of Fig. 4.5 can be read in natural time.

Figure 4.7 depicts $\Pi(\phi)$ for the four SES activities of Fig. 4.5, along with eight "artificial" noises recorded at various stations of the telemetric network which have a similar feature with SES (but do not satisfy the SES recognition criteria; see Section 1.2).

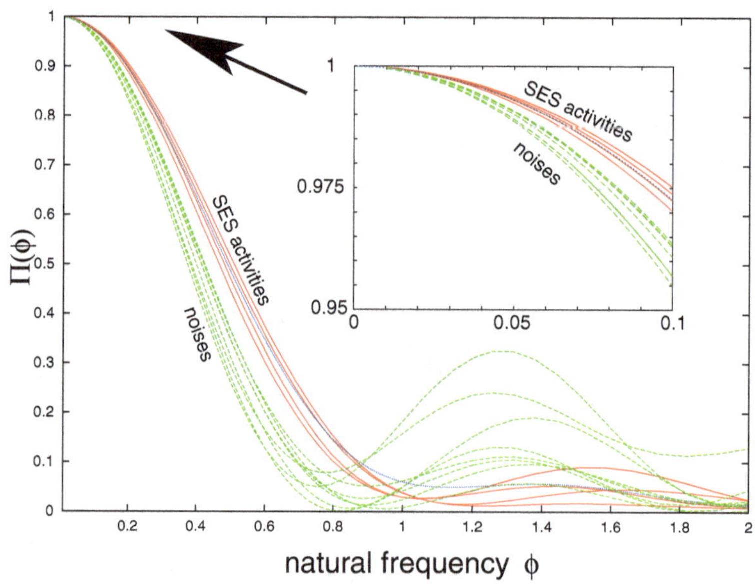

Fig. 4.7 The normalized power spectra $\Pi(\phi)$ for the SES activities (red solid lines) – depicted in Fig. 4.5 – related with the EQs labeled: K, E, and A (in the inset, from the top to the bottom: K1, A, E, K2) along with those of a number of "artificial" noises (green broken lines). The blue dotted curve corresponds to the theoretical estimation of Eq. (2.75), which holds for *critical* dynamics. The inset shows in an expanded scale the behavior of $\Pi(\phi)$ at small ϕ values, i.e., $\phi \to 0$. Taken from Ref. [31].

An inspection of this figure shows the following two facts [31]. First, the curves fall practically into two different classes, labeled "noises" and "SES activities" respectively. This classification, provides a tool for a distinction between "artificial" noises and SES activities (see § 4.2.2).

Secondly, Fig. 4.7 reveals that, for natural frequencies ϕ smaller than 0.5, the $\Pi(\phi)$ values of the SES activities scatter around the dotted curve, which has been estimated from theoretical considerations when approaching a critical point, i.e., Eq. (2.75) of § 2.4.2. In other words, the normalized power spectra in natural time of *all* the SES activities obey a *"universal"* curve.

Note that a possible explanation of the very pronounced "modes" in some "artificial" noises depicted in Fig. 4.7 has been discussed in Ref. [34].

4.2.2 Distinction of SES activities from "artificial" noises based on the normalized power spectrum

Figure 4.8 depicts, for the region of natural frequencies $0 \leq \phi \leq 0.5$, the normalized power spectra $\Pi(\phi)$ of the electric signals mentioned in Fig. 2.8 together with the one corresponding to the "open" states of ICFMC (see Fig. 4.1). The natural time representation of all these electric signals is shown in Fig. 4.9.

Figure 4.8 shows that the curves for the SES activities and "artificial" noises fall practically into two different classes, as already mentioned above (§ 4.2.1), while the ICFMC curve lies just between them and very close to the one that corresponds to the "uniform" distribution (labeled "uniform" in Fig. 4.8); see § 2.1.3. The universal curve for SES activities obeying Eq. (2.75) – which is labeled "theory" in Fig. 4.8 – implies that the variance of χ is $\kappa_1 = \langle \chi^2 \rangle - \langle \chi \rangle^2 = 0.070$ for SES activities (cf. Eq. (2.77), see also the last column in Table 4.1). The κ_1 value that reproduces [32] the ICFMC data is 0.080 ± 0.003 and

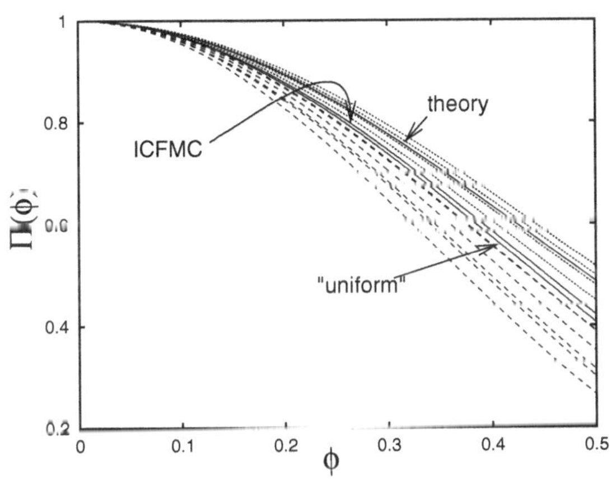

Fig. 4.8 The normalized power spectra $\Pi(\phi)$: SES activities (dotted lines) and "artificial" noises (broken lines) of Fig. 2.8. Three solid curves are also shown: the lower corresponds to the "uniform" distribution (Eq. (2.51) of § 2.4.1), the middle to ICFMC "open" states (see Fig. 4.1 of § 4.1.1), while the uppermost to the theoretical estimation, Eq. (2.75), for SES activities (*critical dynamics*). Reprinted from Ref. [27], Copyright (2009), with permission from TerraPub.

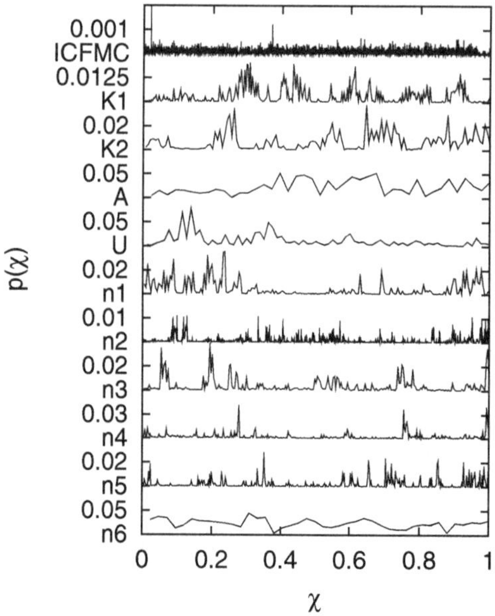

Fig. 4.9 The signals mentioned in Fig. 4.8 read in natural time; it depicts p_k versus χ_k with continuous lines for the sake of reader's convenience and hence $p(\chi)$ versus χ; see Eq. (2.4) (in reality, this should be plotted as in the lower part of Fig. 2.1(a) or as in Fig. 7.2(b)). Taken from Ref. [33]. Excerpts of these signals in the conventional time domain are depicted in Figs. 2.8, 4.1 and 4.2.

$\kappa_u = 1/12 \approx 0.083$ for the "uniform" distribution; see Eq. (2.46). Thus, for the "artificial" noises the variance κ_1 is larger than around 0.083. Hence, the difference $1/12 - \kappa_1 (\equiv \Delta\kappa)$ could be considered as a measure of the deviation of a signal from that of the "uniform" distribution.

By summarizing, SES activities are distinguished from "artificial" noises (AN) according to:

$$\kappa_{1,\mathrm{SES}} < \kappa_u \leq \kappa_{1,\mathrm{AN}} \tag{4.14}$$

where the subscripts designate each class of signals and $\kappa_u \approx 0.083$. Moreover, the SES activities satisfy Eq. (2.77), i.e.,

$$\kappa_{1,\mathrm{SES}} \approx 0.070 \tag{4.15}$$

4.3 Superiority of applying Hurst (R/S) analysis in the natural time domain

4.3.1 Conventional Hurst analysis

A way of studying correlations in a time series is provided by the Hurst analysis [13] known as rescaled range analysis (R/S). This compares the correlations in the time series measured at different time-scales and is similar to the classical fluctuation analysis (FA).

Table 4.1 Summary of the DFA results (when employing E-approximation together with the modification of Eq. (4.20), see § 4.5.2) for the "high"- and the "low"-level states' durations (labeled α_{high} and α_{low}, respectively) along with the κ_1 values for the SES activities and "artificial" noises depicted in Fig. 4.2. Taken from Ref. [34].

Signal	α_{high}	α_{low}	κ_1
K1	0.98±0.08	0.31±0.12	0.063±0.003
K2	0.92±0.10	0.49±0.09	0.078±0.004
A	0.87±0.27	0.34±0.25	0.068±0.004
U	0.98±0.13	0.70±0.15	0.071±0.004
N1	0.68±0.07	0.70±0.08	0.115±0.003
N2	0.79±0.03	0.54±0.04	0.093±0.003
N3	0.78±0.06	0.47±0.08	0.100±0.008
N4	0.76±0.06	0.55±0.06	0.100±0.013
N5	0.68±0.05	0.62±0.05	0.086±0.007
N6	–*)	–*)	0.092±0.004
N7	–*)	–*)	0.083±0.006
N8	–*)	–*)	0.102±0.004
N9	0.78±0.20	0.11±0.20	0.084±0.004

*) For N6, N7 and N8 no reliable slope could be determined in view of the small number of pulses ($N<25$).

Hurst's method fails to determine correlation properties if linear or higher order trends are present in the data, while detrended fluctuation analysis (DFA) (see § 1.4.2) – which is a significant improvement of FA – explicitly deals with monotonous [5] trends in a detrending procedure with remarkable results.

In short (e.g. see Ref. [5]), in Hurst (R/S) analysis, one calculates in each segment n the range R of the 'profile' $y(n)$ (see Eq. (1.9)) given by the difference between maximal and minimal value $\mathrm{R} = \max[y(n)] - \min[y(n)]$. The 'rescaling of range' is done by dividing R by the corresponding standard deviation S of the same segment. The mean $\langle \mathrm{R/S} \rangle$ of all quotients at a particular scale s is analogous to the fluctuation function $F(s)$ already discussed in the description of DFA (see Eq. (1.12)) and for long-range correlated signals shows a power law scaling relationship with s, with an exponent usually called Hurst exponent H (recall Eq. (2.78)). We first note that "*persistence*" usually means the tendency to keep moving in a fixed direction once the random walker has *started* moving in that direction [47].

In a persistent time series the increase in the values of the series is more likely to be followed by an increase and conversely, the decrease is more likely to be followed by a decrease. This occurs when $1/2 < H < 1$.

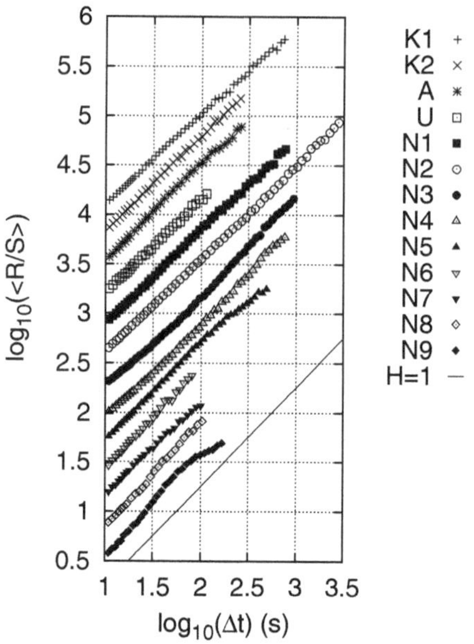

Fig. 4.10 The mean rescaled range ⟨R/S⟩ of the Hurst analysis as a function of Δt for the original time series of the SES activities and "artificial" noises mentioned in Fig. 4.2. For the convenience of the reader, the data points for each time series are vertically displaced after subsequent multiplication by a factor of 2, starting from N9; a solid straight line corresponding to $H = 1$ is also plotted. Taken from Ref. [29].

The results of the (R/S) analysis are given in Fig. 4.10 for the original time series of both the SES activities (the upper four curves) and "artificial" noises mentioned in Fig. 4.2. Since $\langle R/S\rangle \propto (\Delta t)^H$, the value of the Hurst exponent H is found from the slope (labeled H_0 in Table 4.2) of the corresponding log-log plot, when approximating it with a single straight line (note that *all* scaling methods related to the original Hurst analysis that yield the H exponent, assume a *finite* variance and according to the central limit theorem the underlying statistics are *Gaussian*).

An inspection of Fig. 4.10 shows that a value in the range $0 < H < 1/2$ (which means *antipersistent* time series, reflecting that increases in the values of a time series are likely to be followed by decreases, and conversely) cannot be seen.

Furthermore, no case with $H = 1/2$ (purely random changes) can be recognized. In all the cases of Fig. 4.10, the resulting H values lie between approximately 0.9 and 1.0 (Table 4.2), which suggest the *persistent* character of the examined time series (strong memory; see Ref. [32] and references therein). Thus, when Hurst analysis is carried out in the original time series, the H values alone cannot lead to any distinction between SES activities and "artificial" noises.

If we repeat the analysis of Fig. 4.10, but for the *dichotomous* time series (i.e., the converted from the original time series "0–1" dichotomous representation) we find somewhat smaller values (labeled H_d in Table 4.2) approximately in the range 0.75 to 0.90. Thus, the conclusion for the persistent character of the time series still remains.

Table 4.2 Summary of the (R/S) analysis for all the signals mentioned in Fig. 4.2. The symbols H_o and H_d stand for the slopes determined by using either the original time series or the dichotomous representation, respectively. H_{high} and H_{low} stand for the corresponding slopes for the "high"- and the "low"-level states' durations, respectively. Taken from Ref. [34].

Signal	H_o	H_d	H_{high}	H_{low}
K1	0.90±0.02	0.77±0.04	0.85±0.05	0.62±0.05
K2	0.96±0.01	0.81±0.05	0.87±0.09	0.70±0.08
A	0.96±0.02	0.76±0.06	0.82±0.28	0.61±0.21
U	0.95±0.02	0.80±0.06	0.89±0.13	0.72±0.12
N1	0.94±0.01	0.78±0.05	0.70±0.07	0.64±0.06
N2	0.94±0.01	0.84±0.04	0.77±0.03	0.58±0.03
N3	0.97±0.03	0.85±0.04	0.80±0.06	0.57±0.05
N4	0.99±0.03	0.87±0.05	0.72±0.04	0.63±0.04
N5	0.94±0.04	0.79±0.06	0.76±0.04	0.66±0.04
N6	1.06*)±0.02	0.86±0.06	–**)	–**)
N7	0.93±0.02	0.79±0.05	–**)	–**)
N8	1.09*)±0.02	0.86±0.05	–**)	–**)
N9	1.01*)±0.20	0.84±0.25	0.75±0.20	0.55±0.22

*) The value of H should *not* exceed unity (see the text), but here we reproduce the directly computed slope. Note that the computed H_d in the third column never exceeds unity.
**) For N6, N7 and N8 no reliable slope could be determined in view of the small number of pulses ($N < 25$).

By summarizing, the (R/S) Hurst analysis of the SES activities and "artificial" noises reveals a *persistent* character of both time series, but *cannot* distinguish between them.

4.3.2 Hurst analysis of the time series of durations of the "high"- and the "low"-level states. Hurst analysis in natural time

The results of Hurst analysis for the time series of durations of the "high"- and the "low"-level states are shown in Fig. 4.11. The analysis of the former states constitutes, of course, the Hurst analysis in natural time (recall Fig. 2.1(a)). The following common characteristic results for *both* the SES activities and "artificial" noises. The H values are systematically larger for the time series of the "high"-level states' durations when compared to the corresponding values of the "low"-level ones (labeled H_{high} and H_{low}, respectively, in Table 4.2). The *persistent* character ($1/2 < H < 1$) of the time series of the "high"-level states' durations seems to be well-established, while this holds to a lesser degree for the time series of the "low"-level ones (because a few of the corresponding H values, e.g. see A, N3 and N9 in Table 4.2, do not differ significantly from 1/2). Moreover in all cases, H_{high} is greater than H_{low}.

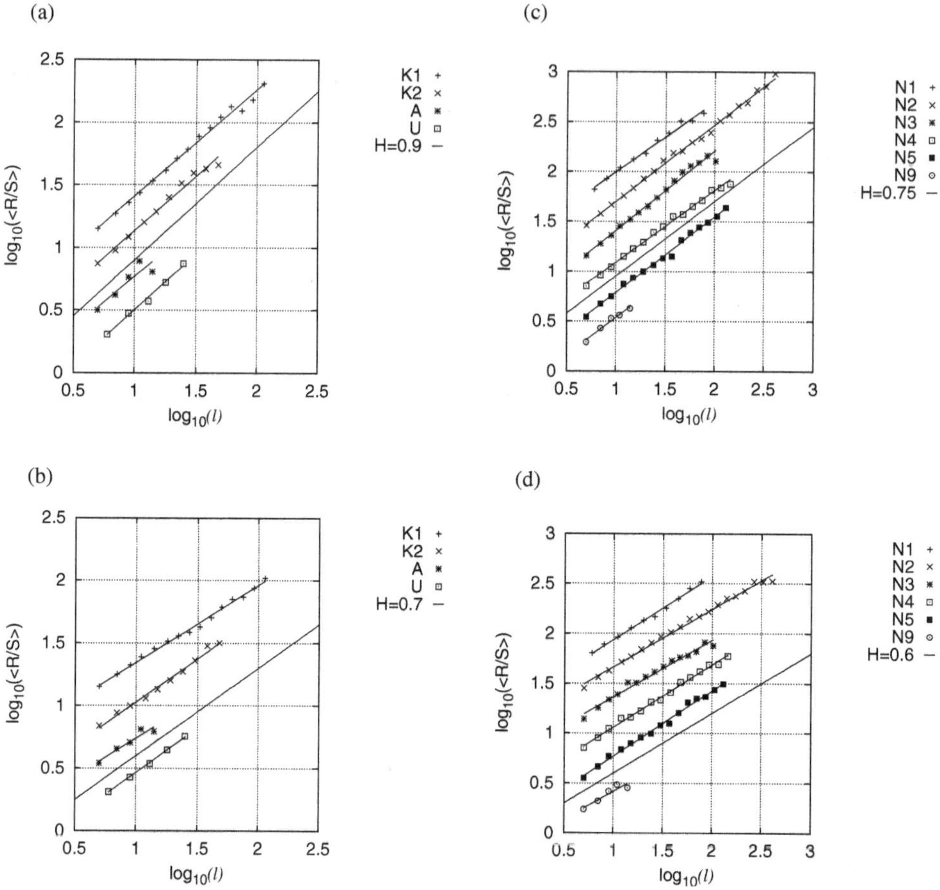

Fig. 4.11 The (R/S) Hurst analysis for the time series of the "high"(panels a and c)- and the "low"(panels b and d)-level states' durations for the SES activities and the "artificial" noises mentioned in Fig. 4.2. The data points for each time series are vertically displaced after subsequent multiplication by a factor of 2, starting from U or N9. For the reader's convenience, apart from the linear least-squares fits, the straight lines with slopes 0.9 in (a), 0.7 in (b), 0.75 in (c) and 0.6 in (d) are also plotted. Taken from Ref. [29].

Hence, the memory of the time series of both the SES activities and "artificial" noises may be mainly attributed to the strong correlation between the "high"-level states' durations.

Note, however, that when comparing the SES activities and "artificial" noises, the H values of their "high"-level states' durations do not differ significantly enough to guarantee a safe distinction between them.

4.4 Superiority of applying detrended fluctuation analysis (DFA) in the natural time domain

4.4.1 DFA of the original time series

Upon using the conventional DFA (§ 1.4.2), we obtain [34] the results depicted in Fig. 4.12 for *both* the SES activities and "artificial" noises mentioned in Fig. 4.2 (cf. recall that the DFA for a long duration SES activity has been already presented in § 1.4.3, see Fig. 1.17). A least squares fit to a single straight line (despite the fact that the data in some cases obviously deviate from such a scheme, see also below) reveals that the slopes of these log-log plots (labeled α_0 in Table 4.3) scatter for all cases around $\alpha \approx 1$, with a plausible uncertainty around 0.15. This reveals long-range temporal correlations. Upon repeating the analysis for their dichotomous time series, slightly different values for each case were obtained (labeled α_d in Table 4.3), and hence the conclusion concerning the strongly persistent character remains the same.

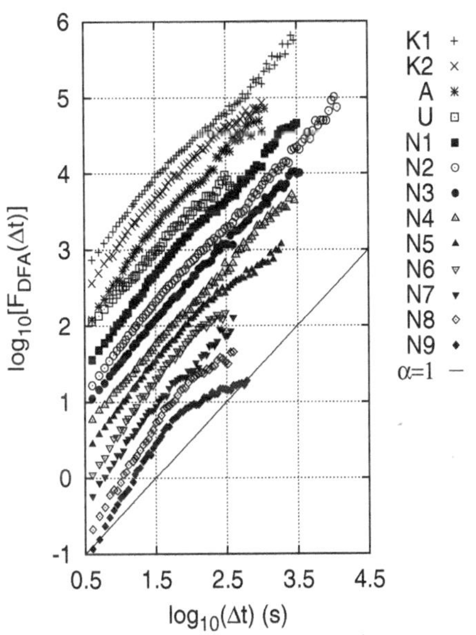

Fig. 4.12 The dependence of F_{DFA} on Δt in the conventional DFA of the original time series (in normalized units) of the SES activities and "artificial" noises mentioned in Fig. 4.2. The data points for each time series are vertically displaced after subsequent multiplication by a factor of 2, starting from N9. For the sake of reader's convenience, a solid straight line corresponding to the slope $\alpha = 1$ is plotted. Taken from Ref. [29].

If the log-log plot in Fig. 4.12 is approximated with two straight lines, the following results were obtained [34]: For *both* SES activities and "artificial" noises, the slope at shorter scales (i.e., $\Delta t \leq 30$ s) was found to lie in the range $\alpha = 1.1$–1.4, labeled α_0^{short} in Table 4.3, while for longer scales a value mostly in the range $\alpha \approx 0.8$–1.0 was determined (labeled α_0^{long} in Table 4.3), without, however, any safe classification between SES activities and "artificial" noises on the basis of the α values alone. The fact that both types of signals exhibit a cross-over at $\Delta t \approx 30$ s and also give almost the same DFA exponent ($\alpha_0^{short} \approx 1.2$)

Table 4.3 Results from the conventional DFA of the signals mentioned in Fig. 4.2. The symbols α_o and α_d stand for the DFA slopes obtained from either the original time series (in normalized units) or the dichotomous representation, respectively, for the whole Δt range. The corresponding slopes when considering either short Δt (smaller than approximately 30 s) or long Δt (larger than approximately 30 s) are also shown, being labeled with a superscript "short" and "long", respectively. Taken from Ref. [34].

Signal	α_o	α_o^{short}	α_o^{long}	α_d	α_d^{short}	α_d^{long}
K1	0.95±0.04	1.19±0.02	0.88±0.02	0.95±0.04	1.21±0.04	0.90±0.02
K2	0.95±0.06	1.22±0.04	0.81±0.02	0.96±0.06	1.23±0.03	0.82±0.02
A	1.06±0.10	1.36±0.05	0.96±0.04	1.08±0.10	1.41±0.05	0.98±0.04
U	0.95±0.04	1.03±0.05	0.81±0.03	0.95±0.04	1.07±0.04	0.79±0.03
N1	1.05±0.05	1.26±0.04	0.98±0.02	1.01±0.05	1.21±0.04	0.95±0.03
N2	1.04±0.03	1.21±0.03	1.01±0.02	0.97±0.03	1.12±0.03	0.94±0.02
N3	1.01±0.04	1.15±0.03	0.97±0.02	0.99±0.04	1.11±0.03	0.95±0.02
N4	1.04±0.04	1.08±0.03	1.02±0.02	1.02±0.04	1.01±0.03	1.02±0.02
N5	0.94±0.10	1.22±0.04	0.79±0.02	0.92±0.10	1.17±0.04	0.78±0.02
N6	1.14±0.11	1.39±0.04	0.89±0.03	1.13±0.11	1.43±0.04	0.86±0.03
N7	1.08±0.09	1.32±0.04	0.96±0.03	1.03±0.09	1.34±0.04	0.82±0.04
N8	1.15±0.12	1.49±0.04	0.78±0.03	1.12±0.12	1.45±0.04	0.76±0.03
N9	0.97±0.20	1.53±0.04	0.55±0.02	0.93±0.20	1.46±0.04	0.52±0.02

can be understood in the context of § 4.1.3 where it is shown that for dichotomous time series such a behavior should be observed at short time scales, i.e., $\Delta t \lesssim \tau_{eff}$.

By summarizing, when the conventional DFA is applied to the original time series of the SES activities and the "artificial" noises, *no* distinction can be achieved.

4.4.2 DFA of the time series of durations of the "high"- and the "low"-level states. Superiority of applying DFA in natural time

We now present the results of DFA for the time series of durations of the "high"- and the "low"- level states which are depicted in Fig. 4.13. Three main points emerge [34]:

First, both the SES activities and "artificial" noises exhibit for the time series of the "high"-level states' durations α values which are systematically larger than the corresponding values of the time series of the "low"-level ones (labeled α_{high} and α_{low}, respectively in Table 4.1).

Second, the α values for the time series of the "high"-level states' durations (which reflects that, *in reality, DFA is applied in natural time*) point to the following difference: for the SES activities (Fig. 4.13(a)) the α_{high} values lie approximately in the range 0.9–1.0, while for the "artificial" noises (Fig. 4.13(c)) the α_{high} values are markedly smaller, i.e., $\alpha_{high} \approx 0.65$–0.8 (Table 4.1). We emphasize that such a difference between SES activities and "artificial" noises is *not* noticed upon comparing their series of the "low"-level states' durations.

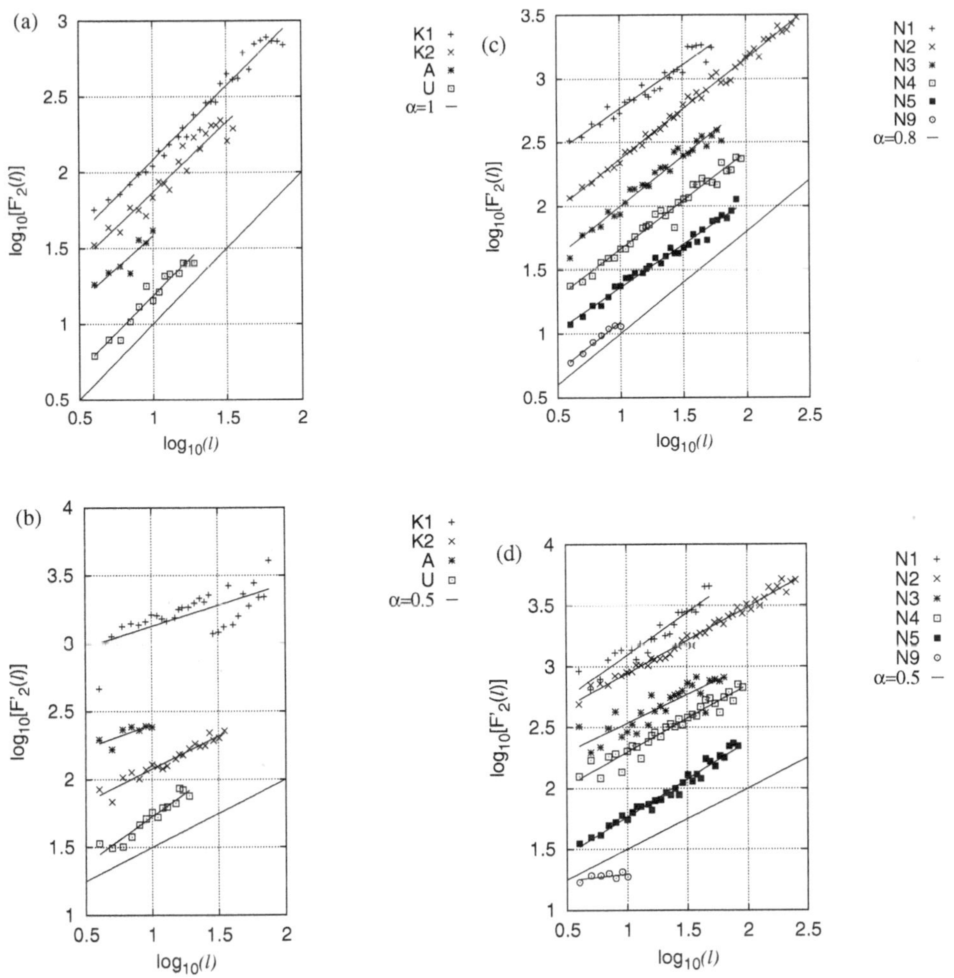

Fig. 4.13 The results of DFA (when employing E-approximation together with the modification of Eq. (4.20); see § 4.5.2) for the time series of the "high" (panels a and c)- and the "low" (panels b and d)-level states' durations (measured in sec, and hence $F_2'(l)$ is also measured in sec) for the SES activities and "artificial" noises mentioned in Fig. 4.2. The data points for each time series are vertically displaced after subsequent multiplication by a factor of 2, starting from U or N9. For the reader's convenience, apart from the linear least-squares fits, the solid straight lines with slopes $\alpha = 1$ in (a), $\alpha = 0.5$ in (b), $\alpha = 0.8$ in (c) and $\alpha = 0.5$ in (d) are also plotted. Taken from Ref. [29].

Third, comparing the α values between the time series of the two states' durations in the SES activities, the following characteristic is found: the α_{low} values for the time series of the "low" level states' durations scatter more or less around 0.5 (see Fig. 4.13(b)), thus being appreciably smaller than the aforementioned values $\alpha_{high} \approx 0.9$–$1.0$ for the series of the "high"-level states' durations (Fig. 4.13(a)).

Hence, *only* in natural time DFA can distinguish SES activities from "artificial" noises leading to an exponent $\alpha \approx 1.0$ for the SES activities, while $\alpha \approx 0.65$–0.8 for "artificial" noises.

4.5 Superiority of applying multifractal detrended fluctuation analysis (MF-DFA) in the natural time domain

4.5.1 Monofractals and multifractals. The necessity for multifractal analysis

Monofractal signals are *homogeneous* in the sense that they have the same scaling properties, characterized locally by a single singularity exponent h_0, throughout the signal. Thus, monofractal signals can be indexed by a single global exponent, e.g., the Hurst exponent $H \equiv h_0$, which suggests that they are *stationary* from the viewpoint of their local scaling properties (e.g., Ivanov et al. [14] and references therein). Since the power spectrum and the correlation analysis (including the conventional DFA, see § 1.4.2) can measure only one exponent, these methods are more suitable for the investigation of monofractal signals.

Concerning the use of these methods, however, the following points should be considered with care. A power spectrum calculation assumes that the signal is stationary and hence when applied to non-stationary time series it can lead to misleading results, as already mentioned in § 1.4.1. (A time series is stationary if the mean, standard deviation, and all higher moments, as well as the correlation functions, are invariant under time translation.) Thus, a power spectrum analysis should be necessarily preceded by a test for the stationarity of the data analyzed. As for the DFA, see § 1.4.2, it can determine the (mono) fractal scaling properties even in non-stationary time series (but see also Refs. [12, 7] on this point), and can avoid, in principle, spurious detection of correlations that are artifacts of non-stationarities.

In several cases, however, the records cannot be accounted for by a single scaling exponent (i.e., do not exhibit a simple monofractal behavior). In general, if a multitude of scaling exponents is required for a full description of the scaling behavior, a multifractal analysis must be applied. Multifractal signals are intrinsically more complex, and inhomogeneous, than monofractals (e.g., Ref. [14] and references therein). A reliable multifractal analysis can be performed by multifractal detrended fluctuation analysis [46, 15], which is summarized below in § 4.5.2. A similar analysis can be also performed by the wavelet transform (e.g., see Ref. [22]; see also § 4.6.1). Both these methods have been used in Refs. [34, 33] to analyze time series of SES activities and "artificial" noises (for the application of these methods to electrocardiograms see § 9.5.1). It was found [34, 33] that the multifractal analysis, when carried out in the conventional time-frame did not lead to any distinction between these two types of signals, but it *does so*, if the analysis is made in natural time. This will be explained below in § 4.5.3 and § 4.6.2.

4.5.2 Multifractal detrended fluctuation analysis. Background

A generalization of the DFA, termed multifractal DFA (hereafter labeled MF-DFA), allows [46, 15] the multifractal characterization of non-stationary time series. Compared to DFA (see § 1.4.2), in MF-DFA the following additional two steps should be taken.

First, we average over all segments to obtain the q-th order fluctuation function $F_q(s)$:

$$F_q(s) \equiv \left\{ \frac{1}{N_s} \sum_{\nu=1}^{N_s} \left[F^2(s,\nu) \right]^{\frac{q}{2}} \right\}^{\frac{1}{q}} \tag{4.16}$$

where

$$F^2(s,\nu) = \frac{1}{s} \sum_{n=(\nu-1)s+1}^{\nu s} \tilde{y}_s(n)^2, \tag{4.17}$$

and the index variable q can take any real value except zero. This is repeated for several scales s.

Second, we determine the scaling behavior of the fluctuation functions by analyzing log-log plots $F_q(s)$ versus s for each value of q. For long-range correlated series, $F_q(s)$ increases for large values of s as a power law:

$$F_q(s) \propto s^{h(q)}, \tag{4.18}$$

where the function $h(q)$ is called *generalized Hurst exponent.*

For *stationary* time series the aforementioned Hurst exponent H (see § 4.3.1) is identical to $h(2)$,

$$h(2) = H. \tag{4.19}$$

For *monofractal* time series, $h(q)$ is independent of q; all *stationary* long-range correlated series can be characterized by the power law decay of their power spectra $S(f) = f^{-\beta}$ with frequency f, and $\beta = 2H - 1$.

Furthermore, Kantelhardt et al. [16], in order to improve the scaling of the DFA fluctuations at short scales s, suggested a modified fluctuation function using randomly shuffled (*shuf*) copies of the original time series. This modification is useful to be incorporated in MF-DFA as well [15] and can be written as:

$$F_q^{mod}(s) = F_q(s) \frac{\overline{F_q^{shuf}(s')}\; s^{0.5}}{\overline{F_q^{shuf}(s)}\; s'^{0.5}} \tag{4.20}$$

for $s' \gg s$, where $\overline{F_q^{shuf}(s')}$ denotes the root mean square fluctuation function averaged over several configurations of the randomly shuffled data taken from the original record and $s' \approx N/20$.

The MF-DFA method requires series of compact support. In order to analyze data with fractal support, Varotsos et al. [34] suggested an additional modification called "*Eu-*

clidean (E-) approximation". In this approximation, instead of $[F^2(s,\nu)]^{q/2}$ in Eqs. (4.16) and (4.20), the "Euclidean distance" $d(s,\nu) \equiv \{[F^2(s,\nu-1)]^q + [F^2(s,\nu)]^q + [F^2(s,\nu+1)]^q\}^{1/2}$ is used.

In Ref. [34], it was shown that when dealing with time series of small length, *both* the above corrections improve significantly the conventional DFA (see fig. 11 of Ref. [34]). The corresponding fluctuation measure is denoted by $F'_q(s)$ and is the one used in Fig. 4.13 as well as for the determination of $h(q)$ in Figs. 4.14, 4.15 and 4.16.

Relation of MF-DFA to standard multifractal analysis. The scaling exponent $\tau(q)$ in the standard multifractal formalism (§ 4.6.1) is connected to the *partition function* $Z_q(s)$ through

$$Z_q(s) \propto s^{\tau(q)} \tag{4.21}$$

It can be shown [15] that $\tau(q)$ is related to the exponent $h(q)$ defined in Eq. (4.18) as follows:

$$\tau(q) = qh(q) - 1. \tag{4.22}$$

4.5.3 Multifractal detrended fluctuation analysis in natural time compared to that in conventional time

The results of the MF-DFA analysis (§ 4.5.2) of the original time series for both the SES activities and "artificial" noises are depicted in Fig. 4.14. An inspection of this figure shows that no obvious common characteristic can be recognized to allow any systematic distinction between SES activities and "artificial" noises. In order to visualize the difficulty of such a distinction, we reproduce in the inset of Fig. 4.14 a case of a SES activity, i.e., A, which, when compared to the artificial noise N4, shows an almost identical dependence of $h(q)$ versus q (for $q < 4$).

When studying the time series of the durations of the "high"- and the "low"-level states alone (Fig. 4.15), the following common feature emerged. In the time series of the "high"-level states (which reflects – if we recall Fig. 2.1(a) – that, *in reality, MF-DFA is applied in natural time*), the $h(q)$ curves for the SES activities (Fig. 4.15(a)) lie systematically higher than those in the case of "artificial" noises (Fig. 4.15(b)). For example, for $q = 2$, the $h(2)$ values for the SES activities lie close to unity, while for the "artificial" noises they scatter approximately in the range 0.65–0.8 (see Fig. 4.16 and the second column in Table 4.4). On the other hand, if we compare the time series of the "low"-level states' durations (although, in general, they have smaller $h(2)$ values than those corresponding to the "high"-level states' durations), *no* general feature can be recognized to distinguish the SES activities from the "artificial" noises. Varotsos et al. [34] emphasized that the "artificial" noises, which are characterized by $\kappa_1 \geq 0.083$ (§ 4.2.2), are accompanied by $h(2)$ values of the "high"-level states' durations smaller than $\approx$0.8 (see Fig. 4.17). We shall return to this point in § 4.7.1.

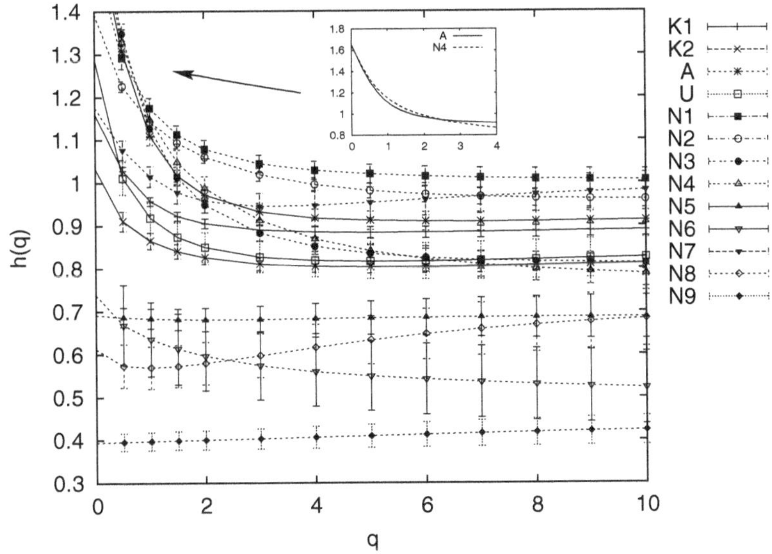

Fig. 4.14 The MF-DFA analysis for the original time series of the SES activities (solid curves) and "artificial" noises (dotted curves) mentioned in Fig. 4.2. The q-dependence of the asymptotic scaling exponent $h(q)$ determined by fits to the log-log plots of $F'_q(s)$ vs s (see § 4.5.2) at the regimes where the fits are straight lines. The corresponding regimes are given in Ref. [29]. For the inset, see the text. Taken from Ref. [29].

Summarizing, when MF-DFA is applied to the original time series of SES activities and "artificial" noises, *no* distinction can be achieved (see Fig. 4.14); *only if* it is applied in natural time can MF-DFA distinguish SES activities from "artificial" noises; see Fig. 4.16 together with the second column in Table 4.4.

4.6 Superiority of applying the wavelet transform in natural time

4.6.1 The wavelet transform, background. Comparison of the estimators of scaling behavior

The main disadvantage of the classical tool of Fourier transform in signal processing is its missing localization property: if a signal changes at a specific time, its transform changes everywhere and a simple inspection of the transformed signal does not reveal the position of the alteration. This originates from the fact that the Fourier transform decomposes a signal in plane waves (trigonometric functions), which oscillate infinitely with the same period and have no local character. Another disadvantage of Fourier analysis lies in the separate description and presentation of time and frequency.

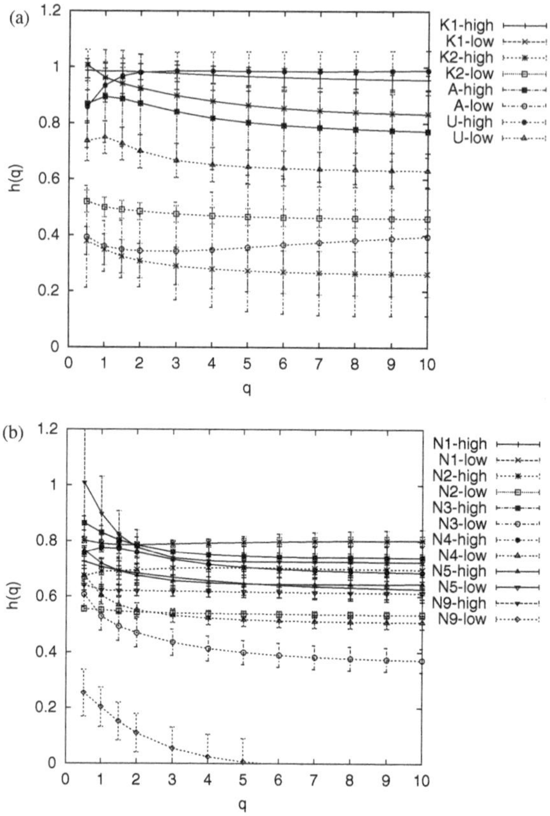

Fig. 4.15 The q-dependence of the asymptotic scaling exponent $h(q)$ obtained from the MF-DFA analysis for the time series of the "high"- and the "low"- level states' durations (labeled high and low, respectively) for: (a) the SES activities and (b) "artificial" noises mentioned in Fig. 4.2. The "high"-level states correspond to the solid lines, and the "low"-level to the dotted lines. Taken from Ref. [29].

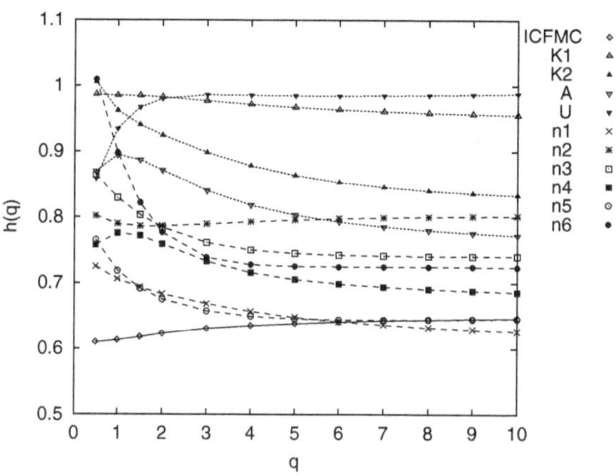

Fig. 4.16 Results of *MF-DFA in natural time* for: the SES activities (dotted lines), "artificial" noises (broken curves) and ICFMC (which corresponds to open states, solid curve). Reprinted from Ref. [27], Copyright (2005), with permission from TerraPub.

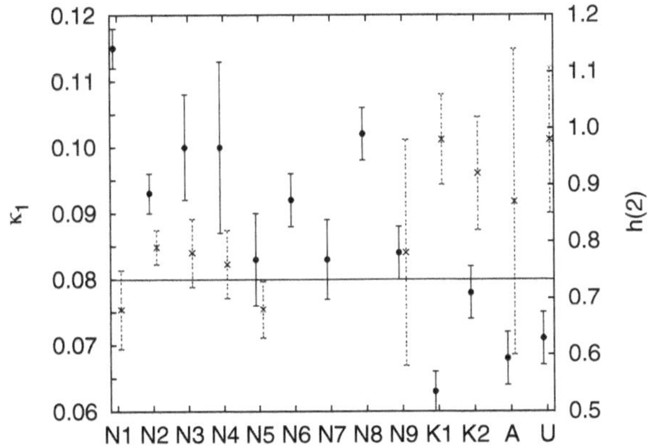

Fig. 4.17 The results of the natural time analysis of the SES activities and the "artificial" noises mentioned in Fig. 4.2. The κ_1 value for ICFMC corresponds to the solid horizontal line at $\kappa_1 = 0.080$. The solid circles correspond to the κ_1 values (left vertical scale). The crosses correspond to the $h(2)$ values (right scale) obtained by MF-DFA (for the noises N6, N7 and N8 no reliable MF-DFA results could be obtained due to the small $N < 25$ number of pulses). Taken from Ref. [34].

Table 4.4 Summary of the results in natural time for the SES activities and the "artificial" noises mentioned in Fig. 4.2 together with the results obtained from the analysis of the closed states for ICFMC. Taken from Ref. [33].

Signal	$h(2)^{*)}$	$h(2)^{**)}$	S	κ_1
K1	0.98±0.08	0.91±0.10	0.067±0.003	0.063±0.003
K2	0.92±0.10	0.94±0.17	0.081±0.003	0.078±0.004
A	0.87±0.27	–	0.070±0.008	0.068±0.004
U	0.98±0.13	1.10±0.27	0.092±0.004	0.071±0.004
ICFMC$_c$	0.86±0.07		0.096±0.003	0.080±0.003
'uniform'			$\ln(2)/2-1/4$	$1/12$
N1 or n1	0.68±0.07	0.86±0.12	0.143±0.003	0.115±0.003
N2 or n2	0.79±0.03	0.81±0.05	0.103±0.003	0.093±0.003
N3 or n3	0.78±0.06	0.69±0.11	0.117±0.010	0.100±0.008
N4 or n4	0.76±0.06	0.84±0.13	0.106±0.010	0.100±0.013
N5 or n5	0.68±0.05	0.77±0.08	0.091±0.011	0.086±0.007
N9 or n6	0.78±0.20	–	0.102±0.007	0.084±0.004

*) From MF-DFA in natural time (§ 4.5.3).
**) From the orthogonal wavelet transform in natural time (§ 4.6.2)

If we use instead a locally confined little wave (*wavelet*), then translation and scaling allows for a "frequency" resolution at arbitrary positions.

Thus, the wavelet transform allows more flexibility (e.g., see Ref. [19]): in simple words, the wavelet, which can be almost any chosen function, can be shifted and dilated to analyze signals. The wavelets can be interpreted as generalized oscillations (small waves) abstractly expressed in a zero mean value (see below). The price of this versatility is that two variables appear in the transform: the location and the scale of the wavelet. If the wavelet ψ is translated to a point t_0 and dilated by a factor l then we calculate the inner (scalar) product of the signal f with the function $\psi_{t_0,l}(t)$. If f shows a big change in a neighborhood of the point t_0 it has a high-frequency spectrum there.

The continuous wavelet transform of a given real function $f(t)$ is defined (e.g., see Ref. [4] and references therein) with a family of test functions $\psi_{t_0,l}(t)$ as the inner product

$$T_\psi[f](t_0,l) = \langle f|\psi_{t_0,l}\rangle \equiv \int f(t)\,\psi_{t_0,l}(t)\,\mathrm{d}t. \tag{4.23}$$

Each test function $\psi_{t_0,l}$ is obtained from a single function $\psi(t)$ (termed analyzing wavelet) by means of a translation and a dilation:

$$\psi_{t_0,l}(t) = \frac{1}{l}\psi\left(\frac{t-t_0}{l}\right) \tag{4.24}$$

where $t_0 \in \mathscr{R}$ and $l \in \mathscr{R}^*_+$ (where $\mathscr{R}$ stands for the set of real numbers and $\mathscr{R}^*_+$ for the positive ones). The function $\psi(t)$ is chosen such that both its spread in time and frequency are relatively limited.

In addition to being well localized both in time and frequency, ψ is required to satisfy the *admissibility* condition which in its weak form implies that ψ must be of zero mean (hence ψ is a band-pass or oscillating function, whence the name "wavelet", e.g. see Ref. [2] and references therein).

In the study of the scaling behavior, the following two features of the wavelet transform play key roles. (a) The wavelet basis is constructed from the dilation (change of scale) operator; thus the analyzing family exhibits a scale-invariant feature. (b) $\psi(t)$ is chosen so as to have a number $n_\psi \geq 1$ of vanishing moments:

$$\int t^k \psi(t)\,\mathrm{d}t \equiv 0,\;\; k = 0,1,\ldots,n_\psi - 1 \tag{4.25}$$

The Fourier transform $\Psi(\omega)$ of $\phi(t)$ satisfies

$$|\Psi(\omega)| \approx \omega^{n_\psi},\;\; \omega \to 0. \tag{4.26}$$

A common way to build admissible wavelets of arbitrary order n_ψ is to successively differentiate a smoothing function, e.g., the Gaussian function:

$$g_{n_\psi}(t) = \frac{\mathrm{d}^{n_\psi}}{\mathrm{d}t^{n_\psi}}\,\mathrm{e}^{-t^2/2} \tag{4.27}$$

The orthogonal wavelet transform. One can show that if ψ is properly chosen, then the family $\left\{2^{j/2}\psi_{j,k}\right\}_{j,k\in\mathscr{Z}}$, with $\psi_{j,k}(t) = 2^{-j}\psi(2^{-j}t-k)$, is an orthonormal basis of L^2 (e.g., Ref. [19]). The term $2^{j/2}$ is just a normalization factor. The orthogonal wavelet coefficients can then be defined by:

$$d_f(j,k) = \langle f|\psi_{j,k}\rangle. \tag{4.28}$$

Orthogonal wavelets that are often used in practice are the Daubechies wavelets, indexed by a parameter $n_D = 1, 2, \ldots$, which corresponds to the order of the wavelet. The Daubechies wavelet with $n_D = 1$ is in fact the *Haar* wavelet [11] (which is discontinuous; it equals 1 at $0 \leq t < 1/2$, -1 at $1/2 < t \leq 1$ and 0 otherwise), but the Daubechies wavelets with $n_D > 1$ are continuous with bounded support, and have n_D vanishing moments.

The Wavelet Transform Modulus Maxima (WTMM) method. This method [22] is based on the local maxima of the modulus of the continuous wavelet transform, i.e., on the local maxima $t_{0,i}$ (over t_0) of the function $|T_\psi[f](t_0, l)|$, where l is a fixed scale. In other words, in practice, instead of averaging over all values of $|T_\psi[f](t_0, l)|$, one averages (within the WTMM) only the local maxima of $|T_\psi[f](t_0, l)|$ and sums up the q-th power of these maxima,

$$Z(q, l) = \sum_{i=1}^{i_{\max}} |T_\psi[f](t_{0,i}, l)|^q \tag{4.29}$$

If scaling behavior is observed, scaling exponents $\tau(q)$ can be defined by:

$$Z(q, l) \propto l^{\tau(q)} \tag{4.30}$$

These $\tau(q)$ exponents are identical [15] to the τ(q) in Eq. (4.21) and related to $h(q)$ as shown in Eq. (4.22). Attention is drawn to the point that usually in WTMM the time series are analyzed directly instead of the profile $y(i)$ defined in § 1.4.2.

4.6.1.1 Comparison of the estimators of scaling behavior

Most of the (non-parametric) techniques for estimating the scaling exponent of time series that display scaling behavior consist essentially in the measurement of a slope in a log-log plot. Abry et al. [3, 2] and Veitch and Abry [44] have advocated the use of orthogonal wavelet-based estimators, which have several advantages. For example, they are blind to eventual superimposed smooth behavior (such as trends) and they are very robust when changing the slope of the underlying probability law.

Various wavelet-based estimators of self-similarity or long-range dependence scaling exponent were compared by Audit et al. [4]. These estimators mainly include the (bi)orthogonal wavelet estimators and the WTMM estimator. Their study focused both on short and long time series and also compared the wavelet-based estimators with DFA that is not wavelet-based. They found, among others, that the WTMM estimator leads to larger mean squared errors (MSE) for short time series of length smaller than 128 (i.e., $N \leq 128$) as compared to the orthogonal estimators but to much smaller MSE for long time series (see table I of Ref. [4]). For time series of size 8192 (i.e., for sizes comparable to those of the original time series of the SES activities and "artificial" noises), the WTMM estimator using the wavelet g_4, see Eq. (4.27), should be used.

Furthermore, for short time series ($N \leq 128$) it was shown [4] that DFA is the best estimator. This justifies why in § 4.4.2 (as well as in Chapter 5) the method of DFA is employed in order to analyze in natural time the SES activities and "artificial" noises which have usually $N \approx 10^2$ pulses (events).

4.6.2 The wavelet-based methods of estimating scaling behavior in natural time compared to that in conventional time

We start with the application of the WTMM method to the (original) time series of SES activities and "artificial" noises mentioned in Fig. 2.8 (see also the caption of Fig. 4.2). Using a g_4 wavelet, see Eq. (4.27), the analysis led to the results shown in Fig. 4.18. Figure 4.18(b) reveals that the curves showing the q dependence of the generalized Hurst exponent $h(q)$ are not classified, thus not allowing any obvious distinction between SES activities and "artificial" noises. The same conclusion is drawn (see Fig. 4.19(a)) if we apply the orthogonal wavelet transform analysis to the original time series of the signals mentioned in Fig. 2.8. This analysis was made with the program provided by Veitch et al. [45] using the Daubechies wavelet $n_D = 1$, after checking several other Daubechies wavelets of higher order, i.e., $n_D > 1$.

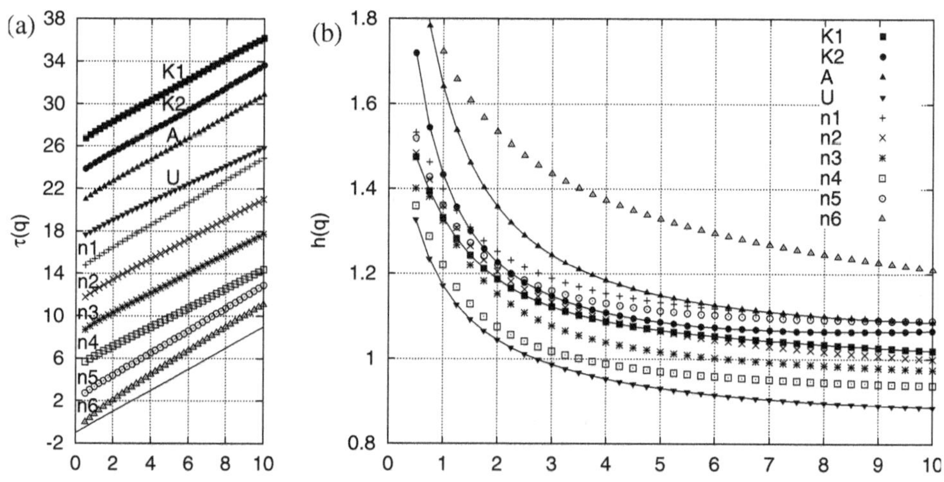

Fig. 4.18 The q dependence of the exponent $\tau(q)$ and the generalized Hurst exponent $h(q)$ (panels a and b, respectively) resulting from the application of WTMM using a g_4 wavelet for the signals mentioned in Fig. 2.8 (see also the caption of Fig. 4.2). For the sake of clarity, the straight line corresponding to a slope $H = 1$ was drawn in (a), while the solid curves in (b) correspond to the four SES activities (bold symbols, while for the "artificial" noises thinner symbols were used). The data points in (a) for each time series are vertically displaced by constant factors. Taken from Ref. [33].

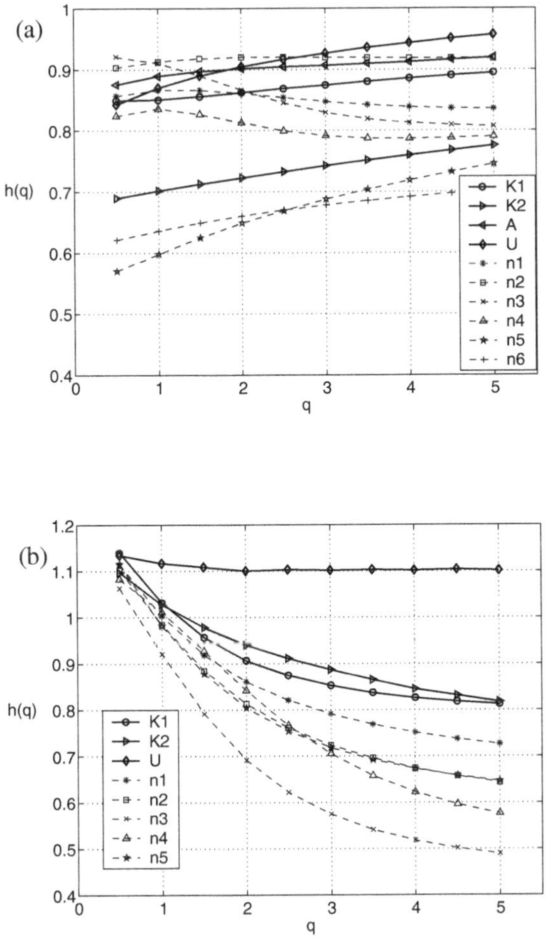

Fig. 4.19 The q dependence of the generalized Hurst exponent $h(q)$ resulting from the orthogonal wavelet transform analysis using a Daubechies1 (i.e., $n_D = 1$) wavelet. The solid curves correspond to the SES activities while the broken to the "artificial" noises: (a) of the original time series mentioned in Fig. 2.8 (see also the caption of Fig. 4.2); the percentage errors in the determination of h(2) are: 5, 15, 8, 4, 6, 3, 4, 12, 19 and 15% for K1, K2, A, U, n1, n2, n3, n4, n5, and n6 respectively.(b) of the time series read in natural time, see Fig. 4.9 (the corresponding errors are presented in Table 4.4); note that the signals A and n6 could not be analyzed in natural time due to the small number of pulses. Taken from Ref. [33].

We now proceed to the application of the wavelet transform to the signals as they are read in natural time, see Fig. 4.9. The results of the orthogonal wavelet transform analysis (note that WTMM could not be reliably applied in view of the small number of pulses), using again the Daubechies $n_D = 1$ (i.e., Haar) wavelet, are depicted in Fig. 4.19(b). An inspection of these $h(q)$ versus q curves, in spite of the large estimation errors seems to show a classification as follows. For q values around 2 or larger the resulting $h(q)$ values for the SES activities are higher than those of the "artificial" noises (see the $h(2)$ values in the third column in Table 4.4). In particular, the results show that the generalized Hurst exponent $h(2)$ for the SES activities is close to unity, while for the "artificial" noises $h(2)$ is markedly smaller. This conclusion is fully compatible with that deduced from the application of MF-DFA in natural time (§ 4.5.3, see also the second column in Table 4.4).

In summary, the wavelet transform analysis allows a distinction between SES activities and "artificial" noises, but *only* if it is applied in natural time leading to $h(2) \approx 1$ for SES activities, while $h(2)$ is markedly smaller for "artificial" noises.

4.7 Combining the normalized power spectrum analysis and multifractal analysis in natural time. The K-means clustering algorithm

4.7.1 Combining the variance κ_1 and the generalized Hurst exponent $h(2)$

Towards this goal, we employ two independent methods: the normalized power spectrum analysis in natural time (leading to the κ_1 values, see § 4.2.2) and the MF-DFA (§ 4.5.3) the application of which in natural time led for $q = 2$ to the $h(2)$ values given in Table 4.4 (see also the columns labeled α_{high} and κ_1 in Table 4.1). Figure 4.20 presents the results for the signals mentioned in Fig. 2.8 (see also the caption of Fig. 4.2) of these two methods applied independently in natural time.

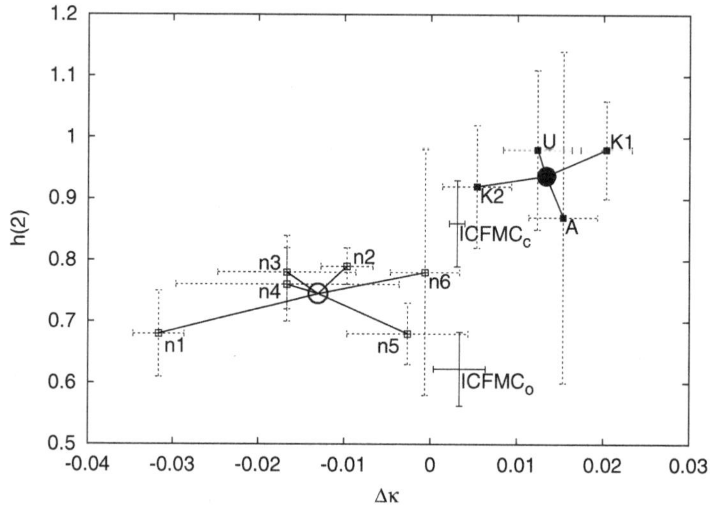

Fig. 4.20 Combined results of the analyses in natural time, see § 4.7.1. Plot of $h(2)$ versus $\Delta\kappa$ (= $1/12 - \kappa_1$): SES activities K1, K2, A, U (filled squares), "artificial" noises n1 to n6 (open squares), and the ICFMC (open states, labeled ICFMC$_o$). The κ_1 values come from the normalized power spectrum analysis (see Fig. 4.8), while the $h(2)$ values were obtained by MF-DFA (Fig. 4.16); all these values are given in Table 4.4. The DFA exponent (≈0.86) of the closed states for ICFMC (labeled ICFMC$_c$) is also inserted [33]. The thick straight lines indicate the two groups resulting from the application of the K-means algorithm explained in § 4.7.2; the full and open circles show the centroids of the two groups. Taken from Ref. [33].

A unified feature seems to emerge. The deviations from the "uniform" behavior quantified by $\Delta\kappa$ (where $\Delta\kappa \equiv 1/12 - \kappa_1$, see § 4.2.2) are interrelated with the $h(2)$ values: First, the SES activities, which correspond to large $\Delta\kappa$ values ($\Delta\kappa > 0$), are characterized by the strongest "memory" (large $h(2)$, close to unity); both their $\Delta\kappa$ and $h(2)$ values are consistent with those expected for a critical behavior (see § 4.7.3 and § 2.4.2). Second, the "artificial" noises simultaneously have smaller $\Delta\kappa$ values ($\Delta\kappa \leq 0$) and weaker "memory" (their $h(2)$ values are markedly smaller than unity). Third, concerning the ICFMC, the values related with the closed states, which have been found [24] to exhibit the stronger "memory" (between the two states, i.e., closed and open, see Fig. 4.1), seem to lie between the aforementioned two regimes.

Finally, Varotsos et al. [33] emphasized that, the randomly "shuffled" series of all the three types of electric signals investigated, lead to $h(2) \approx 0.5$ (simple random behavior) and $\Delta\kappa \approx 0$ (e.g., see the SES activity in § 7.1.1). These two values are internally consistent in the absence of heavy tails, because in the "shuffling" procedure the values are put into random order, thus all correlations (memory) are destroyed (§ 2.5.2.1).

4.7.2 The K-means clustering algorithm

A more elaborated classification of the results depicted in Fig. 4.20, can be obtained by using some clustering algorithm. In Ref. [33] a K-means type was used, which is a least-squares partitioning method allowing users to divide a collection of objects into K groups (e.g., see section 8.8 of Ref. [18]).

The K-means problem consists of dividing a set of multivariate data into non-overlapping groups in such a way as to minimize the sum (across the groups) of the sums of squared residual distances to the group centroid (this statistics is usually called sum of squared errors). In other words, a computer program tries to minimize the sum, over all groups, of the squared within-groups residuals, which are the distances of the objects to the respective group centroid. The groups obtained are such that they are geometrically as compact as possible around their respective centroid.

In Ref. [33] the K-means partitioning program provided by Legendre [17] was used. This program allows users to search through different values of K in a cascade, starting with k_1 groups and ending with k_2 groups, with $k_1 \geq k_2$. In the cascade from a larger to the next smaller number of groups, the two groups whose centroids are the closest in multivariate space are fused and the algorithm iterates again to optimize the sum of squared errors function, reallocating objects to the groups. Varotsos et al. [33] run the program by considering the 10 "objects", i.e., the four SES activities and the six "artificial" noises mentioned in Fig. 2.8 (see also the caption of Fig. 4.2). The $h(2)$ values resulting from MF-DFA in natural time and the κ_1 values reported in Table 4.4 have been used. Studying partitions from $k_1 = 5$ to $k_2 = 2$ groups, the clustering shown in Fig. 4.20 with the thick straight lines was found [33].

This clustering consists of the following two groups ($K = 2$): the first one includes the four SES activities, while the second the six "artificial" noises n1 to n6. The centroid of the first group (solid dot) lies at $\Delta\kappa = 0.013$, $h(2) = 0.9375$, while the centroid of the second at $\Delta\kappa = -0.013$, $h(2) = 0.745$. Note that the $\Delta\kappa$ value ($= 1/12 - \kappa_1$) of the centroid of the group of the four SES activities corresponds to $\kappa_1 = 0.070$, which coincides with the theoretical value obtained for the SES activities in § 2.4.2, see Eq. (2.77).

4.7.3 Comments on the differences in the memory and the variance κ_1 among electric signals of different nature

Let us focus on the tentative origin of the difference in the memory of SES activities and "artificial" noises. In Ref. [34] an attempt was made towards understanding the aforementioned results (§ 4.5.3), which show that the values of the generalized Hurst exponent $h(2)$ of the "high"-level states' durations of the SES activities are close to unity, while those of the "artificial" noises are markedly smaller. Let us consider, at the moment, for the sake of simplicity, the simple case of fBm (which has been proposed [48] to model the SES activities for $H \to 1$ and is the *only* Gaussian self-similar process with self-similarity index $H \neq 0.5$, e.g., see Ref. [21]; see also § 1.5.1.1): the Hurst exponent H has been suggested as a measure of the degree (intensity) of self-similarity or long-range dependence, e.g., see Ref. [26] (see also Refs. [20, 49]). The power law decay of the covariance, Eq. (1.8), characterizes long-range dependence. The higher the H the slower the decay, e.g., see Eq. (1.15). If we now assume that, in general, $h(2)(= H)$ is actually a measure of the intensity of long-range dependence, we may understand that the SES activities, since they exhibit *critical* dynamics (infinite long-ranged correlations), should have a long-range dependence stronger (thus, a higher H) than that of the "artificial" noises. Note that the model of critical behavior discussed in § 2.4.2, which resulted in Eq. (2.77), shows that $\mathscr{E}(Q_k Q_{k+l})$ is independent of l.

As for the fact that the ICFMC curve ($\kappa_1 = 0.080 \pm 0.003$) lies in Fig. 4.8 closer to the "uniform" distribution compared either to the SES activities or the (majority of the) "artificial" noises, this is not unreasonable for a biological system [34] (see Chapter 9, e.g., Fig. 9.11).

4.8 The fluctuation function $F(q) = \langle\chi^q\rangle - \langle\chi\rangle^q$ and the entropy S in natural time

4.8.1 Classification of electric signals based on the function $F(q) = \langle\chi^q\rangle - \langle\chi\rangle^q$ versus q in various types of electric signals

In Ref. [33], it was proposed that a classification of the aforementioned three types of electric signals of dichotomous nature, i.e., ICFMC, SES activities and "artificial" noises, becomes possible if we study, in the range $0 < q \leq 2$, the function $F(q) = \langle \chi^q \rangle - \langle \chi \rangle^q$ versus q.

We recall that Fig. 4.9 shows how the electric signals, mentioned in Fig. 4.8 (see also Figs. 2.8 and 4.2), are read in natural time. The function $\langle \chi^q \rangle - \langle \chi \rangle^q$ versus q, for all these electric signals, is depicted in Fig. 4.21(a), in the range $0 < q \leq 2$. (cf. Eq. (2.38), which was introduced for $n =$ positive integer only). This figure shows that the signals are now classified:

The curves for the SES activities and "artificial" noises, at least in the range $q \in (1,2)$ fall practically into two different classes, while the ICFMC curve lies just between them.

Note that the results, for $q = 2$, exhibit the feature already mentioned in § 4.2.2, i.e., for SES activities, they scatter around the value $\kappa_1 (= \langle \chi^2 \rangle - \langle \chi \rangle^2) = 0.070$, while for the "artificial" noises $\kappa_1 \geq 0.083$, and for ICFMC $\kappa_1 = 0.080 \pm 0.003 \approx \kappa_u$ (see also Fig. 4.8).

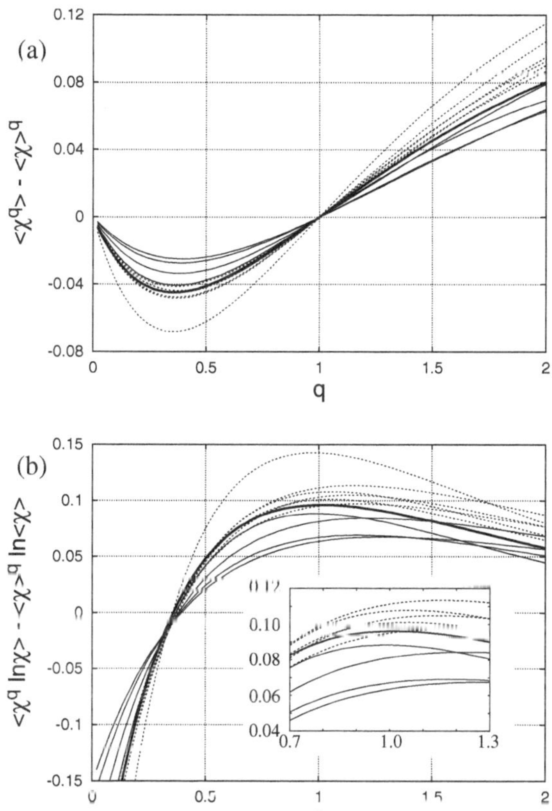

Fig. 4.21 (a) The function $\langle \chi^q \rangle - \langle \chi \rangle^q$ and (b) its derivative with respect to q, i.e., $\langle \chi^q \ln \chi \rangle - \langle \chi \rangle^q \ln \langle \chi \rangle$, versus q. ICFMC: Thick solid line; SES activities: thin solid lines; "artificial" noises: broken lines. The uncertainties for $q = 2$ in (a) and for $q = 1$ in (b) are given in Table 4.4. Taken from Ref. [33].

4.8.2 Classification of electric signals based on the entropy S in natural time

The derivative of the function $F(q) = \langle\chi^q\rangle - \langle\chi\rangle^q$ with respect to q, i.e.,

$$F'(q) = \frac{\mathrm{d}}{\mathrm{d}q}(\langle\chi^q\rangle - \langle\chi\rangle^q) = \langle\chi^q \ln\chi\rangle - \langle\chi\rangle^q \ln\langle\chi\rangle \tag{4.31}$$

is plotted in Fig. 4.21(b) versus q. We may see again a classification. Furthermore, Varotsos et al. [33] drew attention to the region around $q = 1$. The quantity $\langle\chi \ln\chi\rangle - \langle\chi\rangle \ln\langle\chi\rangle$ is just the one defined as entropy S in natural time, i.e., see Eq. (3.1). In addition, Eq. (3.4) states that the entropy S_u of the "uniform" distribution (see § 2.1.3) has the value $S_u = 0.0966$.

Therefore the three types of electric signals seem to be classified as follows (but see also § 4.8.3): The "artificial" noises have an entropy larger than (or equal to) that of the "uniform" distribution, i.e., $S \geq S_u$, while the SES activities exhibit S values smaller than S_u. As for the ICFMC, the S value lies just in the boundary between the SES activities and the "artificial" noises and is very close to S_u. The point that only n5 among the "artificial" noises seems to have a smaller entropy than S_u – see Table 4.4 – is discussed below.

Thus, in short, the entropy $S = \langle\chi \ln\chi\rangle - \langle\chi\rangle \ln\langle\chi\rangle$ of the SES activities is smaller than that (S_u) of the "uniform" distribution, while the "artificial" noises have an entropy larger than (or equal to) S_u.

The following remarks are worth adding. First, when employing the K-means algorithm mentioned in § 4.7.2, if the S values inserted in Table 4.4 are used instead of κ_1, a comparison of partitions into $k_1 = 4$ to $k_2 = 2$ groups also leads to the clustering shown in Fig. 4.20.

Second, for each of the signals depicted in Fig. 4.9, the values of the scaling exponent α (obtained from DFA) and κ_1 do *not* change upon a time reversal. Such a reversal, however, leads to a different S value labeled S_-.

The latter important point has been already treated in Section 3.4 and will be further discussed in Section 4.9. Third, if the number of pulses in a SES activity (or "artificial" noise) is small (e.g. 3–50), the values of *both* κ_1 and S are *smaller* than the actual ones; the extent of this underestimation could be understood on the basis of Fig. 4.22 to which we now turn. Figure 4.22, depicts the expected value for κ_1 and S for a Markovian dichotomous time series (Section 4.1) as a function of the number of the "high" level states N along with their uncertainty of $\pm\sigma$. Recall that the values of κ_1 and S for all the SES activities and "artificial" noises mentioned in Fig. 4.9, are shown in Table 4.4. The fact that only n5 among the "artificial" noises seems to have an entropy somewhat smaller than S_u ($S[\mathrm{n5}] = 0.091 \pm 0.011$) might be understood as follows: for n5, we have $N \approx 400$ (see Table 2.2) for which Fig. 4.22 reveals that the aforementioned value of 0.091 differs from S_u only by an amount smaller than one σ.

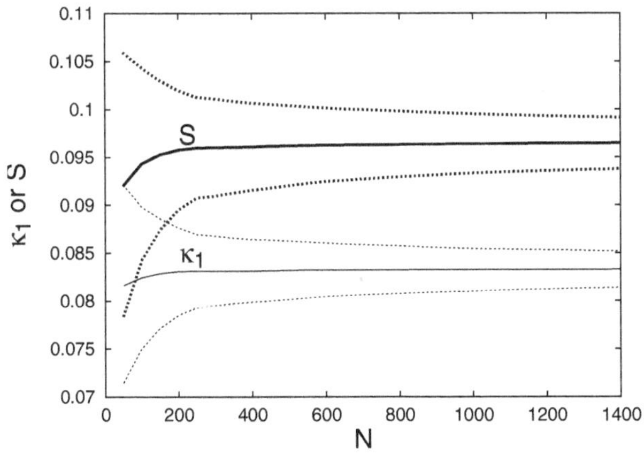

Fig. 4.22 Confidence intervals ($\mu \pm \sigma$) for the variance κ_1 (thin lines) and the entropy S (thick lines) calculated for a Markovian dichotomous time series (note that the durations Q_k are exponentially distributed) versus the number of events N. The solid lines correspond to the average values of κ_1 (thin) and S (thick). Taken from Ref. [28].

4.8.3 Classification of electric signals by the complexity measures using the fluctuations of the entropy in natural time

The values of the complexity measures λ_s, $\lambda_{s,shuf}$ and ν_s, defined in § 3.6.1, for several SES activities and "artificial" noises were calculated in Ref. [38] and the results are shown in Table 4.5. The complexity measures have been calculated only in the short-range because the length of these signals in the natural time domain is on the average $\approx 10^2$ pulses and hence does not significantly exceed the time window length $l \approx 60$ pulses, thus not allowing a reliable calculation of the complexity measures in the longer scale (see § 3.6.1; see also § 9.2.2.1 and § 9.2.7).

Table 4.5 The complexity measures λ_s, $\lambda_{s,shuf}$ and ν_s of SES activities and "artificial" noises along with their S values (note that the latter are compiled from Table 4.4). Taken from Ref. [36].

Signal	λ_s	$\lambda_{s,shuf}$	ν_s	S
K1	1.26	1.27	1.21	0.067±0.003
K2	1.26	1.29	1.30	0.081±0.003
U*)	1.06	1.24	1.17	0.092±0.004*)
A	0.97	1.14	0.97	0.070±0.008
n1	1.25	1.23	1.21	0.143±0.003
n2	1.30	1.31	1.18	0.103±0.003
n3	1.35	1.26	1.24	0.117±0.010
n4	1.36	1.26	1.20	0.106±0.010
n5*)	1.32	1.28	1.12	0.091±0.011*)
n6	1.36	1.01	1.15	0.102±0.007

*) Note that in these two cases the S values are comparable to S_u, and hence their distinction can be made on the basis of the λ_s values which differ markedly.

An inspection of these results reveals that the λ_s values of most "artificial" noises are somewhat *larger* than those in the SES activities. Note that in two cases, i.e., the SES activity U and the "artificial" noise n5, for which the S values are comparable to S_u (thus, no distinction can be made on the basis of the S values alone), the distinction can be achieved on the basis of the λ_s values, $\lambda_s(U) < \lambda_s(n5)$, which differ markedly.

Recapitulating the distinction of similar-looking signals that are emitted from systems of different dynamics, we can now say the following [38]: If the S values differ markedly from S_u (which holds in most SES activities and "artificial" noises), the signals can be distinguished on the basis of the S values alone. On the other hand, if the S values are close to S_u (which holds in all ECG, see Chapter 9, but only in the minority of SES activities and "artificial" noises) the signals can be better classified by using the complexity measures based on the fluctuations δS of the entropy (see also § 3.6.1 and § 9.1.1).

4.9 Using the entropy S_- or the fluctuations of natural time under time reversal

4.9.1 Distinction of SES activities from "artificial" noises based on the entropy in natural time under time reversal

The entropy S_- in natural time under time reversal, defined in § 3.4.1 has been calculated for all the SES activities and "artificial" noises tabulated in Table 4.4 (as well as for some more recent examples) and the results can be found in Ref. [43] (see also Ref. [42]). Here, Table 4.6 compiles the S and S_- values of all these signals along with those of 16 SES activities recorded during the subsequent years. The stations at which the latter SES activities have been recorded are also mentioned in Table 4.6. For the sake of completeness, we also give in Table 4.6, the value of the variance $\kappa_1 = \langle \chi^2 \rangle - \langle \chi \rangle^2$ obtained in each case. An inspection of Table 4.6 reveals the following:

The S values are actually classified, as stated above in § 4.8.2, i.e., $S < S_u$ for the SES activities and $S_u \lesssim S$ for "artificial" noises. On the other hand, this does *not* hold in general for the S_- values.

This is so, since for all the SES activities (with the probable exception of K2) we find that the S_- values are smaller than (or equal to) S_u, but for "artificial" noises no *common* behavior could be found, because S_- is either smaller or larger than S_u.

In other words, no distinction between SES activities and "artificial" noises can be achieved on the basis of S_- values *alone*. This means the following, if we recall that the S value takes into account the sequential order of pulses and hence captures elements of the dynamics hidden in this order [37, 38]: Only when considering the true time arrow (i.e., analyzing in natural time the signal as it was actually recorded in nature) the S value can pinpoint the difference in the dynamics between these two groups of electric signals. Recall that the SES activities are characterized by *critical* dynamics and hence exhibit infinitely ranged long-range correlations, while in "artificial" noises the intensity of the long-range correlations is markedly weaker [33] (see also § 4.7.3). Numerical studies of models which show [41] that *both* S and S_- are smaller than S_u have been already presented in § 3.4.3 and § 3.4.4.

Table 4.6 The values of S, κ_1, S_- for the SES activities and "artificial" noises in Greece analyzed in Ref. [43] (see also table I of Ref. [42]) together with the one labeled E in Fig. 4.5 as well as with 16 more recent SES activities which are the following: M_1 to M_4 were recorded at MYT station, while V_1 at VOL, see fig. 1 of Ref. [41]. The SES activities PAT, shown in Fig. 7.2, and PAT_2, see fig. 2 of Ref. [40], were recorded at PAT station. The signals PIR_1, PIR_2, PAT_3 and PAT_4 correspond to the SES activities depicted in figs. 3(a), 3(b), 3(d) and 3(e) of Ref. [39], respectively. They were recorded at PIR or PAT station. The four additional SES activities recorded at PAT station during 2007 depicted in figs.5(a), 5(b), 5(c) and 5(d) in Ref. [30] are labeled PAT_5, PAT_6, PAT_7 and PAT_8, respectively. Finally, PIR_3 stands for the SES activity (see Fig. 7.22(b)) that was recorded [30] on January 14, 2008, at PIR which preceded the strongest earthquake in Greece during the last 28 years that occurred on February 14, 2008.

Signal	S	κ_1	S_-
K1	0.067±0.003*)	0.063±0.003*)	0.074±0.003
K2	0.081±0.003*)	0.078±0.004*)	0.103±0.003
E	0.071±0.010	0.071±0.006	0.082±0.010
A	0.070±0.008*)	0.068±0.004*)	0.084±0.008
U	0.092±0.004*)	0.071±0.004*)	0.071±0.004
T1	0.088±0.007	0.084±0.007	0.098±0.010
C1	0.083±0.004	0.074±0.002	0.080±0.004
P1	0.087±0.004	0.075±0.004	0.081±0.004
P2	0.088±0.003	0.071±0.005	0.072±0.015
E1	0.087±0.007	0.077±0.017	0.081±0.007
$M_1^{**)}$	0.094±0.005	0.075±0.004	0.078±0.003
$M_2^{**)}$	0.089±0.003	0.076±0.004	0.084±0.003
$M_3^{**)}$	0.089±0.004	0.080±0.005	0.093±0.004
$M_4^{**)}$	0.080±0.005	0.073±0.004	0.086±0.006
$V_1^{**)}$	0.078±0.006	0.074±0.004	0.092±0.005
PAT***)	0.080±0.002	0.072±0.002	0.078±0.002
$PAT_2^{***)}$	0.074±0.002	0.075±0.002	0.078±0.002
$PIR_1^{****)}$	0.070±0.012	0.062±0.010	0.051±0.010
$PIR_2^{****)}$	0.077±0.004	0.076±0.005	0.082±0.004
$PAT_3^{****)}$	0.073±0.007	0.072±0.005	0.081±0.006
$PAT_4^{****)}$	0.085±0.005	0.073±0.007	0.080±0.004
PAT_5	0.067±0.007	0.074±0.007	0.079±0.007
PAT_6	0.071±0.005	0.069±0.003	0.066±0.005
PAT_7	0.072±0.003	0.067±0.003	0.069±0.003
PAT_8	0.070±0.005	0.065±0.005	0.070±0.005
PIR_3	0.086±0.003	0.070±0.005	0.070±0.005

Table 4.6 Continued.

Signal	S	κ_1	S_-
n1	0.143±0.003*)	0.115±0.003*)	0.127±0.004
n2	0.103±0.003*)	0.093±0.003*)	0.122±0.003
n3	0.117±0.010*)	0.100±0.008*)	0.118±0.010
n4	0.106±0.010*)	0.100±0.013*)	0.138±0.010
n5	0.091±0.011*)	0.086±0.007*)	0.120±0.011
n6	0.102±0.007*)	0.084±0.004*)	0.095±0.007
n7	0.116±0.005	0.085±0.005	0.083±0.005
n8	0.117±0.004	0.095±0.007	0.099±0.005
n9	0.110±0.010	0.091±0.005	0.095±0.010
n10	0.112±0.005	0.087±0.007	0.087±0.006
n11	0.122±0.012	0.088±0.007	0.079±0.012
n12	0.104±0.005	0.094±0.005	0.103±0.009
n13	0.124±0.007	0.084±0.007	0.077±0.008
n14	0.124±0.005	0.087±0.005	0.081±0.007

*) From Ref. [33] and mentioned in Fig. 2.8.
**) From Ref. [41].
***) From Ref. [40].
****) From Ref. [39].

In other words, the SES activities can be distinguished from "artificial" noises by considering that for the SES activities *both* S and S_- are smaller than S_u, which is *not* the case for "artificial" noises , i.e.,

$$S, S_- < S_u \text{ for SES activities} \tag{4.32}$$

This happens *in addition to* the fact that for the SES activities the variance κ_1 is $\kappa_1 \approx 0.070$, while for "artificial" noises we have $\kappa_1 \geq \kappa_u \approx 0.083$, see § 4.2.2.

4.9.2 Distinction of SES activities from "artificial" noises on the basis of the fluctuations of natural time under time reversal

In § 2.2.2, it was discussed that a measure of the long-range dependence emerges in natural time if we study the dependence of the fluctuations of the average value of natural time under time reversal

$$\Delta\chi_l^2 \equiv \mathscr{E}[(\langle\chi\rangle - \langle\hat{T}\chi\rangle)^2] = \mathscr{E}\left\{\left[\sum_{k=1}^{l} \frac{k}{l}(p_k - p_{l-k+1})\right]^2\right\}, \tag{4.33}$$

on the window length l that is used for the calculation. In particular, it was shown that Eq. (2.19) holds, i.e,

$$\Delta\chi_l (\equiv \sqrt{\Delta\chi_l^2}) \propto l^{\chi_H} \tag{4.34}$$

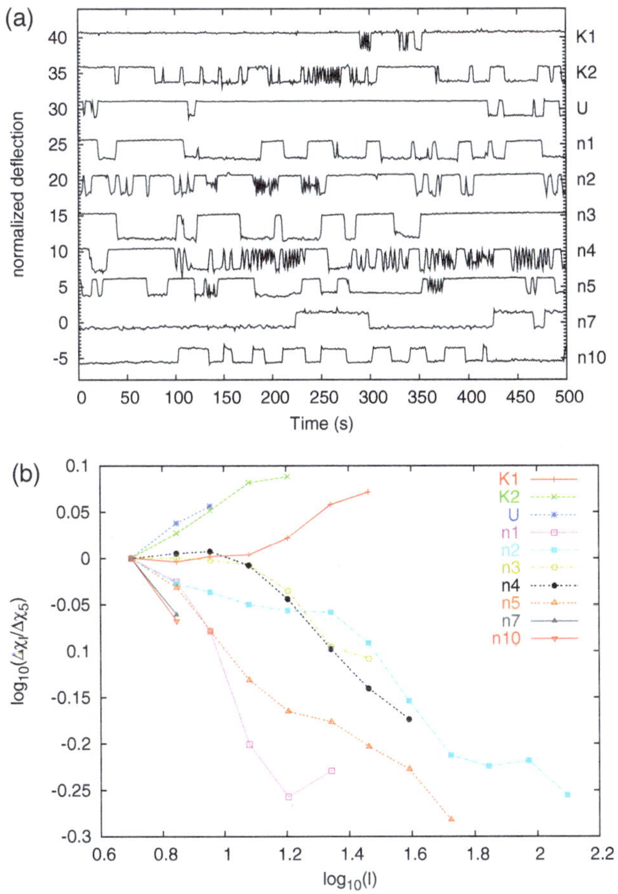

Fig. 4.23 The log-log plot of $\Delta\chi_l$ versus the scale l is depicted in (b) for three SES activities (K1, K2 and U) and seven "artificial" noises (n1-n5, n7 and n10) excerpts of which are shown in (a). The values of $\Delta\chi_l$ are divided by the corresponding values $\Delta\chi_5$ at the scale $l = 5$. Reprinted with permission from Ref. [39]. Copyright (2008), American Institute of Physics.

Hence, the scaling exponent χ_H can be determined from the slope of the $\log\Delta\chi_l$ versus $\log l$ plot. Recall also that in such a plot, we have the interconnection:

$$\chi_H \approx H - 1 \quad \text{for descending curves} \tag{4.35}$$

or

$$\chi_H = H \quad \text{for ascending curves} \tag{4.36}$$

We now show [39] that the aforementioned scale-dependence of the fluctuations of the natural time itself under time reversal provides a useful tool for the discrimination of SES activities from "artificial" noises. We apply this procedure to the time series of the durations of those signals analyzed in Ref. [43] that have enough number of pulses e.g. $\approx 10^2$, excerpts of which are depicted here in Fig. 4.23(a). The relevant results are shown in Fig. 4.23(b). An inspection of this figure interestingly indicates that *all* seven "artificial" noises correspond to *descending* $\Delta\chi_l$ curves versus the scale l, while the three SES activities to *ascending* curves (in a similar fashion as in Figs. 2.3(a) and 2.3(d), respectively) as expected from the fact that the latter exhibit [34] infinitely ranged temporal correlations (having H close to unity), while the former do not.

Hence, the method discussed here, which is based on the fluctuations of the average value of the natural time itself under time reversal, enables the identification of long-range correlations *even for datasets of small size* ($\approx 10^2$), thus allowing the distinction of SES activities from "artificial" noises.

4.10 Summary of the criteria in natural time for the distinction of SES activities from noise

By summarizing the previous Sections of this Chapter, the following three rules are put forward for the distinction between SES activities and "artificial" noises (AN).

First (note that each class of signals below is designated by the relevant subscript):

$$\kappa_{1,\mathrm{SES}} < \kappa_{1,ICFMC}(\approx \kappa_u) \leq \kappa_{1,\mathrm{AN}}, \tag{4.37}$$

where $\kappa_{1,ICFMC} \approx 0.080$ and $\kappa_u \approx 0.083$ and

$$\kappa_{1,\mathrm{SES}} \approx 0.070. \tag{4.38}$$

Second,

$$S_{\mathrm{SES}},\ (S_-)_{\mathrm{SES}} < S_u \leq S_{\mathrm{AN}}, \tag{4.39}$$

where S and S_- stand for the entropy in natural time and that under time reversal, respectively; the value S_u is the one of the "uniform" distribution, i.e., $S_u \approx 0.0966$. The S values themselves are used for the distinction when they differ markedly from S_u. On the other hand, if the S values are found to be close to S_u, which holds for the minority of the SES activities and the AN, the distinction can be better made by using the complexity measure λ_s of the fluctuations δS of the entropy (see § 4.8.3).

Third, if H denotes the generalized Hurst exponent $h(2)$ in natural time,

$$H_{\mathrm{AN}} < H_{\mathrm{SES}}, \tag{4.40}$$

where H_{SES} is close to unity, i.e,

$$H_{\mathrm{SES}} \approx 1.0 \tag{4.41}$$

and $H_{\mathrm{AN}} \leq 0.86$.

The same holds for the DFA exponent in natural time, i.e.,

$$0.86 < \alpha_{SES} \approx 1.0 \tag{4.42}$$

and

$$\alpha_{AN} = 0.65 - 0.80. \tag{4.43}$$

A safe distinction between SES activities and AN should not be solely based on the above three rules but should be used *in conjunction with* the criteria explained in Section 1.2. The basic spirit behind these rules is that SES activities exhibit critical behavior while AN do not. Some types of AN, however, may be also associated with criticality (e.g., when a "man-made" system approaches failure) and hence could in principle be misinterpreted on the basis of the above inequalities.

4.11 Procedure to analyze a long-duration SES activity in natural time

When a short duration SES activity has an obvious dichotomous nature, the procedure to read it in natural time is straightforward, i.e., the one shown in Fig. 2.1(a) where we considered Q_k as being proportional to the duration of the k-th pulse. This is the case, for example, of the SES activity recorded at IOA on April 18, 1995, whose original time series is shown in Fig. 1.11(a) (and see its excerpt in Fig. 4.2(a)), while Fig. 4.6 depicts how this SES activity is read in natural time.

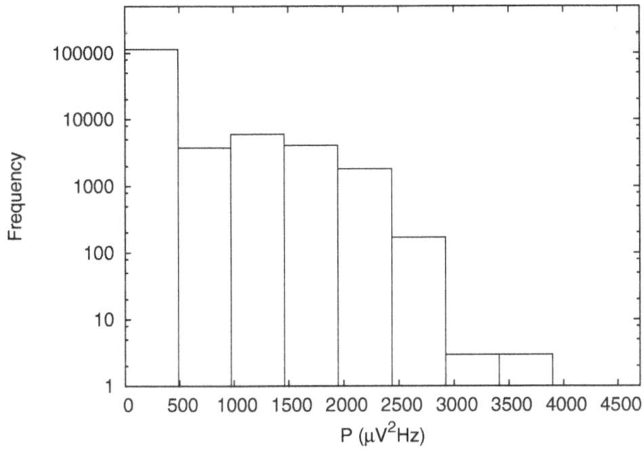

Fig. 4.24 Histogram of the "instantaneous power" P, i.e., the squared amplitude of the signal depicted in channel "e" of Fig. 1.16. Reprinted with permission from Ref. [35]. Copyright (2009), American Institute of Physics.

We now focus on a long-duration SES activity of a non-obvious dichotomous nature which is superimposed on a background that exhibits frequent small MT variations. Let us consider, for example, the SES activity that lasted from February 29 until March 2, 2008 (channel "a" of Fig. 1.16), for which the procedure to subtract the MT background variations has already been presented in § 1.4.3.1. This subtraction results in channel "e" of Fig. 1.16, which provides the time series that should be now analyzed in natural time: To obtain the time series (χ_k, Q_k), the individual pulses of the signal depicted in channel "e" of Fig. 1.16 have to be identified. A pulse starts, of course, when the amplitude exceeds a given threshold and ends when the amplitude falls below it. Moreover, since the signal is not obviously dichotomous, instead of finding the duration of each pulse, one should sum the "instantaneous power" during the pulse duration in order to find Q_k. To this end, we

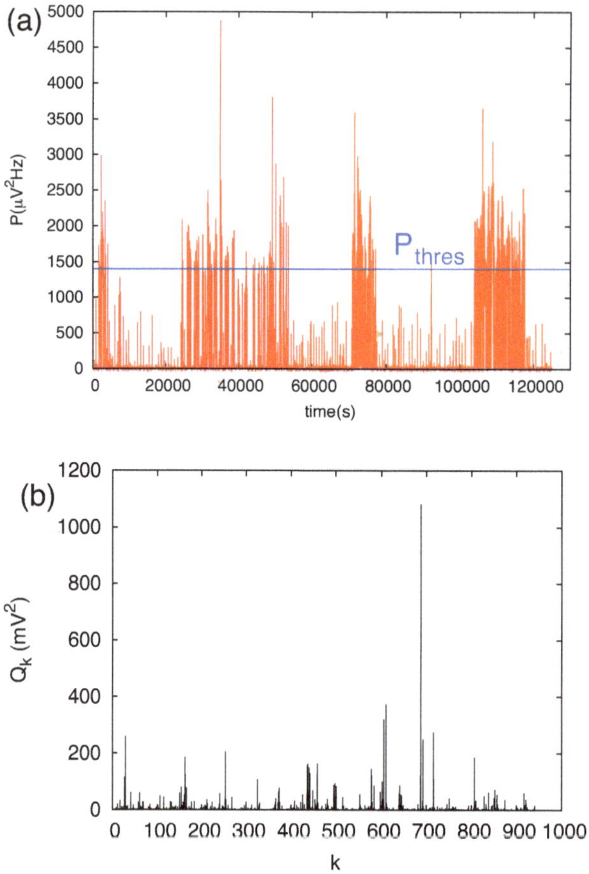

Fig. 4.25 (a) The "instantaneous power" P of the signal depicted in channel "e" of Fig. 1.16 (computed by squaring its amplitude). The solid line parallel to the x-axis marks an example of a threshold P_{thres} ($= 1400\ \mu V^2$ Hz) chosen. (b) The resulting representation of the signal depicted in channel "e" of Fig. 1.16 in natural time, when considering $P_{thres} = 1400\ \mu V^2$ Hz. Reprinted with permission from Ref. [35]. Copyright (2009), American Institute of Physics.

plot in Fig. 4.24 the histogram of the "instantaneous power" P of channel "e" of Fig. 1.16, computed by squaring its amplitude. An inspection of this figure reveals a bimodal feature which signifies the periods of inactivity ($P < 500\ \mu V^2$ Hz) and activity ($P > 500\ \mu V^2$ Hz) in channel "e" of Fig. 1.16. In order to find Q_k, we focus on the periods of activity and select the power threshold P_{thres} around the second peak of the histogram in Fig. 4.24. Let us consider, for example, the case of $P_{thres} = 1400\ \mu V^2$Hz. In Fig. 4.25(a), we depict the "instantaneous power" P of the signal in channel "e" of Fig. 1.16 versus time. Starting from the beginning of the signal, we compare P with P_{thres} and when P exceeds P_{thres} we start summing the P values until P falls below P_{thres} for the first time, $k = 1$. The resulting sum corresponds to Q_1. This procedure is repeated until P falls below P_{thres} for the second time, $k = 2$, and the new sum represents Q_2, etc. This leads to the natural time representation depicted in Fig. 4.25(b). The result depends, of course, on the proper selection of P_{thres}. The latter should be verified by checking whether a small change of P_{thres} around the second peak of the histogram leads to a natural time representation resulting in approximately the same values of the parameters κ_1, S and S_-. By randomly selecting

P_{thres} in the range 500 to 2,000 μV^2 Hz, we obtain that the number of pulses in channel "e" of Fig. 1.16 is $N = 1{,}100 \pm 500$ with $\kappa_1 = 0.070 \pm 0.007$, $S = 0.082 \pm 0.012$ and $S_- = 0.078 \pm 0.006$. When P_{thres} ranges between 1,000, and 1,500 μV^2 Hz, the corresponding values are $N = 1{,}200 \pm 200$ with $\kappa_1 = 0.068 \pm 0.003$, $S = 0.080 \pm 0.005$ and $S_- = 0.074 \pm 0.003$. Thus, we observe that irrespective of the P_{thres} value chosen, the parameters κ_1, S and S_- obey the conditions (4.38) and (4.39) for the classification of this signal as SES activity.

To summarize: natural time analysis allows the distinction between true SES activities and "artificial" (man-made) signals. This type of analysis, however, demands the knowledge of the energy released during each consecutive event. (Note that the determination of this energy is easier to conduct in the case of electric field variations, because the magnetic field variations appear in the form of "spikes" when using coil magnetometers which, as mentioned in § 1.4.4, act as dB/dt detectors.) *If* these electric field variations are of clear dichotomous nature, the energy release is proportional to the duration of each pulse. Otherwise, in the absence of an obvious dichotomous nature, an analysis of the "instantaneous power" similar to that presented above should be carried out to determine the parameters κ_1, S and S_- in natural time.

Acknowledgments We thank Professor P.N.R. Usherwood and Dr. I. Mellor for providing us with the experimental data of ion current through high-conductance locust potential channel.

References

1. Abramowitz, M., Stegun, I.: Handbook of Mathematical Functions. Dover, New York (1970)
2. Abry, P., Flandrin, P., Taqqu, M.S., Veitch, D.: Wavelets for the analysis, estimation and synthesis of scaling data. In: K. Park, W. Willinger (eds.) Self Similar Network Traffic Analysis and Performance Evaluation. Wiley, New York (2000)
3. Abry, P., Veitch, D., Flandrin, P.: Long-range dependence: Revisiting aggregation with wavelets. J. Time Ser. Anal. **19**, 253–266 (1998)
4. Audit, B., Bacry, E., Muzy, G., Arneodo, A.: Wavelet-based estimators of scaling behavior. IEEE Trans. Inf. Theory **48**, 2938-2954 (2002)
5. Bashan, A., Bartsch, R., Kantelhardt, J.W., Havlin, S.: Comparison of detrending methods for fluctuation analysis. Physica A **387**, 5080–5090 (2008)
6. Berezhkovskii, A.M., Weiss, G.H.: Detailed description of a two-state non-Markov system. Physica A **303**, 1-12 (2002)
7. Chen, Z., Ivanov, P.C., Hu, K., Stanley, H.E.: Effect of nonstationarities on detrended fluctuation analysis. Phys. Rev. E **65**, 041107 (2002)
8. Fuliński, A.: Active transport in biological membranes and stochastic resonances. Phys. Rev. Lett. **79**, 4926–4929 (1997)
9. Fuliński, A., Grzywna, Z., Mellor, I., Siwy, Z., Usherwood, P.N.R.: Non-Markovian character of ionic current fluctuations in membrane channels. Phys. Rev. E **58**, 919-924 (1998)
10. Gorczynska, E., Huddie, P., Miller, B., Mellor, I., Vais, H., Ramsey, R., Usherwood, P.: Potassium channels of adult locust (*Schistocerca gregaria*) muscle. Pflügers Archiv European Journal of Physiology **432**, 597–606 (1996)

11. Haar, A.: Zur theorie der orthogonalen funktionensysteme. (erste mitteilung). Mathematische Annalen **69**, 331–372 (1909)
12. Hu, K., Ivanov, P.C., Chen, Z., Carpena, P., Stanley, H.E.: Effect of trends on detrended fluctuation analysis. Phys. Rev. E **64**, 011114 (2001)
13. Hurst, H.E.: Long-term storage capacity of reservoirs. Trans. Am. Soc. Civ. Eng. **116**, 770–808 (1951)
14. Ivanov, P.C., Amaral, L.A.N., Goldberger, A.L., Havlin, S., Rosenblum, M.G., Stanley, H.E., Struzik, Z.R.: From 1/f noise to multifractal cascades in heartbeat dynamics. CHAOS **11**, 641–652 (2001)
15. Kantelhardt, J., Zschiegner, S.A., Koscienly-Bunde, E., Bunde, A., Havlin, S., Stanley, H.E.: Multifractal detrended fluctuation analysis of nonstationary time series. Physica A **316**, 87–114 (2002)
16. Kantelhardt, J.W., Koscielny-Bunde, E., Rego, H.H.A., Havlin, S., Bunde, A.: Detecting long-range correlations with detrended fluctuation analysis. Physica A **295**, 441–454 (2001)
17. Legendre, P.: K-means: Least squares partitioning method. Computer code K-Means available from http://www.bio.umontreal.ca/casgrain/en/labo/k-means.html (2001)
18. Legendre, P., Legendre, L.: Numerical Ecology. Elsevier, Amsterdam (1998)
19. Louis, A.K., Maass, P., Rieder, A.: Wavelets: Theory and Applications. Wiley, New York (1997)
20. Mandelbrot, B.B.: Gaussian Self-Affinity and Fractals. Springer-Verlag, New York (2002)
21. Mercik, S., Weron, K.: Stochastic origins of the long-range correlations of ionic current fluctuations in membrane channels. Phys. Rev. E **63**, 051910 (2001)
22. Muzy, J.F., Bacry, E., Arneodo, A.: The multifractal formalism revisited with wavelets. Int. J. Bifurcation CHAOS **4**, 245–302 (1994)
23. Siwy, Z., Fuliński, A.: Origin of $1/f^{\alpha}$ in membrane channel currents. Phys. Rev. Lett. **89**, 158101 (2002)
24. Siwy, Z., Mercik, S., Weron, K., Ausloos, M.: Application of dwell-time series in studies of long-range correlation in single channel ion transport: analysis of ion current through a big conductance locust potassium channel. Physica A **297**, 79–96 (2001)
25. Talkner, P., Weber, R.O.: Power spectrum and detrended fluctuation analysis: Application to daily temperatures. Phys. Rev. E **62**, 150–160 (2000)
26. Taqqu, M.S., Teverovsky, V., Willinger, W.: Estimators for long-range dependence: An empirical study. Fractals **3**, 785–798 (1995)
27. Varotsos, P.: The Physics of Seismic Electric Signals. TERRAPUB, Tokyo (2005)
28. Varotsos, P.A., Sarlis, N.V., Skordas, E.S.: See (the freely available) EPAPS Document No. E-PLEEE8-68-116309 originally from P.A. Varotsos, N.V. Sarlis and E.S. Skordas, *Phys. Rev. E* **68**, 031106 (2003). For more information on EPAPS, see http://www.aip.org/pubservs/epaps.html.
29. Varotsos, P.A., Sarlis, N.V., Skordas, E.S.: See (the freely available) EPAPS Document No. E-PLEEE8-67-110302 originally from P.A. Varotsos, N.V. Sarlis and E.S. Skordas, *Phys. Rev. E* **67**, 021109 (2003). For more information on EPAPS, see http://www.aip.org/pubservs/epaps.html.
30. Varotsos, P.A., Sarlis, N.V., Skordas, E.S.: Seismic Electric Signals and $1/f$ "noise" in natural time. arXiv:0711.3766v3 [cond-mat.stat-mech] (1 February 2008)
31. Varotsos, P.A., Sarlis, N.V., Skordas, E.S.: Spatio-temporal complexity aspects on the interrelation between Seismic Electric Signals and Seismicity. Practica of Athens Academy **76**, 294–321 (2001)
32. Varotsos, P.A., Sarlis, N.V., Skordas, E.S.: Long-range correlations in the electric signals that precede rupture. Phys. Rev. E **66**, 011902 (2002)
33. Varotsos, P.A., Sarlis, N.V., Skordas, E.S.: Attempt to distinguish electric signals of a dichotomous nature. Phys. Rev. E **68**, 031106 (2003)
34. Varotsos, P.A., Sarlis, N.V., Skordas, E.S.: Long-range correlations in the electric signals that precede rupture: Further investigations. Phys. Rev. E **67**, 021109 (2003)
35. Varotsos, P.A., Sarlis, N.V., Skordas, E.S.: Detrended fluctuation analysis of the magnetic and electric field variations that precede rupture. CHAOS **19**, 023114 (2009)
36. Varotsos, P.A., Sarlis, N.V., Skordas, E.S., Lazaridou, M.S.: See (the freely available) EPAPS Document No. E-PLEEE8-71-134501 originally from P.A. Varotsos, N.V. Sarlis, E.S. Skordas and M.S. Lazaridou *Phys. Rev. E* **71**, 011110 (2005). For more information on EPAPS, see http://www.aip.org/pubservs/epaps.html.
37. Varotsos, P.A., Sarlis, N.V., Skordas, E.S., Lazaridou, M.S.: Entropy in natural time domain. Phys. Rev. E **70**, 011106 (2004)

38. Varotsos, P.A., Sarlis, N.V., Skordas, E.S., Lazaridou, M.S.: Natural entropy fluctuations discriminate similar-looking electric signals emitted from systems of different dynamics. Phys. Rev. E **71**, 011110 (2005)
39. Varotsos, P.A., Sarlis, N.V., Skordas, E.S., Lazaridou, M.S.: Fluctuations, under time reversal, of the natural time and the entropy distinguish similar looking electric signals of different dynamics. J. Appl. Phys. **103**, 014906 (2008)
40. Varotsos, P.A., Sarlis, N.V., Skordas, E.S., Tanaka, H.K., Lazaridou, M.S.: Attempt to distinguish long-range temporal correlations from the statistics of the increments by natural time analysis. Phys. Rev. E **74**, 021123 (2006)
41. Varotsos, P.A., Sarlis, N.V., Skordas, E.S., Tanaka, H.K., Lazaridou, M.S.: Entropy of seismic electric signals: Analysis in the natural time under time reversal. Phys. Rev. E **73**, 031114 (2006)
42. Varotsos, P.A., Sarlis, N.V., Tanaka, H.K., Skordas, E.S.: See (the freely available) EPAPS Document No. E-PLEEE8-71-081503 originally from P.A. Varotsos, N.V. Sarlis, H.K. Tanaka and E.S. Skordas, *Phys. Rev. E* **71**, 032102 (2005). For more information on EPAPS, see http://www.aip.org/pubservs/epaps.html.
43. Varotsos, P.A., Sarlis, N.V., Tanaka, H.K., Skordas, E.S.: Some properties of the entropy in the natural time. Phys. Rev. E **71**, 032102 (2005)
44. Veitch, D., Abry, P.: A wavelet-based joint estimator of the parameters of long-range dependence. IEEE Trans. Inf. Theory **45**, 878–897 (1999)
45. Veitch, D., Abry, P., Chainais, P.: Wavelet estimation tools. Computer codes `Wavelet Estimation Tools` available from http://www.cubinlab.ee.unimelb.edu.au/~darryl/ (2002)
46. Weber, R.O., Talkner, P.: Spectra and correlations of climate data from days to decades. J. Geophys. Res. **106(D17)**, 20,131–20,144 (2001)
47. Weiss, G.H.: Some applications of persistent random walks and the telegrapher's equation. Physica A **311**, 381–410 (2002)
48. Weron, A., Burnecki, K., Mercik, S., Weron, K.: Complete description of all self-similar models driven by Lévy stable noise. Phys. Rev. E **71**, 016113 (2005)
49. Willinger, W., Taqqu, M., Sherman, R., Wilson, D.: Self-similarity through high variability: Statistical analysis of ethernet LAN traffic at the source level. IEEE/ACM Trans. on Networking **5**, 71–86 (1997)

5. Natural Time Investigation of the Effect of Significant Data Loss on Identifying Seismic Electric Signals

Abstract. In many cases of geophysical and/or geological interest, like the case of SES observations, it happens that for substantial parts of the time of data collection high noise prevents any attempt to extract a useful signal so that such data are removed from further analysis. The appearance of such a noise may be either random or periodic. It is the objective of this chapter to examine whether the remaining data allow the identification of the SES activities (*critical dynamics*) characterized by infinitely ranged temporal correlations. The following two cases are treated here. First, the effect of the random removal of data segments of fixed length on the scaling properties of SES activities. Second, the appearance of a periodic noise like in Japan, where the electric field measurements at some sites are seriously contaminated by high noise – due mainly to leakage currents from DC driven trains – during the period 06:00 to 22:00 LT every day, i.e., around 70–80% data loss. Here, we show that, in both cases, the identification of a SES activity becomes possible with probability around 70% even after severe data loss (e.g., 70-80 %). This is achieved by combining natural time analysis and DFA as follows: the remaining data is first represented in natural time and then analyzed in order to deduce the quantities κ_1, S and S_- as well as the exponent α from the slope of the log-log plot of the DFA analysis in natural time. We then examine whether the latter slope has a value close to unity *and/or* the conditions $\kappa_1 \approx 0.070$ and S, $S_- < S_u$ are obeyed.

5.1 Introduction

It is the basic aim of this chapter to investigate how significant data loss affects the scaling behavior of SES activities, which are long-range correlated signals (see § 1.4.3, § 2.4.2 and Chapter 4). The practical importance of this study becomes very clear upon considering that such a data loss is inevitable, mainly due to the following two reasons. First, failure of the measuring system in the field station (including the electric measuring dipoles, the electronics and the data collection system) may occur especially due to lightning. Second, noise-contaminated data segments are often unavoidable due to natural changes such as rainfall, lightning, induction of geomagnetic field variations and ocean-earth tides besides the noise from "artificial" (man-made) sources.

In particular, let us focus on the geoelectrical measurements in Japan. They face the major difficulty that at many sites the recordings are contaminated by high noise due to leakage currents from DC-driven trains and other artificial sources (e.g., see Ref. [2]). Clear SES, however, have been recognized either at noise-free measuring sites [8, 5, 7] or at noisy stations but when the SES happened to occur at midnight, i.e., when the noise level was low [2]. This low noise level occurs from 00:00 to 06:00 and from 22:00 to 24:00 local time (LT) when nearby DC-driven trains cease service, i.e., almost 30% of 24 hours. The answer is attempted here in Section 5.3 provided that the SES activity is of appreciably long duration, i.e., a few days to a few weeks or even more as in the case of the Izu island swarm [5, 7]. This question differs from the one treated in Section 5.2 in which we investigate [4] the effect of the *random* removal of data segments of fixed length on the scaling properties of SES activities. It also differs from the recent study of Ma et al. [1] in which the lengths of the lost or removed data segments are random and may follow a certain type of distribution. In the latter study, a new segmentation approach to generate surrogate signals by randomly removing data segments from stationary signals with different types of long-range correlations has been introduced [1] and will be used here.

5.2 Identification when removing randomly noise-contaminated data segments of fixed length

Let us consider here as an example the SES activity depicted in Fig. 1.11(a) recorded on April 18, 1995, that preceded the M_w6.6 earthquake on May 13, 1995 (see § 7.2.1). This is reproduced here but in normalized units in Fig. 5.1(a).

Following Ma et al. [1], we now describe the segmentation approach used here to generate surrogate signals $\tilde{u}(i)$ by randomly removing data segments of length L from the original signal $u(i)$. The percentage p of the data loss, i.e., the percentage of the data removed, characterizes the signal $\tilde{u}(i)$. The procedure followed is based on the construction of a binary time series $g(i)$ of the same length as $u(i)$. The values of $u(i)$ that correspond to $g(i)$ equal to unity are kept, whereas the data of $u(i)$ when $g(i)$ equals zero are removed. The values of $u(i)$ kept, are then concatenated to construct $\tilde{u}(i)$.

The binary time series $g(i)$ is obtained as follows [1]: (i) We first generate the lengths $l_j = L$ with $j = 1,2,\ldots,M$ of the removed segments, by selecting M to be the smallest integer so that the total number of removed data satisfies the condition $\sum_{j=1}^{M} l_j \geq \mathrm{p}N$. (ii) We then construct an auxiliary time series $a(k)$ with $a(k) = L$ when $k = 1,2,\ldots,M$ and $a(k) = 1$ when $k = M+1,\ldots,N-M(L+1)$ of size $N-M(L+1)$. (iii) We shuffle the time series $a(k)$ randomly to obtain $\tilde{a}(k)$. (iv) We then append $\tilde{a}(k)$ to obtain $g(i)$: if $\tilde{a}(k) = 1$ we keep it, but we replace all $\tilde{a}(k) = L$ with L elements of value '0' and one element with value '1'. In this way, a binary series $g(i)$ is obtained, which has a size equal to the one of the original signal $u(i)$. We then construct the surrogate signal $\tilde{u}(i)$ by simultaneously

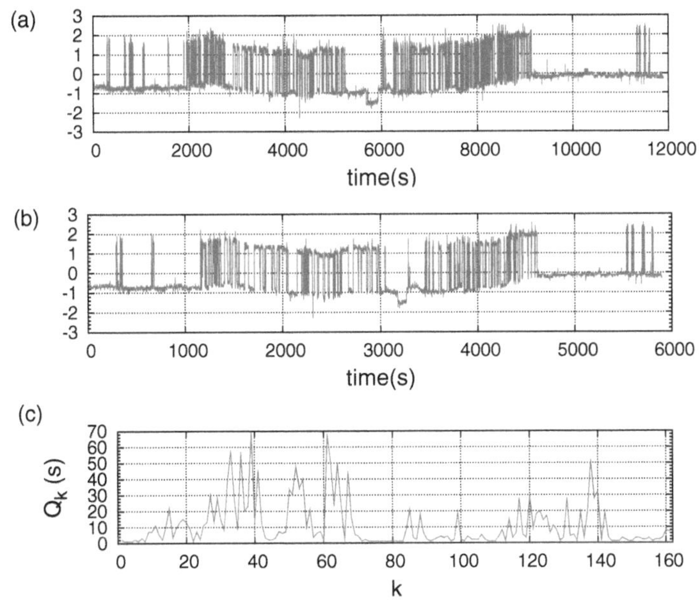

Fig. 5.1 (a): The electric field recordings in normalized units, i.e., by subtracting the mean value μ and dividing by the standard deviation σ, for the SES activity depicted in Fig. 1.11(a), that was recorded on April 18, 1995. (b): Example of a surrogate time series, in normalized units, obtained by removing segments of length $L = 200$ from the signal depicted in (a) with 50% data loss (i.e., p = 0.50). (c):The natural time representation of (b) but drawn with continuous lines for the reader's convenience (in reality, this should be plotted as in the lower part of Fig. 2.1(a) or as in Fig. 7.2(b)). The values obtained from the analysis of (c) in natural time are $\kappa_1 = 0.067(4)$, $S = 0.076(4)$, $S_- = 0.071(4)$ and $a = 0.90(5)$. Reprinted with permission from Ref. [4]. Copyright (2010), American Institute of Physics.

scanning the original signal $u(i)$ and the binary series $g(i)$, removing the i-th element of $u(i)$ if $g(i) = 0$ and concatenating the segments of the remaining data to $\tilde{u}(i)$.

The resulting signal $\tilde{u}(i)$ is subsequently read in natural time. This leads to the quantities κ_1, S and S_- as well as to the DFA exponent α (see § 1.4.2) in natural time. Such an example is given in Fig. 5.1(b), which was drawn on the basis of the SES activity data depicted in Fig. 5.1(a).

Typical DFA plots, obtained for $L = 200$ and p = 30, 50 and 70% are given in Fig. 5.2. For the sake of comparison, this figure also includes the case of no data loss (i.e., p = 0). We notice a gradual decrease of the exponent α upon increasing the data loss, which affects our ability to recognize a signal as SES activity. In order to evaluate this ability to identify SES activities from the natural time analysis of surrogate signals with various levels of data loss, three procedures have been attempted which will be explained separately below. They include the investigation of the following conditions (5.1) and (5.2). The condition (5.1) comes from the relation (4.42) after considering the reasonable experimental error:

$$0.85 \leq \alpha \leq 1.10. \tag{5.1}$$

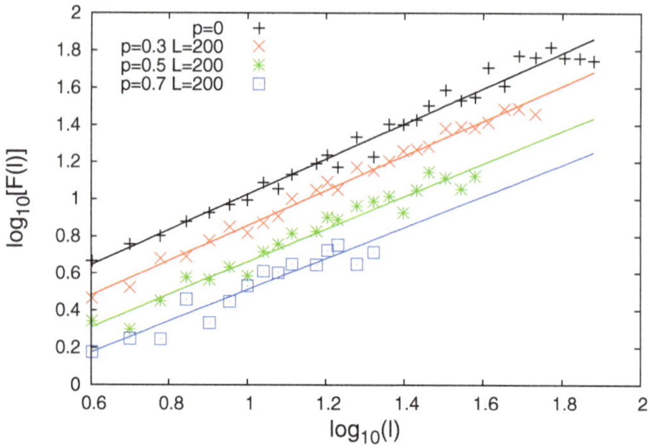

Fig. 5.2 The dependence of the DFA fluctuation function $F(l)$ (see § 1.4.2) versus the scale l in natural time: we increase the percentage of data loss p by removing segments of length $L = 200$ samples from the signal of Fig. 5.1(a). The plus symbols correspond to no data loss (p = 0), the crosses to 30% data loss (p = 0.3), the asterisks to 50% data loss (p = 0.5) and the squares to 70% data loss (p = 0.7). Except for the case p = 0, the data have been shifted vertically for the sake of clarity. The slopes of the corresponding straight lines that fit the data lead to $\alpha = 0.95$, 0.94, 0.88 and 0.84 from the top to bottom, respectively. They correspond to the average values of α obtained from 5000 surrogate time series that were generated with the method described in the text. Reprinted with permission from Ref. [4]. Copyright (2010), American Institute of Physics.

The conditions (5.2) come from Eqs. (4.38) and (4.39) by considering the reasonable experimental error in κ_1:

$$|\kappa_1 - 0.07| \leq 0.01, \;\; S \leq S_u, \;\; S_- \leq S_u. \tag{5.2}$$

In the following, the produced surrogate signals will be investigated whether they obey conditions (5.1) and/or (5.2) using a Monte Carlo comprising a reasonable number of realizations, e.g., of the order of 10^3. The probability that the condition (5.1) is satisfied will be hereafter labeled p_1. By the same token, the probability to satisfy the conditions (5.2) is designated by p_2. Finally, the probability to obey either condition (5.1) or conditions (5.2) will be labeled p_3. Upon considering the number of the Monte Carlo realizations, for example say $M = 10^3$, a plausible estimation error ($3\mathrm{STD}/\sqrt{M}$) at the most around 5% is expected (cf. $1/\sqrt{10^3} \approx 0.032$, and STD stands for the standard deviation of the quantity calculated by Monte Carlo, e.g. see Ref. [12]). The three procedures studied refer to the following investigation(s).

Procedure 1: Investigation whether the exponent α, resulted from the DFA analysis of the natural time representation of a signal, obeys the relation (5.1). If it does, the signal is then classified as SES activity. Figure 5.3(a) shows that for a given amount of data loss (p = const.), upon increasing the length L of the randomly removed segments, the probability p_1 of achieving (after making 5000 attempts for a given value of p and L) the identification of the signal as SES activity is found to gradually increase versus L at small scales and stabilizes at large scales. For example, when considering the case of 70% data loss (magenta color in Fig. 5.3(a)) the probability p_1 is close to 20% for $L = 50$; it increases to $p_1 \approx 30\%$ for $L = 100$ and finally stabilizes around 50% for lengths $L = 300$ to 500. This is essentially consistent with the earlier findings of Ma et al. [1] who noticed that removing the same percentage of data using longer (and fewer) segments has a lesser impact on the scaling behavior compared to removing segments with smaller average length.

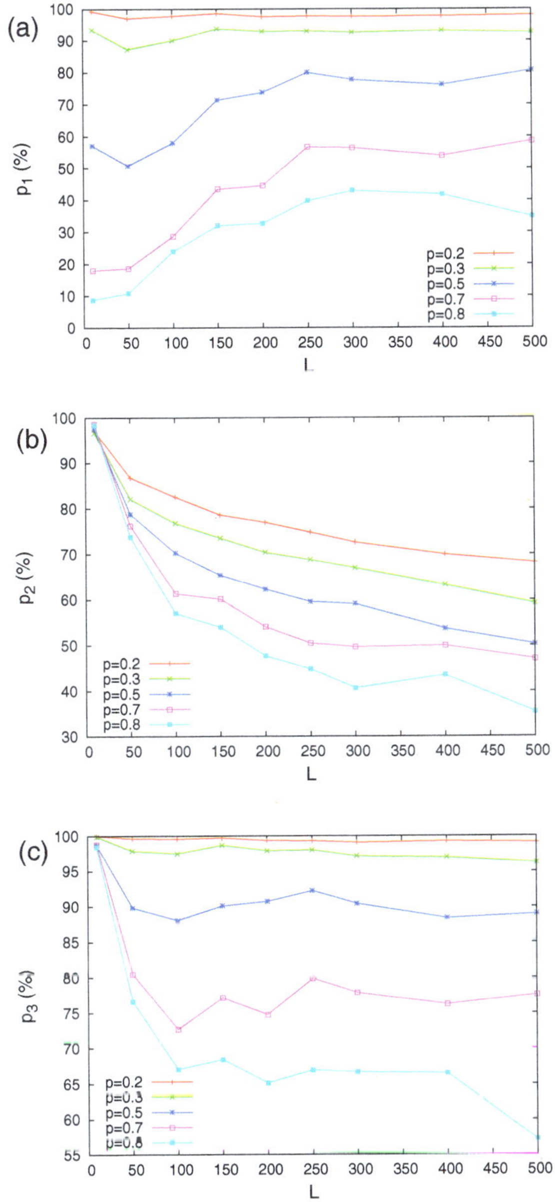

Fig. 5.3 The probabilities p_1, panel (a), p_2, panel (b), and p_3, panel (c), to recognize the signal of Fig. 5.1(a) as true SES activity when considering various percentages of data loss p = 0.2, 0.3, 0.5, 0.7 and 0.8 as a function of the length L of the contiguous samples removed. Note that the removal of large segments leads to better results when using the condition (5.1), see (a), whereas the opposite holds when using the conditions (5.2), see (b). The optimum selection (c) for the identification of a signal as SES activity consists of a proper combination of the aforementioned procedures in (a) and (b); see *Procedure 3* described in the text. The values presented have been obtained from 5000 surrogate time series (for a given value of p and L), which lead to a plausible error 1.4% ($\simeq 1/\sqrt{5{,}000}$). Reprinted with permission from Ref. [4]. Copyright (2010), American Institute of Physics.

Procedure 2: Investigation whether the quantities κ_1, S and S_- (resulted from the analysis of a signal in natural time) obey the conditions (5.2). If they do so, the signal is classified as SES activity. Figure 5.3(b) shows that for a given amount of data loss, the probability p_2 of achieving the signal identification as SES activity – that results after making again 5,000 attempts for each given value of p and L – gradually decreases when moving from the small to large scales. Note that for the smallest length scale investigated, i.e., $L = 10$ (which is more or less comparable – if we consider that $f_{exp} = 1$ sample/sec – with the average duration ≈11 sec of the transient pulses that constitute the signal, see § 4.1.3 and table I of Ref. [9]), the probability p_2 reaches values close to 100% even for the extreme data loss of 80%. This is understood in the context that the quantities κ_1, S and S_- remain almost unaffected when randomly removing segments with lengths comparable to the average pulse's duration. This is consistent with our earlier finding [11] that the quantities κ_1, S and S_- are experimentally stable (Lesche's stability, § 3.3.4) in the sense that they exhibit only slight variations when deleting (due to experimental errors) a small number of pulses. On the other hand, at large scales L, the probability p_2 markedly decreases. This may be understood if we consider that, at such scales, each segment of contiguous L samples comprises on the average a considerable number of pulses the removal of which may seriously affect the quantities κ_1, S and S_-. As an example, for 80% data loss (cyan curve in Fig. 5.3(b)), and for $L = 400$–500, the p_2 value becomes 40%.

Interestingly, a closer inspection of Figs. 5.3(a) and 5.3(b) reveals that p_1 and p_2 play *complementary* roles. In particular, at small scales of L, the probability p_1 increases but p_2 decreases versus L. At large scales, where p_1 reaches (for considerable data loss) its largest value, the p_2 value becomes small. Inspired by this complementary behavior of p_1 and p_2, the following combined procedure was investigated.

Procedure 3: In this procedure, a signal is identified as SES activity when it obeys *either* the condition (5.1) *or* the conditions (5.2). The probability p_3 of achieving such a SES identification, after making 5,000 attempts (for a given value of p and L), is plotted in Fig. 5.3(c). The results are remarkable since, even at significant values of data loss, e.g., p = 70% or 80%, the probability p_3 at scales $L = 100$ to 400 remains relatively high, i.e., $p_3 \approx 75\%$ and 65%, respectively (note also that the value of p_3 reaches values close to 100% at small scales $L = 10$). This is important from a practical point of view, because it states for example the following. Even if the records of a station are contaminated by considerable noise, say 70% of the time of its operation, we have a probability of ≈75% to correctly identify an SES activity from the remaining 30% of the non-contaminated segments. This probability increases considerably, i.e., to $p_3 \approx 90\%$, if only half of the recordings are noisy.

The aforementioned results have been deduced from the analysis of a SES activity lasting around three hours. In cases of SES activities with appreciably longer duration (Section 1.3), e.g., a few to several days detected in Greece [3, 10] or a few months in Japan [7], the results should become appreciably better.

In summary, the identification of a SES activity becomes possible even after significant data loss by employing the following procedure. The remaining data are first read in natural time and then analyzed in order to deduce the quantities κ_1, S and S_- as well as the exponent α from the slope of the log-log plot of the DFA analysis in natural time. We then examine whether this α value is close to unity (cf. Eq. (5.1)) *or* the conditions (5.2) are obeyed. This leads to the following results: even when randomly removing 50% of the data, we have a probability (p_3) around 90%, or higher, to identify correctly a SES activity. This probability becomes somewhat smaller, i.e., 75%, when the data loss increases to 70%.

5.3 Identification upon significant periodic data loss. The case of Japan

As already mentioned in Section 5.1, in geoelectrical field measurements at some sites in Japan, high noise prevails almost during 70% of the 24-hour operational time. Thus, the question arises whether it is still possible to identify a long-duration SES activity upon removing the noisy data segments lasting from 06:00 to 22:00 LT every day.

Let us suppose that we have a long time series of data $s(i)$ (red in Fig. 5.4), with a duration appreciably longer than 24 hours for instance, and we are forced to remove the same segment of these daily data. The portion of the 24-hour data that remain will be hereafter labeled $\mathrm{p_r}$ and the number of data corresponding to one day will be designated by T. Thus, every T samples, $(1-\mathrm{p_r})T$ of them (belonging to the shaded parts of Fig. 5.4) are removed. The remaining segments (drawn in blue in Fig. 5.4) are concatenated to form the new time series $c(i)$ which is subsequently read in natural time.

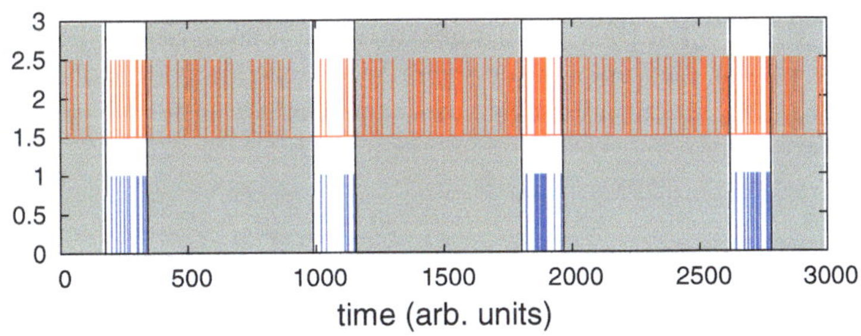

Fig. 5.4 Schematic diagram showing data of fixed length (grey shaded areas) that are periodically removed from a dichotomous time series (red)

Let us now follow a procedure similar to that in Section 5.2 and impose the conditions (5.1) and (5.2). Since we are interested in the low cultural noise night-window, we hereafter focus on $\mathrm{p_r}$ values varying from $p_r = 0.2$ to roughly $p_r = 0.3$.

We recall (Section 1.3) that SES activities of appreciably long duration, i.e, around a few weeks or more, similar to the one observed by Uyeda and coworkers [5, 7] almost two months before the case of the Izu island swarm in Japan (Section 7.4), have not been recorded in Greece. Here, we consider as an example the SES activity depicted in Fig. 1.16 that preceded [10, 6] the most recent major earthquake in Greece. This had almost a 1.5-day duration, which lasted from February 29 to March 2, 2008, and was followed by a magnitude M_w6.4 earthquake at 38.0°N 21.5°E on June 8, 2008 (see § 7.2.6). Its original time series, which is not of an obvious dichotomous nature, is depicted in Fig. 1.16 (channel "a"). We now attempt to answer the following question: if such an SES activity had been recorded in Japan, could its identification have become possible by employing a procedure similar to the *Procedure 3* proposed in the previous Section 5.2? Before applying it, we note that the signal under discussion, as is evident from an inspection of Fig. 1.16 (channel "a"), comprises a number of pulses superimposed on a background which exhibits frequent magnetotelluric (MT) variations. After subtracting these MT variations, as explained in § 1.4.3.1, we find the signal depicted in Fig. 1.16 (channel "e"), which provides the time series that should be considered for further analysis. Recall that its analysis in natural time (i.e., for $\mathrm{p}_r = 0$) was presented in Section 4.11.

To answer the question for the possibility to identify this SES activity in Japan after significant data loss, a Monte Carlo calculation was employed by considering that the first segment to keep starts at some time uniformly distributed during the first 24 hours (i.e, the first 86,400 samples since $f_{exp} = 1$ sample/s). When removing 70% of the data (i.e., $\mathrm{p}_r = 0.3$), we find a probability $p_3 \approx 67\%$ to identify correctly the SES activity (in particular, $p_1 = 0.40$, $p_2 = 0.54$ and $p_3 = 0.67$). This probability becomes somewhat smaller, i.e., $\approx 62\%$, upon increasing the data loss to 80% (in particular, $p_1 = 0.41$, $p_2 = 0.40$ and $p_3 = 0.62$). These values of the probability p_3 would be expected to become markedly larger if the duration of the SES activity were to be similar to the one observed before the Izu island seismic swarm.

Thus, our main conclusion states that when employing two modern techniques, i.e., natural time analysis and DFA, an identification of a long-duration SES activity becomes possible even after removing periodically a significant portion of the data, e.g., even upon removing in Japan the noisy data segments lasting for the period 06:00 to 22:00 LT every day.

References

1. Ma, Q.D.Y., Bartsch, R.P., Bernaola-Galván, P., Yoneyama, M., Ivanov, P.C.: Effect of extreme data loss on long-range correlated and anticorrelated signals quantified by detrended fluctuation analysis. Phys. Rev. E **81**, 031101 (2010)
2. Orihara, Y., Kamogawa, M., Nagao, T., Uyeda, S.: Independent component analysis of geoelectric field data in the northern Nagano, Japan. Proc. Jpn. Acad., Ser. B: Phys. Biol. Sci. **85**, 435–442 (2009)
3. Sarlis, N.V., Skordas, E.S., Lazaridou, M.S., Varotsos, P.A.: Investigation of seismicity after the initiation of a seismic electric signal activity until the main shock. Proc. Japan Acad., Ser. B **84**, 331–343 (2008)

4. Skordas, E.S., Sarlis, N.V., Varotsos, P.A.: Effect of significant data loss on identifying electric signals that precede rupture estimated by detrended fluctuation analysis in natural time. CHAOS **20**, 033111 (2010)
5. Uyeda, S., Hayakawa, M., Nagao, T., Molchanov, O., Hattori, K., Orihara, Y., Gotoh, K., Akinaga, Y., Tanaka, H.: Electric and magnetic phenomena observed before the volcano-seismic activity in 2000 in the Izu Island Region, Japan. Proc. Natl. Acad. Sci. USA **99**, 7352–7355 (2002)
6. Uyeda, S., Kamogawa, M.: Reply to Comment on 'The prediction of two large earthquakes in Greece'. EOS Trans. AGU **91**, 163 (2010)
7. Uyeda, S., Kamogawa, M., Tanaka, H.: Analysis of electrical activity and seismicity in the natural time domain for the volcanic-seismic swarm activity in 2000 in the Izu Island region, Japan. J. Geophys. Res. **114**, B02310 (2009)
8. Uyeda, S., Nagao, T., Orihara, Y., Yamaguchi, T., Takahashi, I.: Geoelectric potential changes: Possible precursors to earthquakes in Japan. Proc. Natl. Acad. Sci. USA **97**, 4561–4566 (2000)
9. Varotsos, P.A., Sarlis, N.V., Skordas, E.S.: See (the freely available) EPAPS Document No. E-PLEEE8-68-116309 originally from P.A. Varotsos, N.V. Sarlis and E.S. Skordas, *Phys. Rev. E* **68**, 031106 (2003). For more information on EPAPS, see http://www.aip.org/pubservs/epaps.html.
10. Varotsos, P.A., Sarlis, N.V., Skordas, E.S.: Detrended fluctuation analysis of the magnetic and electric field variations that precede rupture. CHAOS **19**, 023114 (2009)
11. Varotsos, P.A., Sarlis, N.V., Tanaka, H.K., Skordas, E.S.: Some properties of the entropy in the natural time. Phys. Rev. E **71**, 032102 (2005)
12. Weinzierl, S.: Introduction to Monte Carlo methods. arXiv:hep-ph/0006269v1 (2000)

6. Natural Time Analysis of Seismicity

Abstract. Assuming that a mainshock may be considered as a new phase, the natural time analysis of the seismicity reveals that the normalized power spectrum $\Pi(\phi)$ for small ϕ ($\phi \to 0$) *or* the quantity $\kappa_1 (= \langle \chi^2 \rangle - \langle \chi \rangle^2)$ may be considered as an order parameter for seismicity. The probability distribution $P(\kappa_1)$ of this order parameter is obtained from the calculation of the variance κ_1 when a time window of length l (= number of consecutive events) is sliding through an earthquake catalog. The κ_1 value at which this probability distribution $P(\kappa_1)$ maximizes is designated by $\kappa_{1,p}$. By using $P(\kappa_1)$, we find: first, studying the order parameter fluctuations relative to the standard deviation of its distribution, we observe that (a) the scaled distributions of different seismic areas (as well as that of the worldwide seismicity) fall on a *universal* curve and (b) this curve exhibits an "exponential tail" similar to that observed in certain non-equilibrium systems (e.g. 3D turbulent flow) as well as in several equilibrium critical phenomena, e.g., 2D Ising, 3D Ising, 2D XY. Second, the constant b in the Gutenberg–Richter (G-R) law for EQs, $N(\geq \mathrm{M}) = 10^{a-b\mathrm{M}}$, is determined from the Maximum Entropy Principle which leads to $b \approx 1$ in accordance with the b value obtained from real seismic data. Third, by analyzing either the original earthquake catalog or a shuffled one the following results are obtained for the Southern California Earthquake Catalog (SCEC) as well as for the Japanese Meteorological Agency Earthquake Catalog (Japan). Concerning the $\kappa_{1,p}$ values, we find $\kappa_{1,p} = 0.066$ for the original data, while $\kappa_{1,p} = 0.064$ for the randomly shuffled data (with possible uncertainty of ± 0.001). Both these $\kappa_{1,p}$ values, the difference of which is shown to be associated with temporal correlations between the EQ magnitudes M, differ markedly from the value $\kappa_u = 1/12 (\approx 0.083)$ of the "uniform" distribution, which is interpreted as reflecting that the process's increments' infinite variance contributes significantly to self-similarity. Fourth, upon employing multifractal cascades (generalized Cantor sets) in natural time an interconnection between $\kappa_{1,p}$ and the parameter b of the G-R law is obtained which for $b \approx 1$ leads to $\kappa_{1,p} = 0.064$ that coincides with the $\kappa_{1,p}$ value obtained from the (randomly) shuffled earthquake data of Japan and SCEC. Fifth, by applying DFA to the earthquake magnitude time series of the SCEC and Japan data, we confirm that temporal correlations exist between EQ magnitudes. Sixth, focusing on the order parameter fluctuations of seismicity before and after mainshocks, we find the following. The $P(\kappa_1)$ versus κ_1 plot *before* mainshocks exhibits a significant bimodal feature which is reminiscent of the bimodal feature observed in the pdf of the order parameter when approaching (from below) T_c in equilibrium critical phenomena. Finally, the G-R law or its generalization in the frame of the nonextensive statistical mechanics, *if* combined with natural time, which captures the temporal correlations between EQ magnitudes, can reproduce the features of real seismic data.

6.1 Earthquake scaling laws

It is well known that earthquakes exhibit scaling relations, chief among which are the following three.

First, the well known Gutenberg–Richter (G-R) scaling [33] (hereafter called the G-R law) which states that the (cumulative) number of earthquakes (EQs) with magnitude greater than (or equal to) M, $N(\geq \mathrm{M})$, occurring in a specified area and time is given by

$$N(\geq \mathrm{M}) = 10^{a-b\mathrm{M}}. \tag{6.1}$$

In this relation b is a constant, which varies only slightly from region to region being generally in the range $0.8 \leq b \leq 1.2$ (e.g., see Ref. [53] and references therein) and the constant a gives the logarithm of the number of EQs with magnitude greater than zero [61] being a measure of the intensity of regional seismicity [75] (note that this relation holds both *regionally* and *globally*). For reasons of convenience, we hereafter write the G-R law in the following form

$$N(\geq \mathrm{M}) \propto 10^{-b\mathrm{M}}. \tag{6.2}$$

Considering that the seismic energy E released during an earthquake is related [39] to the magnitude through

$$E \propto 10^{c\mathrm{M}}, \tag{6.3}$$

where c is around 1.5, Eq. (6.2) turns to the distribution of Eq. (2.98), i.e.,

$$P(E) \propto E^{-\gamma} \tag{6.4}$$

for the earthquake energies E, where

$$\gamma = 1 + b/1.5. \tag{6.5}$$

Hence, $b \approx 1$ means that the exponent γ is around $\gamma = 1.6$ to 1.7, see Table 2.1.

Second, a scaling relation (the modified form [76] of Omori's law) describes the temporal decay of aftershock activity and is given in the form (e.g. see Ref. [61]):

$$r(t,\mathrm{M}) = \frac{1}{\tau_0\,[1 + t/c(\mathrm{M})]^p}, \tag{6.6}$$

where $r(t, \mathrm{M})$ is the rate of occurrence of aftershocks with magnitudes greater than M per day, t is the time that has elapsed since the mainshock and τ_0 and $c(\mathrm{M})$ are characteristic times. Note that $p \approx 1$ for large earthquakes (e.g., see Ref. [54]).

Third, the Båth law [15] for aftershocks according to which the difference in magnitude between a mainshock and its largest aftershock is approximately 1.2, a constant independent of the mainshock magnitude.

However, deviations from these scaling laws have been observed and their explanation has attracted a great interest (e.g., see Ref. [43] and references therein). Despite the intense efforts, however, the mechanism behind the complex spatio-temporal behavior of earthquakes still remains a major challenge [12, 44].

It is widely accepted [74, 66, 23] that the aforementioned earthquake scaling laws as well as others (e.g., referring to the distribution $\propto 1/L^2$ of fault lengths L [67], the fractal structure of fault networks [27], the universal law for the distribution of waiting times and seismic rates derived by Bak et al. [11] from the analysis of space-time windows) indicate the existence of phenomena closely associated with the *proximity* of the system to a critical point [35].

In view of this widespread belief, an order parameter for seismicity has been proposed [84] in the frame of natural time, which is explained in § 6.2.1. On the basis of this order parameter, a detailed study of the correlations in real seismic data has been made, a review of which is provided in this Chapter.

6.2 The order parameter and the universal curve for seismicity. The b value of the G-R law from first principles

6.2.1 The order parameter proposed for seismicity

As already mentioned (see Section 6.1), the occurrence of mainshocks can be considered as a critical point (second-order phase transition), but alternative models based on first-order phase transitions have also been proposed, e.g., see Ref. [53] and references therein. (Such a diversity also exists for the brittle rupture which is a phenomenon closely related to earthquakes. Buchel and Sethna [22] have associated brittle rupture with a first-order transition and a similar view has been also expressed in Refs. [42, 89]. On the other hand, Gluzman and Sornette [32] later suggested that it is analogous to a critical point phenomenon.) Both approaches lead to scaling laws or power law distributions for the dynamical variables, because:

Second-order transitions demonstrate scaling near a critical point, whereas *first-order* transitions demonstrate scaling when the range of interactions is large (mean-field condition), as is the case with elastic interactions [53].

Following the wording of Ref. [60], we note that in general:

"A choice of an order parameter is an art, since usually it's a new phase which we do not understand yet, and guessing the order parameter is a piece of figuring out what's going on."

We now proceed to choose the order parameter for seismicity by assuming that a mainshock may be considered as the new phase. We take advantage of the experimental fact (Chapter 1) that several hours to a few months before a mainshock SES are recorded, which probably signals that the system enters into the critical regime; see the pressure-stimulated currents SES generation model described in § 1.6.2 (see also § 2.4.2).

Therefore, we focus our attention on the evolution of the seismicity (in the candidate area) during the period from the SES detection until the mainshock.

If we set the natural time for the seismicity zero at the initiation of the concerned SES activity, we form time series of seismic events in natural time, e.g., see Fig. 2.1(b), for various time windows as the number N of consecutive (small) EQs increases. When computing the normalized power spectrum $\Pi(\phi)$ of the seismicity analyzed in natural time for each of the time windows (see below), we find that, in the range $0 \leq \phi \leq 0.5$, $\Pi(\phi)$ approaches to that given by Eq. (2.75), i.e.,

$$\Pi(\phi) = \frac{18}{5(2\pi\phi)^2} - \frac{6\cos(2\pi\phi)}{5(2\pi\phi)^2} - \frac{12\sin(2\pi\phi)}{5(2\pi\phi)^3}, \tag{6.7}$$

as N increases from 6 to some value usually less than (or around) 40. Simultaneously, the variance κ_1 of natural time becomes equal to 0.070, see Eq. (2.77), i.e.,

$$\kappa_1 = \langle\chi^2\rangle - \langle\chi\rangle^2 = 0.070. \tag{6.8}$$

The coincidence *occurs* usually a few days to around one week before the mainshock. This is not unreasonable, because in Chapter 8 we will show that upon analyzing in natural time the series of avalanches in two dynamical models associated with earthquakes (i.e., the Burridge–Knopoff "train" earthquake model, see § 8.2.2, and the Olami–Feder–Christensen earthquake model, see § 8.3.2), as well as in other critical systems, we find that Eq. (6.8) is fulfilled when the critical point is approached.

In simple words, before a mainshock a sequence of earthquakes occurs, which obeys Eq. (6.7) and this process will be called a *single correlated process*. When the mainshock occurs (the new phase), $\Pi(\phi)$ abruptly increases to approximately unity and κ_1 becomes almost zero. The latter can be visualized in the example depicted in Fig. 6.1, where we plot the κ_1 value versus the number of EQs after the SES detection on April 18, 1995, until the occurrence of the M_w6.6 mainshock on May 13, 1995, at 40.2°N, 21.7°E (see Fig. 1.11(a) and § 7.2.1). This figure shows that the κ_1 value becomes $\kappa_1 \approx 0.070$ at the 12th EQ (see Table 6.1), while upon the occurrence of the mainshock the κ_1 value *abruptly* decreases

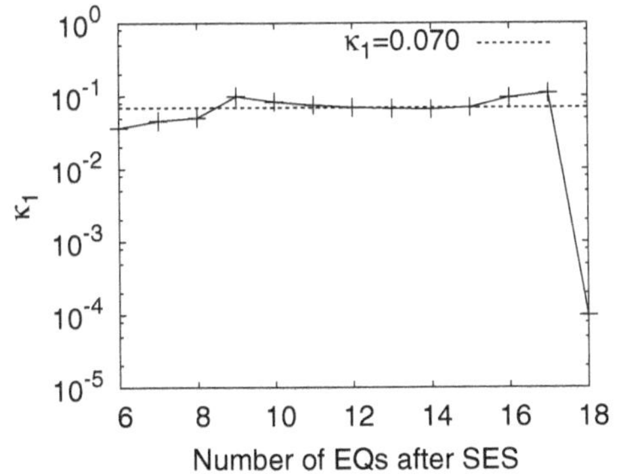

Fig. 6.1 How the variance κ_1 evolves event by event during the following period: from the initiation of the SES activity on April 18, 1995 (see Fig. 1.11(a)), until the occurrence of the M_w6.6 mainshock (numbered 18) on May 13, 1995. All the EQs used in the calculation are tabulated in Table 6.1. Taken from Ref. [84].

to $\kappa_1 \approx 9 \times 10^{-5}$. Such a behavior has been verified [78, 79] for several major EQs (see Chapter 7) and points to the conclusion that κ_1, or $\Pi(\phi)$ for $\phi \to 0$, could be considered as an order parameter.

In what remains, we provide details on the calculations supporting the suggestion that κ_1, or $\Pi(\phi)$ for $\phi \to 0$, may be considered as an order parameter for seismicity using the aforementioned example of the M_w6.6 mainshock. We focus our calculations on the EQs that occurred after April 18, 1995 within the region $N_{39.2}^{40.5}E_{20.3}^{22.0}$ surrounding the epicenter of the mainshock. The earthquakes in this region until the mainshock are tabulated in Table 6.1, where the magnitude M is given for each event. The calculation is carried out as follows. Using, for each EQ, the magnitude given in Table 6.1, we find the corresponding seismic moment M_0(in dyn.cm) through the relation [52]: $\log_{10} M_0 = (0.99 \pm 0.04)\mathrm{M} + (18.1 \pm 0.15)$. The resulting M_0 values of all the events (numbered 1 to 17 in Table 6.1) that preceded the mainshock are plotted in the natural time domain in Fig. 6.2, in a similar fashion as in Fig. 2.1(b). We now calculate the values of $\Pi(\phi)$ for $\phi \in [0,0.5]$, as they

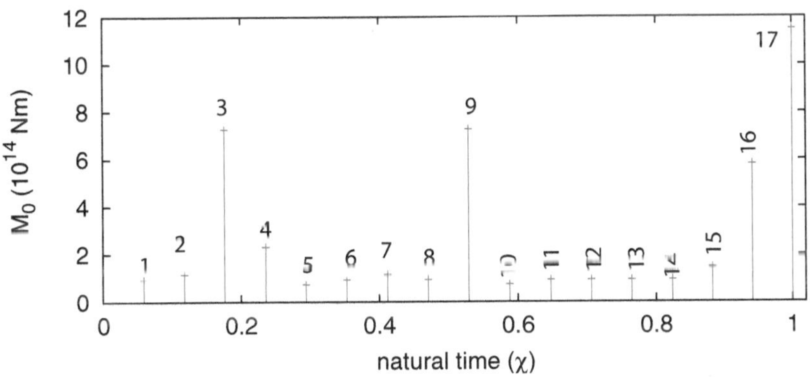

Fig. 6.2 Plot in natural time of the events that occurred after the initiation of the SES activity on April 18, 1995, until the mainshock on May 13, 1995 (the numbers correspond to the EQs of Table 6.1) in a similar fashion as Fig. 2.1(b). Taken from Ref. [83].

Table 6.1 All EQs within $N_{39.2}^{40.5}E_{20.3}^{22.0}$ that occurred after the initiation of the SES activity on April 18, 1995, until the M_w6.6 (from USGS) mainshock at Kozani-Grevena on May 13, 1995. The following data are available from the site of the National Observatory of Athens (NOA): www.gein.noa.gr/services/1950-00.txt (see also Ref. [50]); the reported magnitude M stands either for the local magnitude M_L or for the duration magnitude M_D. The seismic moment of the mainshock was taken from the Appendix of Ref. [52].

No.	Year	Mon.	Day	Hour	min	sec	Lat.	Long.	Depth	M	M_0(N.m)
1	1995	Apr.	27	15	16	55.3	39.50	21.13	10	2.9	9.35×10^{13}
**)	1995	Apr.	28	20	3	16.7	39.19	20.35	17	3.5	3.67×10^{14}
2	1995.	Apr.	30	6	58	24.8	39.79	20.72	29	3	1.17×10^{14}
3	1995	Apr.	30	7	50	32.1	40.44	21.85	3	3.8*)	7.28×10^{14}
4	1995	Apr.	30	21	12	42.6	40.00	20.66	5	3.3	2.33×10^{14}
5	1995	Apr.	30	23	24	54.7	39.81	20.50	10	2.8	7.45×10^{13}
6	1995	Apr.	30	23	46	42.5	39.58	20.58	5	2.9	9.35×10^{13}
7	1995	May	1	1	49	55.5	39.89	20.74	5	3	1.17×10^{14}
8	1995	May	1	22	47	21.1	39.90	21.01	5	2.9	9.35×10^{13}
9	1995	May	2	15	52	18.6	39.55	20.58	5	3.8	7.28×10^{14}
10	1995	May	5	2	58	5.8	39.38	20.35	10	2.8	7.45×10^{13}
11	1995	May	7	5	19	50.3	40.12	20.52	5	2.9	9.35×10^{13}
12	1995	May	10	0	1	4.2	40.34	21.79	10	2.9	9.35×10^{13}
13	1995	May	10	15	23	2.4	39.28	21.69	10	2.9	9.35×10^{13}
14	1995	May	10	18	24	56.3	39.91	20.72	5	2.9	9.35×10^{13}
15	1995	May	11	9	14	24.1	39.94	21.28	10	3.1	1.48×10^{14}
16	1995	May	13	8	42	12.3	40.07	21.75	5	3.7	5.79×10^{14}
17	1995	May	13	8	43	18.7	40.02	21.77	5	4	1.15×10^{15}
18	**1995**	**May**	**13**	**8**	**47**	**17**	**40.18**	**21.71**	**39**	**6.1*****)	$\mathbf{1.25\times10^{19}}$

*) This event is not reported by NOA but comes from USGS with ML_{THE}.
**) This is just in the boundary of the region selected. Note that if the calculation includes this event but disregards the aforementioned(*) one, i.e. $ML_{\mathrm{THE}} = 3.8$, a collapse of the spectra is again observed on May 10, 1995.
***) This is the M_L value, while the moment magnitude is $M_w = 6.6$.

evolve upon the occurrence of each new event by using

$$\Pi(\phi) = \left| \frac{\sum_{k=1}^{N} (M_0)_k \exp\left(i2\pi\phi \frac{k}{N}\right)}{\sum_{n=1}^{N} (M_0)_n} \right|^2, \tag{6.9}$$

which results from Eqs. (2.29) and (2.31) by replacing Q_k with $(M_0)_k$ (see § 2.1.2). In Eq. (6.9), N increases by one when a new EQ takes place, i.e. $N = 1, 2, \ldots 18$ and excerpts of the results are plotted with the crosses in Fig. 6.3. In the same figure, we also plot (solid line, labeled theory), for the sake of comparison, the $\Pi(\phi)$ values calculated from Eq. (6.7). An inspection of this figure shows that since EQ No. 9 (which occurred on May 2, see Table 6.1), see Fig. 6.3(b), the crosses start to gradually approach the solid line. They almost coincide (see Fig. 6.3(e)) upon the occurrence of the EQ No. 12, i.e., on May 10 and hence only 3 days before the mainshock). The corresponding κ_1 value (see Fig. 6.1) is then close to 0.070, thus agreeing with Eq. (6.8). The κ_1 value starts

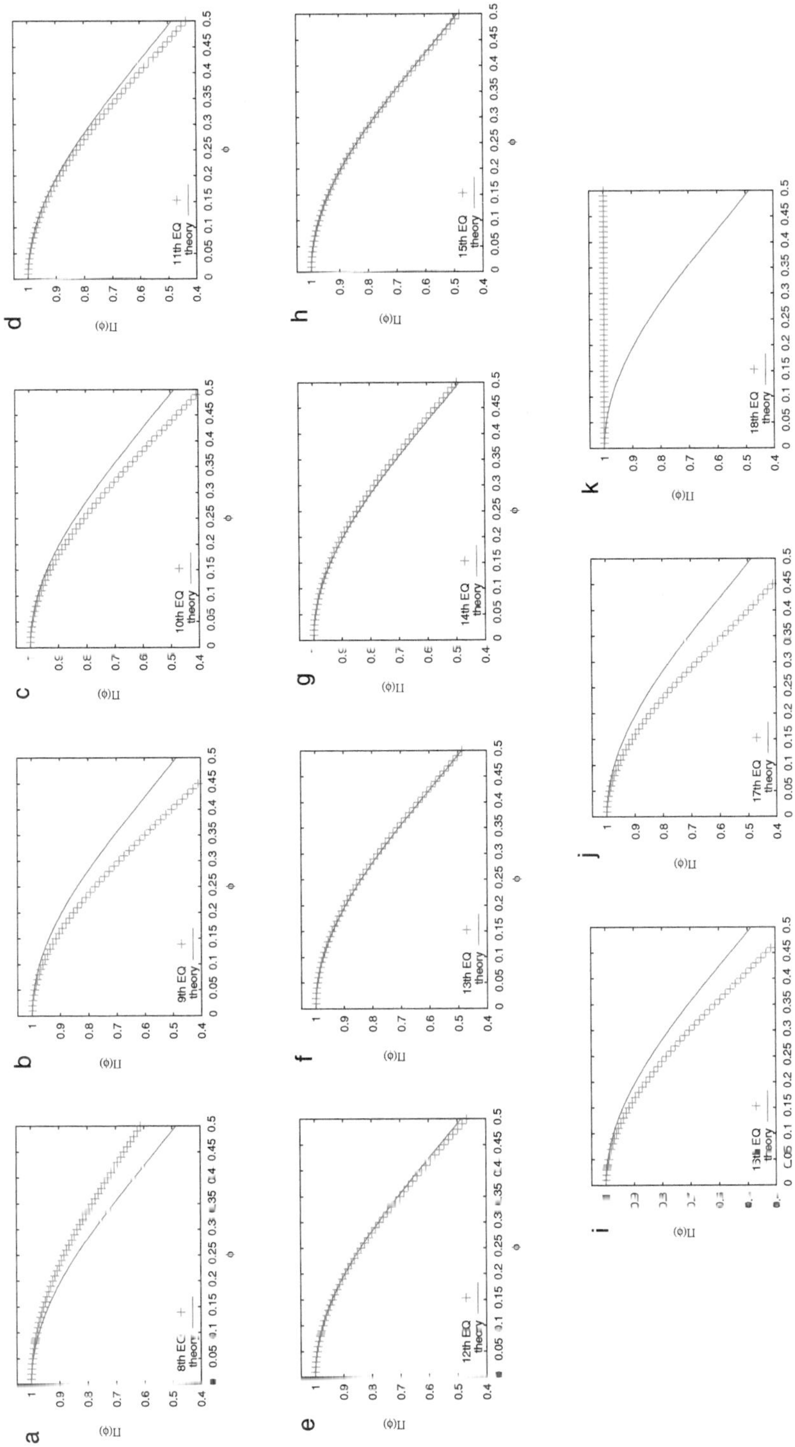

Fig. 6.3 Excerpts of the evolution of $\Pi(\phi)$ (crosses) of the seismicity that occurred after the initiation of the SES activity on April 18, 1995, until the mainshock of May 13, 1995. Figures (a) to (k) correspond to the $\Pi(\phi)$ versus ϕ results (crosses) obtained from Eq. (6.9) upon the occurrence of each of the EQs labeled 8 to 18, respectively, in Table 6.1. The solid lines are calculated from Eq. (6.7). Taken from Ref. [83].

to deviate somewhat from this value only upon the occurrence of the EQs No. 16 and 17 which interestingly took place only a few minutes before the mainshock. Figure 6.3(k) depicts what happens upon the occurrence of the mainshock: The $\Pi(\phi)$ curve then exhibits an *abrupt* change and turns to a straight line almost parallel to the horizontal axis, i.e., $\Pi(\phi) \approx 1$ (note that an abrupt change also appears for κ_1 in Fig. 6.1, see No. 18).

This is exactly the change that motivated us to consider $\Pi(\phi)$ for $\phi \to 0$, or κ_1, as an order parameter for seismicity.

6.2.2 Universal curve for the seismicity in various regions

The properties of the normalized power spectrum for the *long-term seismicity* in natural time can be studied by means of the procedure described in § 2.5.2. First, calculation of $\Pi(\phi)$ was made for an event taking time windows from 6 to 40 consecutive events (for the reasons that will be explained in § 6.4.1, the choice of the precise value of the upper limit, up to 100 or so, is not found decisive [78]). And second, this process was performed for all the events by scanning the whole earthquake catalog. The following data from two different areas, i.e., San Andreas fault system and Japan, have been analyzed. First, the EQs that occurred during the period 1973–2003 within the area $N_{32}^{37}W_{114}^{122}$ using the Southern California Earthquake catalog (hereafter called SCEC). Second, the EQs within $N_{25}^{46}E_{125}^{146}$ for the period 1967–2003 using the Japan Meteorological Agency catalog (hereafter simply called "Japan"). The magnitude thresholds $M \geq 2.0$ and $M \geq 3.5$ have been considered for SCEC and Japan, respectively, for the sake of data completeness [83]. The seismic moments have been obtained by the procedure described later in § 6.2.2.1.

We now study the order parameter fluctuations relative to the standard deviation of its distribution. Thus, we plot in Fig. 6.4 the quantity $\sigma P(X)$ versus $(X - \langle X \rangle)/\sigma$ where X stands for $\Pi(\phi)$ and $\langle \Pi(\phi) \rangle$, σ refer to the mean value and the standard deviation of $\Pi(\phi)$

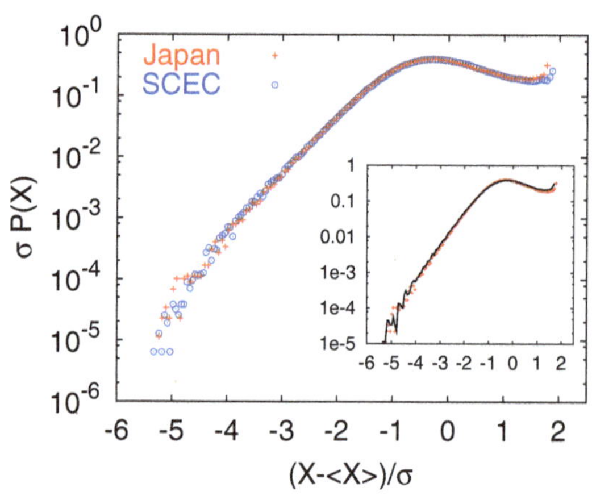

Fig. 6.4 Universality of the probability density function of $\Pi(\phi)$ for EQs in natural time. The log-linear plot of $\sigma P(X)$ versus $(X - \langle X \rangle)/\sigma$, where X stands for the order paramater $\Pi(\phi)$ for $\phi \to 0$. Crosses and circles correspond to Japan ($M \geq 3.5$) and SCEC ($M \geq 2.0$), respectively. The inset depicts the corresponding results for randomly shuffled (black curve) and the original data (red crosses) in Japan. Note that the same graph is obtained for three different regions in Japan (see Fig. 6.5). Taken from Ref. [84].

(recall that the calculations should be done for very small ϕ values, e.g., $\phi = 0.05$, since we assume here $\phi \to 0$, as explained in § 6.2.1). One could alternatively plot $P(\kappa_1)$ versus $(\mu(\kappa_1) - \kappa_1)/\sigma(\kappa_1)$, where $\mu(\kappa_1)$ and $\sigma(\kappa_1)$) refer to the mean value and the standard deviation of κ_1. The results in Fig. 6.4, for *both* SCEC and Japan, fall on the *same* curve. This log-linear plot clearly consists of two segments: The segment to the left shows a decrease of $P(X)$ almost by five orders of magnitude, while the upper right segment has an almost constant $P(X)$. The feature of this plot is strikingly reminiscent of the one obtained by Bak et al. [11] (see their fig. 4) on different grounds, using EQs in California only. More precisely, they measured $P_{S,l}(T)$, the distribution of waiting times T, between EQs occurring within range l whose magnitudes M are greater than $\log_{10} S$. They then plotted $T^{\alpha} P_{S,l}(T)$ versus $TS^{-b} l^{d}$ and found that, for a *suitable choice* of the exponent α (i.e., $\alpha = 1$), the G-R law exponent b (i.e., $b = 1$) and the spatial dimension d (i.e., fractal dimension $d = 1.2$) all the data collapse onto a *single* curve which is similar to that of Fig. 6.4.

Note that Fig. 6.4 was obtained here without considering the waiting time distribution and without the suitable choice of *any* parameter.

After a further inspection of Fig. 6.4, the following points have been clarified.

First, the rapidly decaying part (i.e., the left segment), which is consistent with an almost exponential decaying function over around four orders of magnitude, remains practically unchanged, upon *randomly* shuffling [85] the data. This can be seen in the inset of Fig. 6.4, where for the sake of clarity only the results from the data of Japan (the original as well as the shuffled ones) are depicted.

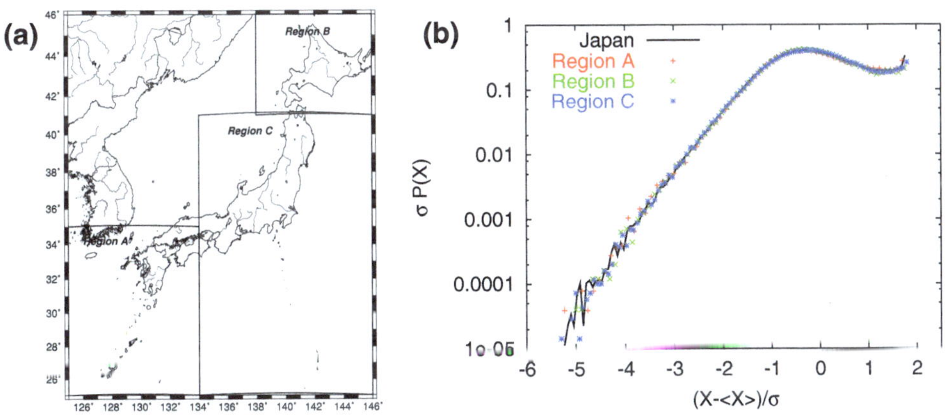

Fig. 6.5 The same as Fig. 6.4 is depicted in (b), but for the regions A (red), B (green) and C (blue) of Japan. A map of these regions is shown in (a). Taken from Ref. [84].

Second, the feature of the plot of Fig. 6.4 is not altered upon changing the seismic region. As an example, Fig. 6.5(b) shows that three different regions A, B, C of Japan (depicted in Fig. 6.5(a)), as well as the whole of Japan, result in almost identical plots.

Third, the "upturn branch" in the upper right part of Fig. 6.4 arises mainly from the presence of aftershocks. It disappears (see the crosses in Fig. 6.6) when, in Japan for example, we delete the EQs with $M \leq 5.7$ (and hence drastically reduce the number of aftershocks), but it does not when deleting EQs with smaller threshold, i.e., $M < 4.0$; the latter can be also visualized in the SCEC example of Fig. 6.6, where we give the results for $M \geq 4.0$ (note that this threshold still allows the presence of a reasonable number of aftershocks).

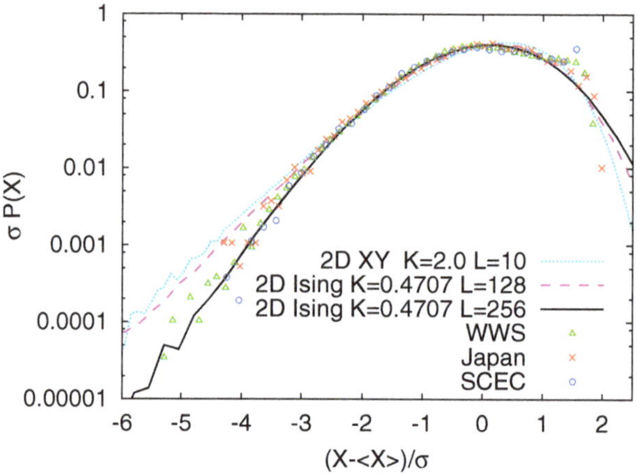

Fig. 6.6 The common feature of fluctuations in different correlated systems. The log-linear plot of $\sigma P(X)$ versus $(X - \langle X \rangle)/\sigma$ for WWS(triangles), Japan (crosses) and SCEC (circles). The magnitude threshold $M > 5.7$ for WWS and Japan (while $M \geq 4.0$ for SCEC) was used see the text. Furthermore, the dotted curve shows the results obtained for the 2D XY model (with [59] inverse Kosterlitz–Thouless transition temperature $K_{KT} \approx 1.2$) ($X = \sqrt{M_x^2 + M_y^2}$), $K = 2.0$ for $L = 10$ ($N = 100$) which has been shown [21] to describe the experimental results for 3D turbulent flow. The results of the 2D Ising model $K = 0.4707$ (while $K_c \approx 0.4407$), either for $L = 128$ (dashed) or $L = 256$ (solid line), are also plotted. Taken from Ref. [84].

Fourth, if we consider the relevant results (1977–2003) for the worldwide seismicity (WWS) by taking a large magnitude threshold, i.e., $M > 5.7$ (so that the data are complete [83]), we find that they fall onto the *same* curve with the results of both Japan and SCEC (see Fig. 6.6 that will be further discussed in § 6.2.3).

If one generates synthetic seismic data either by means of a simple Poisson model or by the G-R law and compare the results to those deduced from actual seismicity data, the following conclusion was drawn [84]: none of these synthetic data can lead to a curve coinciding with the one obtained from the real data. In other words, the scaled distribution deduced within the frame of natural time analysis, reveals *an extra complexity* for the real data compared to the synthetic data even if the latter are produced with b values comparable to the experimental ones. This points to the conclusion that the origin of self-similarity in seismicity *cannot be solely* attributed to the process's increments' "infinite" variance (§ 2.5.1) as it will be further investigated in the remaining Sections (6.3 to 6.5) of this Chapter.

By summarizing, if we analyze the long-term seismicity in natural time and study the order parameter fluctuations relative to the standard deviation of its distribution, we find without making use of *any* adjustable parameter that the scaled distributions of different seismic areas (as well as that of the world-wide seismicity) fall on the *same* curve (*universal*), see Figs. 6.4, 6.5(b) and 6.6.

6.2.2.1 Additional details on the calculations mentioned above

We first recall that the following earthquake catalogs have been used: for the San Andreas fault system, the Southern California Earthquake catalog (SCEC) available from www.data.scec.org/ftp/catalogs/SCSN/, and for Japan, the Japan Meteorological Agency (JMA) catalog. As for the worldwide seismicity (WWS) we used the data available from http://www.globalcmt.org/CMTsearch.html. We also recall that the magnitude thresholds $M \geq 2.0$, $M \geq 3.5$, and $M > 5.7$ have been considered for SCEC, Japan, and WWS, respectively to ensure data *completeness* (i.e., that they obey the G-R law, Eq. (6.1)) for the periods studied.

All the seismic data have been analyzed in natural time in a similar fashion as shown in Fig. 2.1(b). The seismic moment M_0 was obtained from the magnitude M as follows: for Japan, the following approximate formulae, obtained from a fit to fig. 5.3 of Ref. [77], have been used:

$$M_w = 0.701\mathrm{M_{JMA}} + 1.47 \quad \text{for} \quad \mathrm{M_{JMA}} < 5, \tag{6.10}$$

$$M_w = 0.916\mathrm{M_{JMA}} + 0.40 \quad \text{for} \quad 5 \leq \mathrm{M_{JMA}} < 6, \tag{6.11}$$

$$M_w = 1.07\mathrm{M_{JMA}} - 0.509 \quad \text{for} \quad 6 \leq \mathrm{M_{JMA}} < 7.3, \tag{6.12}$$

$$M_w = 1.345\mathrm{M_{JMA}} - 2.56 - 0.0472/(\mathrm{M_{JMA}} - 8.3) \quad \text{for} \quad 7.3 \leq \mathrm{M_{JMA}}, \tag{6.13}$$

where $\mathrm{M_{JMA}}$ stands for the magnitude reported by JMA and M_w stands for the moment magnitude. Then M_0 was obtained [34] through Eq. (1.3), i.e., $M_0 \propto 10^{1.5M_w}$. As for SCEC, we assumed $M_w \approx \mathrm{M}$, where M is the reported magnitude. Finally, for WWS the M_0 values are directly accessible at the aforementioned http address.

6.2.3 Similarity of fluctuations in correlated systems including seismicity

Great interest has been focused on the fluctuations of correlated systems in general and of critical systems in particular [21, 17, 20, 91, 18, 87, 19, 90, 25]. Bramwell, Holdsworth and Pinton (BHP) [21], in an experiment of a closed turbulent flow, found that a normalized form of the probability distribution function (pdf) of the power fluctuations has the same functional form as that of the magnetization (M) of the finite-size 2D (two-dimensional) XY equilibrium model in the critical region below the Kosterlitz–Thouless transition temperature (magnetic ordering is then described by the *order parameter* M). The "normalized" form of the pdf, denoted by $P(m)$, is defined by introducing the reduced magnetization [21] $m = (M - \langle M \rangle)/\sigma$, where $\langle M \rangle$ denotes the mean and σ the standard deviation. For both systems, BHP found that while thc high end ($m > 0$) of the distribution has [21] a Gaussian shape the asymptote of which was later clarified [20] to have a double exponential form, a distinctive exponential tail appears towards the low end ($m < 0$) of the distribution. The latter tail, which will be hereafter simply called, for the sake of convenience, "exponential tail", provides the main region of interest [21], since it shows that the probability for a rare fluctuation, e.g., of greater than six standard deviations from the mean, is almost five orders of magnitude higher than in the Gaussian case. Subsequent independent simulations [17, 20, 91, 90, 25] showed that a variety of highly correlated (non-equilibrium as well as equilibrium) systems, under certain conditions, exhibit approximately the "exponential tail". However, the question of whether earthquakes exhibit an "exponential tail", has not been clarified due to the major difficulty of choosing an order parameter. Since an order parameter for the case of EQs was proposed in § 6.2.1, we examine whether an "exponential tail" appears also in seismicity. We find [84] that this "tail" is identified only *if* we analyze the time series of earthquakes in natural time.

We now compare in Fig. 6.6 the results obtained in § 6.2.2 for seismicity with those obtained in some equilibrium critical systems (e.g., see Ref. [90]). We first recall [90] that the pdf of the order parameter in the *critical* regime depends on the inverse temperature parameter $K = 1/T$ and the length L through a scaling variable $s \equiv L^{1/\nu}(K - K_c)/K_c$, where $K_c = 1/T_c$ and T_c denotes the critical temperature. The quantity s^ν provides the ratio of the lattice size and the *correlation length* ξ (§ 1.5.2) at K. In Fig. 6.6, we include numerical results of the 2D Ising model for $s = 8.72$ ($L = 128$, $K = 0.4707$) and $s = 17.44$ ($L = 256$, $K = 0.4707$). Here, X stands for M. These s values were intentionally selected, because [90] for $s \geq 8.72$ for the 2D Ising model, the normalized forms of the pdfs $P(m;s)$ of a number of critical models (i.e., 2D XY, 2D Ising, 3D Ising, 2D three-state Potts) share the *same* form (up to a constant factor of s), which interestingly exhibits an exponential-like left tail ($m < 0$). An inspection of Fig. 6.6 shows that our 2D Ising results almost coincide (note that this can be safely checked only for the left segment, i.e., $m < 0$) with those of seismicity, i.e., Japan, SCEC and WWS (note that some disparity, which appears in the upper right part of SCEC only, might be attributed to the selection of the magnitude threshold for seismicity; recall the third point mentioned in § 6.2.2). This coincidence (which seems to be better for $s = 17.44$) reveals that the seismicity, irrespective of the seismic area we consider, exhibits – over four orders of magnitude – fluctuations of the order parameter similar to those in several critical systems as well as in 3D turbulent flow [84].

Thus, we conclude that the "universal" curve deduced for seismicity in § 6.2.2, exhibits an "exponential tail" form similar to that observed in certain non-equilibrium systems (e.g., 3D turbulent flow) as well as in several equilibrium critical phenomena (e.g., 2D Ising, 3D Ising, 2D XY).

6.2.4 The pdf of the order parameter of seismicity. The b-value of the Gutenberg–Richter law deduced from first principles

We focus here on a challenging point that emerges [81] from a further elaboration of the results depicted in Fig. 2.5 that have been obtained by the procedure described in § 2.5.4. In particular, we consider randomly shuffled power law distributed energy bursts obeying Eq. (6.4), i.e, $P(E) \propto E^{-\gamma}$, that were subsequently analyzed in natural time. Using a (natural) time window of length $l = 6$ to 40 consecutive events sliding through the whole dataset, the pdf $P(\kappa_1)$ versus κ_1 was constructed for several values of γ as shown in Fig. 2.5. An inspection of this figure shows the following:

First, note that upon increasing the γ value from $\gamma = 1.3$ to 2.0, the feature of the curve changes significantly, becoming *bimodal* at intermediate γ values. Second, we calculate, for each γ value studied, the so-called *differential entropy*, defined as

$$S_I = -\int P(\kappa_1)\ln P(\kappa_1)\, d\kappa_1 \tag{6.14}$$

which is the Shannon information entropy of a continuous probability distribution, e.g., see Ref. [30]. (Note that the Shannon information entropy is *static* entropy and *not* a dynamic one [80]; see Section 3.1.) Finally, we investigate the resulting S_I values versus γ. Such a plot is given in Fig. 6.7, whose inspection reveals that S_I maximizes at a value of γ lying between $\gamma = 1.6$ and $\gamma = 1.7$, which is more or less comparable with the experimental values, see Table 2.1. (In particular for the case of earthquakes, this γ value corresponds to $b \approx 1$, as mentioned in Section 6.1.) This value is almost unaffected by the window length (l) chosen, since it decreases only slightly from $\gamma \approx 1.70$ to $\gamma \approx 1.65$ upon changing l from $l = 6$–40 to $l = 6$–100, see Fig. 6.7.

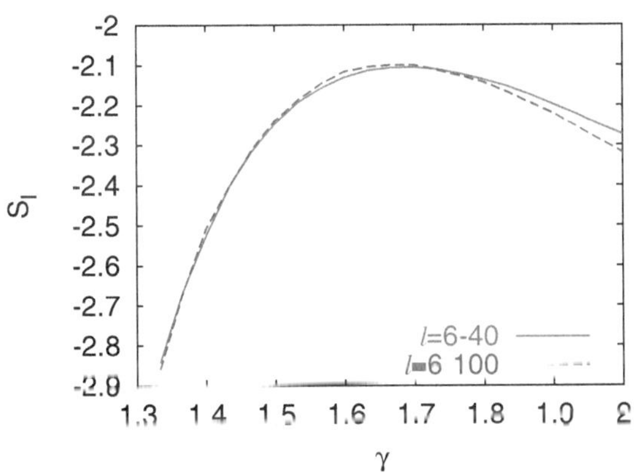

Fig. 6.7 The calculated values of the differential entropy S_I (see the text) versus the exponent γ. Two window length ranges are used and their results differ slightly. Taken from Ref. [81].

In view of the widespread belief (e.g., Ref. [48]) that there is a close analogy between non- equilibrium phase transitions (which is likely to be [84, 82] the case of earthquakes; see Section 6.1 and § 6.2.1) and equilibrium ones (e.g., ferromagnetic materials) – which however are apparently very different problems – our study has been ex-

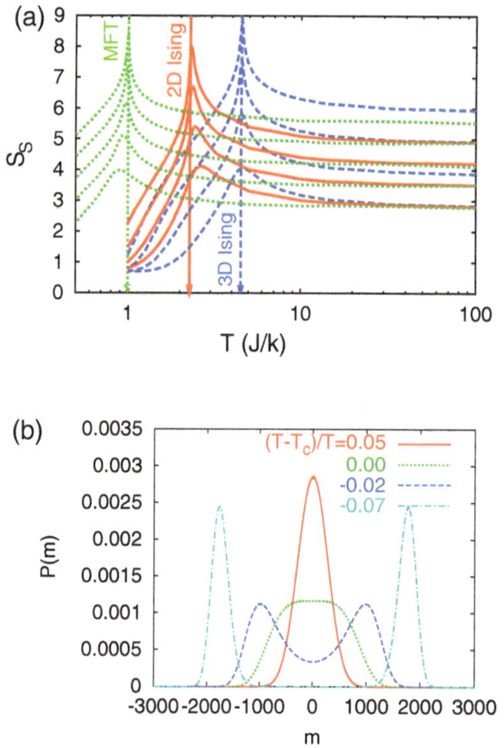

Fig. 6.8 (a) The Shannon entropy $S_S \equiv -\sum_m P(m)\ln[P(m)]$ versus T calculated for the following equilibrium critical models: infinite range model for a ferromagnetic system of $N = 2^n$ spins (MFT, dotted green, $n = 6$, 8, 10, 12 and 14), 2D Ising (solid red, $n = 6$, 8, 10 and 12) and 3D Ising (dashed blue, $n = 6$, 9, 12 and 15); in each case the results are given for various sizes increasing from bottom to top (note that S_S diverges as $\ln N$). The arrows indicate the critical temperatur $T_c$in each case. (b) the probability $P(m)$ of the order parameter m versus m for four values of the quantity $(T - T_c)/T$ for the infinite range model with $N = 4096$. Taken from Ref. [81].

tended to the well-known equilibrium critical systems by investigating the Shannon entropy $S_S \equiv -\sum_m P(m)\ln[P(m)]$, where m denotes the order parameter, versus the temperature. Studying S_S at various temperatures, we find that it maximizes near T_c (for finite sizes S_S diverges proportionally to $\ln N$ as $T \to T_c$). For example, in Fig. 6.8(a) we plot the results for the following models: the infinite range model (a summary of which can be found in the Appendix C of Ref. [81]) of a ferromagnetic system of N spins ($s_i = \pm 1$) (green dotted curves), the 2D Ising model (red solid curves) or the 3D Ising model (blue dashed curves). We now proceed to Fig. 6.8(b), which depicts, as an example, $P(m)$ for the first model at various temperatures above and below the critical temperature T_c for $N = 4096$ spins.

Note that *just* below T_c (see Fig. 6.8(b)) a *bimodal* feature emerges in the probability distribution of the order parameter, which is reminiscent of the one found in Fig. 2.5 (for intermediate γ values). This point is further elaborated in § 6.4.1.

The above inspired us to investigate the slight variation of the γ value (at which S_I maximizes in Fig. 6.7) versus l by means of a procedure analogous to the well-known finite size scaling techniques. Such a technique was actually applied [81] and, after studying for three different l the $P(\kappa_1)$ that arises when S_I is approaching its maximum, the following conclusion was obtained: the value $\gamma_c \approx 1.55$ provides a lower bound for the γ value that maximizes S_I as $l \to \infty$.

Does the above finding in Fig. 6.7 mean that the b or γ value can be determined just by applying the Maximum Entropy Principle in the sense developed by Jaynes [36, 37], who suggested regarding statistical mechanics as a form of *statistical inference* and *start* statistical physics from the principle of maximum entropy inference (MaxEnt)? This is not yet clear, because a widely accepted formalism for *non-equilibrium* statistical mechanics is still lacking. The fact that in some experiments the resulting γ values differ slightly from $\gamma = 1.6$ to 1.7 predicted from Fig. 6.7, could be attributed to the following: Fig. 6.7 is based on randomized data, while the actual data may also exhibit temporal correlations (e.g., the case of aftershocks). In addition, finite size effects [10] might play a significant role.

Thus, in short, when studying the differential entropy associated with the pdf of κ_1 (recall that κ_1 is the order parameter for seismicity; see § 6.2.1), it maximizes when the exponent γ in Eq. (2.98) (or in Eq. (6.4)) lies in the narrow range $\approx$1.6 to 1.7, in agreement with the experimental findings in diverse fields. This for the case of earthquakes immediately reflects that the b value in the Gutenberg–Richter law is $b = 1.5(\gamma - 1) \approx 1$, as actually observed.

6.2.5 Multifractal cascades in natural time and the case of seismicity

Here, we study multiplicative cascades (or generalized Cantor sets [47, 24]) in natural time [81, 56]. In generalized Cantor sets (multiplicative cascades), at the initial stage ($M = 1$) the original region is divided into K segments with possibly variable sizes, but the mass probability from the left to the right is distributed by the constant weights w_i, $i = 1, 2, \ldots K$ with $\sum_i w_i = 1$. The same procedure can then be followed in each segment at the stage $M = 2$, etc. This is what will be hereafter called the *Deterministic Cantor Set* (DCS) in contrast to a procedure in which w_i are assigned randomly (i.e., not from the left to the right) at each segment and stage M. The latter will be called the *Stochastic Cantor Set* (SCS) and will be also studied by means of Monte Carlo simulations. A case of special practical interest is the so-called p-model [47] in which each segment is divided equally into two parts ($K = 2$), with $w_1 = p$ and $w_2 = 1 - p$. This model, in its SCS flavor, was originally proposed to describe turbulence data [47, 49]. Moreover, the DCS case was discussed [51] in relation to power law time sequences in ricepiles.

What is important about DCS is the following. If we consider the original region in the natural time interval $A = (0, 1]$ (note that $\chi \subset A$) and use the obtained mass probabilities as p_k in the sense of Eq. (2.2), then $\Phi(\omega) = \sum_{k=1}^{N} p_k \exp\left(i\omega \frac{k}{N}\right)$ can be factorized and one can obtain easily the properties of DCS in natural time. Under these conditions, for $K = 2$ and equal segments, the following relates $\Phi_{M+1}(\omega)$ at stage $M + 1$ to that $\Phi_M(\omega)$ at stage M:

$$\Phi_{M+1}(\omega) = \left[p + (1-p)\exp\left(i\frac{\omega}{2}\right)\right] \Phi_M\left(\frac{\omega}{2}\right). \tag{6.15}$$

Equation (6.15) can be also generalized for $K > 2$ into

$$\Phi_{M+1}(\omega) = \Phi_M\left(\frac{\omega}{K}\right)\sum_{j=1}^{K} w_j \exp\left[i\frac{(j-1)\omega}{K}\right]. \tag{6.16}$$

Then, we can show that the normalized power spectrum at the stage M, i.e., $\Pi_M(\omega)$, is interconnected to that at the stage $M-1$ through the relation:

$$\Pi_M(\omega) = \Pi_{M-1}\left(\frac{\omega}{K}\right)\left|\sum_{j=1}^{K} w_j \exp\left(i\omega\frac{j-1}{K}\right)\right|^2, \tag{6.17}$$

see also Section I of Ref. [55].

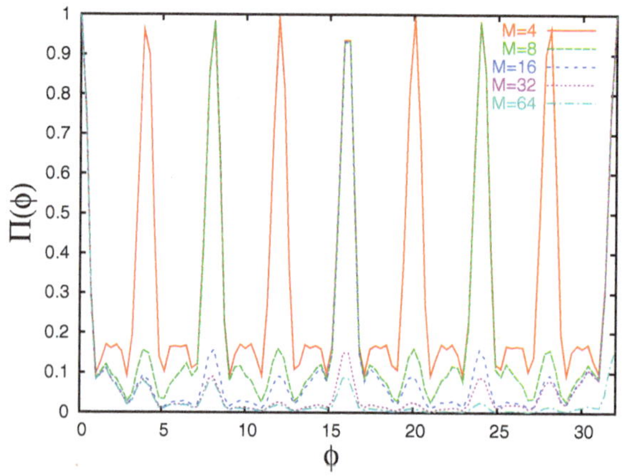

Fig. 6.9 Normalized power spectra $\Pi_M(\phi)$ for the DCS with $K = 2$ and $p = 0.3$ for $M = 4$, 8, 16, 32, etc. Taken from Ref. [81].

Equation (6.15) can be also used for the calculation of κ_1 as $\omega \to 0$. A remarkable property of $\Pi_M(\omega) = |\Phi_M(\omega)|^2$ is that, independent of M, all $\Pi_M(\omega)$ have almost the same shape for natural frequencies ϕ less than 0.5 (see Fig. 6.9). In other words, in the sense discussed above, all these stages share the same characteristic properties but differ in the high natural frequency range. Moreover, the application of Eq. (6.15) for $\Pi_M(\omega)$ as $\omega \to 0$, leads to the following relation for the κ_1 value at stage $M+1$

$$\kappa_{1,M+1} = \frac{\kappa_{1,M} + p(1-p)}{4}, \tag{6.18}$$

which leads to

$$\lim_{M\to\infty} \kappa_{1,M} = \kappa_{1,\infty} = \frac{p(1-p)}{3}. \tag{6.19}$$

Thus, for $p = 0.3$ we obtain $\kappa_{1,\infty} = 0.070$. In Fig. 6.10, we compare such a DCS with the normalized power spectrum given by Eq. (6.7); the results are almost identical in the region $\varphi \in [0, 0.5]$. Note, however that DCS does *not* satisfy the entropy conditions (4.39) which are valid for SES activities (*critical dynamics*) and other cases summarized in Section 7.1. On the other hand the SCS flavor of the same model, which has been proposed to describe

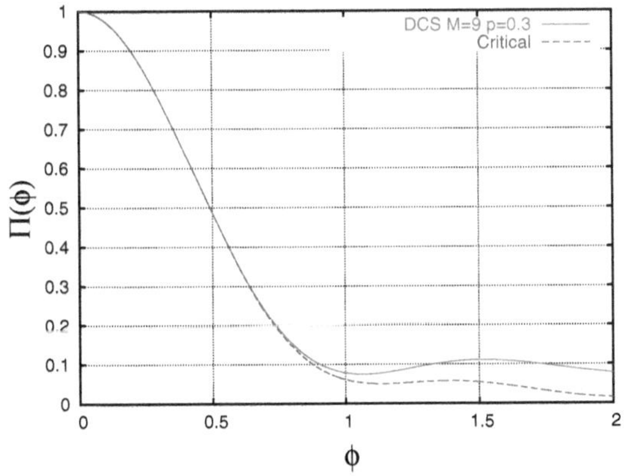

Fig. 6.10 Comparison of the normalized power spectrum of Eq. (6.7) (dashed blue), with that obtained for an $M = 9$ p-model DCS (solid red, $p = 0.3$). The two curves are almost identical in the region $\phi \in [0, 0.5]$. Taken from Ref. [81].

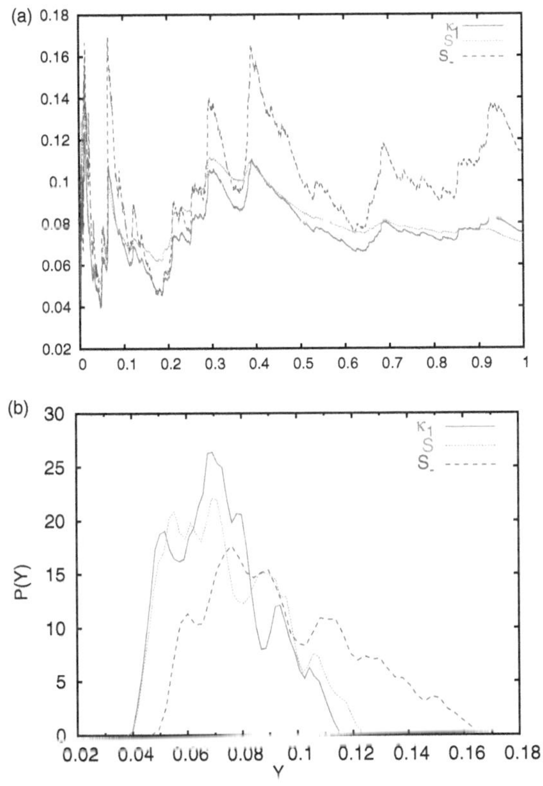

Fig. 6.11 (a) A typical evolution for the SCS p-model, $p = 0.3$; the quantities κ_1 (solid), S (dotted) and S_- (dashed) are plotted versus natural time (b) The pdfs $P(Y)$ of (Y =) κ_1, S, S_- for the case $p = 0.3$ that has been proposed to describe turbulence in Ref. [47]. Taken from Ref. [81].

turbulence [47, 49], also gives an average $\kappa_1 \approx 0.070$ and in addition does satisfy the entropy conditions (4.39) for the majority of the cases treated [81] by Monte Carlo (see Fig. 6.11).

Figure 6.11 summarizes the Monte Carlo study [81] of the stochastic case in which interestingly, the most probable value $\kappa_{1,p}$ of κ_1 is found to be $\kappa_{1,p} \approx 0.070$.

6.2.5.1 Application to seismicity

Shuffled earthquake data are random in time and of course follow the G-R law (Section 6.1). The probability to observe in some area and after some waiting time an EQ of magnitude M greater or equal to M_{thres} is also given by $P(\mathrm{M} \geq M_{thres}) \propto 10^{-bM_{thres}}$. Thus, the frequency $\nu(\mathrm{M})$ of EQs with magnitude M, i.e., the ones having magnitudes within $[\mathrm{M}-\delta\mathrm{M}, \mathrm{M}+\delta\mathrm{M}]$, is just $\nu(\mathrm{M}) \propto 10^{-b\mathrm{M}}$ (note that due to the experimental errors in determining an EQ magnitude a reasonable value of δM is around 0.1). In the light of the p-model, we can now approximate the case of seismicity as follows. Assuming that the largest EQ in some time interval dominates the corresponding energy release in this interval (see Fig. 6.12), if an earthquake of magnitude M_1 dominates the second (segment) time interval, the first segment will be dominated by an earthquake of magnitude $\mathrm{M}_1 - \Delta\mathrm{M}$, having twice the frequency of M_1, i.e., $\nu(\mathrm{M}_1 - \Delta\mathrm{M}) = 2\nu(\mathrm{M}_1)$. Thus, a multiplicative cascade is formed (see Fig. 6.12) with a p value equal to

$$p = 1/(1+10^{c\Delta\mathrm{M}}), \tag{6.20}$$

where c is the constant that interrelates the earthquake energy release with the magnitude see Eq. (6.3), i.e., $E \propto M_0 \propto 10^{c\mathrm{M}}$, where M_0 is the seismic moment of an EQ. Substituting the value of $\Delta\mathrm{M}(=\frac{1}{b}\log_{10}2)$ estimated on the basis of the G-R law, we obtain $p = 1/\left(1+2^{\frac{c}{b}}\right)$, which, in view of Eq. (6.19), leads to the most probable value of κ_1 given by

$$\kappa_{1,p} = \frac{2^{\frac{c}{b}}}{3\left(1+2^{\frac{c}{b}}\right)^2}. \tag{6.21}$$

This interrelates $\kappa_{1,p}$ with the quantity c/b. Typical values of b and c are $b \approx 1$ and $c \approx 1.5$, see Section 6.1, resulting in $\kappa_{1,p} = 0.064$. This value coincides with the value of $\kappa_{1,p}$ obtained [81] for the (randomly) shuffled earthquake data of Japan and SCEC, as it will be explained in Section 6.3.

Thus, in short, the natural time analysis of multiplicative cascades leads to a theoretical interrelation, i.e., Eq. (6.21), between the most probable value $\kappa_{1,p}$ of κ_1 of the (randomly) shuffled earthquake data and the parameter b of the Gutenberg–Richter law. This interrelation, if we just adopt a reasonable value of b, i.e., $b \approx 1$, leads to a $\kappa_{1,p}$ value very close to 0.064 in agreement with the shuffled experimental data of SCEC and Japan.

6.3 Temporal correlations in real seismic data

We now make use of the aspects developed in Section 2.5 on the distinction of the origins of self-similarity and explain how they can be used for the identification of temporal correlations in real seismicity time series. In particular, we will use Eqs. (2.88) and (2.92)

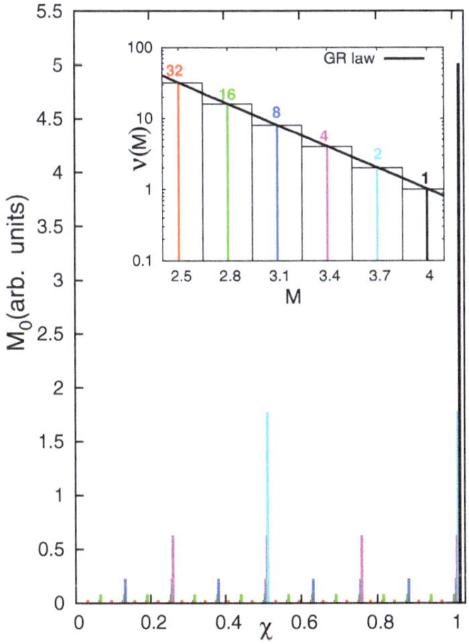

Fig. 6.12 A schematic representation in natural time of the multiplicative cascade model for earthquakes: a series of earthquakes that obey the G-R law (see the inset) with magnitudes varying in the range [2.5,4] arranged in natural time according to their frequency $\nu(\mathrm{M})$. They form a multiplicative cascade, a DCS in this case. Taken from Ref. [55].

(see also Ref. [81]) which give the expectation value $\mathscr{E}(\kappa_1)$ of κ_1 in the actually observed time series, and the expectation value $\mathscr{E}(\kappa_{1,shuf})$ of the randomly shuffled time series, respectively, when a (natural) time window of length l is sliding through the time series $Q_k \geq 0, k = 1, 2, \ldots N$.

For such a window, starting at $k = k_0$, the quantities $p_j = Q_{k_0+j-1} / \sum_{m=1}^{l} Q_{k_0+m-1}$ in natural time are defined and $\mathscr{E}(\kappa_1)$ in the actually observed time series equals (§ 2.5.2) that given by Eq. (2.88), i.e.,

$$\mathscr{E}(\kappa_1) = \kappa_{1,\mathscr{M}} + \sum_{\text{all pairs}} \frac{(j-m)^2}{l^2} \operatorname{Cov}(p_j, p_m), \tag{6.22}$$

where $\kappa_{1,\mathscr{M}}$ is the value of κ_1 corresponding to the time series of the averages $\mu_j \equiv \mathscr{E}(p_j)$ of p_j, i.e., $\kappa_{1,\mathscr{M}} = \sum_{j=1}^{l} (j/l)^2 \mu_j - (\sum_{j=1}^{l} \mu_j j/l)^2$, and $\operatorname{Cov}(p_j, p_m)$ stands for the covariance of p_j and p_m defined as $\operatorname{Cov}(p_j, p_m) \equiv \mathscr{E}[(p_j - \mu_j)(p_m - \mu_m)]$, while the variance of p_j is given by $\operatorname{Var}(p_j) = \mathscr{E}[(p_j - \mu_j)^2]$. The symbol $\sum_{\text{all pairs}}$ stands for $\sum_{j=1}^{l-1} \sum_{m=j+1}^{l}$. Equation (6.22) reveals that $\mathscr{E}(\kappa_1)$ is determined by two factors that involve:

(i) the correlation of the data as reflected in the averages μ_j, e.g., due to decreasing in magnitude aftershocks in an earthquake time series, and

(ii) the covariances' term which sums up the correlations between all natural time lags up to $l-1$.

On the other hand, $\mathscr{E}(\kappa_{1,shuf})$ obtained by randomly shuffling ($shuf$) the original time series is given by Eq. (2.92), i.e.,

$$\mathscr{E}(\kappa_{1,shuf}) = \kappa_u \left(1 - \frac{1}{l^2}\right) - \kappa_u (l+1) \operatorname{Var}(p) \tag{6.23}$$

(note that for the shuffled data $\mathrm{Var}(p_j)$ is independent of j, and hence we merely write $\mathrm{Var}(p) \equiv \mathrm{Var}(p_j)$). If Q_k do not exhibit heavy tails and have finite variance, Eq. (6.23) rapidly converges [81] to $\mathscr{E}(\kappa_{1,shuf}) = \kappa_u$ (see § 2.5.2). Otherwise, $\mathscr{E}(\kappa_{1,shuf})$ differs from κ_u, and the difference

$$\Delta\mathscr{E}(\kappa_{1,shuf}) \equiv \kappa_u\left(1-\frac{1}{l^2}\right) - \mathscr{E}(\kappa_{1,shuf}) = \kappa_u(l+1)\,\mathrm{Var}(p) \tag{6.24}$$

provides a measure of the process's increments' "infinite" variance. By comparing the results obtained from Eqs. (6.22), (6.23) and (6.24) in a time series, we can draw quantitative conclusions on the existence of temporal correlations even if the process's increments' exhibit "infinite" variance.

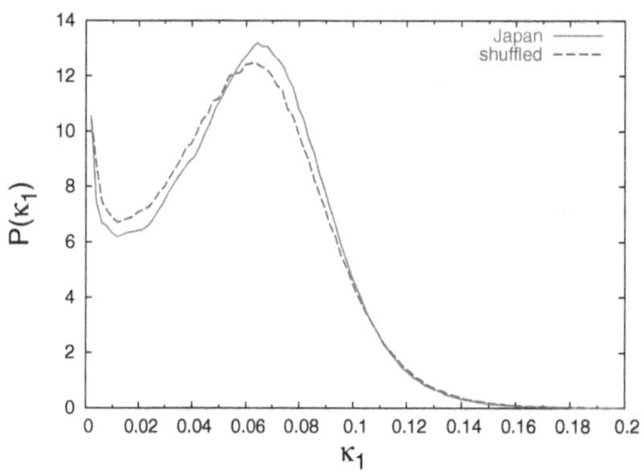

Fig. 6.13 The pdfs of κ_1 when using either the actual seismic catalog of Japan (solid) treated in § 6.2.2 with $M_{\mathrm{JMA}} \geq 3.5$ or the same data in random order (dashed). Taken from Ref. [81].

Let us now use the example of the earthquakes in Japan. Using the EQ catalog of Japan mentioned in § 6.2.2, we give in Fig. 6.13 the two curves $P(\kappa_1)$ versus κ_1 that result when the calculation is made by means of a window of 6–40 consecutive events sliding through either the original catalog or a shuffled one. Both the resulting $\kappa_{1,p}$ values markedly differ from κ_u, and in addition the value of the randomly shuffled data (≈0.064) does not greatly differ from the one (≈0.066) corresponding to the original data. This could be interpreted as reflecting that the self-similarity almost solely originates from the process's increments' "infinite" variance, but we will show here that the method suggested above does reveal clear contribution from temporal correlations as well.

Let us start from the (increased) temporal correlations in the well-known case of earthquake aftershocks. In this case the (modified form of the) Omori law mentioned above in Section 6.1 holds. Using the Southern California Earthquake catalog (with magnitude threshold $M_{thres} = 2.0$, see § 6.2.2.1), we now consider the aftershock series related to the Landers earthquake with magnitude $M_w = 7.3$ (that occurred at 11:57 UT on June 28, 1992, with an epicenter at 34.2°N 116.4°W) and the Hector Mine earthquake with magnitude $M_w = 7.1$ (that occurred at 09:46 UT on October 16, 1999, with an epicenter at 34.6°N 116.3°W). For these two mainshocks, Abe and Suzuki [7] identified the corresponding Omori regimes by examining the best fits of the (modified form of the) Omori law to the data based on the least-squares method. Here, we use the same aftershock dataset and plot

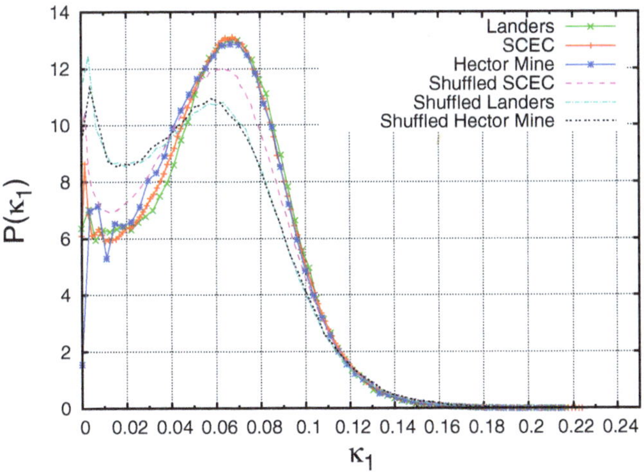

Fig. 6.14 (color) The pdfs of κ_1 when using either the almost 30 year data of SCEC (red +) or the two aftershock series (in the Omori regime [7]) of the Landers (green ×) and Hector Mine (blue ∗) earthquakes, respectively. Note that all these three curves coincide, but they change upon shuffling (turning to dashed magenta, dash-dotted cyan and dotted black, respectively) in a way discussed in the text. Taken from Ref. [81].

in Fig. 6.14 the pdf $P(\kappa_1)$ vs κ_1 by means of a sliding window of 6–40 consecutive events, as above, for Landers (green ×) and Hector Mine (blue ∗) earthquakes, respectively. Beyond these two aftershock series, we plot in Fig. 6.14 the corresponding curve (red +) for all earthquakes that occurred within the area $N_{32}^{37}E_{114}^{122}$ during the period 1973–2003 (called SCEC in § 6.2.2). Interestingly, these three curves more or less coincide and result in a common value of $\kappa_{1,p} \approx 0.066$, which agrees with that determined above from the original data of Japan (Fig. 6.13). Upon shuffling, all these three curves change, but we note that the two aftershock series (dash-dotted cyan and dotted black, which interestingly also almost coincide) exhibit the most noticeable change resulting in $\kappa_{1,p} \approx 0.060$; on the other hand, the shuffled SCEC data (dashed magenta) lead to $\kappa_{1,p} \approx 0.064$ which agrees with the corresponding $\kappa_{1,p}$ determined above from the shuffled data of Japan.

In other words, when focusing on aftershock series, we do observe that $\kappa_{1,p}$ changes markedly upon shuffling, thus pointing to the existence of considerable temporal correlations, as it should.

It seems reasonable that an Omori sequence where the events are clearly interrelated should give increased temporal correlations, and larger changes in $\kappa_{1,p}$, than events in a larger earthquake catalog where there is a possibility of including unrelated events.

In summary, when calculating the κ_1 value in a window $l = 6$ to 40 consecutive events sliding through either the original earthquake catalog or a shuffled one, the following results have been obtained for SCEC as well as for Japan:

Comparing the $\kappa_{1,p}$ values, we find that $\kappa_{1,p} \approx 0.066$ for the original data, while $\kappa_{1,p} \approx 0.064$ for the randomly shuffled ones.

Both these $\kappa_{1,p}$ values (with a plausible uncertainty of ± 0.001) differ markedly from the value $\kappa_u = 1/12$ of the "uniform" distribution. This could be in principle interpreted as reflecting that the self-similarity almost originates from the process' increments "infinite" variance.

Albeit, the existence of temporal correlations is responsible for the difference between the value of $\kappa_{1,p} \approx 0.064$ of the randomly shuffled EQ data from the value of $\kappa_{1,p} \approx 0.066$ of the original EQ data.

To further shed light on the presence of temporal correlations in seismicity data, we considered [56], beyond natural time analysis, the correlation function used in Ref. [46]. As an example, we used the SCEC data with magnitude threshold $M_{thres} = 2.0$ considering the area $N_{32}^{37}W_{114}^{122}$ and the period from 1981–2003. The results also showed the presence of correlations between earthquake magnitudes, thus strengthening the aforementioned conclusion that in natural time analysis the value of $\kappa_{1,p} = 0.064$ of the randomly shuffled data differs from the value of $\kappa_{1,p} = 0.066$ of the original data due to the presence of temporal correlations (arising from the *ordering* of the events in natural time).

Thus, in short, upon employing natural time analysis as well as the correlation function used in Ref. [46], we find that temporal correlations between EQ magnitudes do exist in real seismicity data.

This conclusion will be further strengthened in § 6.4.1 by studying the DFA in the EQ magnitude time series.

6.3.1 Temporal correlations upon changing the magnitude threshold in a catalog

The presence of temporal correlations has been further investigated [56] upon changing the magnitude threshold in the SCEC catalog for the period from 1981–2003. This catalog according to Ref. [69] is complete above $M_{thres} = 1.8$ since 1981 and this is why we selected the period 1981–2003 in this investigation.

For each M_{thres}, the catalog was randomly shuffled and the distribution of $\mathscr{E}(\kappa_{1,shuf})$ was determined. It turned out that $\mathscr{E}(\kappa_{1,shuf})$ exhibits a Gaussian distribution $N(\mu, \sigma)$ with average value μ and standard deviation σ; both parameters μ and σ depend on M_{thres}.

In order to quantify the temporal correlations in the original data as a function of the magnitude threshold, we plot in Fig. 6.15 the z-score ($z \equiv (\mathscr{E}(\kappa_1) - \mu)/\sigma$) of $\mathscr{E}(\kappa_1)$ for the original catalog with respect to the Gaussian distribution of $\mathscr{E}(\kappa_{1,shuf})$. If the z-score differs markedly from zero, this indicates the presence of temporal correlations. Figure

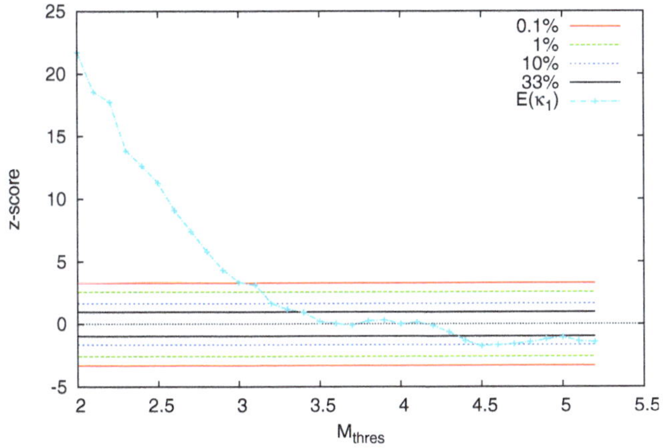

Fig. 6.15 The z-score (cyan) of $\mathscr{E}(\kappa_1)$ of the original SCEC data calculated with respect to the Gaussian distribution $N(\mu,\sigma)$ of $\mathscr{E}(\kappa_{1,shuf})$ (see the text) as a function of M_{thres}. The intervals corresponding to the probability $\mathscr{P}$ to observe the correlation present in the original sequence of events by chance are bounded by the horizontal lines of $\mathscr{P}=10^{-3}$, 10^{-2}, 10^{-1} and 0.33. Taken from Ref. [55].

6.15 reveals the following: a clear descending initial part in the magnitude threshold range $M_{thres}=2$ to 3.1, which indicates a gradual decrease in the statistically *significant* temporal correlations. This result is consistent with the expectation that upon increasing the magnitude threshold, the number of aftershocks involved in the calculation decreases. For the sake of comparison, the values of the aforementioned correlation function used in Ref. [46] have been calculated [56] and interestingly showed the same trend. Note that, in Fig. 6.15, for larger values of M_{thres} *no definite* results can be statistically inferred for the presence of temporal correlations in the catalogs.

6.3.2 The strength of temporal correlations as a function of the EQ inter-occurrence time

It is well known that seismic catalogs exhibit [38] the so-called short-term aftershock incompleteness (STAI). On the other hand, it has been recently shown [45] that correlations between magnitudes are larger for closer in time earthquakes. Thus, it is interesting to investigate in natural time the temporal correlations in a restricted catalog containing not all earthquakes but only those at a time distance (inter-occurrence time) $\delta t < T$ and choosing different values for the parameter T.

This was applied [56] to the aforementioned (§ 6.3.1) SCEC data for $M_{thres}=2.0$ for values of the maximum inter-occurrence time T ranging from half a minute to one day. The resulting catalogs were analyzed in natural time and the value of $\mathscr{E}(\kappa_1)$ for the original data have been determined. Then, the same catalogs were randomly shuffled and the calculation was repeated. Following the discussion of Eqs. (6.22) and (6.24), the relative strength of the temporal correlations with respect to the presence of process's increments' "infinite" variance, can be quantified by the ratio of the change $\Delta\mathscr{E}(\kappa_1)(\equiv \mathscr{E}(\kappa_1)-\mathscr{E}(\kappa_{1,shuf}))$ *upon randomly shuffling* the catalog over the difference $\Delta\mathscr{E}(\kappa_{1,shuf})$ of Eq. (6.24). Since our results are presented for natural time windows $l=6$ to 40, the value $\kappa_u(1-\frac{1}{l^2})$ in Eq. (6.24)

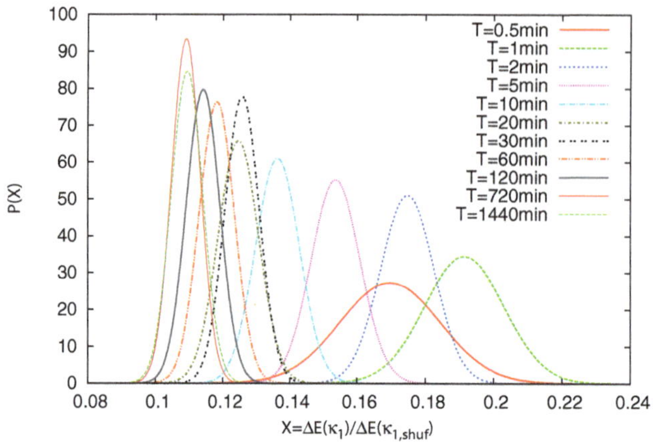

Fig. 6.16 The distributions of X of Eq. (6.25) for various values of the maximum interoccurrence time T ranging from half a minute to one day. Taken from Ref. [56].

can be substituted by its average value $\overline{\kappa_l} = 0.08296$. The study of these restricted catalogs showed that the distribution of

$$X \equiv \frac{\Delta\mathscr{E}(\kappa_1)}{\Delta\mathscr{E}(\kappa_{1,shuf})} = \frac{\mathscr{E}(\kappa_1) - \mathscr{E}(\kappa_{1,shuf})}{\overline{\kappa_l} - \mathscr{E}(\kappa_{1,shuf})} \tag{6.25}$$

can be approximated by Gaussian distributions differing from zero, beyond any statistical doubt, thus reflecting the existence of temporal correlations. These correlations increase (see Fig. 6.16) as T varies from half a day to one minute, thus agreeing with the conclusions of Ref. [45]. When T becomes less than one minute, these correlations diminish (cf. the thick solid red curve corresponding to $T = 0.5$ min with the thick dashed green curve corresponding to $T = 1$min) and this effect could be attributed to STAI: for M equal to the average magnitude of these two catalogs, the appropriate time interval t_{M} to remove [38] STAI is $t_{\mathrm{M}} = 300 \times 10^{(\mathrm{M}-4)/2}$ seconds $= (53 \pm 4)$ seconds.

Thus, natural time analysis leads to results that are compatible with the recent suggestion [46] that correlations between magnitudes are larger for closer in time earthquakes when the maximum inter-occurrence time T varies from half a day to one minute.

6.4 Order parameter fluctuations of seismicity before and after mainshocks

6.4.1 Feature of the pdf of the order parameter for seismicity. DFA of earthquake magnitude time series

In a recent study [44], it has been undoubtedly shown that in the regimes of stationary seismic activity (i.e., during periods at which large aftershock sequences are missing) long-range correlations exist between successive EQ magnitudes. Moreover, a separate study

[13] showed that the fluctuations of seismic activity, defined as the detrended cumulated sum of the magnitude time series, exhibit Family–Vicsek dynamic scaling. In both studies, the sequence index k, i.e., the sequential order in which an EQ had occurred, has been used for the detection of the long-range correlations (e.g. see Fig. 6.17 that will be discussed later). Notice that it is the combination of this index with the released seismic energy during the k-th EQ that constitutes, as mentioned in § 2.1.2, the two quantities which are in fact used in natural time analysis.

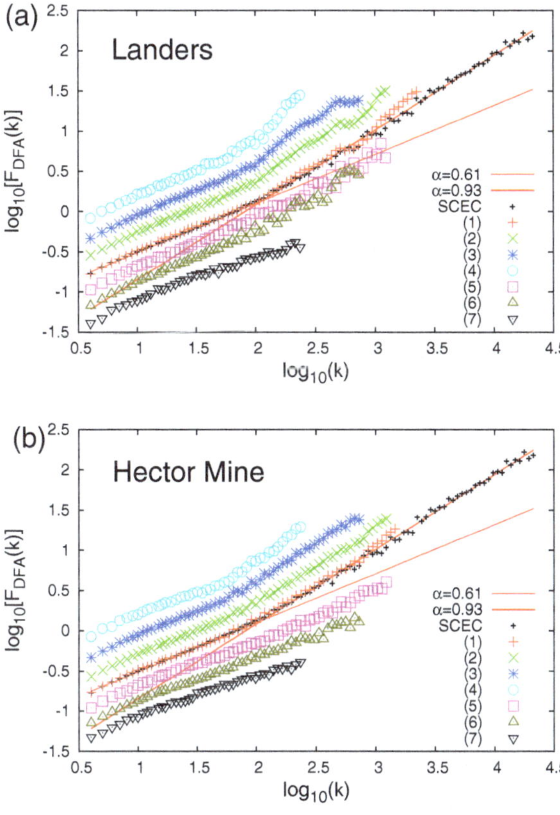

Fig. 6.17 DFA plot of SCEC (black plus) along with the corresponding plots (red crosses (1)) of the aftershock sequence (reported in Ref. [7]) for the Landers EQ (panel a) and the Hector Mine EQ (panel b). The results (2) to (7) depict the DFA plots of $W = 5{,}000$, 3,000, 1,000 EQs immediately *after*, and $W = 5{,}000$, 3,000, 1,000 EQs immediately *before*, the Landers EQ (panel a) and the Hector Mine EQ (panel b), respectively. (These plots have been vertically shifted for sake of clarity.) The straight lines with $\alpha = 0.61$ and 0.93 have been drawn as a guide to the eye. Taken from Ref. [58].

The existence of temporal correlations in seismicity has been already treated on the basis of natural time analysis in Section 6.3. Here, we extend [58] the aforementioned recent study [44] of the DFA for the detection of long-range temporal correlations in earthquake magnitude time series M_k by including the non-stationary periods of seismicity. Recall that DFA has been established as a robust method suitable for detecting long-range power law correlations embedded in non-stationary signals (see § 1.4.2).

Figure 6.17 depicts with black plus symbols the resulting $\log_{10}[F_{DFA}(k)]$ versus $\log_{10}(k)$ for all the 85,862 EQs with $M_k \geq 2.0$ that occurred during the period 1981–2003 within the area $N_{32}^{37}\ W_{114}^{122}$ reported by SCEC. A cross-over is observed at $k \approx 200$ below which the DFA exponent α is close to 0.61 ($\equiv \alpha_{low}$). This value agrees fairly well with the one $\alpha = 0.59(5)$ obtained in Ref. [44] when solely analyzing the periods of *stationary* seismic activity. Thus, the substantially higher value ($\alpha_{high} = 0.93$) obtained for scales longer than $k \approx 200$ now emerges upon the inclusion of the "non-stationary" periods of seismic activity. To further shed light on the origin of this cross-over, we examine the behavior of the magnitude time series after the two most significant earthquakes reported in SCEC already mentioned in Section 6.3, i.e, the Landers EQ (Fig. 6.17(a)) and the Hector Mine EQ (Fig. 6.17(b)). The application of DFA to the aftershock magnitude time series as identified in Ref. [7] (by examining the corresponding Omori law regimes) is shown for both these mainshocks, by the red plus symbols in Figs. 6.17(a) and (b), respectively. Interestingly, an inspection of this figure reveals that the scaling behavior of DFA in both aftershock series is close to that of the whole SCEC. Notice that the cross-over still pertains. Recall that when analyzing both these aftershock data in natural time (Section 6.3), we found that the corresponding pdfs $P(\kappa_1)$ of the *order parameter* κ_1 of seismicity (§ 6.2.1) almost coincide with the $P(\kappa_1)$ for the whole SCEC, see Fig. 6.14. The latter coincidence could be interpreted as a 'return' of the seismic activity to its mean behavior after the completion of the aftershock sequences, thus being in accordance with the previous result of DFA.

We now examine the magnitude time series with lengths $W = 5{,}000$, 3,000 and 1,000 EQs not only just *after* but also just *before* these two EQs in SCEC. The corresponding results of DFA are also given in Figs. 6.17(a) and (b) for the Landers and the Hector Mine EQ, respectively. Upon restricting ourselves to the period just after the mainshocks, the results show (e.g., for $W = 1{,}000$; see the cyan circles in Fig. 6.17) that the high value of the DFA exponent α at longer scales should be attributed to the highly correlated "immediate" aftershocks. We now turn to the study of the magnitude time series solely *before* these two mainshocks. The results of DFA suggest that the α value for scales longer than the cross-over is now significantly *smaller* than in the case of aftershocks, and *much closer to that for scales shorter than the cross-over* (see the squares, triangles and inverted triangles in Fig. 6.17). Thus, the cross-over effect is definitely smoothed in the magnitude time series that end just *before* the mainshocks. A closer inspection of the inverted triangles, i.e., the results obtained from $W = 1{,}000$ EQs just *before* the mainshocks, indicates that the DFA scaling exponent becomes even smaller than α_{low} ($= 0.61$) and the values obtained are $\alpha = 0.53(2)$ and $\alpha = 0.50(2)$ for the Landers and the Hector Mine EQ, respectively.

Thus, the 'foreshocks' appear to exhibit correlations that are somewhat weaker than those already found [44] in the stationary seismicity (recall that Ref. [44] reported $\alpha = 0.59(5)$ for stationary periods).

We now turn to the results of the natural time analysis of the time series with $W = 5{,}000$, 3,000 and 1,000 EQs just *before* and just *after* Landers and Hector Mine EQ. In a seismic catalog comprising W events, the procedure to construct the pdf $P(\kappa_1)$ is the following (in a similar fashion as in § 6.2.2). Starting from the first EQ, we calculate the κ_1 values using $l = 6$ to 40 consecutive events (including the first one). We then proceed to the second EQ, and repeat the calculation of κ_1 and so on. Thus, after sliding event by event

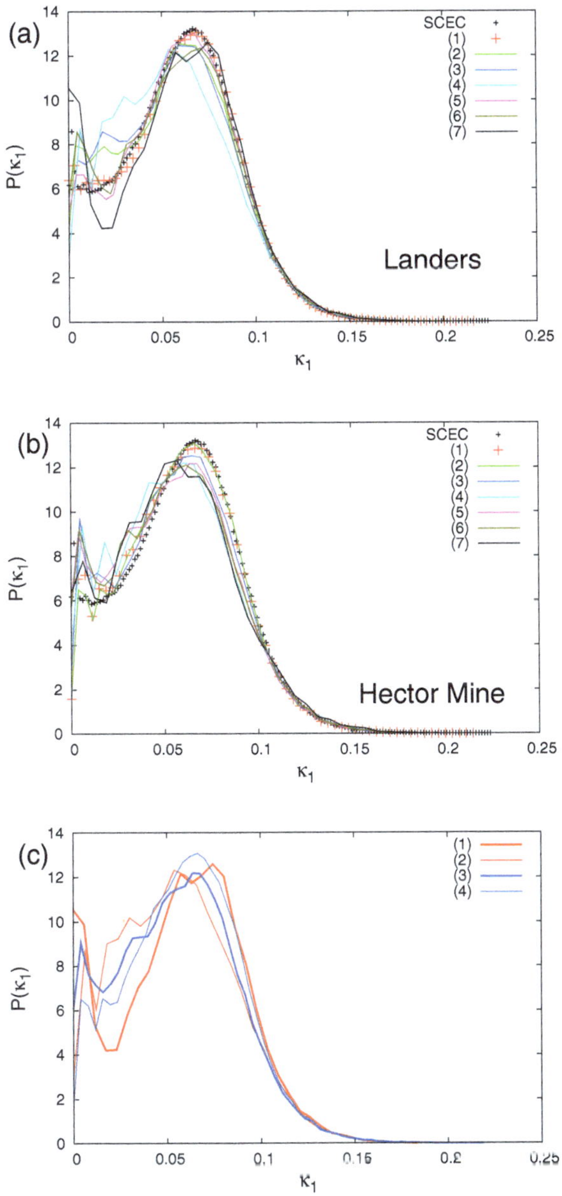

Fig. 6.18 The probability density function $P(\kappa_1)$ versus κ_1 for SCEC (black plus) along with the ones resulting from the aftershock sequence (1) as reported in Ref. [7] for the Landers EQ (panel a) and the Hector Mine EQ (panel b). The results (2) to (7) depict $P(\kappa_1)$ for $W = 5{,}000$, 3,000, 1,000 EQs immediately *after*, and $W = 5{,}000$, 3,000, 1,000 EQs immediately *before*, the Landers EQ (panel a) and the Hector Mine EQ (panel b), respectively. In panel (c), we depict the results for: (1) 1,000 EQs immediately *before* the Landers EQ, (2) 1,000 EQs immediately *after* the Landers EQ, (3) 5,000 EQs immediately *before* the Hector Mine EQ, (4) 5,000 EQs immediately *after* the Hector Mine EQ. Taken from Ref. [58].

through the whole earthquake catalog, the calculated κ_1 values enable the construction of the pdf $P(\kappa_1)$. They are shown in Fig. 6.18(a) and (b) for these two EQs, respectively. In particular, we observe that the pdf $P(\kappa_1)$ versus κ_1 curves differ in general from the corresponding curve obtained from the whole SCEC or from the aftershock time series identified in Ref. [7].

This reveals that either just *before* or just *after* a significant EQ, the seismicity deviates from its mean behavior in natural time.

We now proceed to a comparison between the pdfs $P(\kappa_1)$ of the order parameter κ_1 just *before* and just *after* a significant EQ. In Fig. 6.18(c) when plotting $P(\kappa_1)$ versus κ_1 for $W = 1{,}000$ EQs before and after the Landers EQ with the thick red and the thin red line, respectively, they are found to be markedly different in the following respect:

Before the Landers EQ a significant bimodal feature appears in the $P(\kappa_1)$ vs κ_1 plot. This, which solely emerged from the natural time analysis, is of profound importance as it is strikingly reminiscent of the bimodal feature observed in the pdf of the order parameter when approaching (from below) T_c in equilibrium critical phenomena, e.g., see Fig. 6.8(b). Since κ_1 is the order parameter for seismicity, a similar behavior should be expected before *every* mainshock.

Actually, in Fig. 6.18(c), we depict $P(\kappa_1)$ versus κ_1 for $W = 5{,}000$ events before and after the Hector Mine EQ. The results are shown with the thick blue and the thin blue line, respectively. We again observe that a bimodal feature emerges in the curve *before* the mainshock.

6.4.2 Prediction scheme by quantifying the bimodal feature of the pdf of the order parameter κ_1 for seismicity before mainshocks

Let us now assume [58] that the variability $\beta \equiv \sigma(\kappa_1)/\mu(\kappa_1)$, where $\mu(\kappa_1)$ and $\sigma(\kappa_1)$ stand for the average value and the standard deviation of the κ_1 values considered, constitutes at a first approximation a measure to quantify the presence of the bimodal behavior in $P(\kappa_1)$ versus κ_1 identified above in § 6.4.1. *If* the presence of this bimodal feature actually signifies the occurrence of an impending mainshock, then the quantity β can in principle be considered as a decision variable to predict the occurrence of a large earthquake solely based on the past magnitudes. Such a 'prediction' scheme should *not* be confused, however, with the one achieved when the seismic data are enriched (supplemented) with SES data (Chapter 1), on the basis of which the epicentral area and the magnitude of the impending mainshock can be determined, see § 1.3.5. This is so because, once the latter are available, the natural time analysis of the seismicity that occurs in the future epicentral area *after* the SES detection leads to the identification of the time window of the impending mainshock within a narrow range of a few days to around one week, as it will be explained in Chapter 7.

Hereafter, we proceed as follows. For each EQ of magnitude M_k in SCEC, we estimate $\overline{\kappa_1} = \sum_{l=6}^{40} \kappa_1(l)/35$, i.e., the average value of the $\kappa_1(l)$ calculated upon considering $l = 6$ to 40 consecutive EQs (including the k-th event). Next, we assign this value $\overline{\kappa_1}$ to the k'-th element ($k' = k+40$) of the time series $\kappa(k')(\equiv \overline{\kappa_1})$. This way, $\kappa(k')$ has no information of the event with magnitude $M_{k'}$ which is the 40-th EQ that occurred after M_k. We can now estimate, for various windows of W earthquakes, the time series of the average values

$$\mu_{k'}(W) \equiv \frac{1}{W} \sum_{n=k'-W+1}^{k'} \kappa(n), \tag{6.26}$$

which is equivalent to $\mu(\kappa_1)$ obtained when considering a catalog consisting of the W earthquakes that occurred just before $M_{k'}$. In addition, the time series of standard deviations can be obtained from

$$\sigma_{k'}(W) \equiv \sqrt{\frac{1}{W} \sum_{n=k'-W+1}^{k'} [\kappa(n) - \mu_{k'}(W)]^2}, \tag{6.27}$$

and the variability time series is given by

$$\beta_{k'}(W) = \frac{\sigma_{k'}(W)}{\mu_{k'}(W)}. \tag{6.28}$$

We will now examine whether $\beta_{k'}$ can be used as a decision variable for binary 'predictions'. Following the terminology of Keilis Borok and coworkers [41, 40], the time increased probability (TIP) is turned on when $\beta_{k'}(W) \geq \beta_c$, where β_c is a given threshold in the prediction. If the magnitude $M_{k'}$ is greater than or equal to a target threshold M_{thres}, we have a successful 'prediction'. For the present case of binary predictions, the prediction of events becomes a classification task, with two type of errors: missing an event and giving a false alarm. We therefore choose, following Ref. [31], the receiver operating characteristics (ROC) [29] as the method to analyze here the prediction quality. This is a plot of the hit rate versus the false alarm rate, which is tuned by the threshold β_c. Only if in between the hit rate exceeds the false alarm rate, is the predictor useful. Random predictions generate equal hit and false alarm rate, and hence they lead to the diagonal in ROC plot; see the black straight lines in Figs. 6.19 and 6.20. (If β_c is maximum, both hit rate and false alarm rate are zero, while for very small β_c values both rates tend to unity.) Thus, only when the points lie above this diagonal the predictor is useful. Figure 6.19 depicts the ROC curves, for various values of $M_{thres} = 3.0$ to 4.5, together with the results obtained when using, for example, two *randomly* shuffled copies (green and red circles) of SCEC. The results for various W values are shown, i.e., $W = 70$, 300, 1,000 and 3,000, in Figs. 6.19(a) to (d), respectively.

In all cases, the results are better (i.e., points lying above the diagonal) when deduced from the original SCEC compared to those from the *randomly* shuffled SCEC. This indicates that the predictive power of $\beta_{k'}(W)$ given by Eq. (6.28) stems from *temporal* correlations between EQ magnitudes present in the actual seismicity.

In order to further examine the statistical significance of this 'prediction' scheme, we depict in Fig. 6.20 the results for $W = 1{,}000$ together with the results of 10^2 runs of the

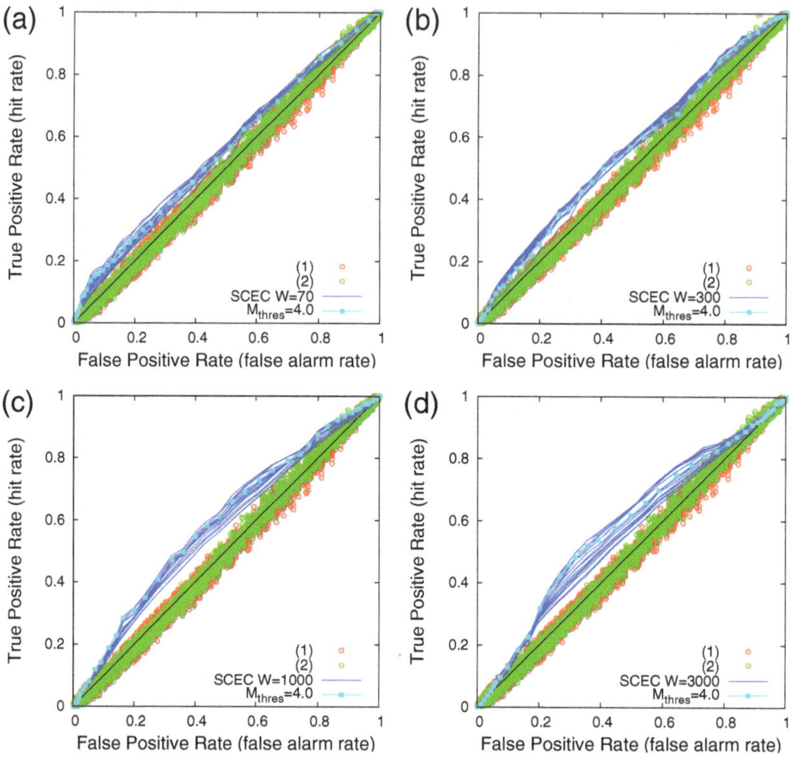

Fig. 6.19 The ROC curves constructed using $\beta_{k'}(W)$ as decision variable, for $W = 70$, 300, 1,000 and 3,000, are depicted in panels (a) to (d), respectively. The blue broken lines correspond to the ROC curves obtained when considering the target thresholds $M_{thres} = 3$ to 4.5 (note that there are only 212 events with $M_k \geq 4.5$). The ROC for $M_{thres} = 4$ is shown in cyan as a guide. The red (1) and green (2) circles correspond to exactly the same analysis, but performed for two independent randomly shuffled copies of SCEC, and fall around the diagonal of chancy predictions because the temporal correlations between consecutive EQs are now lost. Taken from Ref. [58].

same catalog when using as decision variable a uniformly distributed random number in the same range as $\beta_{k'}(1{,}000)$. We observe that none of these runs outperforms $\beta_{k'}(1{,}000)$ for false alarm rates from 20% to 60%. Thus, the decision variable $\beta_{k'}(1{,}000)$ has predictive power which is statistically significant. The inset of Fig. 6.20 depicts the ratio of the hit rate over the false alarm rate versus M_{thres}, which shows that the prediction results become better upon increasing M_{thres}. For example, when $M_{thres} = 4$ (cyan line with squares) the hit rate is approximately 60% when the false alarm rate is 50%. The TIP can be visualized in Fig. 6.21, where the red shaded areas correspond to the periods when the TIP is on (i.e., $\beta_{k'}(1{,}000) \geq 0.35$). The results convincingly outperform chance, but are not spectacular. This (which remains so when using, instead of β, the kurtosis see the black dots in Fig. 6.21) is *not* unreasonable in view of the following fact:

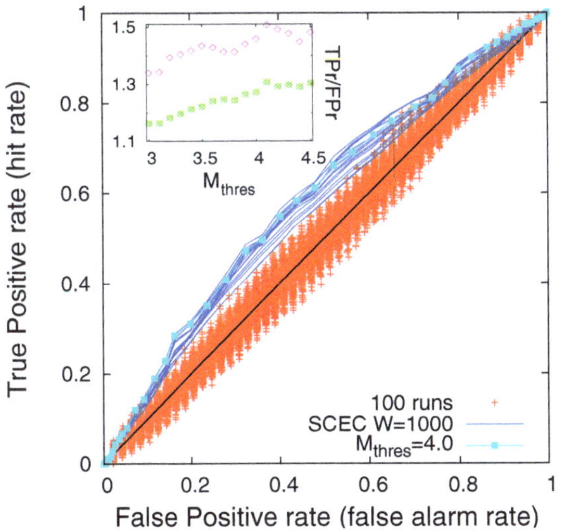

Fig. 6.20 The ROC curves constructed using $\beta_{k'}(W)$ for $W = 1{,}000$ as decision variable (the case of $M_{thres} = 4$ is shown in cyan color as a guide) together with 10^2 repetitions of the prediction scheme using a uniformly distributed random number as decision variable (see the text). The inset depicts the ratio of the True Positive rate (TPr) over the False Positive rate (FPr) versus M_{thres}, for FPr $\approx 25\%$ (magenta diamonds) and FPr $\approx 50\%$ (green stars). Taken from Ref. [58].

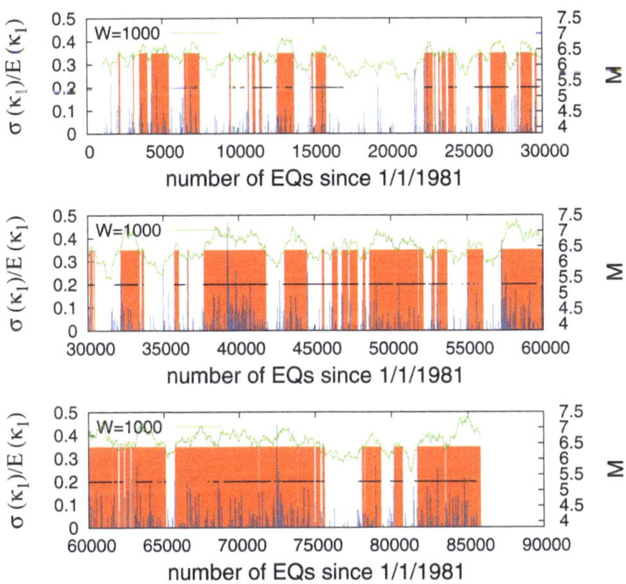

Fig. 6.21 The variability $\beta_{k'}(W)$ (left scale, green color) for $W = 1{,}000$ together with the magnitude time series $M_{k'}$ (right scale, blue impulses) as a function of the sequence index k', i.e., the total number of EQs since January 1, 1981. The red shaded areas, which are formed by impulses when $\beta_{k'}(W) \geq 0.35$, show when the TIP is on. The black dots correspond to the TIP obtained when using, instead of β, the kurtosis. Taken from Ref. [58].

When using a constant natural time window of W events, it may not correspond to the time at which the focal area of the impending mainshock enters into the critical regime, which is captured however by the SES detection *if* available (note that in the case of the SES detection, the results are drastically better as explained in detail in Chapter 7).

6.4.3 Concluding remarks

Combining the results obtained in the preceding § 6.4.1 and § 6.4.2, the main points could be summarized as follows:

We made use of the order parameter κ_1 of seismicity defined in natural time (§ 6.2.1) together with the DFA of the magnitude time series to investigate the period just *before* and just *after* a significant mainshock. The study was focused on two significant EQs that occurred in Southern California in 1992 and 1999, i.e., the Landers and the Hector Mine earhquakes.

Quite interestingly, the natural time analysis of these time series reveals that 'foreshocks' exhibit a behavior characteristic of systems close to their critical point: upon considering the order parameter κ_1 of seismicity the probability distribution function $P(\kappa_1)$ vs κ_1 exhibits a *bimodal feature.*

In an attempt to quantify this bimodal feature, we considered the variability of κ_1, which was then used as decision variable for the 'prediction' of the occurrence of a large earthquake in the next natural time step based solely on the magnitudes of previous earthquakes. These results *outperform* chance but are not spectacular *if* not supplemented with SES detection (see Chapter 7).

In other words, the natural time analysis of seismicity before significant earthquakes reveals that the fluctuations of the order parameter *before* major earthquakes exhibit a bimodal feature which, if quantified properly, may be used as decision variable to predict the occurrence of large earthquakes.

6.5 Nonextensivity and natural time: the case of seismicity

Nonextensive statistical mechanics [4, 72], pioneered by Tsallis [71], provides a theoretical framework for the studies of complex systems in their non-equilibrium stationary states, systems with (multi) fractal and self-similar structures, long-range interacting systems etc. This framework offered recently a generalization of the G-R law. Here, we employ this nonextensive G-R generalization to study the observed seismic data fluctuations. In particular, we combine [57] three modern methods, i.e., the non-extensive generalization of the G-R law together with natural time and detrended fluctuation analysis (DFA, see § 1.4.2). This procedure is applied to synthetic seismic data as well as to real seismic data from two different areas: First, the EQs that occurred during the period 1981–2003 within the area $N_{32}^{37} W_{114}^{122}$ using SCEC. Second, the EQs within $N_{25}^{46} E_{125}^{146}$ for the period 1967–2003 using the Japan Meteorological Agency catalog, simply called "Japan". See Figs. 6.22, 6.23 and 6.24 that will be discussed later. The thresholds $M \geq 2.0$ and $M \geq 3.5$ have been considered for SCEC and Japan, respectively for the sake of data completeness as already mentioned in § 6.2.2.1.

6.5.1 Non extensivity and earthquakes. The generalization of the Gutenberg–Richter law

The first studies on the analysis of EQs in the nonextensivity framework have been made by Abe and coworkers [4, 6, 8]. In this framework, an interesting model for earthquake dynamics has been proposed by Sotolongo-Costa and Posadas (SCP) [68]. It consists basically of two rough profiles interacting via fragments filling the gap between them where the fragments are produced by local breakage of the local plates. In other words, the fundamental idea of this model consists of the fact that the space between faults is filled with the residues of the breakage of the tectonic plates from where the faults originate. In this model the mechanism of earthquake triggering assigns an important role in the fragments: the stress increase between the two fault plates constitutes the main factor that governs the complexity of the fragment–asperity interaction, where eventually the fragments may act as roller bearings, and also as hindering entities of the relative motion of the plates until the growing stress produces their liberation with the subsequent triggering of the earthquake [70]. By using the nonextensive formalism, SCP not only showed the influence of the size distributions of fragments on the energy distribution of earthquakes but also deduced an energy-distribution function which in a particular case leads to the G-R law.

The aforementioned SCP model was revisited by Silva et al. [65] who made two key improvements. The first one made use of a different definition for mean values in the context of Tsallis nonextensive statistics that was achieved in Ref. [3]. In particular, Abe and Bagci [3] considered in depth the two kind of definitions for the expectation value of a physical quantity which both lead to the maximum Tsallis entropy distribution of a similar type. The one is the ordinary definition (note that this was used by SCP) and the other is the normalized q-expectation value employing the escort distribution [16, 5]: Their final conclusion states that the Shore–Johnson theorem [62, 63, 64] for consistent minimum cross-entropy (i.e., relative entropy) principle is shown to select the formalism with the normalized q-expectation value and to exclude the possibility of using the ordinary expectation value from nonextensive statistical mechanics. The second improvement by Silva et al. refers to the introduction of a scaling law, i.e., $\varepsilon \propto r^3$, between the released relative energy ε and the size r of the fragments (this substantially differs from the assumption $\varepsilon \propto r$ used by SCP). Then Silva et al. [65] proceeded as follows. The Tsallis entropy has the form of Eq. (2.101):

$$S_q = k_B \frac{\int p^q(\sigma)(p(\sigma)^{1-q} - 1)\, d\sigma}{q-1} \tag{6.29}$$

where $p(\sigma)$ is the probability of finding a fragment of relative surface σ (which is defined as a characteristic surface of the system), q is a real number usually termed *nonextensive* parameter and k_B is Boltzmann constant which will be hereafter set equal to unity for the sake of simplicity. It is easy to see that Eq. (6.29) recovers the standard Boltzmann–Gibbs entropy in the limit $q \to 1$. The maximum entropy formulation for Tsallis entropy implies that the following two conditions have to be introduced [73, 72]. First, the normalization of $p(\sigma)$:

$$\int_0^\infty p(\sigma)\,\mathrm{d}\sigma = 1 \tag{6.30}$$

Second, the *ad hoc* condition (see also § 6.5.4.1) about the q-expectation value

$$\sigma_q \equiv \langle\sigma\rangle_q = \int_0^\infty \sigma P_q(\sigma)\,\mathrm{d}\sigma, \tag{6.31}$$

where $P_q(\sigma)$ is the escort distribution [16, 5] given by

$$P_q(\sigma) = \frac{p^q(\sigma)}{\int_0^\infty p^q(\sigma)\,\mathrm{d}\sigma}, \tag{6.32}$$

which for $q \to 1$ becomes the definition of the mean value. Silva et al. followed the standard method of conditional extremization of the entropy functional S_q and found an expression for the fragment distribution $p(\sigma)$. Then, assuming the aforementioned energy scale $\varepsilon \propto r^3$, they obtained the energy distribution function $p(\varepsilon)$ for the EQs. Finally, by considering the relationship

$$m = \frac{1}{3}\ln(\varepsilon), \tag{6.33}$$

where m denotes the magnitude, Silva et al. obtained the number $N_{>m}$ of EQs with magnitude larger than m:

$$\log\left(\frac{N_{>m}}{N}\right) = \left(\frac{2-q}{1-q}\right)\log\left[1-\left(\frac{1-q}{2-q}\right)\frac{10^{2m}}{a^{2/3}}\right] \tag{6.34}$$

where N is the total number of the events and a the proportionality constant between ε and r^3.

Equation (6.34) incorporates the characteristics of nonextensivity into the distribution of earthquakes by magnitude, and the G-R law can be deduced as its particular case when considering a significant magnitude *threshold*. Then, Eq. (6.34) reduces to Eq. (6.2) with $b = 2(2-q)/(q-1)$. Thus, Eq. (6.34) can be alternatively termed as a generalized G-R law.

This relation has been found [86] to describe appropriately the energy distribution in a wider detectable range of magnitudes compared to that of the original G-R law. Furthermore, Silva et al. [65] and later Vilar et al. [86] in conjunction with the earlier SCP study [68] led to the conclusion [86] that values of $q \simeq 1.6$–1.7 seem to be universal in the sense that different datasets from different regions of the globe (e.g. California [68], Iberian Peninsula [68], Andalucia [68], Samambaia-Brazil [65], New Madrid (USA) [65], North Anatolian fault, Turkey [65], San Andreas fault [86]) indicate a value lying in this interval. In addition, in a very recent study [70], a comparable q value (i.e., $q = 1.67$) has been found by analyzing the (tectonic) seismicity in Italy, while a somewhat lower value ($q = 1.48$) was reported for the volcanic seismicity in Vesuvius. Finally, we note that very recently [26] an alternative relation has been suggested between the released energy and the surface size of fragments, i.e., $\varepsilon \propto \exp(\sigma^{1/\lambda_0})$, where λ_0 is a constant in contrast to the relation $\varepsilon \propto \sigma^{1/2}$ proposed by SCP [68] and the relation $\varepsilon \propto \sigma^{3/2}$ by Silva et al [65]. This, which has been inspired by the fractal nature of the fragments filling the gaps between adjacent fault plates, leads to a different expression for the distribution of EQs as a function of the magnitude which has a q-exponential form, and the fit with the Iran and California catalogs was found to be good. On the other hand, Eq. (6.34) has no q-exponential form, but it is preferred to be used here since it has been found to describe well the data in a larger number of seismic regions.

6.5.2 Combining nonextensivity with natural time analysis

Recall that by calculating the κ_1 value in a window of length $l = 6$ to 40 consecutive events sliding through either the original earthquake catalog or a shuffled one, the following results have been obtained for SCEC and Japan (see § 6.2.2 and Section 6.3): Concerning the most probable value $\kappa_{1,p}$ of κ_1, we find that $\kappa_{1,p} \approx 0.066$ for the original data while $\kappa_{1,p} \approx 0.064$ for the randomly shuffled data, see Figs. 6.13 and 6.14. Beyond this decrease of the $\kappa_{1,p}$ value, the whole feature of the curve $P(\kappa_1)$ versus κ_1 changes markedly upon shuffling. Both $\kappa_{1,p}$ values, that have a plausible uncertainty of ± 0.001, differ markedly from the value $\kappa_u = 1/12$ of the "uniform" distribution, which indicates a significant contribution from the process's increments' "infinite" variance to self-similarity. In addition, in Section 6.3 the temporal correlations between EQ magnitudes were found to be responsible for the difference between the value of $\kappa_{1,p} \approx 0.064$ of the randomly shuffled data from the value of $\kappa_{1,p} \approx 0.066$ of the original data. This was ascertained in § 6.4.1 by employing also DFA for the analysis of the EQ magnitude time series (see also Fig. 6.23 that will be discussed later).

We now explain the procedure followed in this Section for the generation of synthetic (surrogate) EQ magnitude series. We make use of a simple method [57] to produce long-range correlated (EQ) data (magnitude series) that obey an arbitrary cumulative distribution function $F(x)$. This is based on the well-known random number generator of an arbitrary distribution $F(x)$, described in Ref. [14], as well as on the method suggested in Ref. [28]. Let us first recall that in order to construct [14] a random number generator for the distribution $F(x)(=p)$, we simply need the inverse function $F^{-1}(p)(=x)$. Then by inserting a sequence p_i of (uncorrelated) random numbers uniformly distributed in the region (0,1), we can obtain the (uncorrelated) random numbers $x_i = F^{-1}(p_i)$ which are distributed according to the cumulative distribution function $F(x)$. Here, we shall take advantage of the fact that, *at least* for the exponential distribution (e.g., Eq. (6.1) of the G-R law) or for the distribution function of Eq. (6.34), *if* the sequence p_i is long-range correlated, the same holds for the random numbers $x_i[= F^{-1}(p_i)]$ (see Fig. 6.23). For example, if we want to produce a series of random numbers x_i, having a cumulative distribution function $F(x)$, that have a DFA exponent equal to $\alpha(< 1)$, we can use $x_i = F^{-1}[\Phi_G(z_i)]$, where $\Phi_G(t)$ is the cumulative distribution function of the standard normal (Gaussian) distribution (i.e., with zero mean and unit standard deviation) and z_i is a standard fGn with H (see § 1.5.1.1) equal to α.

Moreover, if we want the generated synthetic data to mimic the temporal correlations of some experimental data y_i, then by using their (experimental) cumulative distribution function $\Phi_y(t)$, we can use $x_i = F^{-1}[\Phi_y(y_i)]$. This simple method for the sake of convenience will be hereafter called cumulative distribution function transformation (CDFT).

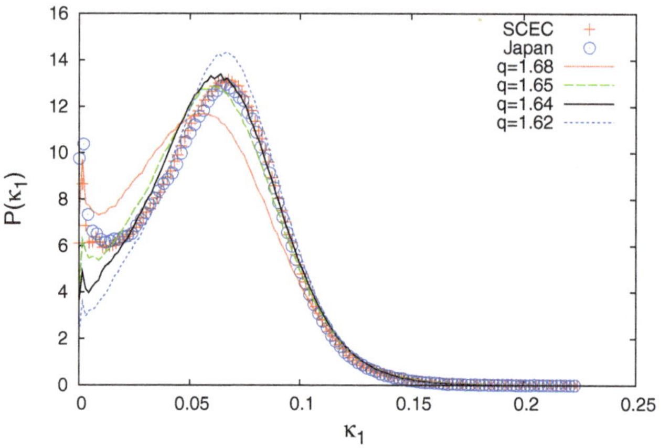

Fig. 6.22 The pdf $P(\kappa_1)$ versus κ_1 resulting from the natural time analysis of *uncorrelated* data obtained from Eq. (6.34) for various q values together with those deduced from real seismic data for SCEC and Japan. Taken from Ref. [57].

Figure 6.22 shows the probability density function $P(\kappa_1)$ versus κ_1 deduced from the natural time analysis of synthetic seismic data with *no temporal correlations* ($\alpha = 0.5$) obeying the nonextensive generalization of the G-R law, i.e., Eq. (6.34). Results are given for four different values of q, i.e., $q = 1.62$, 1.64, 1.65 and 1.68, lying in the universal range $q = 1.6$ to 1.7 (see § 6.5.1). In the same figure, for the sake of comparison, the results obtained from the real seismic data, i.e., SCEC and Japan, are also plotted. An inspection of this figure shows that the results from synthetic data differ markedly from those of the real data. This reveals that, since in natural time analysis the waiting (inter-occurrence) times between EQs do not intervene, temporal correlations do exist in the (magnitude time series of) real seismic data. This is in agreement with the results of § 6.2.2 in which we showed that the G-R law cannot fully account for the complexity observed in the real seismic data.

Thus, as a second step, we investigate whether synthetic data obeying Eq. (6.34) can reproduce the real situation but when *inserting* long-range temporal correlations. To quantify the long range temporal correlations in the real seismic data, we depict in Fig. 6.23 the DFA plots (as in § 6.4.1) for the original magnitude time series of SCEC (red pluses) and Japan (blue circles). The thin and the thick straight lines result from a linear least-squares fit to the short ($\log_{10}(k) \leq 2$) and long ($\log_{10}(k) \geq 2.5$) scales, respectively, for SCEC (red, dotted lines) and Japan (blue, short-dashed lines). The values of the slope α at the short scales are $\alpha = 0.61(2)$ and $0.57(2)$ for SCEC and Japan, respectively. These values are comparable, as mentioned (§ 6.4.1), to those obtained [44] by analyzing the

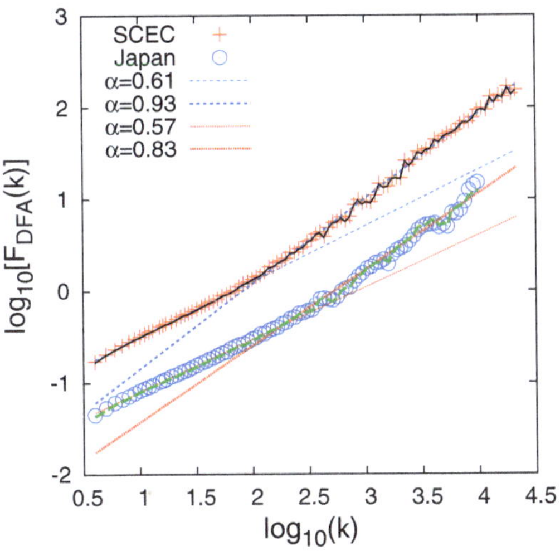

Fig. 6.23 The DFA of the original magnitude time series for SCEC (red pluses) and Japan (blue circles). The thin and thick straight lines correspond to the linear least-squares fit at the short and long scales, respectively. The existence of a cross-over at $k \approx 200$ indicates an extra complexity in the case of earthquake time series. For this reason, synthetic time series obeying the G-R law (Eq. (6.1)) with $b = 1.08$ have been produced by CDFT, the DFA of which are shown with the thick black (solid) and green (long-dashed) broken lines for SCEC and Japan, respectively. The DFA of Japan has been displaced for the sake of clarity. Taken from Ref. [57].

seismic records in regimes of *stationary* seismic activity in Northern and Southern California. At longer scales, a cross-over is evident in Fig. 6.23 (see also Fig. 6.17) at $k \approx 200$ above which the slopes are found to be $\alpha = 0.93(3)$ and 0.83(3) for SCEC and Japan, respectively.

The aforementioned DFA behavior (i.e., smaller α value at short scales and larger α at long scales) of the real seismic data was then reproduced by synthetic (obtained from CDFT of the original) seismic data coming from the G-R law with $b \approx 1.08$, the DFA plots of which are shown in Fig. 6.23 with the broken lines (black and long-dashed green for SCEC and Japan, respectively). Figure 6.24 depicts $P(\kappa_1)$ versus κ_1 for the real seismic data of SCEC, Fig. 6.24(a), and Japan, Fig. 6.24(b), along with those obtained from synthetic G-R data by CDFT (red, dotted lines). There exists a good agreement between synthetic and real data. This agreement implies that the temporal correlations between EQ magnitudes have been successfully incorporated as well as the fact that only EQs above the magnitude *completeness* threshold (§ 6.2.2.1) have been considered.

To proceed one step further, synthetic seismic data were deduced by using, instead of the G-R law, Eq. (6.34) and CDFT. In particular, for SCEC we obtained the results depicted in Fig. 6.24(a) after adopting $q = 1.65$, 1.66, 1.67 and $q = 1.68$ and inserting (by means of CDFT) long-range temporal correlations between EQ magnitudes comparable to those found in real data, i.e., $\alpha = 0.61$ and 0.93 for short and long scales, respectively. They are shown by the green (long-dashed), black (solid), blue (short-dashed) and cyan (dashed dotted) lines for the four values of q, respectively. We observe that the $q = 1.67$ curve is closer to the real data but some differences still remain. As for Japan (see Fig. 6.24(b)), the synthetic long-range correlated data that come from Eq. (6.34) with $q = 1.66$ (green, long-dashed) with $\alpha = 0.57$ and 0.83 for the short and long scales, respectively, exhibit much better agreement with the real ones.

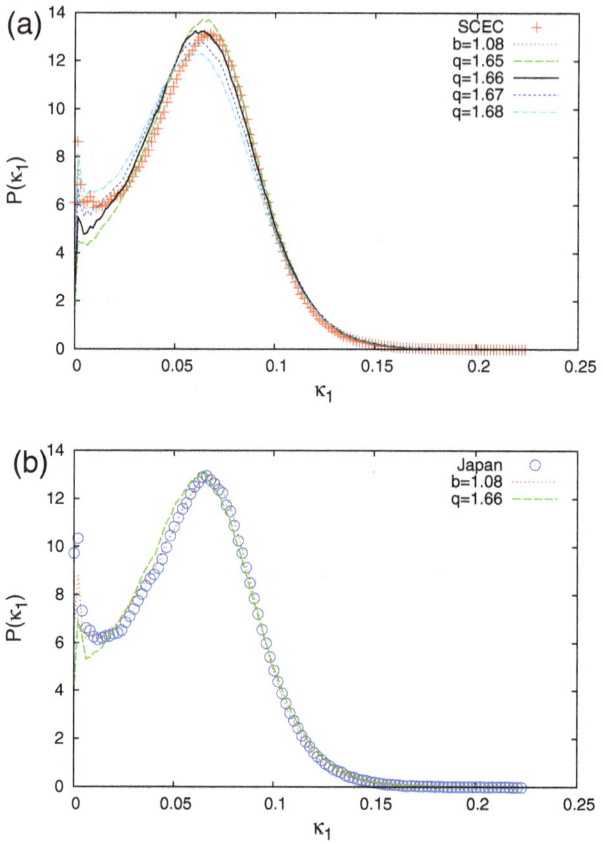

Fig. 6.24 (a) The probability density function $P(\kappa_1)$ versus κ_1 for SCEC (plus) together with G-R distributed data with $b = 1.08$ (red, dotted line) having the same temporal correlations between EQ magnitudes as the original data (see the black broken line in Fig. 6.23). The green (long-dashed), black (solid), blue (short-dashed) and cyan (dashed-dotted) lines depict $P(\kappa_1)$ versus κ_1 resulting from Eq. (6.34) for $q = 1.65$, 1.66, 1.67 and 1.68, respectively, when taking *also* into account the temporal correlations between EQ magnitudes of the original data by CDFT. (b) The same as in (a) for Japan (circles), but here only the case of $q = 1.66$ (green, long-dashed line) is shown. Taken from Ref. [57].

This agreement between synthetic and real data can be considered as satisfactory if we recall that there exists a considerable deviation between them in Fig. 6.22 where the results have been obtained from Eq. (6.34) by *ignoring long-range temporal correlations between earthquake magnitudes.*

6.5.3 Discussion of the results obtained from the combination of nonextensivity with natural time analysis

Recapitulating the results obtained in § 6.5.2, as well as those discussed in Section 6.3, we can say that:

Natural time analysis of seismic data for both SCEC and Japan reveals that long-range temporal correlations between earthquake magnitudes do exist.

This finding, which agrees with the results obtained by independent analyses of real seismic data in Refs. [46, 45] through a different procedure, also corroborates with a recent theoretical study by Woodard et al. [88] of SOC systems. The latter study shows that the memory of past events (avalanches) is stored in the system profile and that the existence of these correlations contradicts the notion that a SOC time series is simply a *random* superposition of events with sizes distributed as a power law (as has been claimed by several previous studies). This is the notion which was initially interpreted as stating that in SOC systems an event 'can occur randomly anywhere at any time and cannot "know" how large it will become', thus incorrectly concluding that EQ prediction is impossible, which was proven in Ref. [88] to be a misconception (see also Section 8.1).

6.5.4 Conclusions from the combination of nonextensivity with natural time analysis of earthquakes

Summarizing, we can say the following. In this Section, we investigated the nonextensive generalization of the G-R law, i.e., Eq. (6.34), but see also § 6.5.4.1. We considered only values of the nonextensive parameter q that have been found in the recent literature to fit well with the real seismic data. The results obtained when combining this generalized law with natural time analysis as well as with DFA, show the following.

(1) The results of the natural time analysis of the synthetic seismic data obtained from either G-R law or its nonextensive generalization, deviate markedly from those of the real seismic data for both SCEC and Japan. This unambiguously reveals that long-range temporal correlations between magnitudes exist in the real datasets.

(2) DFA applied to the magnitude time series of the real seismic data demonstrate independently the existence of temporal correlations. The DFA exponent is around 0.6 for short scales but $\alpha = 0.8$–0.9 for longer scales (note that the cross-over is noticed around $k \approx 200$ earthquakes).

(3) Inspired from point 2, temporal correlations, with different α values (i.e., $\alpha \approx 0.6$ and 0.8–0.9 for short and long scales, respectively) were inserted to synthetic seismic data coming from either the G-R law or its nonextensive generalization of Eq. (6.34). The natural time analysis of the correlated synthetic seismic data deduced from the G-R law leads to results that agree well with those obtained from the real seismic data of Japan and SCEC, thus confirming the importance of temporal correlations between the magnitudes of successive earthquakes. As for the synthetic seismic data deduced from Eq. (6.34) by inserting long-range temporal correlations, a satisfactory agreement with real data has been obtained for the case of Japan for $q = 1.66$, while for SCEC some differences still remain.

The present results show that the nonextensive parameter q does *not* capture the effect of long-range temporal correlations between the magnitudes of successive earthquakes. Thus, published claims (*not* by the pioneers of the field of nonextensive statistical mechanics) that q is a measure of temporal organization do not hold.

On the other hand, either the generalization of the G-R law or the G-R law itself, when combined with natural time analysis (which focuses on the *sequential order* of the energies of the events that appear in nature) does enable a satisfactory description of the real seismic data fluctuations.

6.5.4.1 More recent developments

Very recently, the applicability of the nonextensivity framework has been discussed by Abe [1] (see also Refs. [9, 2]) who argued that discreteness of basic physical variables is, at least, essential for generalized statistical mechanics with non-logarithmic entropy, as the one in Eq. (6.29), to be thermodynamically applicable to classical systems.

Furthermore, it has been pointed out (Abe, personal communication) that the q-average formalism (i.e., the normalized q-expectation value; see Eq. (6.31)) may result in conceptual difficulties and hence should be avoided. In other words, the ordinary average must be used and not the escort average as given by Eq. (6.31). In this case, Eq. (6.34) along with the resulting interconnection between b and q will change.

Acknowledgments Acknowledgments are due to Professor Sumiyoshi Abe for bringing to our attention his recent work on nonextensive Statistical Mechanics.

References

1. Abe, S.: Essential discreteness in generalized thermostatistics with non-logarithmic entropy. EPL **90**, 50004 (2010)
2. Abe, S.: Reply to the Comment by B. Andresen. EPL **92**, 40006 (2010)
3. Abe, S., Bagci, G.B.: Necessity of q -expectation value in nonextensive statistical mechanics. Phys. Rev. E **71**, 016139 (2005)
4. Abe, S., Okamoto, Y. (eds.): Non Extensive Statistical Mechanics and its Applications. Springer, Berlin (2001)
5. Abe, S., Suzuki, N.: Itineration of the internet over nonequilibrium stationary states in Tsallis statistics. Phys. Rev. E **67**, 016106 (2003)
6. Abe, S., Suzuki, N.: Law for the distance between successive earthquakes. J. Geophys. Res. **108(B2)**, 2113 (2003)
7. Abe, S., Suzuki, N.: Aging and scaling of earthquake aftershocks. Physica A **332**, 533–538 (2004)
8. Abe, S., Suzuki, N.: Scale-free statistics of time interval between successive earthquakes. Physica A **350**, 588–596 (2005)
9. Andresen, B.: Comment on "Essential discreteness in generalized thermostatistics with non-logarithmic entropy" by Abe Sumiyoshi EPL **92**, 40005 (2010)
10. Ausloos, M., Lambiotte, R.: Brownian particle having a fluctuating mass. Phys. Rev. E **73**, 011105 (2006)
11. Bak, P., Christensen, K., Danon, L., Scanlon, T.: Unified scaling law for earthquakes. Phys. Rev. Lett. **88**, 178501 (2002)
12. Balankin, A.S.: Dynamic scaling approach to study time series fluctuations. Phys. Rev. E **76**, 056120 (2007)
13. Balankin, A.S., Morales Matamoros, D., Patiño Ortiz, J., Patiño Ortiz, M., Pineda León, E., Samayoa Ocha, D.: Scaling dynamics of seismic activity fluctuations. EPL **85**, 39001 (2009)

14. Barnett, R.M., *et al.*: Review of particle physics. Phys. Rev. D **54**, 1–708 (1996)
15. Båth, M.: Lateral inhomogeneities of the upper mantle. Tectonophysics **2**, 483–514 (1965)
16. Beck, C., Schlögl, F.: Thermodynamics of Chaotic Systems: An Introduction. Cambridge University Press, Cambridge, U.K. (1993)
17. Bramwell, S.T., Christensen, K., Fortin, J.Y., Holdsworth, P.C.W., Jensen, H.J., Lise, S., López, J.M., Nicodemi, M., Pinton, J.F., Sellitto, M.: Universal fluctuations in correlated systems. Phys. Rev. Lett. **84**, 3744–3747 (2000)
18. Bramwell, S.T., Christensen, K., Fortin, J.Y., Holdsworth, P.C.W., Jensen, H.J., Lise, S., López, J.M., Nicodemi, M., Pinton, J.F., Sellitto, M.: Bramwell et al. reply. Phys. Rev. Lett. **87**, 188902 (2001)
19. Bramwell, S.T., Christensen, K., Fortin, J.Y., Holdsworth, P.C.W., Jensen, H.J., Lise, S., López, J.M., Nicodemi, M., Pinton, J.F., Sellitto, M.: Bramwell et al. reply. Phys. Rev. Lett. **89**, 208902 (2002)
20. Bramwell, S.T., Fortin, J.Y., Holdsworth, P.C.W., Peysson, S., Pinton, J.F., Portelli, B., Sellitto, M.: Magnetic fluctuations in the classical XY model: The origin of an exponential tail in a complex system. Phys. Rev. E **63**, 041106 (2001)
21. Bramwell, S.T., Holdsworth, P.C.W., Pinton, J.F.: Universality of rare fluctuations in turbulence and critical phenomena. Nature **396**, 552–554 (1998)
22. Buchel, A., Sethna, J.P.: Statistical mechanics of cracks: Fluctuations, breakdown, and asymptotics of elastic theory. Phys. Rev. E **55**, 7669–7690 (1997)
23. Carlson, J.M., Langer, J.S., Shaw, B.E.: Dynamics of earthquake faults. Rev. Mod. Phys. **66**, 657–670 (1994)
24. Chen, H., Sun, X., Chen, H., Wu1, Z., Wang, B.: Some problems in multifractal spectrum computation using a statistical method. New J. Phys. **6**, 84 (2004)
25. Clusel, M., Fortin, J.Y., Holdsworth, P.C.W.: Criterion for universality-class-independent critical fluctuations: Example of the two-dimensional Ising model. Phys. Rev. E **70**, 046112 (2004)
26. Darooneh, A.H., Mehri, A.: A nonextensive modification of the Gutenberg-Richter law: q-stretched exponential form. Physica A **389**, 509–526 (2010)
27. Davy, P., Sornette, A., Sornette, D.: Some consequences of a proposed fractal nature of continental faulting. Nature **348**, 56–59 (1990)
28. Eichner, J.F., Kantelhardt, J.W., Bunde, A., Havlin, S.: Statistics of return intervals in long-term correlated records. Phys. Rev. E **75**, 011128 (2007)
29. Fawcett, T.: An introduction to ROC analysis. Pattern Recogn. Lett. **27**, 861–874 (2006)
30. Garbaczewski, R.: Differential entropy and time. Entropy **7**, 253–299 (2005)
31. Garber, A., Hallerberg, S., Kantz, H.: Predicting extreme avalanches in self-organized critical sandpiles. Phys. Rev. E **80**, 026124 (2009)
32. Gluzman, S., Sornette, D.: Self-consistent theory of rupture by progressive diffuse damage. Phys. Rev. E **63**, 066129 (2001)
33. Gutenberg, B., Richter, C.F.: Seismicity of the Earth and Associated Phenomena. Princeton Univ. Press, Princeton, New York (1954)
34. Hanks, T.C., Kanamori, H.: Moment magnitude scale. J. Geophys. Res. **84(B5)**, 2348–2350 (1979)
35. Holliday, J.R., Rundle, J.B., Turcotte, D.L., Klein, W., Tiampo, K.F., Donnellan, A.: Space-time clustering and correlations of major earthquakes. Phys. Rev. Lett. **97**, 238501 (2006)
36. Jaynes, E.T.: Information theory and statistical mechanics. Phys. Rev. **106**, 620–630 (1957)
37. Jaynes, E.T.: Probability Theory: The Logic of Science. Cambridge University Press, New York (2003)
38. Kagan, Y.Y.: Short-term properties of earthquake catalogs and models of earthquake source. Bull. Seismol. Soc. Am. **94**, 1207–1228 (2004)
39. Kanamori, H.: Quantification of earthquakes. Nature **271**, 411–414 (1978)
40. Keilis-Borok, V.I., Kossobokov, V.G.: Premonitory activation of earthquake flow: algorithm M8. Phys. Earth Planet. Inter. **61**, 73–83 (1990)
41. Keilis-Borok, V.I., Rotwain, I.M.: Diagnosis of time of increased probability of strong earthquakes in different regions of the world: algorithm CN. Phys. Earth Planet. Inter. **61**, 57–72 (1990)
42. Kun, F., Herrmann, H.J.: Transition from damage to fragmentation in collision of solids. Phys. Rev. E **59**, 2623–2632 (1999)
43. Lennartz, S., Bunde, A., Turcotte, D.L.: Missing data in aftershock sequences: Explaining the deviations from scaling laws. Phys. Rev. E **78**, 041115 (2008)

44. Lennartz, S., Livina, V.N., Bunde, A., Havlin, S.: Long-term memory in earthquakes and the distribution of interoccurrence times. EPL **81**, 69001 (2008)
45. Lippiello, E., de Arcangelis, L., Godano, C.: Influence of time and space correlations on earthquake magnitude. Phys. Rev. Lett. **100**, 038501 (2008)
46. Lippiello, E., Godano, C., de Arcangelis, L.: Dynamical scaling in branching models for seismicity. Phys. Rev. Lett. **98**, 098501 (2007)
47. Meneveau, C., Sreenivasan, K.R.: Simple multifractal cascade model for fully developed turbulence. Phys. Rev. Lett. **59**, 1424–1427 (1987)
48. Miguel, M.C., Zapperi, S.: Fluctuations in plasticity at the microscale. Science **312**, 1151–1152 (2006)
49. O'Neil, J., Meneveau, C.: Spatial correlations in turbulence: Predictions from the multifractal formalism and comparison with experiments. Phys. Fluids A **5**, 158–172 (1993)
50. Papanastassiou, D., Latoussakis, J., Stavrakakis, G.: A revised catalogue of earthquakes in the broader area of Greece for the period 1950-2000. Bulletin of the Geological Society of Greece **34**, 1563–1566 (2001)
51. Pastor-Satorras, R.: Multifractal properties of power-law time sequences: Application to rice piles. Phys. Rev. E **56**, 5284–5294 (1997)
52. Roumelioti, Z., Kiratzi, A., Theodoulidis, N., Papaioannou, C.: S-wave spectral analysis of the 1995 Kozani-Grevena (NW Greece) aftershock sequence. Journal of Seismology **6**, 219–236 (2002)
53. Rundle, J.B., Turcotte, D.L., Shcherbakov, R., Klein, W., Sammis, C.: Statistical physics approach to understanding the multiscale dynamics of earthquake fault systems. Rev. Geophys. **41**, 1019 (2003)
54. Saichev, A., Sornette, D.: Power law distributions of seismic rates. Tectonophysics **431**, 7–13 (2007)
55. Sarlis, N.V., Skordas, E.S., Varotsos, P.A.: See (the freely available) EPAPS Document No. E-PLEEE8-80-014908 originally from N.V. Sarlis, E.S. Skordas and P.A. Varotsos, *Phys. Rev. E* **80**, 022102 (2009). For more information on EPAPS, see http://www.aip.org/pubservs/epaps.html.
56. Sarlis, N.V., Skordas, E.S., Varotsos, P.A.: Multiplicative cascades and seismicity in natural time. Phys. Rev. E **80**, 022102 (2009)
57. Sarlis, N.V., Skordas, E.S., Varotsos, P.A.: Nonextensivity and natural time: The case of seismicity. Phys. Rev. E **82**, 021110 (2010)
58. Sarlis, N.V., Skordas, E.S., Varotsos, P.A.: Order parameter fluctuations of seismicity in natural time before and after mainshocks. EPL **91**, 59001 (2010)
59. Schultka, N., Manousakis, E.: Finite-size scaling in two-dimensional superfluids. Phys. Rev. B **49**, 12,071–12,077 (1994)
60. Sethna, J.P.: Order parameters, broken symmetry, and topology. In: L. Nagel, D. Stein (eds.) 1991 Lectures in Complex Systems, Santa Fe Institute Studies in the Sciences of Complexity, Proc. Vol. XV. Addison-Wesley, New York (1992)
61. Shcherbakov, R., Turcotte, D.L., Rundle, J.B.: A generalized Omori's law for earthquake aftershock decay. Geophys. Res. Lett. **31**, L11613 (2004)
62. Shore, J.E., Johnson, R.W.: Axiomatic derivation of the principle of maximum entropy and the principle of minimum cross-entropy. IEEE Trans. Inf. Theory **IT-26**, 26–37 (1980)
63. Shore, J.E., Johnson, R.W.: Properties of cross-entropy minimization. IEEE Trans. Inf. Theory **IT-27**, 472–482 (1981)
64. Shore, J.E., Johnson, R.W.: Comments on and correction to 'axiomatic derivation of the principle of maximum entropy and the principle of minimum cross-entropy' (Jan 80 26-37) (corresp.). IEEE Trans. Inf. Theory **IT-29**, 942–943 (1983)
65. Silva, R., França, G.S., Vilar, C.S., Alcaniz, J.S.: Nonextensive models for earthquakes. Phys. Rev. E **73**, 026102 (2006)
66. Sornette, D.: Critical Phenomena in Natural Science, 2nd edn. Springer, Berlin (2004)
67. Sornette, D., Davy, P.: Fault growth model and the universal fault length distribution. Geophys. Res. Lett. **18**, 1079–1082 (1991)
68. Sotolongo-Costa, O., Posadas, A.: Fragment-asperity interaction model for earthquakes. Phys. Rev. Lett. **92**, 048501 (2004)
69. See the document `SCSN/README.old` included in `SCSN_catalogs.tar.gz` available at `http://www.data.scec.org/ftp/catalogs/SCSN/`

70. Telesca, L.: Nonextensive analysis of seismic sequences. Physica A **389**, 1911–1914 (2010)
71. Tsallis, C.: Possible generalization of Boltzmann-Gibbs statistics. J. Stat. Phys. **52**, 479–487 (1988)
72. Tsallis, C.: Introduction to Nonextensive Statistical Mechanics. Springer, Berlin (2009)
73. Tsallis, C., Mendes, R.S., Plastino, A.R.: The role of constraints within generalized nonextensive statistics. Physica A **261**, 534–554 (1998)
74. Turcotte, D.L.: Fractals and Chaos in Geology and Geophysics, 2nd edn. Cambridge University Press, Cambridge (1997)
75. Turcotte, D.L., Malamud, B.D., Guzzetti, F., Reichenbach, P.: Self-organization, the cascade model, and natural hazards. Proc. Natl. Acad. Sci. USA **99**, 2530–2537 (2002)
76. Utsu, T.: A statistical study of the occurrence of aftershocks. Geophys. Mag. **30**, 521 (1961)
77. Utsu, T.: Seismology. Kyoritsu (in Japanese), Tokyo (2001)
78. Varotsos, P.A., Sarlis, N.V., Skordas, E.S.: Spatio-temporal complexity aspects on the interrelation between seismic electric signals and seismicity. Practica of Athens Academy **76**, 294–321 (2001)
79. Varotsos, P.A., Sarlis, N.V., Skordas, E.S.: Seismic Electric Signals and seismicity: On a tentative interrelation between their spectral content. Acta Geophys. Pol. **50**, 337–354 (2002)
80. Varotsos, P.A., Sarlis, N.V., Skordas, E.S., Lazaridou, M.S.: Entropy in natural time domain. Phys. Rev. E **70**, 011106 (2004)
81. Varotsos, P.A., Sarlis, N.V., Skordas, E.S., Tanaka, H.K., Lazaridou, M.S.: Attempt to distinguish long-range temporal correlations from the statistics of the increments by natural time analysis. Phys. Rev. E **74**, 021123 (2006)
82. Varotsos, P.A., Sarlis, N.V., Skordas, E.S., Tanaka, H.K., Lazaridou, M.S.: Entropy of seismic electric signals: Analysis in the natural time under time reversal. Phys. Rev. E **73**, 031114 (2006)
83. Varotsos, P.A., Sarlis, N.V., Tanaka, H.K., Skordas, E.S.: See (the freely available) EPAPS Document No. E-PLEEE8-72-058510 originally from P.A. Varotsos, N.V. Sarlis, H.K. Tanaka and E.S. Skordas, *Phys. Rev. E* **72**, 041103 (2005). For more information on EPAPS, see http://www.aip.org/pubservs/epaps.html.
84. Varotsos, P.A., Sarlis, N.V., Tanaka, H.K., Skordas, E.S.: Similarity of fluctuations in correlated systems: The case of seismicity. Phys. Rev. E **72**, 041103 (2005)
85. Varotsos, P.A., Sarlis, N.V., Skordas, E.S., Tanaka, H.K.: A plausible explanation of the b-value in the Gutenberg-Richter law from first principles. Proc. Japan Acad., Ser. B **80**, 429–434 (2004)
86. Vilar, C.S., França, G.S., Silva, R., Alcaniz, J.S.: Nonextensivity in geological faults? Physica A **377**, 285–290 (2007)
87. Watkins, N.W., Chapman, S.C., Rowlands, G.: Comment on "universal fluctuations in correlated systems". Phys. Rev. Lett. **89**, 208901 (2002)
88. Woodard, R., Newman, D.E., Sánchez, R., Carreras, B.A.: Persistent dynamic correlations in self-organized critical systems away from their critical point. Physica A **373**, 215–230 (2007)
89. Zapperi, S., Ray, P., Stanley, H.E., Vespignani, A.: First-order transition in the breakdown of disordered media. Phys. Rev. Lett. **78**, 1408–1411 (1997)
90. Zheng, B.: Generic features of fluctuations in critical systems. Phys. Rev. E **67**, 026114 (2003)
91. Zheng, B., Trimper, S.: Comment on "universal fluctuations in correlated systems". Phys. Rev. Lett. **87**, 188901 (2001)

7. Identifying the Occurrence Time of an Impending Mainshock

Abstract. Natural time enables the determination of the occurrence time of an impending major earthquake since it can identify when a complex system approaches a critical point. Considering that the detection of a SES activity signifies that the system enters the critical regime, the small earthquakes that occur (in the region candidate to suffer the mainshock) after the SES detection are analyzed in natural time. It was found that the variance κ_1 of natural time becomes equal to 0.070 (which manifests the approach to the critical point) usually a few days to around one week *before* the mainshock. This, which exhibits spatial as well as magnitude threshold invariance, has been observed to date for *all* major earthquakes that occurred in Greece since the introduction of the natural time concept in 2001 (note that it has been also ascertained in retrospect for the two major earthquakes in Greece during the previous decade, i.e., in the 1990s). For example, the occurrence time of the M_w6.9 earthquake on February 14, 2008, which is the strongest earthquake in Greece during the last 28 years, was announced as imminent on February 10, 2008. The procedure has been also ascertained in the case of the volcanic-seismic swarm activity in 2000 in the Izu island region in Japan as well as of the M_s7.1 Loma Prieta earthquake in California in 1989.

7.1 Determination of the time-window of the impending mainshock by analyzing in natural time the seismicity after the initiation of the SES activity

We first recall (see Eq. (2.75) or Eq. (6.7)) that the relation

$$\Pi(\omega) = \frac{18}{5\omega^2} - \frac{6\cos\omega}{5\omega^2} - \frac{12\sin\omega}{5\omega^3}. \tag{7.1}$$

for $\omega \to 0$, simplifies to

$$\Pi(\omega) \approx 1 - 0.070\omega^2 \tag{7.2}$$

which shows that the second-order Taylor expansion coefficient of $\Pi(\omega)$, labeled κ_1, is equal to 0.070. The quantity κ_1 equals (see Eq. (2.37)) to the variance $\langle\chi^2\rangle - \langle\chi\rangle^2$ of natural time χ, i.e.,

$$\kappa_1 = \langle\chi^2\rangle - \langle\chi\rangle^2 = 0.070. \tag{7.3}$$

This has been shown for SES activities (§ 2.4.2) as well as for the time series of avalanches in a number of dynamical models (see Table 8.1), including the "train" Burridge–Knopoff earthquake model (§ 8.2.2) and the Olami–Feder–Christensen earthquake model (§ 8.3.2), when the system approaches the critical point. Furthermore, since it has been observed for several EQs that, when analyzing the seismicity that occurs after the SES activity, the resulting κ_1 value slowly approaches to 0.070 just before the mainshock and abruptly changes to vanishingly small when the main shock occurs, it was proposed (see § 6.2.1) that κ_1 (or $\Pi(\omega)$ for $\omega \to 0$) may be considered as an *order parameter for seismicity* [54].

In addition, we recall that the entropy S in natural time as well as the entropy S_- under time reversal, have been found(see Eq. (4.32)) to obey the following conditions [55, 51, 50] for SES activities

$$S, S_- < S_u. \tag{7.4}$$

These also hold for long-range correlated fBm time series with $\alpha_{DFA} \approx 1$ (see § 3.4.3) as well as for an on–off intermittency model when the critical value is approached from *below* (see § 3.4.4) . Note that it has been suggested that [23] "The Californian earthquakes are long-range correlated according to the persistence of a fractal Gaussian intermittent noise with $H = 1$ known as $1/f$ or pink noise" as well as that [7]: the intermittent criticality model as being more appropriate for earthquakes.

In view of the above and based on our fundamental premise that mainshock occurrence is a critical phenomenon, the conditions (7.1) to (7.3) and (7.4) have been used to study the evolution of seismicity in natural time before a mainshock occurrence. To obtain the order parameter κ_1 or $\Pi(\omega)$ for $\omega \to 0$ (as well as the quantities S and S_-), however, it is necessary to decide the initiation time of seismicity analysis. We decided to start the analysis immediately after the SES initiation since it signals, as mentioned in § 6.2.1, that the system enters the critical stage (recall that the SES emission marks *cooperative* orientation of the electric dipoles and hence the establishment of long-range correlations; see § 1.6.2 and § 2.4.2).

Once a SES activity has been recorded, the area to suffer the mainshock can be estimated, as explained in § 1.3.5, on the basis of the so-called selectivity map of the station at which the SES was recorded and in addition by considering the ratio of the two SES components. Thus, we have in principle some area (see also the discussion in § 7.2.3), labeled A, in which we count the small EQs, e_i , that occur after the initiation of the SES activity. In order to check the spatial invariance of the results, the study was also repeated for a smaller area. This procedure, which for the sake of convenience will be hereafter, called *preliminary* procedure, was used during the period 2001–2008 in a series of publications (e.g., see Refs. [45, 54, 51, 50, 35, 48]) to determine the occurrence time of the impending mainshock by means of the natural time analysis of the seismicity subsequent to a SES activity. Since there has been, however, some room for subjective judgment to identify the approach to critical stage, because the time variation of parameters was traced only on a single subarea, a more objective procedure, which for reasons of brevity will be hereafter called "*updated*" procedure, has been developed [21], in 2008, and considers the natural time analysis of the seismicity in *all* the possible subareas, instead of a single smaller area, of the larger area under discussion.

7.1.1 The *preliminary* procedure to determine the occurrence time of the impending mainshock

The actual procedure was carried out as follows. We set the natural time zero at the initiation time of the SES activity, and then formed time series of seismic events in natural time for the area A, each time when a small EQ (above a magnitude threshold $M \geq M_{thres}$) occurred; in other words, when the number of the events increased by one. The normalized power spectrum in natural time $\Pi(\omega)$ for $\omega \to 0$ (or the variance κ_1) for each of the time series was computed for the pairs (χ_k, Q_k) and compared with that of Eq. (7.1) for $\omega \in [0, \pi]$. We also calculated the evolution of the quantities S and S_- to ascertain Eq. (7.4) was also satisfied. The actual criteria for recognizing a *true* coincidence of the observed time series with that of critical state were as follows [45, 35, 51, 50, 48]:

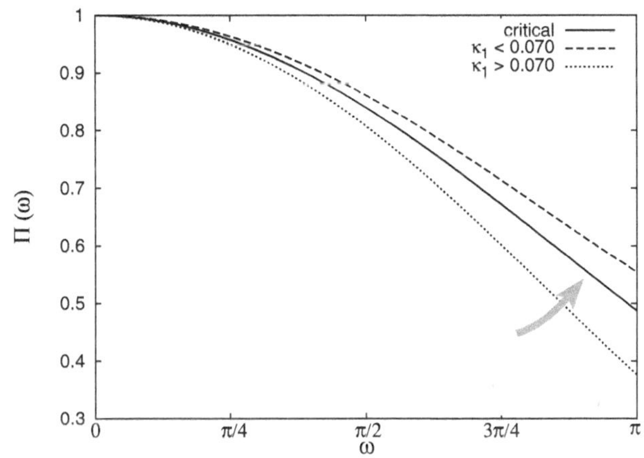

Fig. 7.1 Schematic diagram showing the normalized power spectrum $\Pi(\omega)$ in natural time for $\omega \in [0, \pi]$. Solid line is $\Pi(\omega)$ obtained from Eq. (7.1) which holds for critical stage ($\kappa_1 = 0.070$), whereas two other lines are for $\kappa_1 > 0.070$ and $\kappa_1 < 0.070$. The grey arrow indicates how the $\Pi(\omega)$ curve approaches the critical from below.

First, the 'average' $\langle D \rangle$ distance between the curves of $\Pi(\omega)$ of the evolving seismicity and Eq. (7.1) for $\omega \in [0, \pi]$ should be smaller than 10^{-2} (note that this was regarded as showing that $\langle D \rangle = 0$). This was a practical criterion for stopping calculation.

Second, the final approach of the evolving $\Pi(\omega)$ to that of Eq. (7.1) must be by approaching from *below* as shown by the grey arrow in Fig. 7.1. This alternatively means that before major EQs, the κ_1 value gradually changes with time and *finally* approaches from *above* that of the critical state ($\kappa_1 = 0.070$, see Eq. (7.3)). This rule was found empirically [45].

Third, both values S and S_- should be smaller than $S_u(= 0.0966)$ at the coincidence (see Eq. (7.4)).

Finally and fourth, since the process concerned is supposed to be self-similar (critical dynamics), the time of the occurrence of the *true* coincidence should not vary, in principle, upon changing (within reasonable limits) the magnitude threshold M_{thres} and the size of area A.

We clarify, however, that if higher magnitude threshold is used, the description of the real situation approaching criticality is expected to become less accurate due to '*coarse graining*' [43, 49] since the number of events is finite.

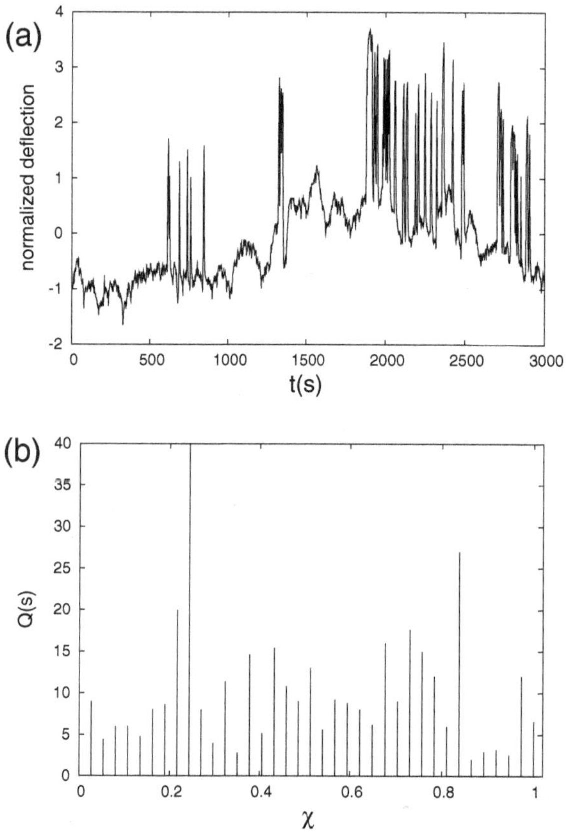

Fig. 7.2 (a) A SES activity recorded on February 13, 2006 at PAT station(sampling rate $f_{exp} = 1$ Hz). The actual electric field E of the SES pulses is 6 mV/km (see Ref. [49]), but here the signal is presented in normalized units, i.e., by subtracting the mean value and dividing by the standard deviation. (b) How the SES activity in (a) is read in natural time. Taken from Ref. [50].

It has been observed [45, 35, 51, 50, 48] that the aforementioned *true* coincidence appears usually a few days (up to around one week) *before* the occurrence of the mainshock. As an example, we report a SES activity recorded at a station located in central Greece (close to Patras city, PAT; see Fig. 1.2) on February 13, 2006. It is depicted in Fig. 7.2(a) and comprises 37 pulses, the durations Q_k of which vary between 1 s and 40 s (see Fig. 7.2(b)). Beyond the application of the four criteria of Section 1.2, a natural time analysis of this SES activity (labeled PAT in Table 4.6) was made which led [50] to the following values: $\kappa_1 = 0.072 \pm 0.002$, $S = 0.080 \pm 0.002$, $S_- = 0.078 \pm 0.002$ which obey the conditions (4.38) and (4.39), i.e., $\kappa_1 \approx 0.070$ and $S, S_- < S_u$, that have to be obeyed for SES activities. In addition, the Detrended Fluctuation Analysis (DFA) (§ 1.4.2) in natural time of this SES activity, resulted in an exponent $\alpha = 1.07 \pm 0.36$, which agrees with the finding $\alpha \approx 1$ in several other SES activities (see § 4.4.2 and Eq. (4.42)). If we repeat the computation by *shuffling* the durations Q_k randomly (and hence their distribution is conserved), the corresponding quantities, designated by adding a subscript "shuf", have the following values: $\kappa_{1,shuf} \approx \kappa_u$ and $S_{shuf} \approx S_{-,shuf} \approx S_u$. This points to the conclusion that the self-similarity of SES activities results from the process's memory only (see § 4.7.1 and § 2.5.5). All these results showing that the signal recorded on February 13, 2006, is a true SES activity were submitted [50] for publication on February 25, 2006 (see Table 7.1). Actually, on April 3, 2006, a strong seismic activity started with an earthquake of magnitude $M_s(\text{ATH}) = 5.3$ and lasted until April 19, 2006 with earthquakes of magnitude up to 5.9 in a region 80 to 100 km west of PAT station, i.e., around 37.6°N 20.9°E (see also table I of Ref. [49]). We will now explain how the occurrence time of the initiation of this earthquake activity has been specified [49] by following the *preliminary* procedure:

First, after the recording of this SES activity, the area to suffer the impending mainshock was estimated as follows: We considered that the epicenters of the EQs that have been preceded, up to that time, by SES activities at PAT station lie approximately within the area $\text{N}_{37.5}^{38.6}\,\text{E}_{19.8}^{23.3}$, i.e., this was the selectivity map (§ 1.3.4) of PAT station. Then, by using the additional information of the ratio of the two SES components (§ 1.3.5), we selected from the selectivity map the region A: $\text{N}_{37.6}^{38.6}\,\text{E}_{20.9}^{22.6}$ as candidate that might have emitted the SES activity under discussion.

Second, we now study in natural time the seismicity that evolved after the recording of the relevant SES activity at PAT, thus we put natural time zero for seismicity at the initiation time of this SES activity, i.e., at 19:04 UT on February 13, 2006. The study is made in the areas A: $\text{N}_{37.55}^{38.64}\,\text{E}_{20.85}^{22.64}$ as well as in its smaller area B: $\text{N}_{37.55}^{38.34}\,\text{E}_{20.85}^{22.15}$. We now form time series of seismic events in natural time for various time windows as the number N of consecutive (small) EQs increases. We then compute the normalized power spectrum of seismicity in natural time $\Pi(\phi)$ (for $\phi \to 0$, e.g. $\phi \in [0, 0.5]$) for each of the time windows. We clarify that the seismic moment M_0 was estimated from the relation [5] $\log_{10}(M_0) = 1.5M_w + \text{const.}$ by using $M_w = 1.09M_L - 0.21$, i.e., the least-squares fit proposed in Ref. [19], and the values of the local magnitude M_L were taken from the GI-NOA catalog. In short, the relation $\log_{10}(M_0) = 1.64M_L + \text{const.}$ has been used. Excerpts of the results of these computations which refer to the values deduced during the period March 27 to April 1, 2006, are depicted in red in Fig. 7.3. In this figure, Fig. 7.3(a) corresponds to the area A with magnitude threshold $M_{thres} = 3.0$ (defined by means of the local magnitude M_L and of the 'duration' magnitude M_D), while Fig. 7.3(b) to the area B with $M_{thres} = 2.8$. In the same figure, we plot in blue the normalized power spectrum obeying Eq. (7.1). The date and the time of the occurrence of each small earthquake (with magnitude exceeding (or equal to) the aforementioned threshold) that occurred in each of the areas A and B, is also written in red in each panel.

Table 7.1 All EQs with $M_s(\mathrm{ATH}) \geq 6.0$ within $\mathrm{N}_{36}^{41}\,\mathrm{E}_{19}^{27}$ since 2001 along with the relevant SES activities. The cases in parentheses refer to EQs for which the expected magnitude (on the basis of the SES amplitude) was $M_s(\mathrm{ATH}) \approx 6.0$, but the actual magnitude turned out to be somewhat smaller. The last column gives, in each case, the relevant documentation publicized *before* the mainshock occurrence, when available. The EQs grouped together refer to almost the same epicentral location.

EQ			Related SES activities		
Date D/M/Y	Epicenter °N–°E	Magnitude M_s(ATH)– M_w(USGS)	Station	Date D/M/Y	Publication
26/7/2001	39.05–24.35	5.8–6.5	VOL	17/3/2001	Ref. [41] submitted on 25 March 2001
14/8/2003	38.79–20.56	6.4–6.2	PIR	8/8/2003	Ref. [35]
31/1/2005	37.41–20.11	6.2–5.7	PIR	17/10/2004	Ref. [52]
17/10/2005	38.13–26.59	6.0–5.5	MYT	21/3 and 23/3/2005	Ref. [51] submitted on 16 April 2005 and Ref. [43]
(17/10/2005	38.14–26.59	5.9–5.8	MYT	21/3 and 23/3/2005	Ref. [51] submitted on 16 April 2005 and Ref. [43])
20/10/2005	38.15–26.63	6.1–5.9	MYT	21/3 and 23/3/2005	Ref. [51] submitted on 16 April 2005 and Ref. [43]
18/10/2005	37.58–20.86	6.1–5.7	PIR	17/9/2005	Ref. [42] submitted on 22 October 2005 and Ref. [43]
8/1/2006	36.21–23.41	6.9–6.7	PIR	17/9/2005	Ref. [42] submitted on 22 October 2005 and Ref. [43]
(3/4/2006	37.59–20.95	5.3–5.0	PAT	13/2/2006	arXiV:0602603v1 [25 February 2006] and Ref. [50])
(4/4/2006	37.58–20.93	5.7–5.3	PAT	13/2/2006	arXiV:0602603v1 [25 February 2006] and Ref. [50])
(11/4/2006	37.64–20.92	5.7–5.4	PAT	13/2/2006	arXiV:0602603v1 [25 February 2006] and Ref. [50])
(11/4/2006	37.68–20.91	5.9–5.5	PAT	13/2/2006	arXiV:0602603v1 [25 February 2006] and Ref. [50])
(12/4/2006	37.61–20.95	5.9–5.6	PAT	13/2/2006	arXiV:0602603v1 [25 February 2006] and Ref. [50])
25/3/2007	38.34–20.42	6.0–5.7	PAT	8/2/2007	arXiv:0703683v1 [26 March 2007]
(29/6/2007	39.25–20.26	5.7–5.2	PAT	23 and 24/4/2007	arXiv:0703683v5 [15 May 2007])
6/1/2008	37.11–22.78	6.6–6.2	PAT	7/11/2007	arXiv:0711.3766v1 [23 November 2007] and Ref. [21]
14/2/2008	36.50–21.78	6.7–6.9	PIR	14/1/2008 and 21-26/1/2008	arXiv:0711.3766v3 [1 February 2008] and Ref. [21]
14/2/2008	36.22–21.75	6.6–6.5	PIR	14/1/2008 and 21-26/1/2008	arXiv:0711.3766v3 [1 February 2008] and Ref. [21]
20/2/2008	36.18–21.72	6.5–6.2	PIR	14/1/2008 and 21-26/1/2008	arXiv:0711.3766v3 [1 February 2008] and Ref. [21]
8/6/2008	37.98–21.51	7.0–6.4	PIR	29/2-2/3/2008	arXiv:0802.3329v4 [29 May 2008] and Ref. [21]
21/6/2008	36.03–21.83	6.0–5.6	PIR	5/6/2008	
14/10/2008	38.85–23.62	6.1–5.2	missed		
(13/12/2008	38.72–22.57	5.7–5.2	PAT	9/10/2008	arXiv:0711.3766v5 [7 December 2008])
16/2/2009	37.13–20.78	6.0–5.5	PIR	12/12/2008	arXiv:0707.3074v3 [5 February 2009] and Ref. [47]
3/11/2009	37.39–20.35	6.1–5.8	PIR	24/10/2009	–
(18/1/2010	38.41–21.95	5.7–5.5	PAT	24/10/2009	arXiV:0904.2465v8 and v9[14 and 27 November 2009])
(22/1/2010	38.42–21.97	5.6–5.2	PAT	11/11/2009	arXiV:0904.2465v8 and v9[14 and 27 November 2009])
(9/3/2010	38.87–23.65	5.6–	LAM	27-30/12/2009	arXiV:1003.1383v1 [6 March 2010])

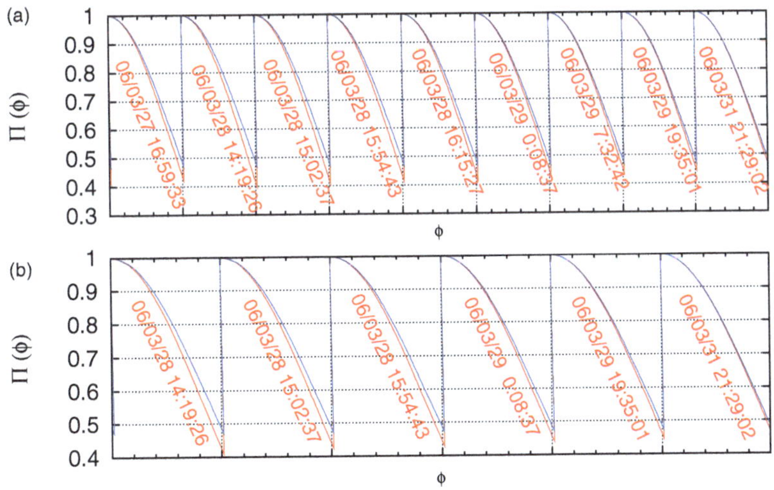

Fig. 7.3 The normalized power spectrum(red) $\Pi(\phi)$ of seismicity as it evolves event by event (whose date and time of occurrence are written in each panel) after the initiation of the SES activity on February 13, 2006. The two excerpts presented here refer to the period March 27 to March 31, 2006, and correspond to: (a) the area A with $M_{thres} = 3.0$ and (b) the area B with $M_{thres} = 2.8$. In each case the spectrum for small ϕ values, e.g. $\phi \in [0, 0.5]$ (for the reasons discussed in Section 2.4) is depicted (separated by the vertical dotted lines), whereas the $\Pi(\phi)$ of Eq. (7.1) is depicted by blue color. The minor horizontal ticks for ϕ are marked every 0.1. Taken from Ref. [49].

An inspection of Fig. 7.3 reveals that the red line approaches the blue line as N increases and a *coincidence* occurs at the last small event which had $M_L = 3.0$ and occurred at 21:29 UT on March 31, 2006, i.e., roughly two days before the first strong EQ (00:50 UT on April 3, 2006). To ensure that this coincidence is a *true* one, we also calculate the evolution of the quantities κ_1, S and S_- and the results are depicted in Fig. 7.4 for both magnitude thresholds 2.8 and 3.0 for each of the areas A and B.

We now examine whether the aforementioned criteria for a coincidence to be considered as *true* are obeyed: First, concerning the 'average' distance $\langle D \rangle$ see Fig. 7.5, where we plot $\langle D \rangle$ versus the conventional time for the aforementioned two areas and the two magnitude thresholds (hence four combinations were studied in total). In order to better visualize the details of this figure, its four consecutive segments are enlarged and separately depicted in Fig. 7.6(a) to (d). Note that in Fig. 7.5 or Fig. 7.6(d), upon the occurrence of the aforementioned last small event of March 31, 2006, in both areas A and B and both magnitude thresholds (i.e., $M_{thres} = 2.8$ and 3.0) their $\langle D \rangle$ values become smaller than 10^{-2}. Second, a few events *before* the coincidence leading to the strong EQ, the evolving $\Pi(\phi)$ has been found to approach that of Eq. (7.1), i.e., the blue one in Fig. 7.3, from *below* (note that this reflects that during this approach the κ_1 value decreases as the number of events increases see Fig. 7.4(a)). In addition, both values S and S_- are smaller than S_u at the coincidence; see Fig. 7.4(b) and 7.4(c), respectively. Finally, since the process concerned is self-similar

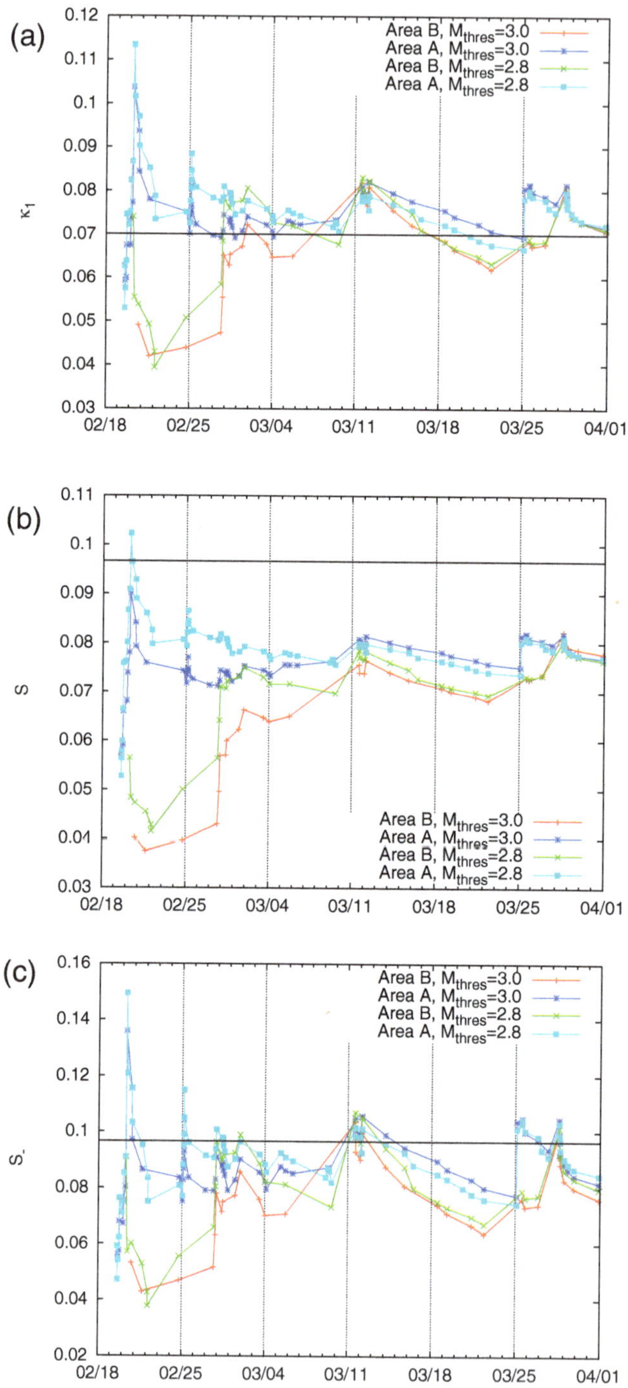

Fig. 7.4 Evolution of the quantities κ_1, S and S_- for seismicity after the initiation of the SES activity on February 13, 2006, depicted in Fig. 7.2(a). They are shown in (a), (b) and (c), respectively for two magnitude thresholds, i.e., $M \geq 2.8$ and $M \geq 3.0$, for both areas A and B. After the event at 14:19 UT of March 28, 2006 the four curves (corresponding to the four combinations, i.e., resulting from the two areas and the two magnitude thresholds) *almost collapse on the same curve*. This points to the *scale invariance* when approaching the critical point (see the text). Taken from Ref. [49].

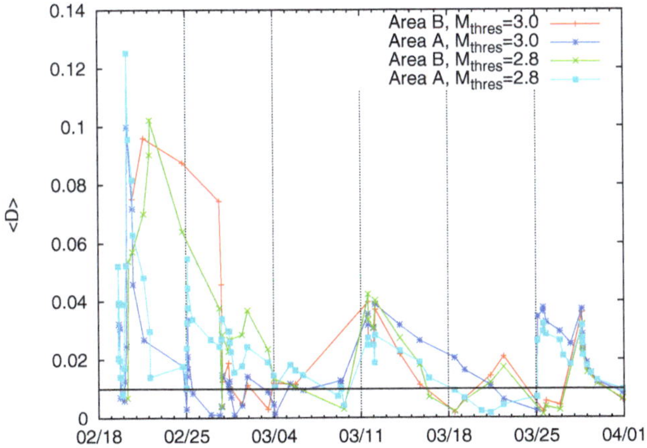

Fig. 7.5 The average distance $\langle D\rangle$ versus the conventional time. The calculation of $\langle D\rangle$ is made upon the occurrence of every consecutive earthquake when starting the calculation after the SES activity of February 13, 2006 (depicted in Fig. 7.2(a)), for each of the two areas A and B by considering two magnitude thresholds 2.8 and 3.0. Taken from Ref. [49].

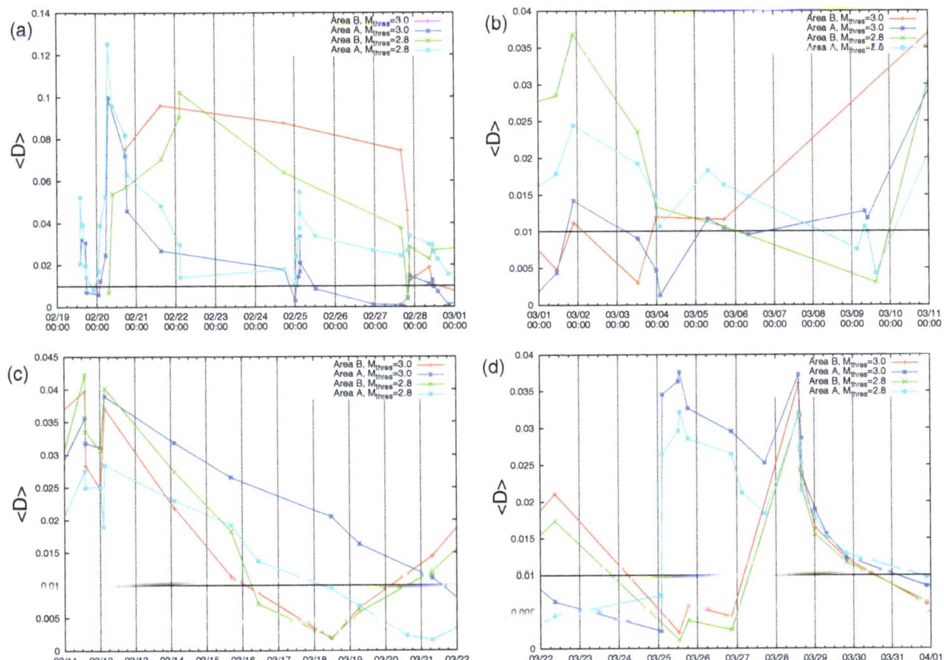

Fig. 7.6 Four consecutive segments of Fig. 7.5. Note that in (d), after 14:19 UT of March 28, 2006 the four curves (corresponding to the four combinations resulting from the two areas and the two magnitude thresholds) *almost collapse on the same curve*. This points to the *scale invariance* when approaching the critical point (see the text). Taken from Ref. [49].

(*critical* dynamics), the occurrence time of the (true) coincidence should *not* change, in principle, upon changing either the (surrounding) area or the magnitude threshold used in the calculation. This was actually checked in this example since we considered two areas and two magnitude thresholds. Hence, this coincidence can be considered as *true*, while other coincidences that occurred earlier (i.e., before March 31, 2006) have been found *not to be true ones* since they violate one or more of the aforementioned conditions. Let us briefly summarize:

The occurrence time of the initiation of the strong seismic activity, that lasted from April 3 to April 19, 2006 at an epicentral region 80 to 100 km west of PAT, has been specified within a narrow range around 2 days. This is so, because the normalized power spectrum in natural time of the evolving seismicity after the SES activity of February 13, 2006, collapses on the one expected for critical dynamics at 21:29 UT on March 31, 2006, i.e., almost two days *before* the occurrence time of the 5.3 earthquake of April 3, 2006, obeying the conditions for a *true* coincidence.

Additional examples for the *preliminary* procedure will be presented in § 7.2.1 and § 7.2.4.

7.1.2 The *updated* procedure to determine the occurrence time of the impending mainshock

The basic idea behind the new approach suggested in Ref. [21] is the following. When area A reaches criticality, one expects in general that all its subareas have also reached criticality simultaneously. At that time, therefore, the evolution of seismicity in each of these subareas is expected to result in κ_1 value close to 0.070. Assuming equi-partition of probability among the subareas, the distribution of the κ_1 values of all subareas should be peaked at around 0.070 exhibiting magnitude threshold invariance. Before the criticality is reached, the κ_1 values will *not* show such a behavior.

We recall that, as mentioned above in Section 7.1, once a SES activity has been recorded, we identify in principle an area, labeled A, in which we count the small EQs, e_i, that occur after the initiation of the SES activity. Each EQ e_i is characterized by its epicentral location $\mathbf{x}(e_i)$, the conventional time of its occurrence $t(e_i)$, and its magnitude $\mathrm{M}(e_i)$ or the equivalent seismic moment $M_0(e_i)$. The index $i = 1, 2, \ldots$ increases by one each time a new EQ with M larger or equal to some threshold M_{thres} occurs within the area A. Thus, a set of events denoted as $A_{M_{thres}} = \{e_i\colon \mathbf{x}(e_i) \in \mathrm{A},\ \mathrm{M}(e_i) \geq M_{thres}\}$ is formed each time until the mainshock occurs. Here, the number of EQs in $A_{M_{thres}}$ is denoted by $|A_{M_{thres}}|$. Since, in forming $A_{M_{thres}}$, we place the EQs in sequence of their occurrence time, $A_{M_{thres}}$ is a *time-ordered* set.

In practice, in order to check whether criticality as described above has been approached at the occurrence of a new event i within the predicted area A, we should construct all the possible proper subsets of $A_{M_{thres}}$ that *necessarily* include the event i and examine whether their κ_1 values reveal a probability distribution peaked at 0.070. A subset is qualified as a *proper* subset ($P_{M_{thres}}$) *iff* it includes all EQs that took place inside its corresponding rectangular subarea denoted by $R(P_{M_{thres}})$. This is a simplification because other geometries, e.g., circular, could be also envisaged. It is worthwhile to clarify, however, that even in the frame of this simplification:

> The accuracy in the determination of the epicentral coordinates of the EQs involved in the computation, may somewhat affect – as intuitively expected – the results as it will be further commented on in § 7.2.5.1.

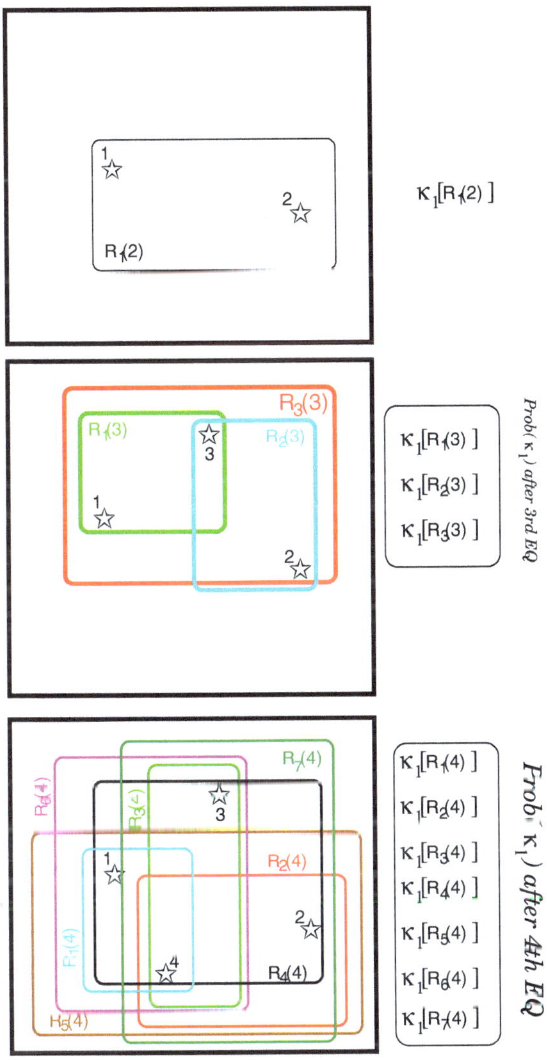

Fig. 7.7 The area A (thick black rectangle) and its rectangular subareas $R_j(i)$, corresponding to the proper subsets immediately after the occurrence of the second EQ "2" (upper panel), the third EQ "3" (middle panel) and the fourth EQ "4" (bottom panel). The location of each EQ is shown by an open star. Right column shows the κ_1 values that can be obtained for each subarea. Taken from Ref. [21].

Let us now consider the schematic example shown in Fig. 7.7, in which four EQs have occurred (area A is indicated by a black line rectangle in each panel) in a sequence indicated by the numbers $i = 1,2,3$ and 4. Colored rectangles depict proper subareas $R(P_{M_{thres}}) = R_j(i)$ just after the occurrence of each EQ. Figure 7.7 shows that the number of subareas j increases by an integer larger than or equal to one, when a new EQ occurs. For each of these proper subsets (which form the $\varepsilon[A_{M_{thres}}]$ ensemble at each time instant), one can compute the κ_1 values and then construct their distribution denoted by Prob(κ_1) hereafter. Just after the occurrence of the second event a single proper subset can be defined, thus only $\kappa_1[R_1(2)]$ is available. Later, just after the occurrence of the third event, three proper subsets of $A_{M_{thres}}$ can be defined as shown in Fig. 7.7. Recall that the necessary condition for a proper subset at a given time instant is that it includes the last event (the third EQ in this case). Therefore, $\kappa_1[R_1(2)]$ obtained before the third event is not included for the construction of the distribution Prob(κ_1) at the instant of the third event. By the same token, after the occurrence of the fourth event, seven proper subsets result. Thus, we can now calculate κ_1 for each of these 7 subsets and construct the Prob(κ_1) versus κ_1 graph to examine whether it maximizes at $\kappa_1 \approx 0.070$ (i.e., if it obeys Eq. (7.3)). In actual cases, the number of EQs, depending on the threshold magnitude, are usually tens to a few hundreds and the number of subareas varies from hundreds up to a few tens of thousands.

In the new approach, the κ_1 values of all these subareas and the largest area A, are treated on *equal* footing, which reflects that the adopted largest area A may be a proper subarea of an even larger area in which the mainshock actually occurs. This is a useful notion when the selectivity map of the concerned station is incomplete or a portion of it is adopted for some reason as in the case of the M_w6.4 EQ on June 8, 2008 (see Table 7.1), that will be discussed later in § 7.2.6.

By summarizing, upon the recording of a SES activity, one can estimate (through the procedure explained in § 1.3.5) an area A within which the impending mainshock is expected to occur. Analyzing in natural time the subsequent seismicity (as it evolves event by event) in *all* the possible subareas of A, the probability density function of κ_1 is obtained until it maximizes at $\kappa_1 \approx 0.070$ exhibiting also magnitude threshold invariance. This usually occurs a few days to around one week *before* the mainshock, thus it enables the prediction of the occurrence time of major EQs with time window of the order of a week or so. Examples of this procedure will be presented in § 7.2.2, § 7.2.3, § 7.2.5 and § 7.2.6.

Note also that, as shown later in § 8.4.3, in the mean field case of a *growing* sandpile (§ 8.4.2) even when studying a *single* realization and select random subseries of the process described by Eq. (8.21) to be analyzed in natural time, the pdf deduced for κ_1 maximizes at $\kappa_1 \approx 0.070$; see Fig. 8.17.

7.2 What happened before all earthquakes in Greece with $M_s(ATH) = 6.0$ or larger since 2001. The cases of the major earthquakes with magnitude $M_w 6.4$ or larger since 1995

Since the introduction of *natural time* [45] in 2001 a number of earthquakes (EQs) with magnitude $M_w = 6.0$ or larger occurred in Greece. In this Section, we report what was observed before these EQs, which are included in Table 7.1 (note that predictions of earlier EQs – which have been undoubtedly shown to clearly outperform chance in a debate published in a Special Issue of Geophysical Research Letters, i.e., Vol. 23, No. 11, May 27, 1996, under the title: Debate on "VAN" – were compiled in Ref. [35]). Particular attention is focused on the five major EQs (see Fig. 7.8) with $M_w \geq 6.4$, i.e., the $M_w 6.5$ at 39.05°N 24.35°E on July 26, 2001, the $M_w 6.7$ at 36.21°N 23.41°E on January 8, 2006, the $M_w 6.9$ at 36.50°N 21.78°E on February 14, 2008, the $M_w 6.5$ at 36.22°N 21.75°E on February 14, 2008, and the $M_w 6.4$ at 37.98°N 21.51°E on June 8, 2008 (note that M_w is taken from [26]). In addition, our attention here is focused on the two major EQs with $M_w > 6.4$ of the previous decade (which are *also* plotted in Fig. 7.8), i.e., the $M_w 6.6$ at 40.14°N 21.67°E on May 13, 1995 and the $M_w 6.5$ at 38.4°N 22.3°E on June 15, 1995, the data of which have been analyzed in natural time in retrospect.

During the last fifteen year period, in accordance with the recommendation of the European Advisory Committee for earthquake prediction of the Council of Europe (see p. 101 of Ref. [35]), the following policy was adopted: *if* the expected EQ magnitude M_s(ATH) estimated from the amplitude of the SES activity was larger than (or equal to) 6.0, quick report on the relevant information was submitted to international journals (e.g., see Refs. [52, 51, 50]) *before* the EQ occurrence. The symbol M_s(ATH) stands for the magnitude defined by

$$M_s(\mathrm{ATH}) = M_L + 0.5, \tag{7.5}$$

where M_L denotes the local magnitude reported by GI-NOA (www.gein.noa.gr/services/monthly-list.html).

In Table 7.1, we include *all* EQs with $M_s(\mathrm{ATH}) \geq 6.0$ that occurred in Greece within the area $N_{36}^{41} E_{19}^{27}$ since 2001. In addition, this Table also includes in parentheses the data for the cases in which the expected magnitude (documented on the basis of the SES amplitude) was $M_s(\mathrm{ATH}) \approx 6.0$, but the actual EQ magnitude turned out to be somewhat smaller. For each EQ, we tabulate the date and the station at which the relevant SES activity was recorded along with the publication at which this preseismic information was documented. For the reader's convenience, we also give the submission date of each publication in cases where this documentation was made *before* the EQ occurrence. We emphasize that, in this documentation, it has been confirmed that the SES activity reported in each case, was classified as such since it obeys *both* the criteria described in Section 1.2 as well as the criteria in natural time summarized in Section 4.10.

In § 7.2.1 to § 7.2.6, we restrict ourselves to the description on what happened before the major earthquakes in Greece with magnitude $M_w \geq 6.4$ since 1995.

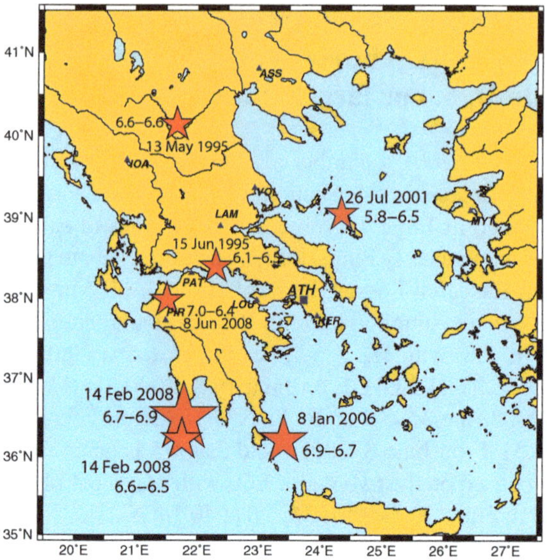

Fig. 7.8 Map showing the location of the VAN stations (triangles) operating in Greece. The location of the central station GLY (which is a suburb of Athens, ATH, rectangle), to which the data of all stations are transferred telemetrically in real time is also shown. The epicenters of the five major EQs with $M_w \geq 6.4$ since 2001(see Table 7.1) along with the two ones in 1995 are indicated by red stars.

7.2.1 The major Grevena-Kozani M_w6.6 earthquake on May 13, 1995

An International Workshop was held by the Royal Society (London, May 11–12, 1995, e.g., see Lighthill [12, 13]) under the title: "A critical review of VAN" just before the occurrence of the M_w6.6 earthquake in Greece on May 13, 1995. This EQ was highly unexpected, because it occurred in an "aseismic" area. The relevant prediction had been forwarded to the chairman of the Workshop (Sir James Lighthill) well in advance (see below). Furthermore, one week after the Workshop, another prediction was sent to the chairman that was related with the catastrophic M_w6.5 Eratini-Egion earthquake of June 15, 1995, which will be discussed in § 7.2.2. These two EQs were the largest events that occurred during 1983–1995 in Greece and their predictions, which attracted a strong interest in the international literature (e.g., Masood [15, 14], Kerr [8], Monastersky [16]), can be found in the Proceedings of the Workshop published several months later (see Varotsos et al. [39]; copies of these predictions are also reproduced here). The chairman included the following conclusion in the Proceedings [13]:

"The earthquakes occurring after the meeting (on 13 May in northern Greece and on 15 June in Egion, which were the two largest in Greece for over a decade) are carefully related to the corresponding VAN predictions (those received by myself, for example – along with other interested scientists – on 2 May and on 20 May 1995). It is noteworthy that the distinguished seismologist, Professor H. Kanamori, was influenced partly by these events, as well as by the proceedings of the review meeting (which he had attended in an initially neutral spirit), to give the views he has expressed above in 'A seismologist looks at VAN', suggesting that for the larger earthquakes in Greece the VAN group appears to have usefully identified SES precursors."

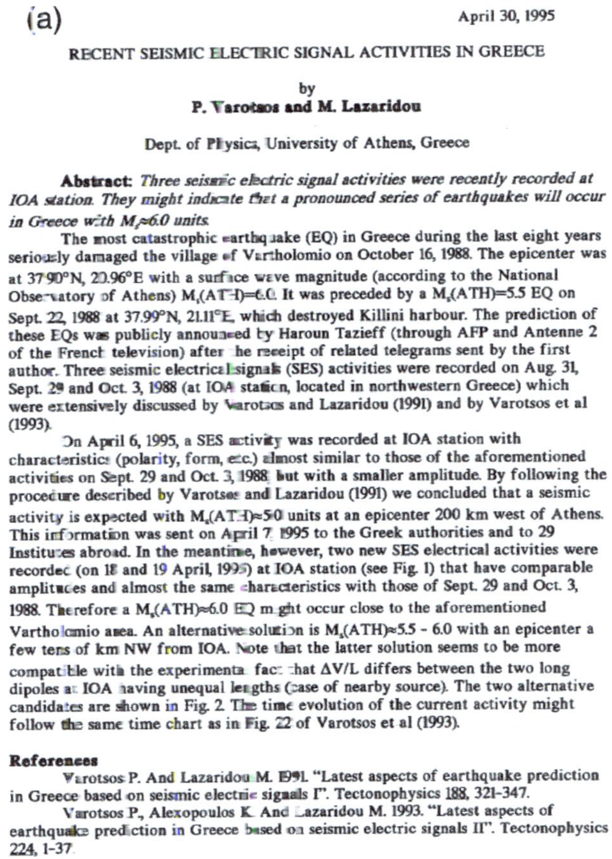

(a) April 30, 1995

RECENT SEISMIC ELECTRIC SIGNAL ACTIVITIES IN GREECE

by

P. Varotsos and M. Lazaridou

Dept. of Physics, University of Athens, Greece

Abstract: *Three seismic electric signal activities were recently recorded at IOA station. They might indicate that a pronounced series of earthquakes will occur in Greece with $M_s \approx 6.0$ units.*

The most catastrophic earthquake (EQ) in Greece during the last eight years seriously damaged the village of Vartholomio on October 16, 1988. The epicenter was at 37.90°N, 20.96°E with a surface wave magnitude (according to the National Observatory of Athens) M_s(ATH)=6.0. It was preceded by a M_s(ATH)=5.5 EQ on Sept. 22, 1988 at 37.99°N, 21.11°E, which destroyed Killini harbour. The prediction of these EQs was publicly announced by Haroun Tazieff (through AFP and Antenne 2 of the French television) after the receipt of related telegrams sent by the first author. Three seismic electrical signals (SES) activities were recorded on Aug. 31, Sept. 29 and Oct. 3, 1988 (at IOA station, located in northwestern Greece) which were extensively discussed by Varotsos and Lazaridou (1991) and by Varotsos et al (1993).

On April 6, 1995, a SES activity was recorded at IOA station with characteristics (polarity, form, etc.) almost similar to those of the aforementioned activities on Sept. 29 and Oct. 3, 1988 but with a smaller amplitude. By following the procedure described by Varotsos and Lazaridou (1991) we concluded that a seismic activity is expected with M_s(ATH)≈5.0 units at an epicenter 200 km west of Athens. This information was sent on April 7, 1995 to the Greek authorities and to 29 Institutes abroad. In the meantime, however, two new SES electrical activities were recorded (on 18 and 19 April, 1995) at IOA station (see Fig. 1) that have comparable amplitudes and almost the same characteristics with those of Sept. 29 and Oct. 3, 1988. Therefore a M_s(ATH)≈6.0 EQ might occur close to the aforementioned Vartholomio area. An alternative solution is M_s(ATH)≈5.5 - 6.0 with an epicenter a few tens of km NW from IOA. Note that the latter solution seems to be more compatible with the experimental fact that ΔV/L differs between the two long dipoles at IOA having unequal lengths (case of nearby source). The two alternative candidates are shown in Fig. 2. The time evolution of the current activity might follow the same time chart as in Fig. 22 of Varotsos et al (1993).

References

Varotsos P. And Lazaridou M. 1991. "Latest aspects of earthquake prediction in Greece based on seismic electric signals I". Tectonophysics 188, 321-347.

Varotsos P., Alexopoulos K. And Lazaridou M. 1993. "Latest aspects of earthquake prediction in Greece based on seismic electric signals II". Tectonophysics 224, 1-37.

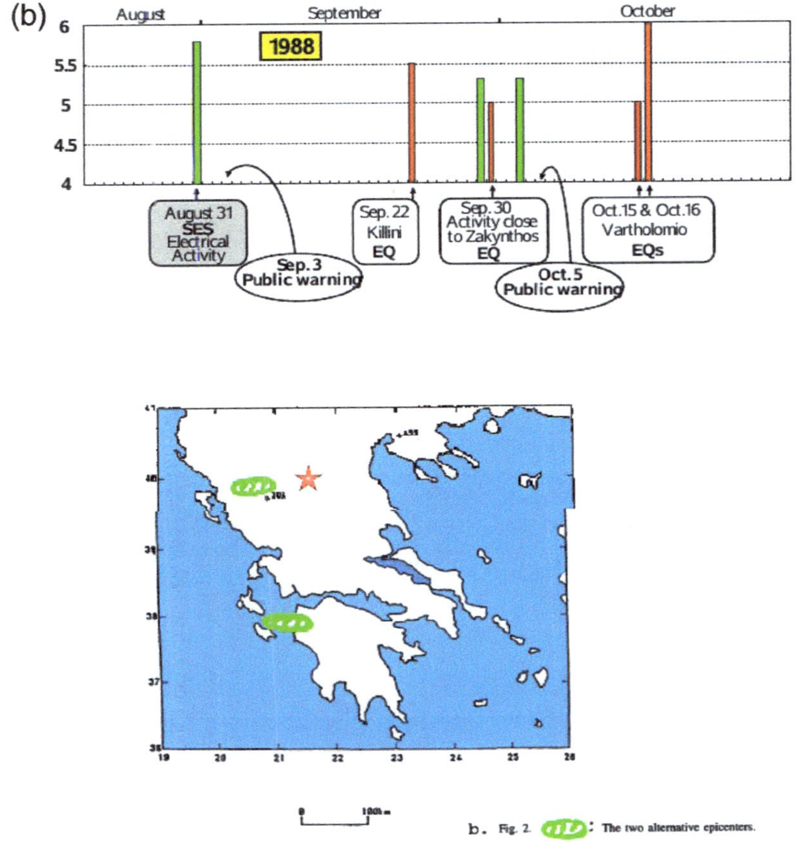

Fig. 7.9 The prediction issued on April 30, 1995, related to the M_w6.6 EQ that occurred on May 13, 1995. A third page was also attached, which was just a copy of the recordings of the SES activities at IOA. For the convenience of the reader, the actual epicenter has been added on the prediction map with a red star.

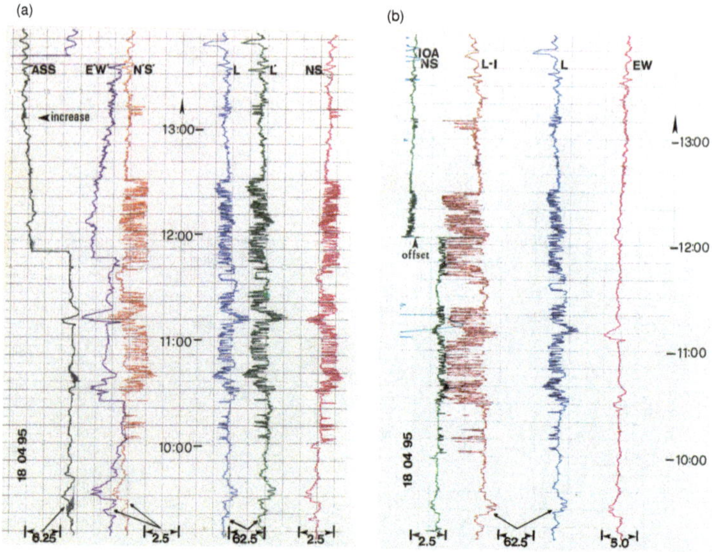

Fig. 7.10 SES activity recorded at IOA on April 18, 1995 (raw data collected by the real-time telemetric network; the scales are in mV). All dipoles are installed at IOA (see the text), except the one labeled ASS (given to distinguish the MT disturbances). The arrow labeled "increase" indicates the direction of increasing ΔV measured in mV. Taken from Ref. [37].

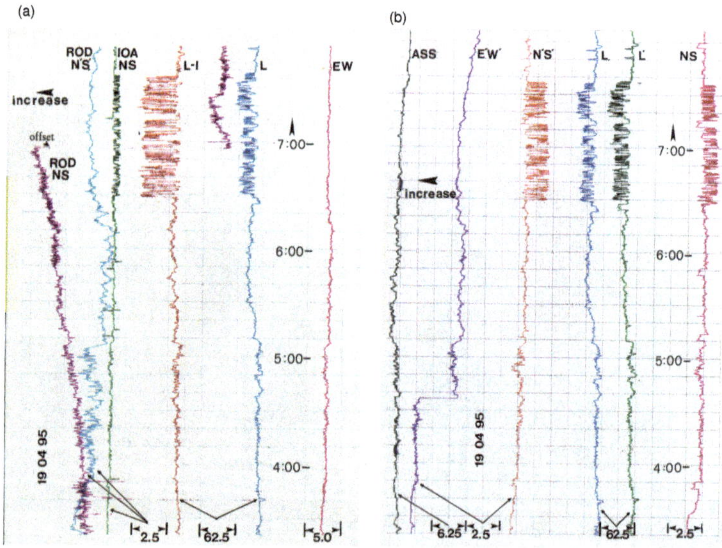

Fig. 7.11 SES activity at IOA on April 19, 1995. They are photocopies from the recordings at the central station (GLY) of the real-time telemetric network. All channels correspond to IOA, except those labeled ROD or ASS, which refer to other stations. The arrow, labeled "increase", indicates the direction of increasing value of ΔV (e.g., see p. 324 of Varotsos and Lazaridou [38]). All the scales are in mV. Reprinted from Ref. [40], Copyright (2005), with permission from TerraPub.

We now proceed here to a description of what happened before the M_w6.6 EQ at 08:47 UT on May 13, 1995 (this EQ is labeled 'K'; see Fig. 4.5(b)).

The SES data and the prediction issued. On April 30, 1995, a three-page prediction was issued. The first page is reproduced in Fig. 7.9(a). It was a short paper under the title "Recent Seismic Electric Signal activities in Greece", the abstract of which stated: *"Three SES activities were recently recorded at IOA station. They might indicate that a pronounced series of EQs will occur in Greece with M_s(ATH) $\approx$ 6.0 units."* The two strongest SES activities (Fig. 1.11(a),(b)) were recorded on April 18 and April 19 (and were classified as such since they obey the criteria mentioned in Section 1.2). The second page reproduced in Fig. 7.9(b) contained the probable time-chart that will be followed as well as a map indicating the two candidate epicentral areas. The prediction text (Fig. 7.9(a)) stated that the *epicentral area located close to IOA was more probable*. The third page was a photocopy of the SES data, as collected through the real-time telemetric network; see Figs. 7.10 and 7.11 to which we now turn. (Recall that Fig. 1.11 depicts data collected with datalogger, see Section 1.1).

Figure 7.10 shows the intense SES activity recorded at IOA on April 18, 1995. It was mainly recorded on the NS short dipole array and on the 3 long dipoles. Figure 7.10(a) shows the recordings of the following 5 dipoles (see the map in Fig. 1.3): Two NS short dipoles ($L = 100$ and 184 m), one EW dipole ($L = 50$ m), and two long dipoles, labeled L and L$'$ (see Fig. 1.3(b)). Figure 7.10(b) depicts the recordings at the following 4 dipoles: one EW short dipole ($L \approx 50$ m), one short dipole ($L \approx 50$ m, labeled IOA, NS) which is almost parallel to the long dipoles that connect IOA with Perama village, and two long dipoles, labeled L-I and L$'$ in Fig. 1.3(b) (L$'$ coincides with that also depicted in Fig. 7.10(a)). The corresponding SES activity recorded on April 19, 1995, is given, as mentioned, in Fig. 7.11.

In addition, in Figs. 1.11(a) and 1.11(b), we have presented the SES recordings collected at two of the IOA sites, i.e., "B" and "C" (see the map in Fig. 1.3(c)), along with the variations of the two horizontal components of the magnetic field that have already been discussed in § 1.3.6.

Evaluation of the prediction. The prediction issued on April 30, 1995 shown in Fig. 7.9(a) discriminated between the two candidate epicenters, depicted in the prediction map of Fig. 7.9(b). One candidate epicenter was in western Greece (i.e., close to the Vartholomio-Killini area), while the second alternative was close to IOA. The prediction text clarified that the second solution seemed to be more compatible with the experimental facts. The predicted magnitude (for the latter solution) was (verbatim) *"M_s(ATH) $\approx$ 5.5–6.0 with an epicenter a few tens of km NW from IOA."* The actual epicenter (USGS [26]) was 40.14°N, 21.67°E, i.e., lying at a distance $\Delta r = 80$–90 km far from the predicted area. The fact that the actual magnitude M_s(ATH)$\approx$6.6 exceeded the predicted one M_s(ATH) $\approx$ 6.0 by $\Delta M = 0.6$, is consistent with what was naturally expected for larger epicentral distances than predicted "a few tens of km" As for the time-window, the EQ occurred on May 13, 1995, i.e., during the fourth week after the SES initiation, in accordance to the expected time chart; see § 1.3.1 case (b).

In summary, this prediction obeyed the tolerances (with respect to the time-window, epicenter and magnitude) for *successful* prediction. The latter is considered as such *if* $\Delta r \lesssim 100$ km, $\Delta M (= 3\sigma) \leq 0.7$ and in addition Δt obeys the expected (time) limits (§ 1.3.1).

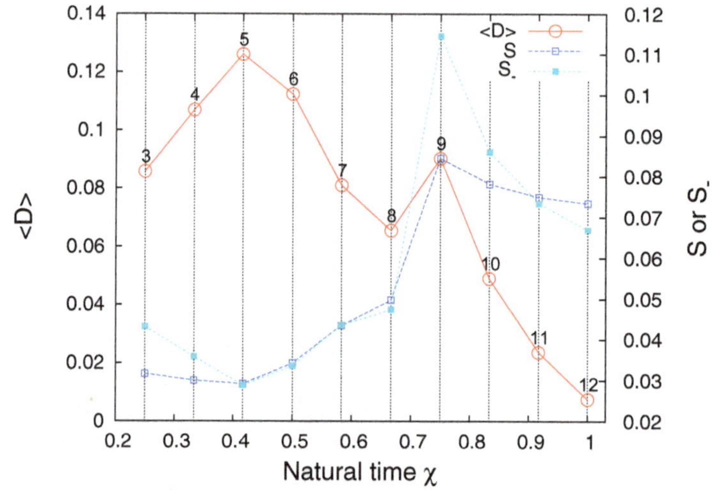

Fig. 7.12 Determination of the occurrence time of the major M_w6.6 EQ on May 13, 1995 (see also Fig. 7.13). The average distance $\langle D \rangle$ (red circles, left scale), the entropy S (open squares, right scale) and the entropy under time reversal S_- (filled squares, right scale) of the seismicity versus the natural time χ. The distance $\langle D \rangle$ drastically decreases only a few days before the occurrence of the mainshock, and the entropies S, S_-, become smaller than $S_u(= 0.0966)$ satisfying condition (7.4). The numbers correspond to the earthquakes listed in Table 6.1.

We now explain, following the *preliminary* procedure (§ 7.1.1), how the occurrence time of this EQ could have been identified in advance. We consider all EQs within the area A: $N_{39.2}^{40.5} E_{20.3}^{22.0}$ that occurred after the SES activity at IOA on April 18, 1995. These have already been listed in Table 6.1 and their analysis in natural time, explained in detail in § 6.2.1, resulted in the evolution of $\Pi(\phi)$, event by event, depicted in Fig. 6.3 (crosses). A careful inspection of this figure (in conjunction with that of Fig. 6.1) reveals that a coincidence is observed upon the occurrence of the EQ No. 12 on May 10, i.e., only 3 days before the mainshock. This is a *true* coincidence, because (see § 7.1.1): first, the average distance $\langle D \rangle$ between the curves of $\Pi(\phi)$ of the evolving seismicity and Eq. (7.1), as shown by the red circles in Fig. 7.12 is less than 10^{-2} at the coincidence; second, Fig. 7.13 – which is an excerpt of Fig. 6.3 depicting only the cases from Fig. 6.3(b) to Fig. 6.3(e) – shows that the evolving $\Pi(\phi)$, i.e., the red crosses, approach the blue curve, i.e., that of Eq. (7.1), from *below* upon the occurrence of the EQs No. 9, 10 and 11 (see Table 6.1) and the *coincidence* occurs at the event No. 12. Third, the criterion of Eq. (7.4) is obeyed, see Fig. 7.12. Finally, the occurrence time of the coincidence does not vary upon

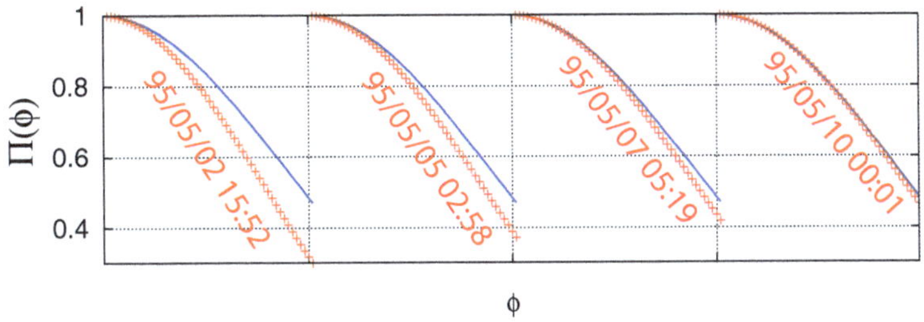

Fig. 7.13 Determination of the occurrence time of the major M_w6.6 EQ on May 13, 1995 (see also Fig. 7.12). The normalized power spectrum (red crosses) $\Pi(\phi)$ of the seismicity within the area $N_{39.2}^{40.5} E_{20.3}^{22.0}$ as it evolves event by event (whose date and time (UT) of occurrence are written in each panel) after the initiation of the SES activity on April 18, 1995. The excerpt presented here corresponds to Figs. 6.3(b) to 6.3(e). In each case only the normalized power spectrum in the window $0 < \phi < 0.5$ is depicted (separated by the vertical dotted lines), whereas the $\Pi(\phi)$ of Eq. (7.1) is depicted by the blue solid line.

changing *either* the magnitude threshold from $M_{thres} = 2.8$ to $M_{thres} = 2.9$ *or* the area from $N_{39.2}^{40.5} E_{20.3}^{22.0}$ to $N_{39.5}^{40.4} E_{20.5}^{22.0}$.

Thus, in short, applying the *preliminary* procedure, the occurrence time of this EQ could have been identified around 3 days in advance.

7.2.2 The major Eratini-Egion M_w6.5 earthquake on June 15, 1995

SES data and the prediction issued. This is the EQ labeled E in Fig. 4.5(b) and its prediction, as already explained in § 7.2.1, has been forwarded to the chairman (Sir James Lighthill) of the International Workshop held by the Royal Society (London, May 11–12, 1995).

Figure 4.5(a) shows the strong SES activity (labeled E) that was recorded on April 30, 1995, at the station VOL (Fig. 1.2). The operation of this station had started only six months before and hence the selectivity map, as well as the calibration of this station, was still unknown. No SES activity (simultaneous to that at VOL) was recorded at the other operating stations.

On the basis of the aforementioned SES activity, a two-page prediction was forwarded to the Government with some delay (caused by the occurrence of the aforementioned major EQ on May 13, 1995), i.e., on May 19, 1995. A photocopy is shown in Fig. 7.14, and its abstract clarifies that *a new strong EQ might hit Greece at a different epicentral area.* Despite the fact that the SES activity was recorded at a station not yet calibrated, the SES amplitude (10 mV/km) allowed the estimation, that the expected magnitude would be comparable to that of the EQ on May 13, 1995, i.e., around 6.6. Since the selectivity map of VOL was still unknown, the epicenter was estimated as follows: in addition to the short dipole arrays, the SES activity was recorded at two long dipoles (having almost the same direction, i.e., SSW and SW in respect to the Volos city, and lengths $L_1 = 5$ km and $L_2 \approx 22$ km)

THE UNIVERSITY OF ATHENS

DEPARTMENT OF PHYSICS
SOLID STATE PHYSICS SECTION
Knossou st. 36, Ano Glyfada
Athens 165 61, GREECE

P. Varotsos
Professor of Physics
Chairman of the Department of Physics

Athens, May 19, 1995

CONTINUATION OF THE SEISMIC ELECTRIC SIGNAL ACTIVITIES IN GREECE

by **P. Varotsos and M. Lazaridou**

Abstract: *The 6.6 earthquake of May 13, 1995 was preceded by seismic electrical activities recorded at IOA. A similar seismic electrical activity was subsequently recorded at* ***VOL.*** *It seems probable that a new strong earthquake (EQ) might hit Greece. This EQ should occur at a different epicentral area but with comparable magnitude. The present prediction is not equally reliable with the previous one, as VOL station is not yet calibrated (because it is operating during the last 6 months).*

On April 27 and 30, 1995 a prediction was issued based on seismic electrical signal (SES) activities recorded at IOA on April 18 and 19, 1995. This prediction was actually followed by a M_s=6.6 earthquake with a USGS epicenter at 40.0°N, 21.6°E (labelled with asterisk in Fig. 1). This area was previously considered to be *aseismic* because such EQ had not occurred there for a period more than 1000 years.

Since last September an experimental station is operating at VOL. The inspection of its records indicated that an SES activity was recorded on April 30 (i.e. **not** simultaneous with those earlier recorded at IOA) on both short and long dipole arrays. Although this station is not yet calibrated we might guess that it is precursor of an EQ similar to that of May 13, because its amplitude (10 mV/km) is comparable to those at IOA. The epicenter should be different due to the following facts:

(i) The two long dipoles installed at VOL (with lengths 22 km and 5 km) show comparable $\Delta V/L$-values thus indicating a non nearby source (and hence the neighbouring area of VOL should be excluded).

(ii) As the SES activity was not recorded at IOA, the areas belonging to IOA selectivity map (large hatched area in Fig. 1) should be excluded. As the epicenter of May 13 seems to belong to IOA selectivity map, the regions lying in its immediate vicinity (and especially those in its western side) should also be excluded.

(iii) As the SES activity was not recorded at ASS, the area belonging to ASS selectivity map (Fig. 1) i.e. that surrounded by the regions a, b, and c (and that of Chalkidiki peninsula and the neighbouring sea) should be excluded. (Note however that the VER area still remains as a candidate area as it is not well verified that it belongs to the ASS selectivity map. Unfortunately VER is out of operation).

(iv) As the SES activity was not recorded at KER, the area lying in the vicinity of Athens (and those in Peloponese) should be excluded. GOR is not operating and its immediate vicinity cannot be excluded.

(v) The central Aegean sea (recorded at ASS or at KER) should be excluded. However the area around Skiros and Alonisos islands cannot be excluded (they are of smaller probability).

By summarising: The new EQ might occur in the remaining part [i.e. after deleting the areas due to the points (i) to (v)] of continental Greece. The spectrum of the SES activity at VOL is **strikingly similar** to those recorded at IOA (on April 18, 19) thus indicating that the EQ of May 13 and the expected EQ might belong to the **same** tectonic process which, according to our opinion, is still going on. The time evolution might follow Fig. 22 of Varotsos et al (1993).

Fig. 7.14 The prediction issued on May 19, 1995, related to the M_w6.5 EQ that occurred on June 15, 1995 that was sent on May 20, 1995 at several institutes abroad.

with the same $\Delta V/L$ value; the latter fact indicated that the impending focal area should lie at a distance r appreciably larger than the dipole lengths, i.e. $r \gg L_1, L_2$ and hence $r/L_2 \gg 1$. As the ratio r/L_2 had to be, at least, around 4–5, the epicenter should lie at a distance more than ~100 km from VOL. Furthermore as the SES activity was not recorded at the other four stations operating at that time, i.e. IOA, ASS, KER and PIR (see Fig. 1.2), we excluded as candidate epicenters the seismic areas belonging to their selectivity maps. We also excluded the area around the epicenter of the M_w6.6 earthquake which had just occurred on May 13, 1995, because the latter was preceded by SES activities at IOA.

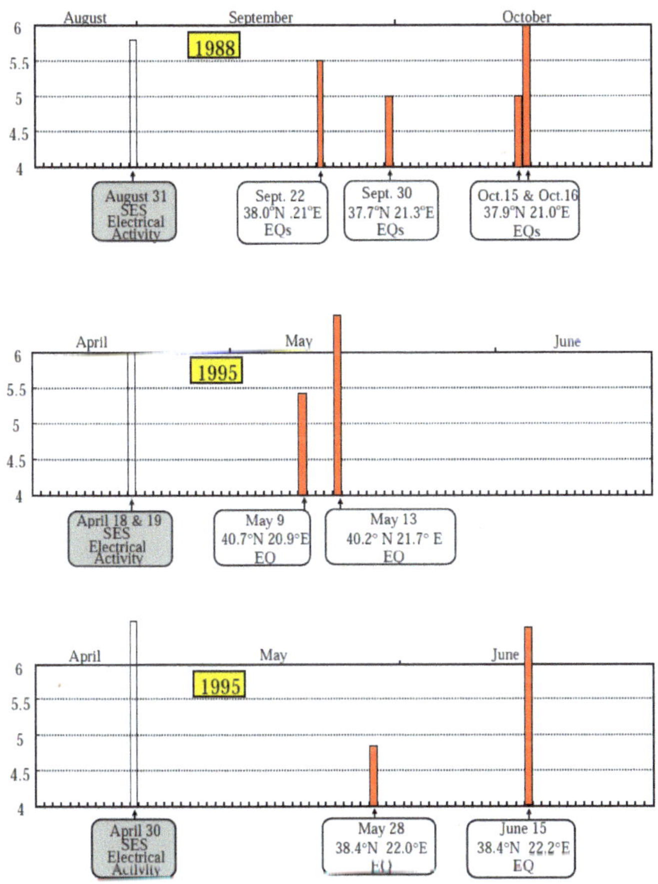

Fig. 7.15 Time evolution of the SES activities related with the 2 big EQs in Greece in 1995. For the sake of comparison, the case of Killini-Vartholomio destructive EQs in 1988 (i.e., fig. 28A of Varotsos and Lazaridou [38]) is also given. Open bars and full bars correspond to SES activities and EQs, respectively.

Thus, the prediction of the epicenter was summarized in the text of the prediction as follows: *"The new EQ might occur in the remaining part... of continental Greece"*. More precisely, the following areas were excluded from continental Greece: central western Greece, Chalkidiki area (including Thessaloniki), the area within a radius of at least ~100 km around VOL, the Peloponnese, the neighboring area around Attica (i.e. Athens)

Table 7.2 All EQs within $N_{37.5}^{39.7} E_{21.5}^{25.0}$ that occurred after the initiation of the SES activity at VOL on April 30, 1995, until the M_w6.5 mainshock on June 15, 1995. Taken from Ref. [46].

No	Year	Month	Day	Hour	min	sec	Lat.	Lon.	Depth	M_L
1	1995	4	30	19	4	41	38.82	21.45	9	2.9
2	1995	5	2	8	26	56	38.20	21.76	32	2.7
3	1995	5	4	16	11	49	38.33	22.05	5	2.9
4	1995	5	6	1	44	12	37.70	21.46	10	2.5
5	1995	5	6	17	44	59	38.51	21.50	24	2.6
6	1995	5	6	23	10	21	38.44	21.80	5	2.6
7	1995	5	8	5	11	9	38.32	22.14	21	4.0
8	1995	5	9	12	48	34	38.32	22.09	10	2.5
9	1995	5	10	15	23	2	39.28	21.69	10	2.9
10	1995	5	12	7	25	13	39.12	24.48	31	3.6
11	1995	5	13	11	53	1	39.56	22.53	10	3.2
12	1995	5	13	13	31	55	38.52	22.04	5	3.3
13	1995	5	15	20	15	13	38.13	21.66	9	2.8
14	1995	5	16	5	15	44	38.97	23.18	33	3.6
15	1995	5	16	10	1	30	38.93	21.77	5	3.0
16	1995	5	17	23	5	25	39.73	21.89	5	2.9
17	1995	5	17	23	10	52	39.70	21.91	5	3.0
18	1995	5	17	23	20	30	39.74	21.97	5	3.1
19	1995	5	18	4	48	27	38.30	22.18	22	3.2
20	1995	5	19	23	19	49	38.24	21.87	11	2.7
21	1995	5	19	23	59	26	38.12	22.65	34	2.8
22	1995	5	20	20	32	33	38.41	21.79	9	2.9
23	1995	5	22	17	35	27	39.54	22.43	5	3.0
24	1995	5	23	2	56	49	39.51	22.25	10	2.7
25	1995	5	25	16	41	31	39.08	23.50	10	2.9
26	1995	5	25	20	32	11	39.74	21.57	35	3.0
27	1995	5	26	1	28	47	38.36	22.63	10	2.6
28	1995	5	26	7	9	25	38.36	22.00	5	2.9
29	1995	5	26	21	30	35	38.43	21.81	6	2.7
30	1995	5	28	16	14	44	38.90	25.04	49	3.2
31	1995	5	28	19	56	41	38.38	21.96	5	4.1
32	1995	5	28	20	9	14	38.40	21.90	5	3.0
33	1995	5	28	21	51	1	38.28	22.67	10	3.0
34	1995	5	29	13	3	3	37.61	22.78	5	2.8
35	1995	5	30	9	6	31	38.50	21.74	5	3.1
36	1995	5	31	12	25	42	39.21	22.88	10	3.0
37	1995	5	31	21	43	30	39.39	22.63	29	3.0
38	1995	6	1	14	4	53	38.13	21.74	5	3.2
39	1995	6	2	14	47	46	39.20	23.14	32	3.1
40	1995	6	4	18	47	35	38.50	22.25	5	2.6
41	1995	6	5	15	4	40	38.88	21.51	5	2.9
42	1995	6	5	16	50	24	38.86	21.47	5	2.9
43	1995	6	5	18	34	46	38.98	21.47	12	2.7
44	1995	6	5	18	35	31	38.97	21.47	7	2.7
45	1995	6	6	20	12	14	38.80	21.58	5	2.9
46	1995	6	12	20	27	7	38.21	22.22	39	2.9
47	1995	6	13	2	48	39	38.29	22.47	10	2.6
48	1995	6	14	11	8	41	38.04	21.54	28	2.5
EQ	**1995**	**6**	**15**	**0**	**15**	**51**	**38.37**	**22.15**	**26**	**5.6**

and, of course, the area of northern Greece around the major earthquake of May 13, 1995. Of the remaining small part of continental Greece, the area lying in the vicinity of GOR (this is a site shown in Fig. 1.1 lying in the vicinity of LAM; see Fig. 1.2) was the more probable. Recall that the region to the north of GOR, close to VOL, was already excluded in view of the same $\Delta V/L$ value collected at the long dipoles of VOL.

The actual epicenter of the mainshock at 00:15UT on June 15 was at 38.4°N, 22.3°E (USGS [26]) being consistent with the prediction, since it lies less than 40–50 km almost south of GOR. The actual EQ magnitude was $M_w = 6.5$, thus being also consistent with the predicted value 6.6.

As for the prediction of time, the last row of the prediction text indicated that the time evolution of seismicity might follow fig. 22 of Varotsos et al. [36]. What actually happened is shown in the lowest time-chart of Fig. 7.15 and the comparison to the predicted time chart (i.e., the upper one in Fig. 7.15), reveals a striking agreement. Note that a smaller EQ with $M_s(\mathrm{ATH}) = 4.8$ occurred on May 28, 1995, at 38.4°N, 22.0°E, i.e. practically at the same area where the mainshock occurred almost two weeks later.

We now apply the *updated* procedure (§ 7.1.2) for the determination of the occurrence time of this EQ since the *preliminary* procedure can be found elsewhere [45, 46]. We consider all EQs (see Table 7.2) that occurred within the area A: $N_{37.5}^{39.7} E_{21.5}^{25.0}$ after the initiation of the SES activity recorded at VOL on April 30, 1995, and their M_0 values are estimated using the relation $\log_{10}(M_0) = 1.64M_L + \text{const.}$ as in § 7.2.1. The computation of κ_1 is extended, as mentioned in § 7.1.2, to all possible subareas of the area A and then the plot of the probability distribution $\mathrm{Prob}(\kappa_1)$ versus κ_1 is constructed after the occurrence of each small event since April 30, 1995. Excerpts of these results that correspond to the period June 1 to June 12, 1995, are shown in Figs. 7.16(a) to 7.16(c) for three magnitude thresholds, i.e., $M_{thres} = 2.5$, 2.6 and 2.8. An inspection of these figures reveals that:

Upon the occurrence of the $M_L = 2.9$ event at 20:27 UT on June 12, 1995, the probability distribution $\mathrm{Prob}(\kappa_1)$ maximizes at $\kappa_1 = 0.070$ for all three magnitude thresholds (see the arrows in Figs. 7.16(a) to 7.16(c)), thus signaling the impending mainshock that occurred almost two days later at 00:15 UT on June 15, 1995.

7.2.3 The major Aegean M_w6.5 earthquake on July 26, 2001

This is the major earthquake labeled A in Fig. 4.5. This figure also depicts the preceding SES activity which had duration of around two hours and was recorded at the station VOL on March 17, 2001. It was clearly detected at several short- and long-measuring dipoles located in a zone with spatial dimensions (a few tens km) × (several km), see Fig. 1.4. A copy from the recordings of the real time telemetric network is given in Ref. [35] as well as in Ref. [41], while Fig. 7.17(a) depicts the digital recordings from the long-dipole

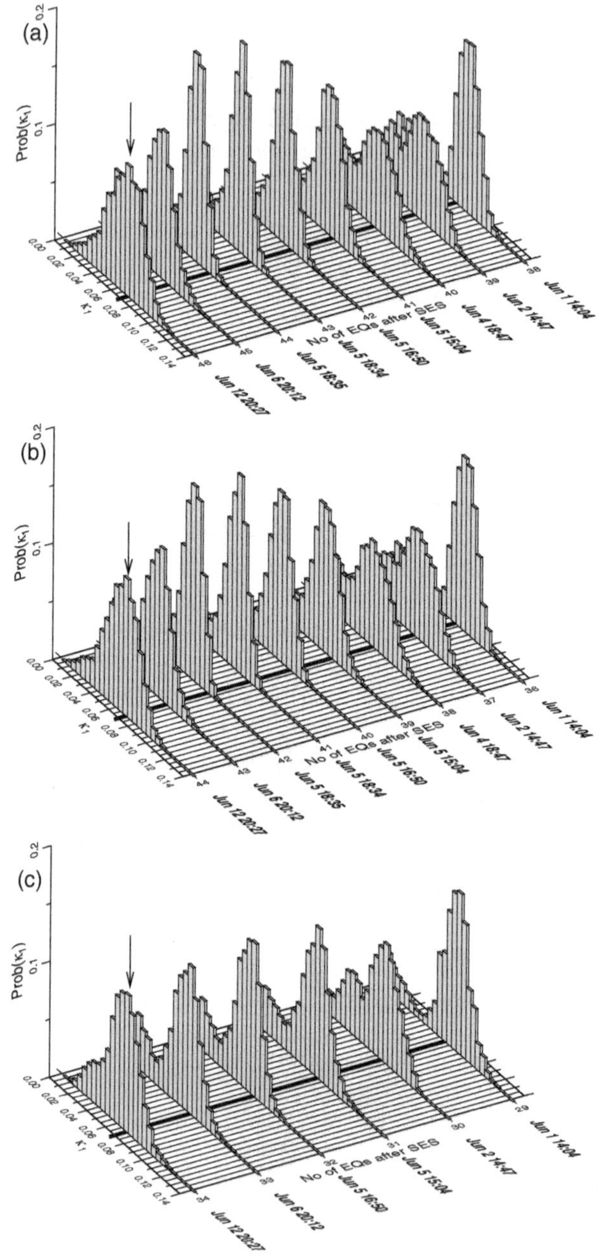

Fig. 7.16 Determination of the occurrence time of the M_w6.5 EQ on June 15, 1995. Prob(κ_1) versus κ_1 when considering the seismicity within the area $N_{37.5}^{39.7}$ $E_{21.5}^{25.0}$ since the initiation of the SES activity recorded at VOL on April 30, 1995. Excerpts for the period June 1 to June 12, 1995, are shown for (a) $M_{thres} = 2.5$, (b) $M_{thres} = 2.6$ and (c) $M_{thres} = 2.8$. The thick horizontal line corresponds to $\kappa_1 = 0.070$. The arrows show the maximum of Prob(κ_1) vs κ_1 observed at $\kappa_1 = 0.070$ on June 12, 1995, for all thresholds $M_{thres} = 2.5$, 2.6 and 2.8.

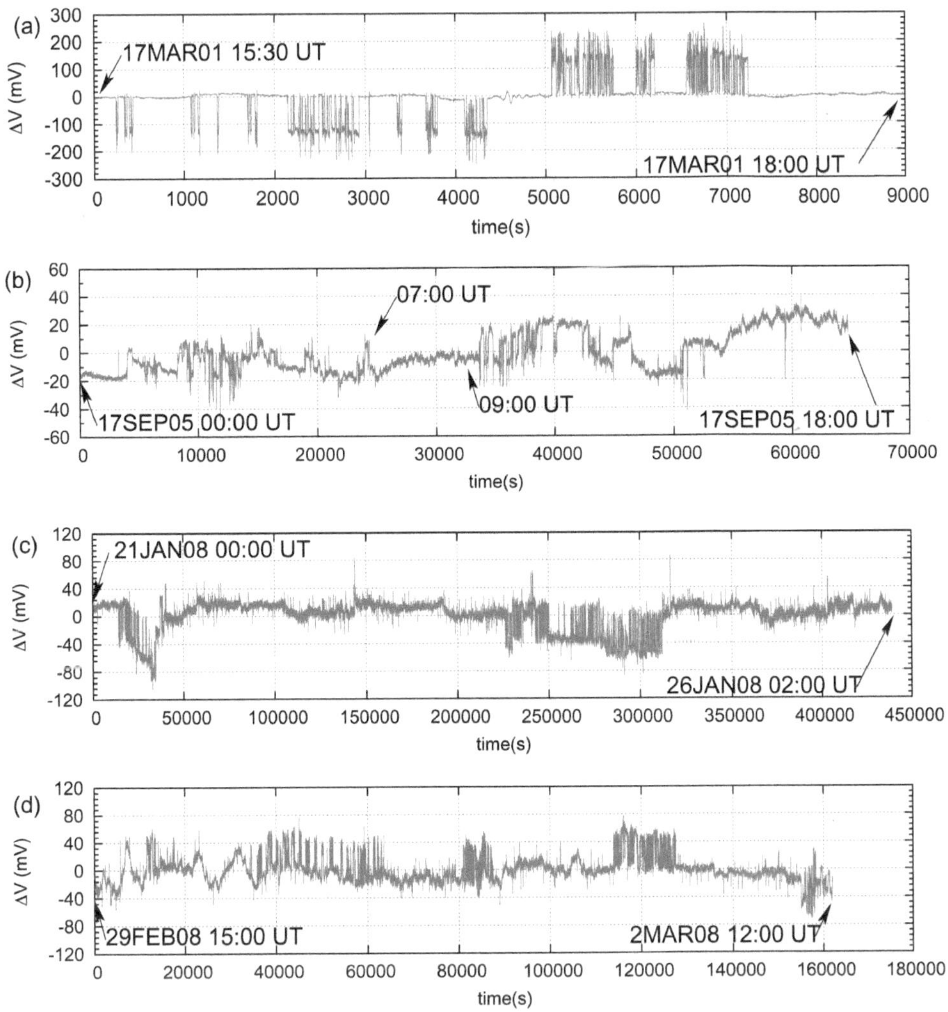

Fig. 7.17 SES activities before the major EQs with $M_w \geq 6.4$ since 2001. The short-duration SES activity (a) was recorded at VOL in 2001, while the three long-duration SES activities of (b), (c) and (d) at PIR in 2005 (b) and in 2008 (c, d) (see the text).

V − $S_{\Sigma B}$; see the map in Fig. 1.4. The digital recordings from all the measuring dipoles can be found in Ref. [41].

The epicenter of the impending seismic activity was estimated to be within the region marked with the broken line in Fig. 7.18. The procedure through which the SES activity was identified, as well as the expected epicenter and magnitude ($M \approx 6.5$) were determined, has been described in detail in Ref. [41] that was submitted for publication on March 25, 2001, i.e., almost four months before the EQ occurrence. Such a lead time seems to be in principle too long (note that a tentative explanation in terms of tectonics

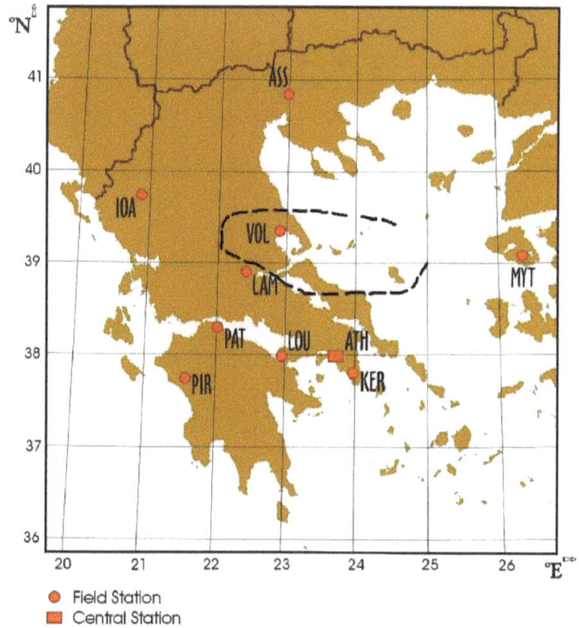

Fig. 7.18 The area 'bordered' by the broken curve (surrounding VOL) was the predicted area in Ref. [41] for the epicenter of the impending EQ related to the SES activity depicted in Fig. 7.17(a). Taken from Ref. [41].

and geodynamics of that seismic area has been discussed in Ref. [3]) but interestingly conforms with natural time analysis of the subsequent seismicity to which we now turn.

Since the *preliminary* procedure can be found elsewhere [45, 46], we present here the *updated* procedure (§ 7.1.2). We consider all EQs that occurred after the initiation of the SES activity at VOL on March 17, 2001, within the area A: $N_{38.5}^{39.5} E_{22.2}^{25.6}$, which includes the predicted area 'bordered' by the broken line in Fig. 7.18. The natural time analysis of seismicity (by using, as in § 7.2.1, the relation $\log_{10}(M_0) = 1.64 M_L + \text{const.}$, where M_L is taken from GI-NOA) was made, as explained in § 7.1.2, for all possible subareas of the area A and the resulting κ_1 values lead to the probability distribution Prob(κ_1) of κ_1 shown in Fig. 7.19. An inspection of this figure shows that:

Upon the occurrence of the $M_L = 3.0$ EQ at 16:35 UT on July 25, 2001, which took place just eight hours before the mainshock (that occurred at 00:21 UT on July 26, 2001), Prob(κ_1) vs κ_1 exhibits a maximum at $\kappa_1 = 0.070$ marked with arrows in Fig. 7.19(a), (b) and (c) for three magnitude thresholds, i.e., $M_{thres} = 2.8$, 2.9 and 3.0, respectively.

Concerning the actual magnitude of this mainshock, i.e, $M_w = 6.5$, it is comparable to the one estimated in advance [41].

As for its actual epicenter (see Table 7.1), it actually lies *within the predicted area* 'bordered' by the broken line in Fig. 7.18 and close to its eastern side.

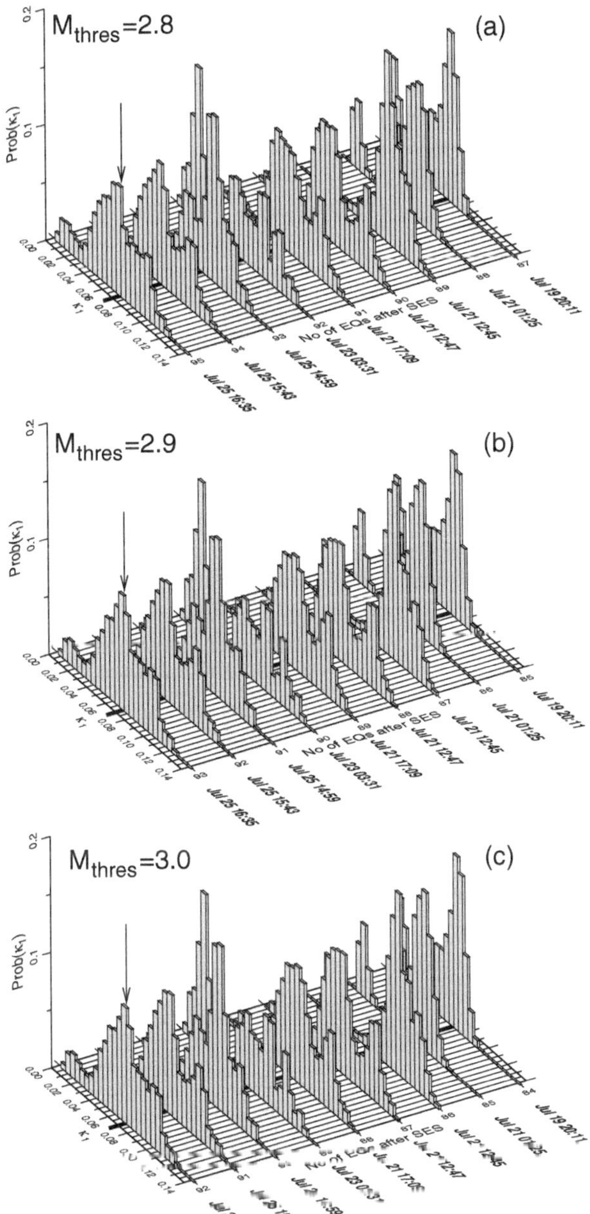

Fig. 7.19 Determination of the occurrence time of the major $M_w6.5$ EQ on July 26, 2001. Prob(κ_1) versus κ_1 when considering the seismicity within the area $N^{39.5}_{38.5} E^{25.6}_{22.2}$ since the initiation of the SES activity recorded at VOL on March 17, 2001. The period July 19 to July 25, 2001, is shown for (a) $M_{thres} = 2.8$, (b) $M_{thres} = 2.9$, and (c) $M_{thres} = 3.0$. The thick horizontal line corresponds to $\kappa_1 = 0.070$. The arrows show the maximum of Prob(κ_1) vs κ_1 observed at $\kappa_1 = 0.070$ upon the occurrence of the $M_L = 3.0$ event at 16:35 UT on July 25, 2001.

Note that the predicted area is smaller than the one, i.e., $N^{39.5}_{38.5}\,E^{25.6}_{22.2}$, considered in the computation of κ_1 in the natural time analysis of seismicity. This could be understood in the following context. The (predicted) area in Fig. 7.18 is solely based on the SES characteristics governed by the electrical inhomogeneities in the Earth's crust, and hence does not necessarily coincide with the area considered in the updated procedure that involves the preceding small EQs that finally establish long-range temporal correlations. The same argument holds for the case of the EQ discussed in § 7.2.2.

7.2.4 The major $M_w6.7$ earthquake in southern Greece on January 8, 2006

Two intense SES activities, with a duration of several hours each, were recorded [42] at PIR station on September 17, 2005. They are shown together in Fig. 7.17(b), where we see that the first lasts until around 07:00 UT, while the second one starts after 09:00UT.

Almost one month later, a $M_w5.7$ EQ occurred in Western Greece on October 18, 2005, with an epicenter at 37.58°N 20.86°E (Table 7.1). USGS and Harvard reported that this EQ was mainly of thrust type, which, however, seemed to deviate from an earlier conclusion of Uyeda et al. [28] who had found that, for the EQs in the transform fault zone west of Kefallinia, the station PIR was mainly sensitive to strike–slip type EQs. In view of this deviation, doubts were raised whether any of the two SES activities of Fig. 7.17(b) were actually correlated with the EQ of October 18, 2005. As a result four days later, i.e., on October 22, 2005, a paper was submitted [42] raising the possibility that the two SES activities in Fig. 7.17(b) were in fact a one-day long-duration SES activity probably correlated with an impending strong EQ (*not* from the aforementioned area studied by Uyeda et al. [28]). Actually, at 11:34 UT on January 8, 2006, the $M_w6.7$ EQ occurred in southern Greece with an epicenter at 36.3°N 23.3°E, i.e., in an area different from the one studied earlier by Uyeda et al. [28].

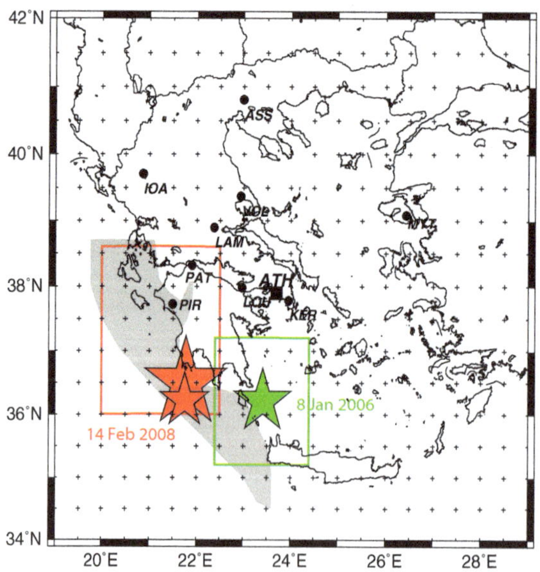

Fig. 7.20 A map showing the areas discussed in § 7.2.4, § 7.2.5 and § 7.2.6. The corresponding determination of the occurrence time for the $M_w6.9$ EQ on February 14, 2008 (red star), and the $M_w6.7$ EQ on January 8, 2006 (green star), was made by considering the seismicity within the red rectangle $N^{38.6}_{36.0}\,E^{22.5}_{20.0}$ and the green rectangle $N^{37.2}_{35.2}\,E^{24.4}_{22.4}$, respectively (see the text). The shaded area shows the PIR selectivity map updated in 2008 that was used later (§ 7.2.6) for the determination of the occurrence time of the major $M_w6.4$ EQ on June 8, 2008. Solid dots show the measuring stations of the telemetric network.

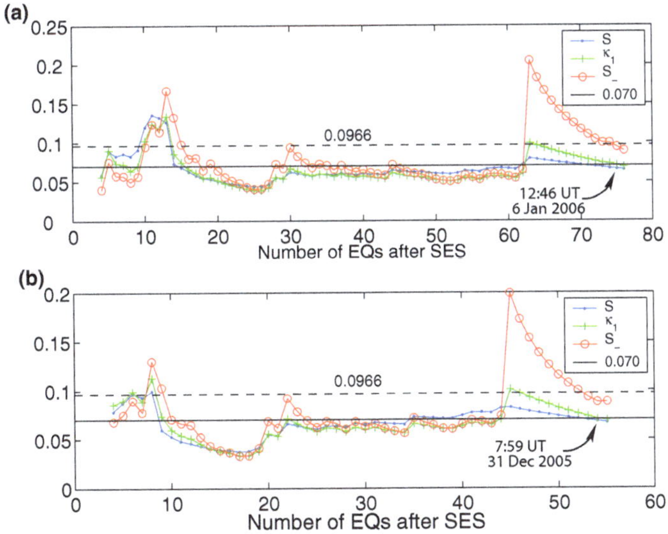

Fig. 7.21 Determination of the occurrence time of the major M_w6.7 EQ on January 8, 2006. The variance κ_1 (green), the entropy S (blue) and the entropy under time reversal S_- (red) of the seismicity within the green rectangular region of Fig. 7.20, as it evolves event by event after the long-duration SES activity recorded at PIR on September 17, 2005 (Fig. 7.17(b)): (a): For all small seismic events reported by GI-NOA, i.e., $M_{thres} = 2.6$ and (b): For seismic events with $M_{thres} = 3.0$. The horizontal solid line corresponds to $\kappa_1 = 0.070$ while the broken to $S_u = 0.0966$. Taken from Ref. [43].

We now follow the *preliminary* procedure explained in § 7.1.1. We set the natural time for seismicity zero at the initiation time of the SES activity recorded at PIR on September 17, 2005 (Fig. 7.17(b)) and form time series of seismic events in natural time for various time windows as the number of consecutive (small) EQs increases. We consider [42] all the small EQs (i.e., with $M_L \geq 2.6$) that occurred before the mainshock, within the region $N_{35.2}^{37.2} E_{22.4}^{24.4}$ surrounding the epicenter (see the green rectangular area in Fig. 7.20) according to the EQ catalog of GI-NOA (the corresponding M_0 values have been estimated from the relation $\log_{10}(M_0) = 1.64M_L + \text{const.}$ as in § 7.2.1). For each of the time windows, the following quantities have been computed: κ_1, $\langle D \rangle$, S and S_- and the results are plotted in Fig. 7.21(a). An inspection of this figure shows that κ_1 approaches the value 0.070 from *above* at 12:46 UT on January 6, 2005, i.e., almost two days before the occurrence of the mainshock. Furthermore, both S and S_- values at the coincidence are smaller than the value $S_u = 0.0966$ in accordance to Eq. (7.4). In addition, we confirmed that $\langle D \rangle$ is smaller than 10^{-2}. Finally, upon changing the magnitude threshold (i.e., taking $M_{thres} = 2.8$, instead of $M_{thres} = 2.6$) and studying a smaller region, i.e., $N_{35.7}^{36.9} E_{22.6}^{24.2}$, the occurrence time of the coincidence remains the same. Thus, we conclude that the conditions mentioned in § 7.1.1 for a *true* coincidence are obeyed. Despite this fact, and in order to shed more light on a point already tackled in § 7.1.1, we repeated the same calculation, but by imposing an even larger magnitude threshold, i.e., $M_{thres} = 3.0$. We then obtained the results depicted in Fig. 7.21(b) showing that the critical point is approached a week before the mainshock (note that *no* EQ with $M_L \geq 3.0$ occurred during that week). The difference in the results is understood in the context already mentioned in § 7.1.1: if higher magnitude threshold is used, the description of the real situation approaching criticality becomes less accurate due to *'coarse graining'* [43] since the number of events is finite.

In summary, the natural time analysis of the seismicity subsequent to the long-duration SES activity at PIR enables the determination of the occurrence time of the M_w6.7 EQ on January 8, 2006, within a narrow range of around 2 days up to 1 week.

7.2.5 The two major M_w6.9 and M_w6.5 earthquakes in southwestern Greece on February 14, 2008

In this case, both short- and long-duration SES activities have been recorded (Table 7.1). The short one came first [44] and it was recorded (see Fig. 7.22(b)) on January 14, 2008, at the station PIR. Almost one week later, a long duration SES activity of the same polarity and amplitude was recorded also at PIR (Fig. 7.17(c)). The natural time analysis of the former (labeled PIR_3 in Table 4.6), which is of clear dichotomous nature, led [44] to the following parameters: $\kappa_1 = 0.070 \pm 0.005$, $S = 0.086 \pm 0.003$, $S_- = 0.070 \pm 0.005$, which obey the conditions in order to classify this signal as SES activity (note that it also satisfies the criteria mentioned in Section 1.2).

After this classification, the study of the seismicity in natural time was immediately started in the area A: $N^{38.6}_{36.0}E^{22.5}_{20.0}$ (determined by means of the procedure described in § 1.3.5) as publicized on February 1, 2008, by Varotsos et al. [44] (this area is marked with the red rectangle in Fig. 7.20). The corresponding M_0 values have been again estimated using the relation $\log_{10}(M_0) = 1.64M_L + \text{const.}$ as in § 7.2.1. We now draw attention to the difficulty arisen if the *preliminary* procedure (§ 7.1.1) is applied to the present case. Within the area $N^{38.6}_{36.0}E^{22.5}_{20.0}$ studied since the initiation of this SES activity on January 14, 2008, two EQs with magnitudes $M_s(\text{ATH}) \approx 5.5$ occurred on February 4, 2008, close to PAT associated with the SES activity at PAT on January 10, 2008; see Fig. 7.22(a). This results in the fact that the κ_1 value becomes very small, i.e., $\kappa_1 \approx 0$, at any small area surrounding the epicenters of these two EQs (see § 6.2.1; see also Ref. [54]). On the other hand, in the *updated* procedure (§ 7.1.2) the computation of κ_1 is extended to all possible subareas of the area $N^{38.6}_{36.0}E^{22.5}_{20.0}$. Then the plot of the probability distribution Prob(κ_1) versus κ_1 (shown in Fig. 7.23 for $M_{thres} = 3.2$) constructed after the occurrence of each small event exhibited a bimodal feature. The one mode, corresponding to nearly zero κ_1 values, results from the subareas that contain the aforementioned two EQs of magnitude 5.5. The other mode, maximized at $\kappa_1 = 0.070$, comes from subareas which do not include these two EQs. It is the latter mode that upon the occurrence of a small event at 04:07 UT on February 12, 2008; see the case marked with an arrow in Fig. 7.23, signifies the approach to the critical point. Two days later, i.e., at 10:09 UT on February 14, 2008, the M_w6.9 earthquake occurred at 36.5°N 21.8°E inside the area $N^{38.6}_{36.0}E^{22.5}_{20.0}$ specified in advance [44]. Two hours later, i.e., at 12:08 UT, a M_w6.5 EQ followed almost at the same epicenter.

At that period, beyond the *updated* procedure, the *preliminary* one was simultaneously applied. The latter procedure, upon avoiding the difficulty described above (i.e., by exclud-

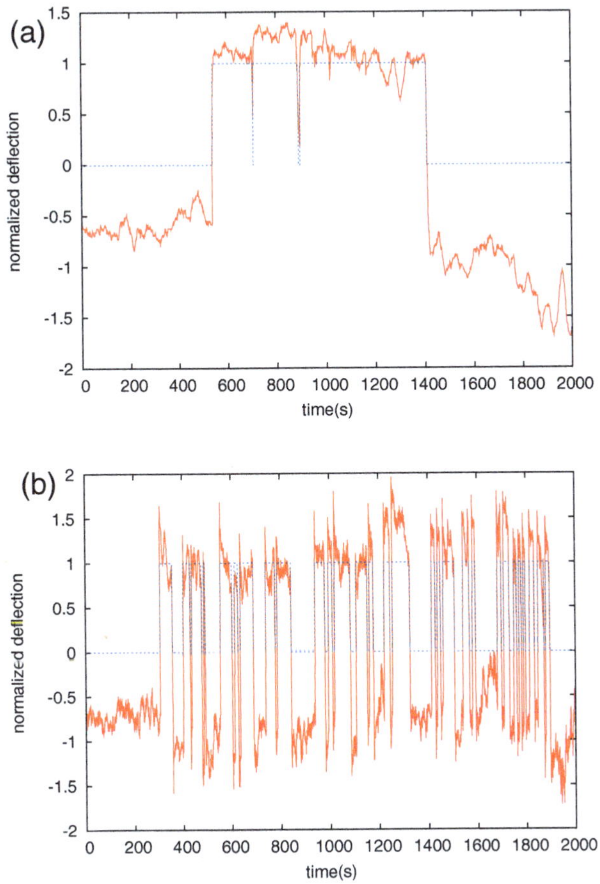

Fig. 7.22 The short duration SES activities recorded on January 10, 2008, at PAT (a) and on January 14, 2008, at PIR (b) in normalized units (i.e., by subtracting the mean value and dividing the results by the standard deviation) along with the dichotomous representation marked by the dotted (blue) line. Taken from Ref. [44].

ing the influence of the aftershocks around the two M_s(ATH) $\approx$ 5.5 EQs that had already occurred close to Patras on February 4, 2008), had led to the conclusion that the critical point was approached somewhat earlier, i.e., on February 10, 2008 (note that the difference in the results of the two procedures can be understood on the basis of the discussion in § 7.1.1 concerning the *'coarse graining'* when using different magnitude thresholds). This explains why we were able to publicly announce on February 10, 2008, that the major EQ is imminent, as described in detail by Uyeda and Kamogawa [30, 31].

The M_w6.9 earthquake on February 14, 2008, according to USGS [26], is the strongest one in Greece since 1983. As explained above, all the parameters of this earthquake, i.e., the epicentral area (see the red rectangle in Fig. 7.20), the magnitude (recall that *only* when the expected M is larger than 6.0, a prediction is publicized) and the occurrence time were specified and announced in advance.

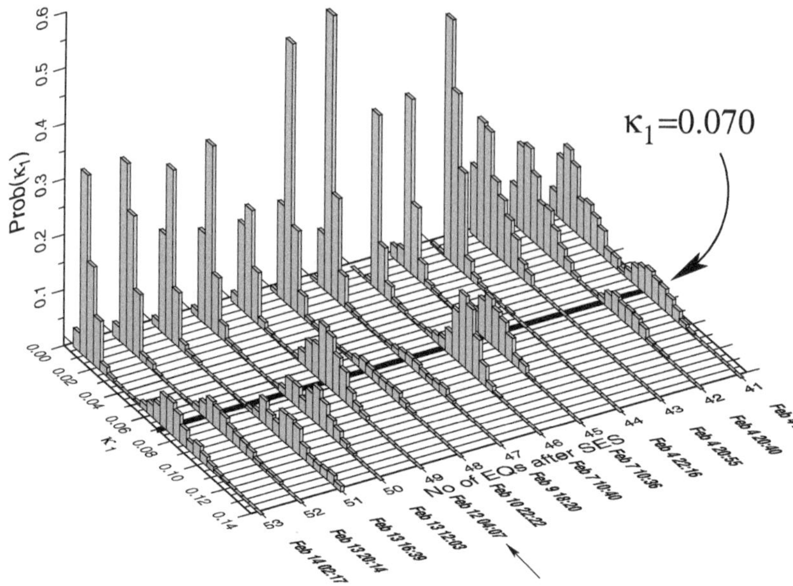

Fig. 7.23 Determination of the occurrence time of the major EQs on February 14, 2008. Study of the Prob(κ_1) versus κ_1 for the seismicity ($M_{thres} = 3.2$) that occurred within the area $N_{36.0}^{38.6} E_{20.0}^{22.5}$ after the short duration SES activity at PIR on January 14, 2008, depicted in Fig. 7.22(b). Taken from Ref. [21].

7.2.5.1 The experimental error in the epicentral location of small EQs and its influence on the determination of the occurrence time of an impending mainshock

The results depicted in Fig. 7.23 have been obtained upon adopting a reasonable experimental error in the determination of the epicentral coordinates of the small EQs involved in the aforementioned computation. In particular, two small EQs have been assumed to occur at different locations *iff* their reported epicentral coordinates differ more than $0.02^\circ \times 0.02^\circ$. In other words, the number of the possible subareas inside the estimated area A: $N_{36.0}^{38.6} E_{20.0}^{22.5}$ was counted after using a grid with "cells" having dimensions of $0.02^\circ \times 0.02^\circ$ and considering the reported epicentral coordinates of the small EQs. On the other hand, if we assume that the EQ epicentral coordinates, that have been reported with two decimals, are accurate and construct a grid based on these coordinates (adaptive grid), the population of the resulting possible subareas of the area $N_{36.0}^{38.6} E_{20.0}^{22.5}$ becomes markedly larger, thus leading to a somewhat different result. Namely, based on the latter assumption the computation was repeated and led to the results depicted in Figs. 7.24(a), 7.24(b) and 7.24(c) for $M_{thres} = 2.8$, 2.9 and 3.0, respectively. They show that a maximum of Prob(κ_1) versus κ_1 at $\kappa_1 \approx 0.070$ is simultaneously observed in all the three magnitude thresholds upon the occurrence of a $M_L = 3.4$ event at 10:40 UT on February 7, 2008, with epicenter at 38.37°N 20.32°E. This date, which is almost one week before the M_w6.9 mainshock of February 14, 2008, differs from the one (i.e., February 12, 2008) of the maximum observed in Fig. 7.23.

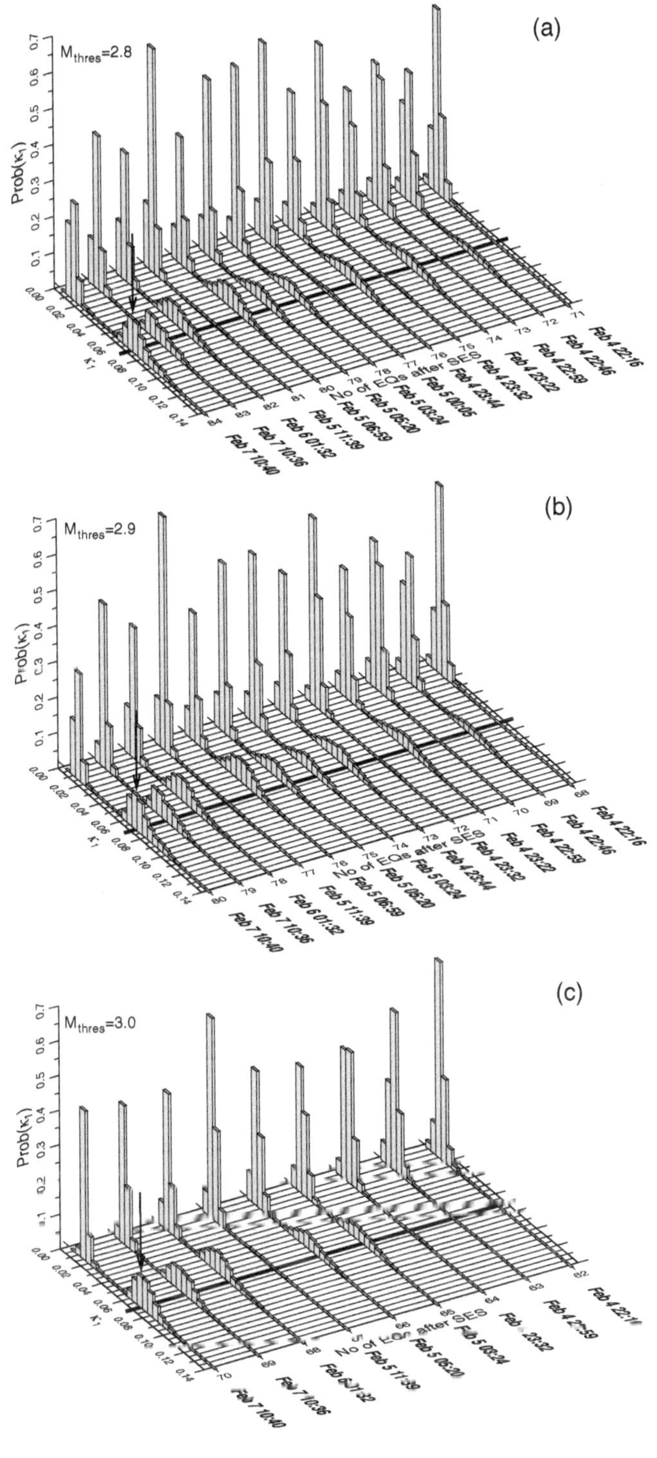

Fig. 7.24 Determination of the occurrence time of the major EQs on February 14, 2008, when considering the *assumption* discussed in § 7.2.5.1. Study of the Prob(κ_1) versus κ_1 for the seismicity for (a) $M_{thres} = 2.8$, (b) $M_{thres} = 2.9$, and (c) $M_{thres} = 3.0$ that occurred within the area $N_{36.0}^{38.6} E_{20.0}^{22.5}$ after the SES activity at PIR on January 14, 2008, depicted in Fig. 7.22(b). The simultaneous maxima of Prob(κ_1) versus κ_1 at $\kappa_1 \approx 0.070$, marked with arrows, are observed upon the occurrence of a $M_L = 3.4$ EQ at 10:40 UT on February 7, 2008, with epicenter at 38.37°N 20.32°E.

In other words, we conclude that the date at which the maximum of Prob(κ_1) versus κ_1 at $\kappa_1 \approx 0.070$ is observed, depends somewhat on the accuracy considered in the epicentral coordinates of the small earthquakes involved in the computation.

This accuracy depends of course on several factors (including the density of the seismological network operating in the area investigated) and should be considered with care in each case separately. Since, however, the estimation of this accuracy is far outside of the scope of the present monograph, in all the other examples treated here, we assumed that the epicentral coordinates as reported are accurate.

We also note that a random experimental error (≈0.2 to 0.3) in the EQ magnitude, does not seem to affect the date of a *true* coincidence, as shown by Uyeda et al. [32] when applying the *preliminary* procedure.

7.2.6 $M_w6.4$ earthquake in the Peloponnese on June 8, 2008

This major EQ was preceded by that long-duration SES activity-lasted from February 29 to March 2, 2008 (see Fig. 7.17(d) which just reproduces the upper channel of Fig. 1.16). After subtracting the MT background with the procedure described in § 1.4.3.1, the signal was analyzed in natural time (see Section 4.11) and classified as an SES activity (note that it also obeys the criteria mentioned in Section 1.2).

The investigation of the subsequent seismicity was conducted at first (see Ref. [20]) in the area $N_{37.0}^{38.6} E_{20.0}^{22.0}$, which is somewhat smaller than the PIR selectivity map known at that time. This was in an attempt to avoid as much as possible the influence of aftershocks of the $M_w6.9$ EQ at 36.5°N 21.8°E on February 14, 2008. This policy was considered justified, based on the notion that a criticality approach would take place in proper subareas simultaneously. At the same time, an attempt was also made to extend the area A to include the shaded area along the Hellenic Arc as shown in Fig. 7.20. This extension was based on the recent pieces of information for PIR selectivity map, including the occurrences of the aforementioned $M_w6.9$ EQ on February 14, 2008 (see § 7.2.5), associated with the SES activity of Fig. 7.17(c) and the $M_w6.7$ EQ at 36.3°N 23.2°E on January 8, 2006 (see § 7.2.4) following the SES activity of Fig. 7.17(b) at PIR [42]. In the study for the extended PIR selectivity map area (shaded region in Fig. 7.20), we raised the magnitude threshold to $M_{thres} = 3.9$, 4.0 and 4.1, because the extended area along the Hellenic Arc is highly seismic and there were too many (more than half a thousand) events to handle for $M_{thres} = 3.2$. This study showed that upon the occurrence of a M_s(ATH) $= 5.1$ EQ at 35.5°N 22.4°E at 23:26 UT on May 27 (practically May 28), 2008, the probability Prob(κ_1) exhibits a pronounced maximum at $\kappa_1 \approx 0.070$ marked by a vertical arrow in Fig. 7.25(a) drawn for $M_{thres} = 3.9$. Similar maxima at $\kappa_1 \approx 0.070$ appeared simultaneously for $M_{thres} = 4.0$ and $M_{thres} = 4.1$ (see Figs. 7.25(b) and 7.25(c), respectively), thus indicating that the critical point has been approached. This was reported on May 29, 2008, in Ref. [22]. Actually, at 12:25 UT on June 8, 2008, a $M_w6.4$ EQ occurred at 38.0°N 21.5°E, i.e., inside the candidate area $N_{37.0}^{38.6}E_{20.0}^{22.0}$ (see Ref. [20] publicized on March 20, 2008). It caused extensive damage (four people

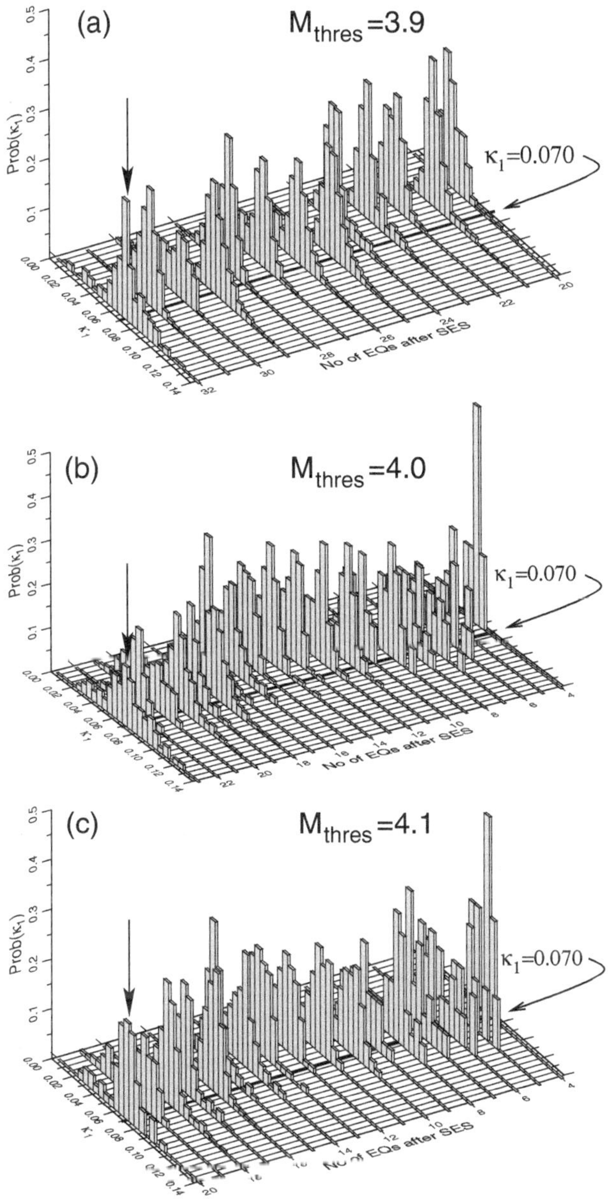

Fig. 7.25 Determination of the occurrence time of the major M_w6.4 EQ on June 8, 2008. Prob(κ_1) versus κ_1 of the seismicity $M_{thres} = 3.9$ (a), $M_{thres} = 4.0$ (b) and $M_{thres} = 4.1$ (c) within the shaded area shown in Fig. 7.20 subsequent to the long duration SES activity recorded at PIR during February 29 to March 2, 2008; see Fig. 7.17(d). The vertical arrows mark the maxima of Prob(κ_1) vs κ_1 at $\kappa_1 \approx 0.070$ that occurred at 23:26 UT on May 27, 2008 (practically May 28), and has been followed by the M_w6.4 on June 8, 2008. Taken from Ret. [22].

were killed while several hundred houses were seriously damaged). The magnitude 6–7 class EQ expected from the amplitude of the SES activity, as mentioned in the last paragraph of the Appendix of Ref. [21](which had been submitted for publication on March 21, 2008, i.e., after the completion of the analysis in natural time of the SES activity depicted in Fig. 7.17(d)), was reasonably well supported by the actual EQ magnitude [26], i.e., $M_w = 6.4$.

Thus, in short, all the parameters of the M_w6.4 earthquake that occurred at 12:25 UT on June 8, 2008, i.e., the epicentral area, the magnitude and the occurrence time, were specified and announced well in advance.

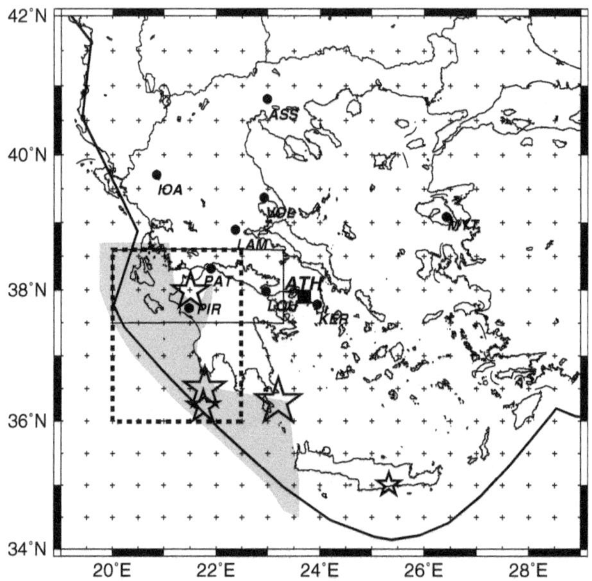

Fig. 7.26 The shaded area shows the up to that date (29 May 2008) addition to the PIR selectivity map. Solid dots show the measuring stations, while the stars denote the epicenters of the EQs discussed in Ref. [21] that were preceded by SES recorded at PIR. The rectangle with solid lines corresponds to the area $N_{37.5}^{38.6}E_{20.0}^{23.3}$ which is the preliminary selectivity map of PAT while the one with broken lines to $N_{36.0}^{38.6}E_{20.0}^{22.5}$, which is also shown (in red) in Fig. 7.20. Taken from Ref. [21].

7.3 Summary of all SES predictions issued along with all earthquakes of magnitude $M_w \geq 6.0$ in Greece since 2001

Table 7.1, as mentioned in § 7.2, compiles the information on what happened before *all* EQs with $M_s(\mathrm{ATH}) \geq 6.0$ that occurred in Greece within the area $N_{36}^{41}E_{19}^{27}$ since 2001. We clarify that this Table also compiles *all* the predictions issued since 2001 considering that a prediction is issued *only* when the expected magnitude (on the basis of the SES amplitude) is $M_s(\mathrm{ATH}) \geq 6.0$ (see § 7.2).

An inspection of Table 7.1 along with the contents of § 7.2.1 to § 7.2.6, which explain what happened before each of the major EQs with $M_w \geq 6.4$, leads to the following main conclusions:

(a) Concerning the natural time results of both the most significant SES activities and their subsequent seismicities until the corresponding mainshock since 2001: the results (see Table 7.1) reveal that in *all* cases but one (i.e., the one in 2003 in which the body wave magnitude mb = 5.6) with $M_w \geq 6.0$, natural time analysis enabled the classification of the relevant SES activity. This was documentated publicly well *before* the EQ occurrence.

(b) The cases mentioned in (a), include *all* five major EQs with $M_w \geq 6.4$ related to four mainshocks (Fig. 7.8). In each of these mainshocks, the occurrence time was identified within a narrow range, a few days to around one week or so, by analyzing in natural time the seismicity after the initiation of the SES activity. The same holds for the two major EQs during the previous decade (1990–2000, see Fig. 7.8) as shown by natural time analysis carried out in retrospect.

7.4 The volcanic-seismic swarm activity in 2000 in the Izu Island region, Japan

SES experimentation has been carried out by Uyeda and coworkers (e.g., Uyeda [27], Uyeda et al. [33, 29, 32], Orihara et al. [18]). The study has been made in two stages: In Stage 1 (1987–1995), only long dipole ($L = 1$–10 km) networks were used (Kinoshita et al. [9]; Takahashi et al. [24]; Nagao et al. [17]). In Stage 2, i.e., since 1996, short ($L \approx 100$ m) dipoles have been also installed. Several precursory changes similar to those observed in Greece have been recorded. They have been summarized by Uyeda et al. [33, 34] as well as in pp. 34–37 of Ref. [35].

Below we focus on the natural time analysis of the preseismic electrical anomalous changes and the seismicity observed in the 2000 swarm in Izu Island region, Japan.

This study by Uyeda et al. [32] is important because the nature of both seismic and electrical activities is vastly different from the Greek cases, i.e., the number of EQs subsequent to the initiation of the electrical disturbance was almost two orders of magnitude larger and the duration of electrical activity was around one order of magnitude longer than in Greek cases. Moreover, the swarm in the Izu Island region was considered closely related to volcanic/magmatic activity in contrast to the Greek cases.

In this Section we closely follow Uyeda et al. [32].

7.4.1 Natural time analysis of the precursory electric signals

The data collected. In the Izu Island region, a map of which is given in Fig. 7.27, electrical measurements were carried out in Niijima Island by means of 16 measuring electric dipoles (long and short ones) with sampling rate $f_{exp} = 0.1$ Hz. Anomalous electrical changes were recorded [29] at two of these electric dipoles. Niijima Island is usually electrically almost noise free (Figs. 7.28(a) and 7.28(d)), the long (≈6 km) dipole "Wak-Air" connecting Wak (Wakago Village) and Air (Airport) and the short (≈30 m) dipole "Wak" in Wakago Village started to show innumerable visually clear unusual changes from 2 months before the onset of the swarm activity (i.e., on April 26, 2000) as illustrated in Figs. 7.28(b) and 7.28(c). Figure 7.29 shows the 3-year records of daily spectrum intensity at 0.01 ± 0.003 Hz after reducing noises common to "Air-Boe" dipole which showed no unusual changes by taking the intensity ratio of "Wak-Air" and "Air-Boe" dipoles. These two dipoles are almost in the same NS direction (see Fig. 7.27). They showed similar noises, mainly due to geomagnetic variations [29], while only "Wak-Air" dipole showed the unusual changes. In Fig. 7.29, it is clear that the anomalous changes were enhanced after the swarm activity started until the monitoring was interrupted in July and August 2000 by power failure caused by EQ shaking and typhoons.

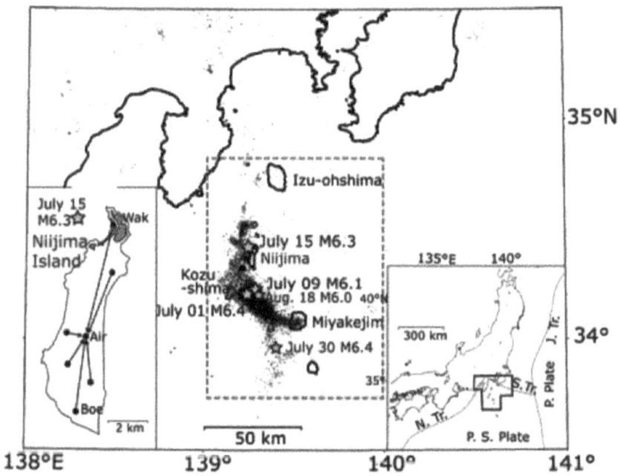

Fig. 7.27 Index map of the Izu Island region. Dots are $M_{JMA} \geq 0$ EQs from 1 June to September 30, 2000. Stars are $M_{JMA} \geq 6$ EQs. Right inset is a map of Japan with plate boundaries. P. plate, Pacific Plate; P. S. Plate, Philippine Sea Plate; N. Tr., Nankai Trough; S. Tr., Sagami Trough. Left inset shows the long dipole configuration of Niijima Island. A short dipole (not shown) is also installed at the far end of each long dipole centered at Air (Airport). Shaded parts near Wak are basaltic exposures. Broken rectangle shows the region of seismicity study $N^{34.8}_{33.7}E^{140}_{139}$. Taken from Ref. [32]. Copyright (2009), American Geophysical Union. Reproduced/modified by permission of American Geophysical Union.

These anomalous changes observed on almost perpendicularly oriented "Wak-Air" long and "Wak" short dipoles cannot be attributed to any source of "artificial" noise in this island of small population and no industry. Furthermore, the observed changes cannot be related with electrode noises, because the two dipoles were independent without a common electrode. As already mentioned in § 1.3.4, it is not uncommon that SES-sensitive sites are locally highly selective which means most sites are insensitive and a sensitive site is found only after a painstaking search through repeatedly moving temporary observation network, e.g. see Refs. [36, 10]. Moreover, as pointed out by Uyeda et al. [32], young basaltic rocks are exposed only at Wak area on the Island which otherwise exclusively consists of less conductive rhyolitic rocks (see the inset in Figure 7.27), suggesting highly heterogeneous underground electrical structure typical of a volcanic zone. According to volcanological studies [11, 25], Niijima Island was formed by rhyolitic activity in the Late Pleistocene and the basaltic exposure in Wakago area is less than a few thousand years old, the last basaltic magma phreatic activity being in the 9th century. It might be speculated [32] that the basaltic exposure is connected to the underground magma body, providing possible electrical channel for the transmission of electrical signals. In order to check these conjectures, which seem to be supported by a detailed geoelectrical modeling by Huang and Lin [6], a thorough electromagnetic exploration of the island is needed.

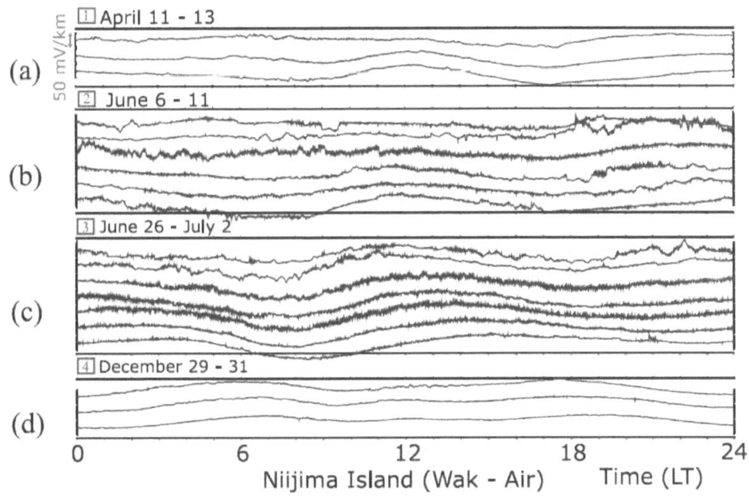

Fig. 7.28 Examples of typical 24-h records of the Wak-Air long dipole potential difference [29] (50 mV/km scale is indicated on the vertical axis). (a) Before 26 April. Records showed mainly smooth variations only. (b) During 2 months before the onset (26 June) of the seismic swarm activity. Numerous anomalous changes occurred. (c) Just after 26 June. Anomalous changes were more conspicuous. (d) After the cessation of the swarm activity, records resumed usual quietness. Time windows 1, 2, 3, and 4 are indicated in Fig. 7.29 (top). Taken from Ref. [32]. Copyright (2009), American Geophysical Union. Reproduced/modified by permission of American Geophysical Union.

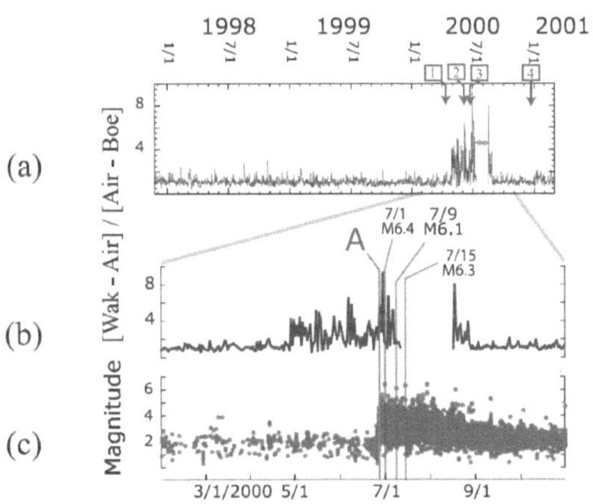

Fig. 7.29 Three-year record [29]. Time change of the 0.01 Hz spectral intensity ratio of geoelectric potential difference at Wak-Air and Air-Boe dipoles, Niijima Island. Anomalous changes started about two months before the seismic swarm (26 June to 29 August). The gap in data was caused by the system failure due to shaking and typhoons in July and August 2000. The numbers 1, 2, 3, and 4 correspond to those in Fig. 7.28. A is the date of the "true" coincidence. Taken from Ref. [32]. Copyright (2009), American Geophysical Union. Reproduced/modified by permission of American Geophysical Union.

Uyeda et al. [32] proceeded to the natural time analysis of the observed anomalous electric signals as follows. They first subtracted the MT background changes by applying a procedure similar to that explained in § 1.4.3.1 and the remaining signal was subsequently analyzed by applying natural time analysis as described earlier in Section 4.11. They found that these electrical disturbances had common characteristic properties with the SES activities in Greece. Thus, Uyeda et al. [32] concluded that they may well be called a SES activity.

7.4.2 Natural time analysis of Izu 2000 seismicity subsequent to the initiation of the SES activity

Uyeda et al. [32] applied the *preliminary* procedure explained in § 7.1.1. By setting natural time zero at the initiation time of the SES activity, analysis of the time series of seismic events in the rectangular region from N33.7° to N34.8° and from E139° to E140° as marked by broken lines (Fig. 7.27) was conducted using the JMA Catalog. In other words, the time series of seismic events in natural time was formed for increasingly longer time windows as the number N of consecutive EQs increased. Then, they computed $\Pi(\phi)$ for each of the time windows and examined its behavior. Specifically, the investigation was made for the period from 15:33 (LT) on April 30 (which was the occurrence time of the first EQ with magnitude greater than 3.5 after the initiation of the SES activity) until just before the occurrence of the first magnitude 6 class EQ very close to Niijima Island (July 1, 2000).

Uyeda et al. [32] used the magnitude in the JMA catalog (M_{JMA}) and employed Eqs. (6.10) to (6.13) to calculate the moment magnitude M_w. Then, the relation [5] $M_0 \propto 10^{1.5M_w}$ was used to obtain the values of the seismic moment M_0, as indicated in Fig. 2.1(b). The spatiotemporal evolution of the seismicity for magnitude threshold

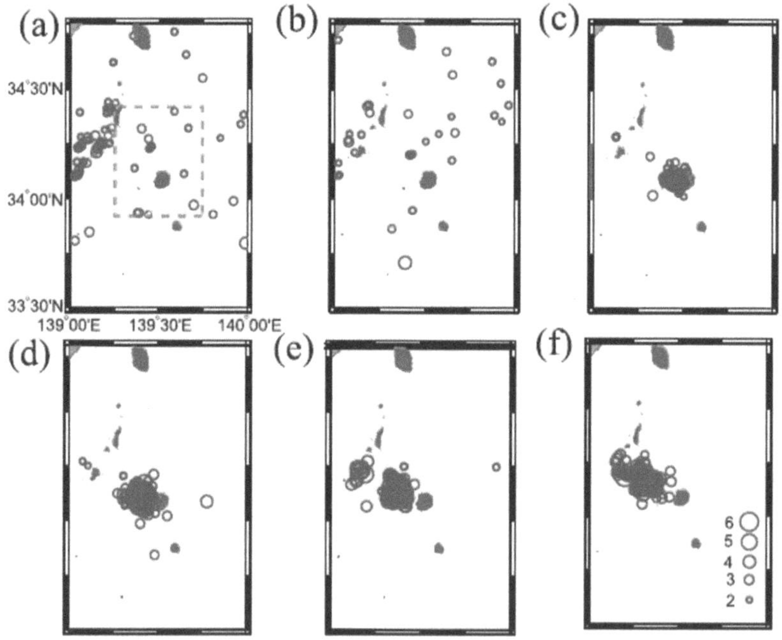

Fig. 7.30 Seismicity for $M_{JMA} \geq 2.0$ in the region $N_{33.7}^{34.8}E_{139}^{140}$ in the study period in 2000: (a) 1 January to 25 April, (b) 26 April to 25 June, (c) 26 June to 14:45 LT on 27 June, i.e., almost until the "true coincidence" (see Fig. 7.31), (d) after the "true coincidence" until 16:47 LT on 28 June, (e) 16:47 LT on 28 June until 15:31 LT on 29 June, and (f) after this, until the first magnitude 6 class EQ on July 1. Inset rectangle shows the smaller study area. Taken from Ref. [32]. Copyright (2009), American Geophysical Union. Reproduced/modified by permission of American Geophysical Union.

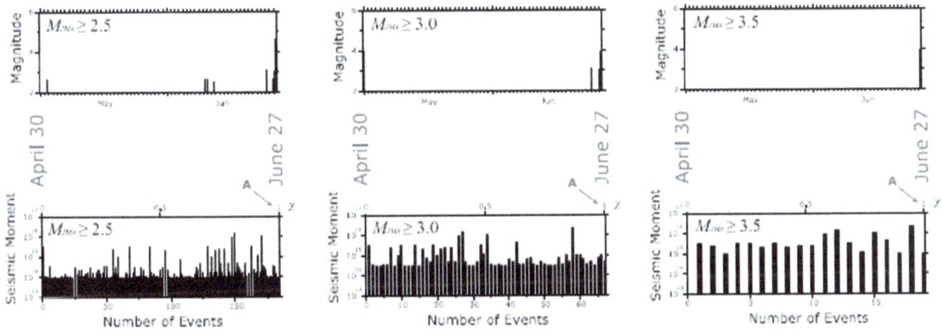

Fig. 7.31 The seismicity after 30 April until the "true coincidence" on 27 June for magnitude thresholds (a) $M_{JMA} \geq 2.5$, (b) $M_{JMA} \geq 3.0$, and (c) $M_{JMA} \geq 3.5$ for (top) conventional time and (bottom) natural time. In the bottom panel, the number of events (instead of χ) is given in the horizontal axis for the reader's convenience. Taken from Ref. [32]. Copyright (2009), American Geophysical Union. Reproduced/modified by permission of American Geophysical Union.

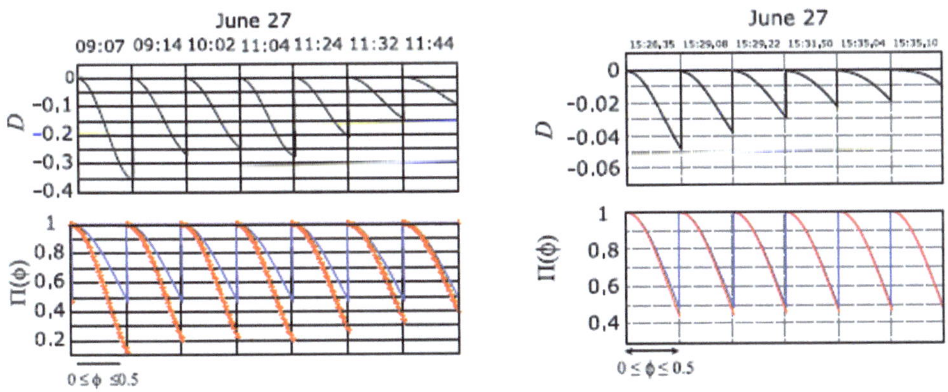

Fig. 7.32 (bottom) Time evolution of $\Pi(\phi)$ for $0 \leq \phi \leq 0.5$ of the seismic activity for $M_{JMA} \geq 3.0$ when the calculation was started on 30 April. $\Pi(\phi)$ curves (red) fall on the theoretical $\Pi(\phi)$ curves (blue) calculated from Eq. (7.1) as critical stage is approached. (top) The difference D between the two curves. (a) Examples for the morning hours of 27 June. (b) At the last six events which occurred at 15:26:35, 15:29:08, 15:29:22, 15:31:50, 15:35:04, and 15:35:10 LT on 27 June until the "true coincidence". Taken from Ref. [32]. Copyright (2009), American Geophysical Union. Reproduced/modified by permission of American Geophysical Union.

$M_{JMA} \geq 2.0$ in the studied region is shown in Figs. 7.30(a) to 7.30(f). The readings of the seismicity in natural and conventional time frames until the coincidence marked A on June 27 are shown in Fig. 7.31 for three different magnitude thresholds. Figures 7.30(a) to 7.30(f) show how nonlinearly the two time frames are interconnected. One may notice that the natural time covered in Figs. 7.30(a) to 7.30(f) is practically from June 26 to June 27, indicating that important changes took place in a short period even before the bulk of the swarm activity (see Fig. 7.29). Figure 7.32 (bottom) clearly shows that for magnitude threshold 3.0 as an example, the computed $\Pi(\phi)$ curve approaches the critical $\Pi(\phi)$ curve from *below* on June 27, 2000, a few days before the first $M \geq 6$ earthquake of July 1,

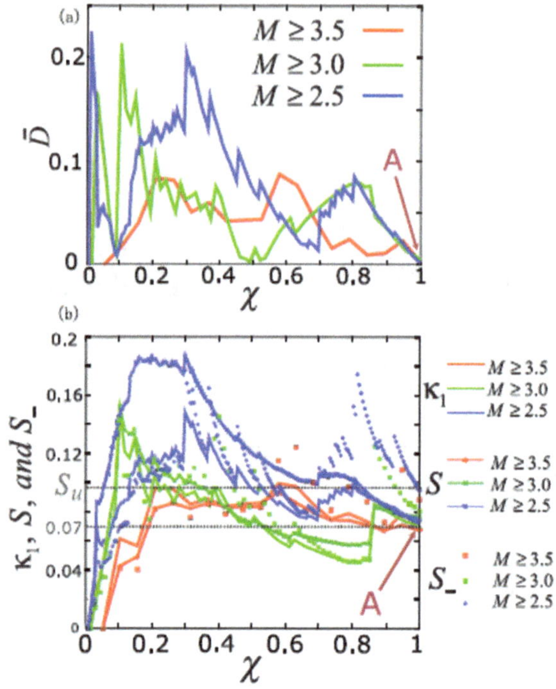

Fig. 7.33 $\overline{D} \equiv \langle D \rangle$, κ_1, S, and S_- of the evolving seismic activity versus the natural time χ. The calculation was started on 30 April and continued until the true coincidence A on 27 June just 4 days before the first $M \geq 6$ class EQ on 1 July. Three magnitude thresholds ($M_{JMA} \geq 2.5$), ($M_{JMA} \geq 3.0$), and ($M_{JMA} \geq 3.5$) are considered. (a) $\overline{D}$ is plotted and (b) the quantities κ_1, S, and S_- are shown with the symbols depicted. Taken from Ref. [32]. Copyright (2009), American Geophysical Union. Reproduced/modified by permission of American Geophysical Union.

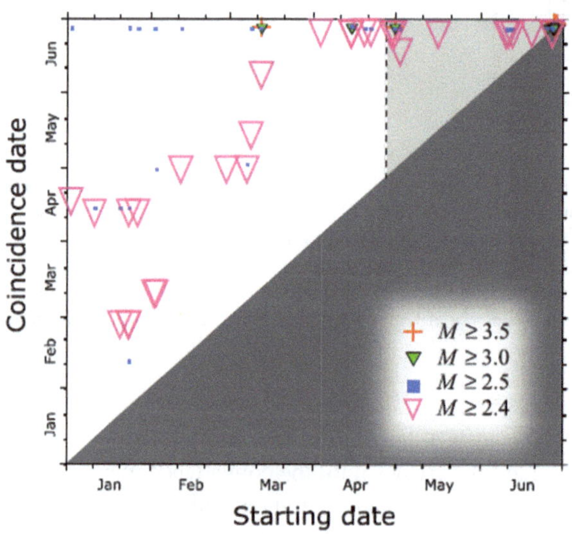

Fig. 7.34 Coincidence date versus the starting date of calculation for four magnitude thresholds. The SES activity started on April 26. The shaded triangular area is irrelevant because the coincidence has to be only in the unshaded area. Taken from Ref. [32]. Copyright (2009), American Geophysical Union. Reproduced/modified by permission of American Geophysical Union.

2000. The approach of the two curves is more clearly demonstrated in the upper panel of Fig. 7.32, in which D is plotted.

Moreover, Fig. 7.33 depicts $\langle D \rangle$, κ_1, S, and S_{-}, as they evolved event by event during the whole period (April 30 to June 27). This figure also shows that all three different magnitude thresholds resulted in approximately the same time of coincidence on June 27, supporting the self-similar structure of the process concerned [32].

As to the spatial self-similar nature of the process, a similar calculation was made for a smaller region depicted in Fig. 7.30(a). The results showed the same behavior. Thus, the coincidence on June 27 is considered as *true* coincidence since all the conditions mentioned in § 7.1.1 are obeyed. It may be added here that in fact, Uyeda et al. [32] made the calculations until the last event before the first $M \geq 6$ class EQ of July 1 and there was another case with $\kappa_1 = 0.070$. But this second case was discarded because it did not satisfy the entropy criterion for true coincidence, i.e., the condition (7.4).

Figure 7.34 shows the coincidence dates (vertical axis) when the calculations were started on the dates shown on the horizontal axis for four magnitude thresholds. The calculation with M2.4 threshold was added here in order to check if the *true* coincidence A recognized by the abovementioned three threshold calculations satisfies the magnitude threshold invariance even for M2.4 threshold. Figure 7.34 clearly shows that *true* coincidence is reached at a time close to the date of the first $M \geq 6$ shock, i.e., late June, only when the calculation was started around the initiation date of the SES activity, which is indicated by a vertical broken line in Fig. 7.34. It was found that the self-similarity condition for M2.4 threshold was useful for identifying *true* coincidence. One may wonder if the uncertainty in magnitude (or moment) determination bothers this kind of analysis.

Hence, Uyeda et al. [32] have conducted simulation test giving 0.2–0.3 random error of magnitude and concluded that the date of the *true* coincidence is not affected.

7.4.3 Main conclusions from the study of the Izu 2000 case

Uyeda et al. [32], after analyzing in natural time both the SES activity started on April 26, 2000, as well as the subsequent seismicity, as explained in § 7.4.1 and § 7.4.2 respectively, obtained the following main conclusions:

First, before the first magnitude 6 class EQ on July 1, one *true* coincidence was observed on June 27. Thus, the analysis in the natural time domain of the seismicity led to an estimation on the date of the impending large EQ of July 1, 2000, with a narrow time window of the order of a few days.

Second, it has been demonstrated that starting the calculation more than 2 weeks earlier than the initiation time of the SES activity does *not* result in *true* coincidence, whereas starting the calculation at later time does so. This is consistent with Greek cases in which natural time zero was set at the time of SES activity initiation.

7.5 Results from California: the $M_s7.1$ Loma Prieta earthquake on October 18, 1989

This is the best-known case in the USA for which clear precursory electromagnetic variations have been reported. Almost one month before this earthquake, i.e., on September 12, 1989, magnetic field variations were recorded at a site just 7 km from the earthquake epicenter [4, 1] similar to those accompanying the SES activities in Greece for earthquakes with M6.5 or larger [56] (see § 1.3.6).

Table 7.3 The seismic data (reported from the Northern California Earthquake Data Center, http://www.ncedc.org/ncedc/catalog-search.html, as they appeared on January 8, 2010) analyzed in natural time. The magnitude M corresponds either to M_L or M_D. It is converted to seismic moment according to $M_w = \mathrm{M}$. Taken from Ref. [53].

Number	Magnitude M	Date	Time(UT)	Latitude	Longitude
1	2.7	1989/9/16	18:41:24	37.33	−121.70
2	3.2	1989/9/28	15:42:37	36.57	−121.11
3	2.7	1989/10/1	12:21:37	38.15	−121.90
4	3.0	1989/10/1	13:10:24	38.14	−121.93
5	3.2	1989/10/1	13:19:27	38.16	121.93
6	3.1	1989/10/1	22:08:35	36.56	−121.15
7	3.1	1989/10/1	22:09:17	36.56	−121.15
8	2.7	1989/10/2	11:20:19	38.15	−121.91
9	2.6	1989/10/6	15:53:36	37.32	−122.11
10	3.3	1989/10/8	12:36:46	36.44	−121.01
11	2.7	1989/10/9	11:51:24	37.63	−121.70
12	2.7	1989/10/9	12:06:02	37.29	−122.09
13	3.1	1989/10/9	12:42:03	37.63	−121.69
14	2.8	1989/10/13	12:22:11	36.63	−121.08
15	7.0	1989/10/18	00:04:15	37.04	−121.88

Following Ref. [53], in order to determine the occurrence time of the impending mainshock, we analyze in natural time all the earthquakes (see Table 7.3) that occurred after September 12, 1989, which is the date of the initiation of the aforementioned (SES like) precursory magnetic field change, within the area A: $\mathrm{N}^{38.5}_{36.2}\mathrm{W}^{120.7}_{122.7}$ surrounding the Loma Prieta earthquake epicenter. The seismic data used here are from the Northern California Earthquake Data Center and the relevant epicenters are depicted in Fig. 7.35. We set the natural time zero at the initiation time of the magnetic field change, and then formed time

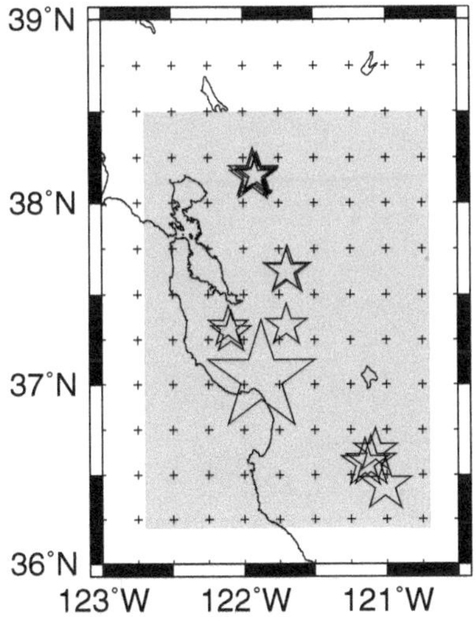

Fig. 7.35 The area $N_{36.2}^{38.5}W_{122.7}^{120.7}$ (shaded) surrounding the epicenter of the Loma Prieta earthquake (largest star) in which the seismicity after the initiation on September 12, 1989, of the precursory magnetic field variations [4, 1] is analyzed in natural time. Taken from Ref. [53].

series for the area A each time a small earthquake (with magnitude M exceeding a certain threshold M_{thres}, i.e., $M \geq M_{thres}$) occurred. The quantity κ_1 for each of the time series was computed for the pairs (χ_k, Q_k). The quantity Q_k was taken as the seismic moment M_{0k} of the k-th event (see Fig. 2.1(b)), calculated from the relation $\log_{10} M_0 \approx 1.5 M_L + \text{const.}$ (H. Kanamori, personal communication).

Applying the *updated* procedure (§ 7.1.2), in order to check whether criticality has been approached at the occurrence of a new event k within the area A, we construct all the possible subareas of $A_{M_{thres}}$ that necessarily include the event k and examine whether their κ_1 values reveal a probability distribution Prob(κ_1) maximized at 0.070. We considered only earthquakes with $M > 2.5$ in order to have homogeneous and complete catalog (see Ref. [2]). In other words, we take $M_{thres} = 2.6$. The results are depicted in Fig. 7.36(a), which shows how Prob(κ_1) versus κ_1 evolves upon the occurrence of each event before the October 18, 1989, M_s7.1 Loma Prieta earthquake. We see that Prob(κ_1) maximizes at $\kappa_1 = 0.070$ upon the occurrence of a 2.8 event at 12:22 UT on October 13, 1989, i.e., almost 5 days before the main shock. Upon repeating the calculation for larger magnitude thresholds, i.e., $M_{thres} = 2.7$ and 2.8, see Figs. 7.36(b) and 7.36(c), respectively, we find again that the maximum of Prob(κ_1) versus κ_1 is observed at $\kappa_1 = 0.070$ on October 13, 1989.

In summary, we analyzed in natural time the small earthquakes that occurred after the initiation on September 12, 1989, of the (SES-like) magnetic field variations in the area surrounding the epicenter of the M_s7.1 Loma Prieta earthquake. We find that Prob(κ_1) versus κ_1 exhibits a maximum at $\kappa_1 = 0.070$, for $M_{thres} = 2.6$, 2.7 and 2.8, on October 13, 1989, i.e., *five days before* the occurrence of the mainshock.

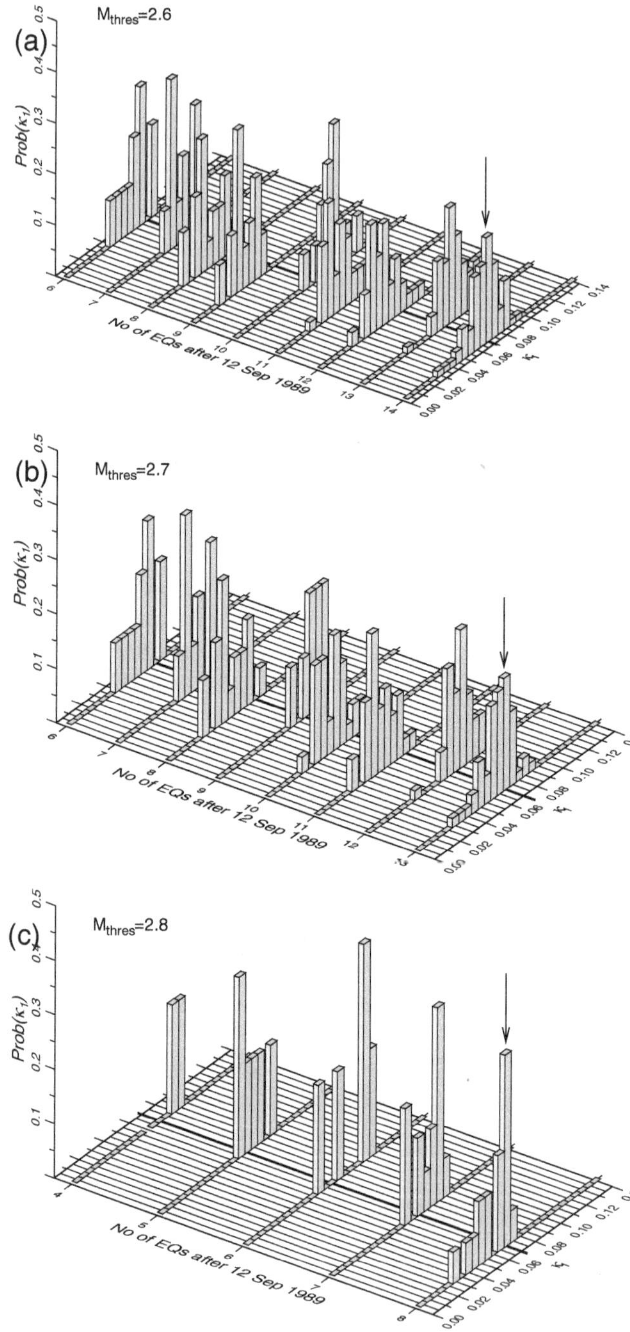

Fig. 7.36 Detemination of the occurrence time of the M_s7.1 Loma Prieta EQ on October 18, 1989. Prob(κ_1) versus κ_1 for the seismicity in the area $N^{38.5}_{36.2}W^{120.7}_{122.7}$, for $M_{thres} = 2.6$ (a), $M_{thres} = 2.7$ (b) and $M_{thres} = 2.8$ (c) subsequent to the initiation on September 12, 1989, of the precursory (SES like) magnetic field variations reported in Refs. [4, 1]. The last event corresponds to the magnitude 2.8 earthquake that occurred at 12:22 UT on October 13, 1989 with an epicenter at 36.63°N 121.08°W (see Table 7.3). Taken from Ref. [53].

References

1. Bernardi, A., Fraser-Smith, A.C., McGill, P.R., Villard, O.G.: ULF magnetic field measurements near the epicenter of the Ms 7.1 Loma Prieta earthquake. Physics of The Earth and Planetary Interiors **68**, 45–63 (1991)
2. Davidsen, J., Grassberger, P., Paczuski, M.: Networks of recurrent events, a theory of records, and an application to finding causal signatures in seismicity. Phys. Rev. E **77**, 066104 (2008)
3. Dologlou, E., Hadjicontis, V., Mavromatou, C.: Electrical precursors of earthquakes in Aegean sea during the last decade (1997-2007). Natural Hazards and Earth System Science **8**, 123–128 (2008)
4. Fraser-Smith, A.C., Bernardi, A., McGill, P.R., Ladd, M.E., Helliwell, R.A., Villard, O.G.: Low-frequency magnetic-field measurements near the epicenter of the Ms-7.1 Loma Prieta earthquake. Geophys. Res. Lett. **17**, 1465–1468 (1990)
5. Hanks, T.C., Kanamori, H.: Moment magnitude scale. J. Geophys. Res. **84(B5)**, 2348 (1979)
6. Huang, Q., Lin, Y.: Selectivity of Seismic Electric Signal (SES) of the 2000 Izu earthquake swarm:a 3D FEM numerical simulation model. Proc. Jpn. Acad., Ser. B **86**, 257–264 (2010)
7. Jaume, S.C., Sykes, L.R.: Evolving towards a critical point: A review of accelerating seismic moment/energy release prior to large and great earthquakes. Pure Appl. Geophys. **155**, 279 (1999)
8. Kerr, R.A.: Quake prediction tool gains ground. Science **270**, 911–912 (1995)
9. Kinoshita, M., Uyeshima, M., Uyeda, S.: Earthquake prediction research by means of telluric potential monitoring, Progress Rep. 1. Bull. Earthq. Res. Inst. **64**, 255–311 (1989)
10. Kondo, S., Uyeda, S., Nagao, T.: The selectivity of the Ioannina VAN station. J. Geodynamics **33**, 433–461 (2002)
11. Koyaguchi, T.: Evidence for two-stage mixing in magmatic inclusions and rhyolitic lava domes on Niijima island, Japan. Journal of Volcanology and Geothermal Research **29**, 71–98 (1986)
12. Lighthill, J.: A brief look back at the Review Meeting's Proceedings. In: Sir J. Lighthill (ed.) The Critical Review of VAN: Earthquake Prediction from Seismic Electric Signals, pp. 349–356. World Scientific, Singapore (1996)
13. Lighthill, J.: A brief look forward to future research needs. In: Sir J. Lighthill (ed.) The Critical Review of VAN: Earthquake Prediction from Seismic Electric Signals, pp. 373–376. World Scientific, Singapore (1996)
14. Masood, E.: Court charges open splite in Greek earthquake experts. Nature **377**, 375 (1995)
15. Masood, E.: Greek earthquake stirs controversy over claims for prediction method. Nature **375**, 617 (1995)
16. Monastersky, R.: Electric signals may herald earthquakes. Science News **148**, 260 (1995)
17. Nagao, T., Uyeda, S., Asai, Y., Kono, Y.: Anomalous changes in geoelectric potential preceding four earthquakes in Japan. In: Sir J.Lighthill (ed.) The Critical Review of VAN: Earthquake Prediction from Seismic Electric Signals, pp. 292–300. World Scientific, Singapore (1996)
18. Orihara, Y., Noda, Y., Nagao, T., Uyeda, S.: A possible case of SES selectivity at Kozu-shima island. J. Geodynamics **33**, 425–432 (2002)
19. Roumelioti, Z., Kiratzi, A., Benetatos, C.: The instability of M_w and M_L comparison for earthquakes in Greece for the period 1969 to 2007. Journal of Seismology **14**, 309–337 (2010)
20. Sarlis, N.V., Skordas, E.S., Lazaridou, M.S., Varotsos, P.A.: Investigation of the seismicity after the initiation of a Seismic Electric Signal activity until the main shock. arXiv:0802.3329v2 [cond-mat.stat-mech] (20 March 2008)
21. Sarlis, N.V., Skordas, E.S., Lazaridou, M.S., Varotsos, P.A.: Investigation of seismicity after the initiation of a Seismic Electric Signal activity until the main shock. Proc. Japan Acad., Ser. B **84**, 331–343 (2008)
22. Sarlis, N.V., Skordas, E.S., Lazaridou, M.S., Varotsos, P.A.: Investigation of the seismicity after the initiation of a Seismic Electric Signal activity until the main shock. arXiv:0802.3329v4 [cond-mat.stat-mech] (29 May 2008)
23. Scafetta, N., West, B.J.: Multiscaling comparative analysis of time series and a discussion on "earthquake conversations" in California. Phys. Rev. Lett. **92**, 138501 (2004)
24. Takahashi, I., Nagao, T., Uyeda, S.: On some geoelectric potential changes in Naha, Okinawa, Japan and their possible relationship with nearby earthquakes (in Japanese with English abstract). Bull. Inst. Oceanic Res. and Develop. **20**, 31–40 (1999)

25. Tsukui, M., Saito, K., Hayashi, K.: Frequent and intensive eruptions in the 9th century, Izu Islands, Japan: Revision of volcano- stratigraphy based on tephras and historical document (in Japanese with English abstract). Bull. Volcanol. Soc. Jpn. **51**, 327–338 (2006)
26. USGS: (2010). See the United States Geological Survey (USGS) earthquake search web page http://neic.usgs.gov/neis/epic/epic.html for the relevant seismic catalogs
27. Uyeda, S.: VAN method of short-term earthquake prediction shows promise. EOS Trans. AGU **79**, 573–580 (1998)
28. Uyeda, S., Al-Damegh, E., Dologlou, E., Nagao, T.: Some relationship between VAN Seismic Electric Signals (SES) and earthquake parameters. Tectonophysics **304**, 41–55 (1999)
29. Uyeda, S., Hayakawa, M., Nagao, T., Molchanov, O., Hattori, K., Orihara, Y., Gotoh, K., Akinaga, Y., Tanaka, H.: Electric and magnetic phenomena observed before the volcano-seismic activity in 2000 in the Izu Island Region, Japan. Proc. Natl. Acad. Sci. USA **99**, 7352–7355 (2002)
30. Uyeda, S., Kamogawa, M.: The prediction of two large earthquakes in Greece. EOS Trans. AGU **89**, 363 (2008)
31. Uyeda, S., Kamogawa, M.: Reply to a Comment on 'The prediction of two large earthquakes in Greece'. EOS Trans. AGU **91**, 163 (2010)
32. Uyeda, S., Kamogawa, M., Tanaka, H.: Analysis of electrical activity and seismicity in the natural time domain for the volcanic-seismic swarm activity in 2000 in the Izu Island region, Japan. J. Geophys. Res. **114**, B02310 (2009)
33. Uyeda, S., Nagao, T., Orihara, Y., Yamaguchi, T., Takahashi, I.: Geoelectric potential changes: Possible precursors to earthquakes in Japan. Proc. Natl. Acad. Sci. USA **97**, 4561–4566 (2000)
34. Uyeda, S., Nagao, T., Tanaka, H.: A report from the RIKEN international frontier research project on earthquakes. Terr. Atmos. Ocean. Sci. **15**, 269–310 (2004)
35. Varotsos, P.: The Physics of Seismic Electric Signals. TERRAPUB, Tokyo (2005)
36. Varotsos, P., Alexopoulos, K., Lazaridou, M.: Latest aspects of earthquake prediction in Greece based on Seismic Electric Signals, II. Tectonophysics **224**, 1–37 (1993)
37. Varotsos, P., Eftaxias, K., Lazaridou, M., Nomicos, K., Sarlis, N., Bogris, N., Makris, J., Antonopoulos, G., Kopanas, J.: Recent earthquake prediction results in Greece based on the observation of Seismic Electric Signals. Acta Geophys. Pol. **44**, 301–327 (1996)
38. Varotsos, P., Lazaridou, M.: Latest aspects of earthquake prediction in Greece based on Seismic Electric Signals. Tectonophysics **188**, 321–347 (1991)
39. Varotsos, P., Lazaridou, M., Eftaxias, K., Antonopoulos, G., Makris, J., Kopanas, J.: Short term earthquake prediction in Greece by Seismic Electric Signals. In: Sir J. Lighthill (ed.) The Critical Review of VAN: Earthquake Prediction from Seismic Electric Signals, pp. 29–76. World Scientific, Singapore (1996)
40. Varotsos, P., Sarlis, N., Bogris, N., Makris, J., Kapiris, P., Abdulla, A.: A comment on the $\Delta V/L$-criterion for the identification of Seismic Electric Signals, pp. 1–45. TERRAPUB, Tokyo (1999)
41. Varotsos, P., Sarlis, N., Skordas, E.: A note on the spatial extent of the Volos SES sensitive site. Acta Geophys. Pol. **49**, 425–435 (2001)
42. Varotsos, P.A.: Recent Seismic Electric Signals (SES) activities in Greece. Acta Geophys. Pol. **54**, 158–164 (2006)
43. Varotsos, P.A.: What happened before the last five strong earthquakes in Greece. Proc. Jpn. Acad., Ser. B: Phys. Biol. Sci. **82**, 86–91 (2006)
44. Varotsos, P.A., Sarlis, N.V., Skordas, E.S.: Seismic Electric Signals and $1/f$ "noise" in natural time. arXiv:0711.3766v3 [cond-mat.stat-mech] (1 February 2008)
45. Varotsos, P.A., Sarlis, N.V., Skordas, E.S.: Spatio-temporal complexity aspects on the interrelation between Seismic Electric Signals and seismicity. Practica of Athens Academy **76**, 294–321 (2001)
46. Varotsos, P.A., Sarlis, N.V., Skordas, E.S.: Seismic Electric Signals and seismicity: On a tentative interrelation between their spectral content. Acta Geophys. Pol. **50**, 337–354 (2002)
47. Varotsos, P.A., Sarlis, N.V., Skordas, E.S.: Detrended fluctuation analysis of the magnetic and electric field variations that precede rupture. CHAOS **19**, 023114 (2009)
48. Varotsos, P.A., Sarlis, N.V., Skordas, E.S., Lazaridou, M.S.: Fluctuations, under time reversal, of the natural time and the entropy distinguish similar looking electric signals of different dynamics. J. Appl. Phys. **103**, 014906 (2008)

49. Varotsos, P.A., Sarlis, N.V., Skordas, E.S., Tanaka, H.K., Lazaridou, M.S.: See (the freely available) EPAPS Document No. E-PLEEE8-74-190608 originally from P.A. Varotsos, N.V. Sarlis, E.S. Skordas, H.K. Tanaka and M.S. Lazaridou *Phys. Rev. E* **74**, 021123 (2006). For more information on EPAPS, see http://www.aip.org/pubservs/epaps.html.
50. Varotsos, P.A., Sarlis, N.V., Skordas, E.S., Tanaka, H.K., Lazaridou, M.S.: Attempt to distinguish long-range temporal correlations from the statistics of the increments by natural time analysis. Phys. Rev. E **74**, 021123 (2006)
51. Varotsos, P.A., Sarlis, N.V., Skordas, E.S., Tanaka, H.K., Lazaridou, M.S.: Entropy of seismic electric signals: Analysis in the natural time under time reversal. Phys. Rev. E **73**, 031114 (2006)
52. Varotsos, P.A., Sarlis, N.V., Skordas, E.S., Tanala, H.K., Lazaridou, M.S.: Additional evidence on some relationship between Seismic Electric Signals and earthquake source parameters. Acta Geophys. Pol. **53**, 293–298 (2005)
53. Varotsos, P.A., Sarlis, N.V., Skordas, E.S., Uyeda, S., Kamogawa, M.: Natural time analysis of critical phenomena. the case of seismicity. EPL **92**, 29002 (2010)
54. Varotsos, P.A., Sarlis, N.V., Tanaka, H.K., Skordas, E.S.: Similarity of fluctuations in correlated systems: The case of seismicity. Phys. Rev. E **72**, 041103 (2005)
55. Varotsos, P.A., Sarlis, N.V., Tanaka, H.K., Skordas, E.S.: Some properties of the entropy in the natural time. Phys. Rev. E **71**, 032102 (2005)
56. Varotsos, P.A., Sarlis, N.V., Skordas, E.S.: Electric fields that "arrive" before the time derivative of the magnetic field prior to major earthquakes. Phys. Rev. Lett. **91**, 148501 (2003)

8. Natural Time Analysis of Dynamical Models

Abstract. We apply here the natural time analysis to the time series of the avalanches in several SOC models as well as to other dynamical models. First, in a simple deterministic SOC system introduced to describe avalanches in stick–slip phenomena that belongs to the same universality class as the "train" model for earthquakes introduced by Burridge and Knopoff, we find that the value $\kappa_1 = 0.070$ can be considered as quantifying the extent of the organization of the system at the onset of the critical stage. Second, in the conservative case of the Olami Feder–Christensen (OFC) earthquake model, the value $\kappa_1 = 0.070$ is accompanied by an abrupt exponential increase of the avalanche size which is indicative of the approach to a critical behavior. In the non-conservative case of OFC, in the later part of the transient period, coherent domains of the strain field gradually develop accompanied by κ_1 values close to 0.070. Furthermore, there is a non-zero change ΔS of the entropy in natural time under time reversal, thus reflecting predictability in the OFC model. Third, an explanation for the validity of the condition $\kappa_1 = 0.070$ for critical systems on the basis of the dynamic scaling hypothesis is forwarded. Fourth, when quenching the 2D Ising model at temperatures close to but below T_c, which is qualitatively similar with the pressure stimulated currents SES generation model, and set $Q_k = |M_k|$ (where M_k stands for the evolution of the magnetization per spin), we find $\kappa_1 = 0.070$. Fifth, in a deterministic version of the original Bak–Tang–Wiesenfeld sandpile model, the value $\kappa_1 \approx 0.070$ is reached during the transient to the self-organized criticality. Finally, natural time analysis of the avalanches observed in laboratory experiments on three-dimensional ricepiles and on the penetration of the magnetic flux into thin films of high T_c superconductors, leads to κ_1 values around $\kappa_1 = 0.070$. A further investigation of the experiment on ricepiles reveals that the *sequential order* of the avalanches captured by the natural time analysis is of profound importance for establishing the SOC state and constitutes the basis for the observation of the result $\kappa_1 \approx 0.070$.

8.1 Is self-organized criticality (SOC) compatible with prediction? Recent aspects. The models analyzed here in natural time

The SOC concept, that has been originally introduced by Bak, Tang and Wiesenfeld [6] using as an example the sandpile model, was an attempt to explain the ubiquity of scale invariance in nature (see also Refs. [44, 74]). Systems, in general, are termed self-organized critical if they reach a stationary state (after a transient during which the system acquires criticality [32, 33]) characterized by power laws without the need for fine-tuning an external parameter, for example the temperature or pressure. There is more or less a general tendency [85] on confining the term self-organized critical to those systems that are slowly driven and that display fast, avalanche-like dissipation events. In other words, in SOC systems, the competition between a driving force that very slowly injects energy and the dynamics of local thresholds can drive the system into a *critical state* where a minor perturbation can trigger an avalanche of any size and duration [6, 7, 66]. In particular, in the original sandpile model, the random, slow addition of "blocks" in a two-dimensional lattice (along with a local conservation law) drives the system into a critical state, where power law distributed avalanches maintain a steady regime far from equilibrium.

The fact that avalanches were taken [6] as uncorrelated in the original SOC sandpile model, has been used as an argument that is not possible to predict the occurrence of large avalanches (relevant claims are cited in Refs. [66, 65]). A belief has been expressed that power law distributed avalanches are inherently unpredictable, which came from the concept of SOC, but interpreted in the way that, at any moment, any small avalanche can eventually cascade to a large event.

However, prediction is possible, because the system is not *at*, but *close to*, the critical state [21, 66].

This, became clear from the accumulated theoretical and experimental evidence, which could be summarized as follows (see also § 6.5.3). First, some cellular automaton SOC models have been analyzed for the predictability of very large avalanches (responsible for the cut-off on the power law distribution) [62] and in addition precursors of large events have been identified [41, 67] in dissipative or hierarchical lattices. Second, the prediction of extreme avalanches in self-organized critical sandpiles have been studied in recent detailed numerical studies [32] which showed that: (a) particularly large events in a close to SOC system can be predicted on the basis of past observations; (b) the predictive power stems from temporal correlations which are pure finite size effects, i.e., it disappears in the infinite system size limit as all avalanches become independent of each other; (c) under variation of the system size, predictability persists if the magnitude used to define extreme events is scaled linearly in the maximal possible avalanche size. It was also clarified [33] that SOC seems to be an unsuitable mechanism for the explanation of the extreme events that occur in clusters. Third, experimental work has recently demonstrated [66] the possibility of avalanche prediction in the classical SOC paradigm, i.e., a pile of grains: by knowing the position of every grain in a two-dimensional pile, avalanches of moving grains do follow a distinct power law distribution, but large avalanches are found [66] to

Table 8.1 Compilation of the 14 cases described in this monograph in which the condition $\kappa_1 = 0.070$ has been ascertained.

No.	Case	Class
1	SES activities (Section 4.2)	Field experiments
2	Seismicity preceding major EQs (Sections 7.1 to 7.5)	"
3	Ricepiles (§ 8.5.1)	Laboratory measurements
4	Magnetic flux avalanches in high T_c superconductors (§ 8.5.2)	"
5	Burridge & Knopoff "train" EQ model (§ 8.2.2)	Dynamical models
6	Olami–Feder–Christensen EQ model (conservative case, § 8.3.2) and "foreshocks" in the non-conservative case (§ 8.3.3)	"
7	Dynamic scaling hypothesis with $z = 2.0$–2.4 (Section 8.4, § 8.4.3)	"
8	Deterministic version of the original sandpile SOC model (§ 8.4.2)	"
9	Generalized stochastic directed SOC model (§ 8.5.2)	"
10	2D Ising model quenched close to, but below, T_c (§ 8.4.1)	Other models
11	Pressure stimulated currents model for SES generation (§ 2.4.2)	"
12	fBm time series with DFA exponent $\alpha = 1$ (§ 3.4.3)	"
13	Stochastic Cantor set: p-model describing turbulence (§ 6.2.5)	"
14	Power law distributed uncorrelated energy bursts with $\gamma = 1.87$ (§ 2.5.4, Fig. 2.6)	"

be preceded on the average by continuous detectable variations in the internal structure of the pile.

To answer the aforementioned question on predictability, in this Chapter we will analyze in natural time the time series of avalanches in some dynamical models including typical SOC examples like sandpiles, as well as in laboratory measurements on ricepiles and magnetic flux penetration in high T_c superconductors which are believed to be SOC systems.

Table 8.1 includes all 14 cases discussed in this monograph where the condition $\kappa_1 \approx 0.070$ has been ascertained, thus strengthening the conjecture that: *if* a system acquires criticality, the condition $\kappa_1 = 0.070$ holds (but *not* the inverse as for example case No. 14).

8.2 Natural time analysis of the Burridge & Knopoff "train" earthquake model

8.2.1 The earthquake model proposed by Burridge & Knopoff. The "train" model. Introduction

An earthquake is a stick–slip dynamical instability of a pre-existing fault driven by the motion of a tectonic plate [71, 72]. A relatively simple dynamical model that contains much of the essential physics of earthquake faults is the so-called spring-block model originally proposed [14] by Burridge and Knopoff (BK). It consists of an assembly of blocks, each of which is connected via elastic springs to the nearest neighboring blocks.

The blocks are also connected to the driving plate by elastic springs and rest on a surface with a velocity-weakening stick–slip friction force (note that the friction force decreases as the velocity is increased). When the force acting on a block overcomes the static friction with the surface, the block slips. Then a redistribution of forces takes place in the neighbors that eventually trigger new displacements. An EQ event is defined as a cluster of blocks that move (slip) due to the initial slip of a single block. A numerical study in one dimension had already been made by BK, and later Carlson, Langer and others [18, 19] proceeded to more extensive studies of the one-dimensional and two-dimensional BK models focusing on the magnitude distribution of EQ events. Spatiotemporal correlations of the 2D BK model have been studied [55] by considering also long-range inter-block interactions.

In the BK model studied by Carlson, Langer and others, each block is connected, as mentioned above, to the driving element. To model the dynamics of EQs, Burridge and Knopoff in their original work [14] also studied the case of a chain of blocks (situated on a rough surface with friction) connected by elastic springs and pulled only at one end with a constant small velocity. The dynamics of the model is as follows. All the blocks are initially at rest. As the driver pulls the first block, the latter remains stuck until the elastic force overcomes the static friction. When this occurs, the first block will move a little. Such small events (or EQs) will continue and increase the elastic force on the second block. When the elastic force on the second block overcomes the friction force, an event involving the two blocks will occur. The dynamics continues with events involving three, four, five or all the blocks in the system. This model is usually called the "train" model since it has some similarity with a train, where the driving force is applied only at one end of the chain (e.g., Ref. [75]). The dynamics here is governed by coupled ordinary differential equations which makes its study very time-consuming. To make this system more amenable to computer simulations, de Sousa Vieira [76] introduced a continuous cellular automaton that exhibits SOC and belongs to the same universality class as the "train" model. This *deterministic* one-dimensional model, for the avalanches in stick–slip phenomena, which is very close to the case of an array of connected pendulums first discussed by Bak et al. [6], is defined as follows (see Refs. [76, 26, 70]). Consider a one-dimensional system, where a continuous (force) variable $f_l \geq 0$ is associated with each site l, $l = 1, 2, \ldots, L$. Initially all f_l have the same value f_0 which lies below a threshold f_{th}. One can set $f_{th} = 1.0$ without loss of generality. The basic time step consists of varying the force on the first site according to $f_1 = f_{th} + \delta f$; the system then relaxes with a conservative redistribution of the forces at the site $f_l \geq f_{th}$ (toppling site) and its nearest neighbors according to $f_l = \Psi(f_l - f_{th})$ and $f_{l\pm1} = f_{l\pm1} + \Delta f/2$, where Δf is the change of force at the overcritical site and $\Psi(x)$ a periodic nonlinear function. This condition mimics the redistribution of forces when the block l is displaced (stick–slips) by Δx_l during an 'earthquake' in the "train" model [76]. The relaxation continues until all sites have $f_l < f_{th}$ for all l. The size of the 'earthquake' corresponds to the number of topplings, s, required for the system to relax, and is considered here as the appropriate value of Q_k in natural time. Then, the driving force at the first site sets in again. This is complemented by open boundary conditions; i.e., the force is 'lost' at $l = 1$ and $l = L$. The nonlinear periodic function used here (which means that, when considering that the force supposed mimics the net effect of the two forces in the "train" model, i.e., the elastic and the friction forces, the periodicity of the elastic force dominates over the form of the friction force) is similar to the one used in Refs. [76, 26], i.e., a sawtooth function $\Psi(x) = 1 - ax + [ax]$, where $[\ldots]$ denotes the integer part of ax and a is a number. It was shown [76] that such a system evolves to a SOC state where the avalanche distributions are scale-free, limited only by the overall system size.

8.2.2 Natural time analysis of the "train" model

In Fig. 8.1, we present the results obtained from the deterministic one-dimensional SOC system described above in § 8.2.1 that belongs, as mentioned, to the same universality class as the "train" model for EQs. The same parameters as in Ref. [26], i.e., $L = 1024$, $a = 4$, $f_0 = 0.87$ and $\delta f = 0.1$, have been used. In Fig. 8.1(a), the number of topplings s is plotted in red versus the avalanche number i for the first 160,000 avalanches which shows in fact how these series of avalanches can be read in natural time. The blue curve in Fig. 8.1(a), shows how the quantity κ_1 evolves avalanche by avalanche. There, we also plot in green the total force $X(i)$ of the system after each avalanche, computed from $X(i) = \sum_{l=1}^{L} f_l(i)$, whose stabilization provides [26] a measure of the approach to SOC. An inspection of Fig. 8.1(a) reveals that (after the transient and hence) when the system enters into the

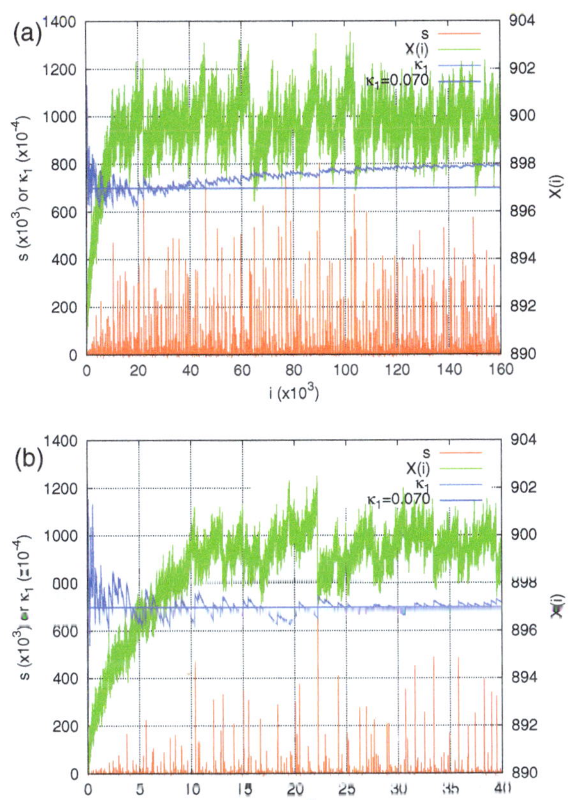

Fig. 8.1 The results of the model discussed in the text for 160,000 (a) and 40,000 (b) avalanches as read in natural time, for $L = 1024$, $a = 4$, $f_0 = 0.87$ and $\delta f = 0.1$. The avalanche size s is depicted by red color, and the variance κ_1 by blue color. The thick blue line corresponds to $\kappa_1 = 0.070$. The total force of the system after each avalanche $X(i)$ is plotted with green color and read in the right scale. For an extension of this figure to 10^6 avalanches see Fig. 8.2. Taken from Ref. [82].

critical state, the κ_1 value fluctuates around 0.070 (designated by the thick blue line). The latter becomes clear in Fig. 8.1(b), which reproduces Fig. 8.1(a) but in an enlarged time scale for the first 40,000 avalanches and shows that for $i > 5{,}000$ (i.e., just when the system enters into the SOC state) κ_1 scatters around 0.070. This behavior has been verified for a wide range of parameters L, a, f_0 and δf just before the SOC state is reached. Note that, once the statistically steady SOC state is established, the κ_1 value gradually increases reaching the corresponding value of $\kappa_u = 1/12$ of a "uniform" distribution (see § 2.1.3). This can be seen in Fig. 8.2 which has been plotted for 10^6 avalanches. The model discussed here leads to a power law with a realistic b value of the Gutenberg–Richter law. In particular, de Sousa Vieira [76] concluded that the distribution of avalanche sizes s is a power law with an exponent $\tau \approx 1.54$ that corresponds to $b \approx 0.81$. This lies in the range (0.8 to 1.2) of the b values found experimentally (see Section 6.1). In spite of this agreement, however, we note that the BK model cannot account for the observed spatiotemporal complexity of seismicity, e.g. the Omori law for aftershocks [55].

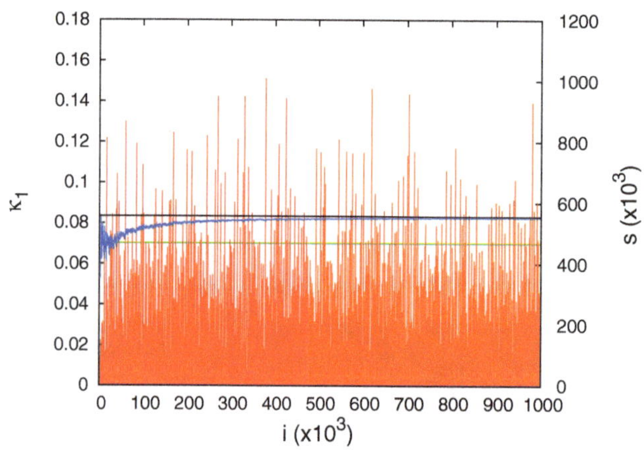

Fig. 8.2 The results of the model discussed in the text for 10^6 avalanches. The parameters (as well as the symbols) here are the same as in Fig. 8.1 (except that the total force $X(i)$ is not plotted, for the sake of simplicity). The horizontal green line corresponds to $\kappa_1 = 0.070$, and the black one to $\kappa_u = 1/12$.

In the focal region of a future earthquake the stress gradually changes *before* failure. It is commonly accepted that, after the mainshock occurrence, the stress value reduces to a smaller value, a fact, however, which is not fully captured by the simple BK model considered here. In other words, in the steady SOC state of this model the system has an average f_l value, $\bar{f}_l$, around $\bar{f}_l = 0.8785$ that remains almost constant (i.e., practically within 0.0055) after the occurrence of any avalanche (cf. $X(i)$ in Fig. 8.1). Our computations reveal (see Fig. 8.3) that when considering a reasonable decrease, e.g., by a few percent, of $\bar{f}_l$, the system exits the steady SOC state and then returns to it through a transient in which the κ_1 value scatters around 0.070, similarly to that depicted in Fig. 8.1. This can be seen in Fig. 8.3, which depicts the results that show what happens with κ_1 when reducing each f_l by 1% (a), 2% (b) and 10% (c) of its value at SOC. The κ_1 value is given here in red while magenta corresponds to one standard deviation $\pm\sigma$. A reasonable reduction of f_l may be around a few percent at the most (see especially Fig. 8.3(a) and 8.3(b)). The results have been obtained by means of the Monte Carlo procedure described

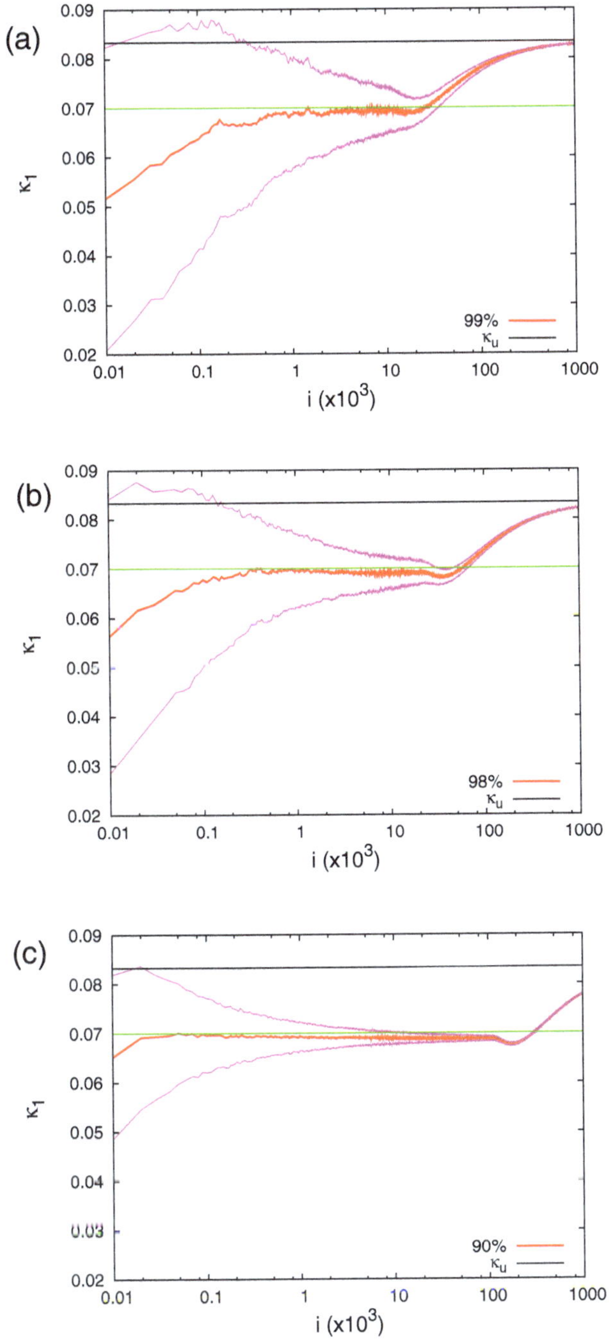

Fig. 8.3 Results showing the behavior of the average value of κ_1 (red) when decreasing $\bar{f}_l$ to (a) 99%, (b) 98% and (c) 90% of its value at SOC. The results have been obtained as follows. We considered 10^3 systems with the initial f_l values randomly scattered around $f_0 = 0.87$. Each system was driven to SOC and in order to obtain a reliable series f_{lSOC}, $l = 1, 2, \ldots L$, the first 10^7 avalanches were ignored in natural time analysis. Then, each of these f_l values was reduced to 99%, 98% and 90%, respectively, of its value at SOC, i.e., f_{lSOC}, and natural time analysis was initiated ($i = 0$). The magenta curves depict the one standard deviation($\pm\sigma$) interval. The horizontal green line corresponds to $\kappa_1 = 0.070$, and the black one to $\kappa_u = 1/12$. Taken from Ref. [82].

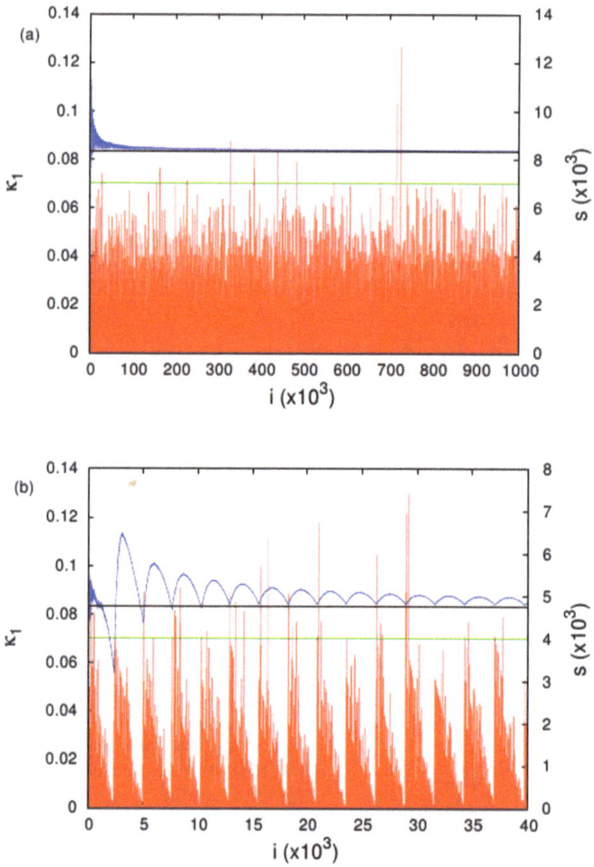

Fig. 8.4 The results obtained when using, instead of the periodic function $\Psi(x)$, the strictly non-increasing function $\phi(x) = \frac{(2-\delta f)^2/a}{x+(2-\delta f)/a} - 1$, introduced in Ref. [57]. The parameters $L = 1024$, $a = 4$, $f_0 = 0.87$ and $\delta f = 0.1$ are the same as those in Figs. 8.1 and 8.2. Results are given for 10^6(a) and 40,000(b) avalanches. The quantities shown here are the same as those plotted in Fig. 8.2. The horizontal green line corresponds to $\kappa_1 = 0.070$, and the black one to $\kappa_u = 1/12$. As shown in (b), the avalanche size $s(\equiv Q_k)$ is quasi-periodic, leading to κ_1 values "oscillating" close to (but mostly higher than) κ_u.

in the caption of this figure. Hence, the value $\kappa_1 = 0.070$ *can be considered as quantifying the extent of the organization of the complex system at the onset of the critical stage.*

We emphasize that such a behavior is not observed for a variant of the model which *does not* exhibit SOC [76], e.g., when using, instead of a periodic function $\Psi(x)$, the strictly non-increasing function $\phi(x)$ introduced by Nakanishi [57] (see Fig. 8.4). This figure shows the results, in a similar fashion to those depicted in Fig. 8.1, obtained from the model when using, instead of the periodic function $\Psi(x)$, the strictly non-increasing function introduced by Ref. [57]. In this case the behavior of κ_1 is found to be distinctly different from that of the SOC model depicted in Fig. 8.1 as well as in Fig. 8.2.

In summary, natural time analysis was made for a one-dimensional SOC model introduced to describe avalanches in stick–slip phenomena. It belongs to the same universality class as the "train" model for earthquakes suggested by Burridge and Knopoff. We found that the value $\kappa_1 = 0.070$ can be considered as quantifying the extent of the organization of the complex system at the onset of the critical stage.

8.3 Natural time analysis of the Olami–Feder–Christensen (OFC) earthquake model

8.3.1 The Olami–Feder–Christensen model. Introduction

The OFC model originated by a simplification of the Burridge & Knopoff spring-block model [14] by mapping it into a non-conservative cellular automaton, simulating the earthquake's behavior and introducing dissipation in the family of SOC systems. In the spring-block model, which as mentioned in § 8.2.1 consists of a two-dimensional array of blocks in a flat surface, each block is connected (by elastic springs) with its neighbors, and in the vertical direction, to a driving plate which moves horizontally at velocity v. When the force acting on a block overcomes the static friction with the surface, the block slips. In the OFC model the force on a block is stored in a site of a square lattice, and the static friction threshold is assumed to have the same value over all blocks. If force input occurs in discrete steps instead of continuous and if thresholds are random but not quenched, quasi-periodicity emerges combined with power laws [65].

The criticality of the OFC model has been debated [21, 54]. Also, the SOC behavior of the model is destroyed upon introducing some small changes in the rules of the model, e.g., replacing open boundary conditions with periodic boundary conditions [64], introducing frozen noise in the local degree of dissipation [56] or in its threshold value [43], including lattice defects [23]. Despite these findings as well as others which show [61], that it is insufficient to account for certain aspects of the *spatiotemporal* clustering of seismicity, the OFC model appears to show many features found in real earthquakes. As far as earthquake predictability [62] or Omori law [39, 36] are concerned, the OFC model appears to be closer to reality than others [85]. In addition, for certain values of the local degree 'α' of dissipation (i.e., if 'α' is chosen above 0.17, see also below), the OFC model exhibits avalanche size distribution that agrees well [52] with the Gutenberg–Richter (G-R) law; see Eq. (6.1). These are some of the reasons why the OFC model is considered to be the prime example [5] for a supposedly SOC system for earthquakes but the question of whether real earthquakes are described or not by SOC models of this type, or whether other kinds of mechanisms, e.g., Refs. [51, 50], need to be involved, remains unsolved [9, 86, 39, 36, 65]. Note also that an analysis of the OFC model in the nonextensivity framework (Section 6.5) has been made by Caruso et al. [20] and further discussed in Ref. [69].

Description of the Olami–Feder–Christensen (OFC) model. The OFC model [59] runs as follows: we assign a continuous random variable $z_{ij} \in (0,1)$ to each site of a square lattice, which represents the local 'energy'. Starting with a random initial configuration taken from a uniform distribution in the segment (0,1), the value z_{ij} of all sites is simultaneously increased at a uniform loading rate until a site ij reaches the threshold value $z_{thres} = 1$ (i.e., the loading Δf is such that $(z_{ij})_{max} + \Delta f = 1$). This site then topples which means that z_{ij} is reset to zero and an 'energy' αz_{ij} is passed to every nearest neighbor. If this causes a neighbor to exceed the threshold, the neighbor topples also, and the avalanche continues until all $z_{kl} < 1$. Then the uniform loading increase resumes. The number of topplings defines the size s of an avalanche or "earthquake". This is the quantity that is used as Q_k in the natural time analysis. The coupling parameter α can take values from zero to 0.25. Smaller α means more dissipation, and $\alpha = 0.25$ corresponds to the conservative case.

The parameter α is the *only* parameter of the model, apart from the system size L, the edge length of the square lattice. Except from the initial condition the model is deterministic. The model can be supplemented by open boundary conditions (OBC) in which the sites at the boundary distribute energy to the outer sites, which cannot topple, thus energy is removed at the boundary. Another possibility, is to use free boundary conditions (FBC). In this case, α varies locally

$$\alpha_{ij} = \frac{1}{n_{ij} + K}. \tag{8.1}$$

where n_{ij} is the actual number of nearest neighbors of the site ij. For sites in the bulk $n_{ij} = 4$, for sites at the edges $n_{ij} = 3$ and for the four sites at the corners $n_{ij} = 2$. The symbol K denotes the elastic constant of the upper leaf springs measured relatively to that of the other springs between blocks [36]. Obviously the OFC model is non-conservative for $K > 0$ for which $\alpha_{ij} < 0.25$ in the bulk. Finally, periodic boundary conditions (PBC) can be imposed but these destroy [64] criticality. Except in the case of PBC, the sites at the boundary receive energy only from three or two neighbors, and therefore topple on average less often than sites in the interior, which leads to the formation of "patches" of sites with similar energy. This patch formation proceeds from the boundaries inward [53, 29]. Due to the dynamics of the model, there occur avalanches of all sizes. The mechanism producing these avalanches are different on different scales [29]. Large avalanches are mainly patch-wide avalanches, while smaller avalanches occur between patches and constitute a series of 'foreshocks' or 'aftershocks' [39]. Also, avalanches at different distances from the boundaries have different sizes.

As already mentioned, there has been no agreement as to whether the model is indeed critical for all values of the coupling or only in the conservative case [21, 22, 54]. In particular, detailed analytical studies [13, 24] for a *random-neighbor* version of the OFC model concluded that *only* in the conservative limit the model becomes critical (this conclusion was also shared by de Carvalho and Prado [21]). Furthermore, using a variety of arguments and large-scale computer simulations, the most exhaustive analyses [10, 35, 85, 5] coincide to the conclusion that the spatially extended version of the non-conservative OFC model is *not* critical.

Thus, the state of the art is [9] that the OFC model is *not* truly scale-invariant except for its conservative limit.

8.3.2 Natural time analysis of the Olami–Feder–Christensen model

8.3.2.1 Results in the transient and the stationary regime of the OFC model

We first present the results [68] for the transient regime of the OFC model using the quantity $f = \sum(\Delta f)$, which represents the total increase of z_{ij} due to the external force loading in each site. Since the loading rate is assumed uniform in time, *f plays a role analogous to that of the conventional time T*, i.e., $T \equiv f$.

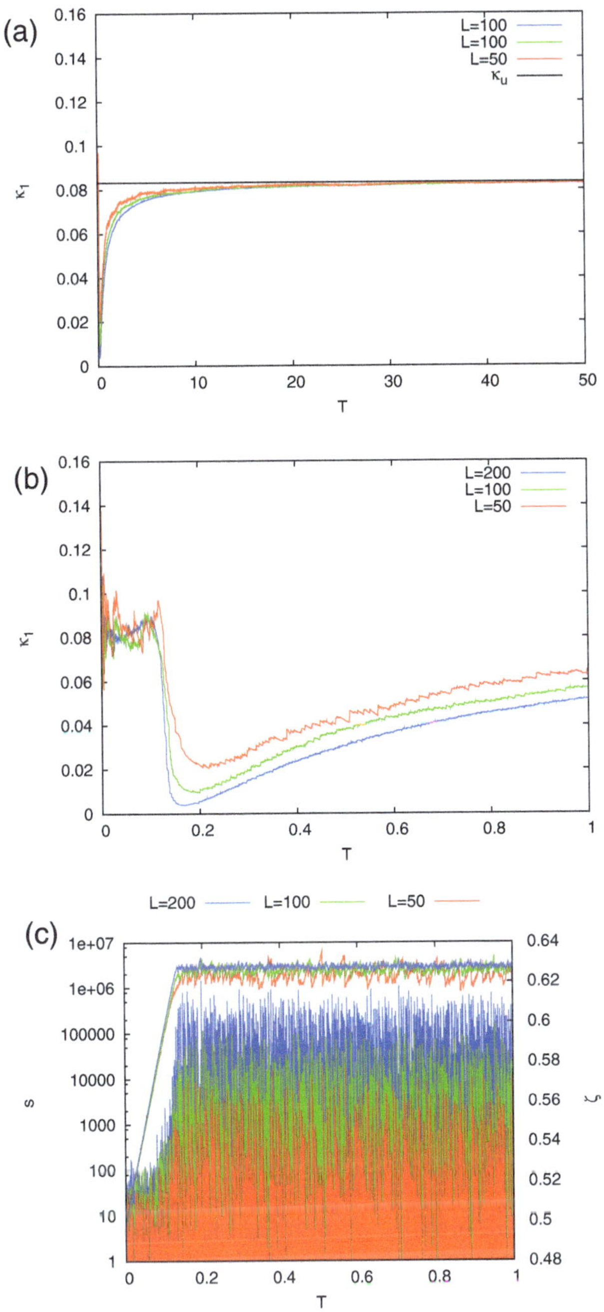

Fig. 8.5 The conservative case of the OFC model for various L (= 50, 100, 200) and OBC: (a) κ_1 versus the 'time' T; (b) an excerpt of (a) showing the transient regime; (c) the avalanche size s (left scale) and the mean energy ζ (right scale) versus the 'time' T.

We find that the conservative and non-conservative cases of the OFC model display a qualitatively different behavior.

In the former case, as can be seen in Figs. 8.5(a),(b) which depict the evolution of κ_1 versus the 'time' T, the quantity κ_1 exhibits a *single* transient consisting of an abrupt decrease, from a value larger than κ_u down to $\kappa_1 \approx 0$ (for larger L see Fig. 8.6), and then κ_1 gradually increases up to the value $\kappa_u = 1/12$. The latter value reflects that the system has reached a steady state, thus the κ_1 value approaches that of the "uniform" distribution. Note that, as the number of avalanches taken into account in the κ_1-calculation increases, the contribution of the avalanches in the transient regime to the κ_1 value becomes gradually smaller compared with that of the avalanches in the stationary regime. In addition, we note that the number of avalanches corresponding to the minimum value of κ_1 was found to scale with L^2. This is reminiscent of the scaling found in Ref. [22] when plotting the mean 'energy' per site $\zeta = \sum z_{ij}/L^2$ versus the number of avalanches. The use of the 'time' T which is *intensive* and *not extensive* quantity, as does the number of avalanches, simplifies the study of this transient. By investigating the κ_1 versus T curves for various L and examining their behavior close to $\kappa_1 \approx 0$, we find (see Fig. 8.6) that all these curves

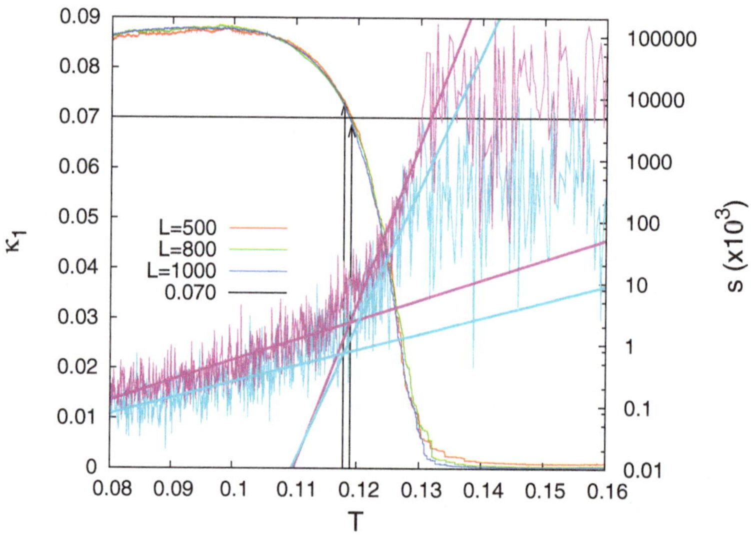

Fig. 8.6 The collapse of various κ_1 vs T curves for the conservative case of the OFC model onto a single curve upon varying the size of the system $L = 500$ (red), 800 (green), 1000 (blue). In the left vertical scale we depict κ_1, while in the right the avalanche size s for $L = 500$ (cyan) and $L = 1000$ (magenta). The cyan (magenta) line corresponds to the maxima observed every 30 (100) avalanches. The thick straight lines correspond to their exponential fits for $T \in [0.08, 0.11]$ and $T \in [0.12, 0.131]$ (note that $T = 0.131$ is approximately the value at which the linear increase of ζ ceases, see Fig. 8.5(c)), and exhibit a cross-over during which the slope increases by one order of magnitude. The arrows show the values of κ_1 corresponding to the points at which the two linear fits (of the same color) intersect. The horizontal black line corresponds to $\kappa_1 = 0.070$.

collapse onto a single curve. Figure 8.5(c) depicts the size s of avalanches versus T along with the mean energy ζ. We observe that the almost abrupt decrease of κ_1 is due to the drastic (exponential) increase by several orders of magnitude of the avalanche size s when ζ approaches its steady state value. This exponential increase is better visualized in Fig. 8.6 where, for the sake of clarity, we depict for $L = 500(1000)$ the maximum avalanche size deduced every 30(100) avalanches versus T (cf. the two values 30 and 100 are considered to account for the fact that the larger system exhibits more avalanches for the same increase in T). One can recognize roughly three linear regions (only two of which are fitted with straight lines in Fig. 8.6, for the sake of clarity) in this log-linear plot (right scale). The first one corresponds to the region $T \in [0.08, 0.11]$ during which the (maximum) avalanche size increases by almost one order of magnitude (see the lower thick solid lines in Fig. 8.6). The second stage corresponds to an almost abrupt later increase by almost five orders of magnitude during $T \in [0.12, 0.131]$, which is linear in the log-linear plot, see the steeper thick solid lines in Fig. 8.6. After $T = 0.131$, which is the value at which the initial linear increase of ζ ceases, see Fig. 8.5(c), the (maximum) avalanche size does not exhibit any obvious trend, thus making unnecessary the plot of the corresponding fits in Fig. 8.6.

The non-conservative case gives a more complicated feature, see Fig. 8.7, because the aforementioned single transient of the conservative case now splits into two parts. Figures 8.7(a),(b) depict the evolution of κ_1 versus T for $\alpha = 0.24$ for various lattice sizes. In Fig. 8.7(b), which is an excerpt of Fig. 8.7(a), an abrupt decrease of κ_1 is observed at $T \approx 0.3$, accompanied by a peak of ζ (see Fig. 8.7(c)) centered at $T \approx 0.16$, which for large L does *not* depend on L. This κ_1 decrease is followed by an increase – coming from a decrease of s (see Fig. 8.7(c)) – and κ_1 reaches a maximum which is subsequently followed by a gradual decrease down to a minimum. This second minimum is observed at $T \approx 1$ for $L = 50$, $T \approx 5$ for $L = 100$ and $T \approx 25$ for $L = 200$, thus indicating that it scales somewhat faster than L^2, which deviates from the finite-size scaling found for the single minimum of the transient in the conservative case. Note that the finite-size scaling observed during the first part of the transient could be attributed to an almost one order of magnitude exponential increase of s when T varies approximately in the range $[0.10, 0.20]$ (see Fig. 8.7(c) and Fig. 8.8(a)) which is similar to the one observed in the conservative case when $T \in [0.08, 0.11]$ (see Fig. 8.8(b)). The dissipation, however, does not allow the emergence of the second much more significant exponential increase of s observed in the conservative case (see the steeper thick solid lines in Fig. 8.6 and the insets of Fig. 8.8). This might be the reason for which the simple finite-size scaling found for $T \in [0.10, 0.20]$ ceases at later times T. The apparent similarity of the first transient stages observed for $T \in [0.08, 0.11]$ and $T \in [0.10, 0.20]$ for the conservative (e.g. Fig. 8.8(b)) and non-conservative cases (e.g. Fig. 8.8(a)), respectively, could be understood as follows. Figure 8.8 shows the T-dependence of the percentage P_{vis} (red solid lines) of the sites "visited" by the OFC toppling rule, which *'diffuses'* energy to the nearest neighbors. We observe that in both cases a similar exponential increase starts when P_{vis} approaches the value of the site *percolation* threshold p_c ($p_c = 0.59274598(4)$; see Ref. [49]) for the two-dimensional square lattice. Thus, when the "visited" sites begin to percolate through the square lattice, the two cases exhibit a similar behavior irrespective of the energy conservation. It seems that the interplay between the diffusive character of the OFC toppling rule and the geometrical phenomenon of percolation is dominating at this stage. Recall

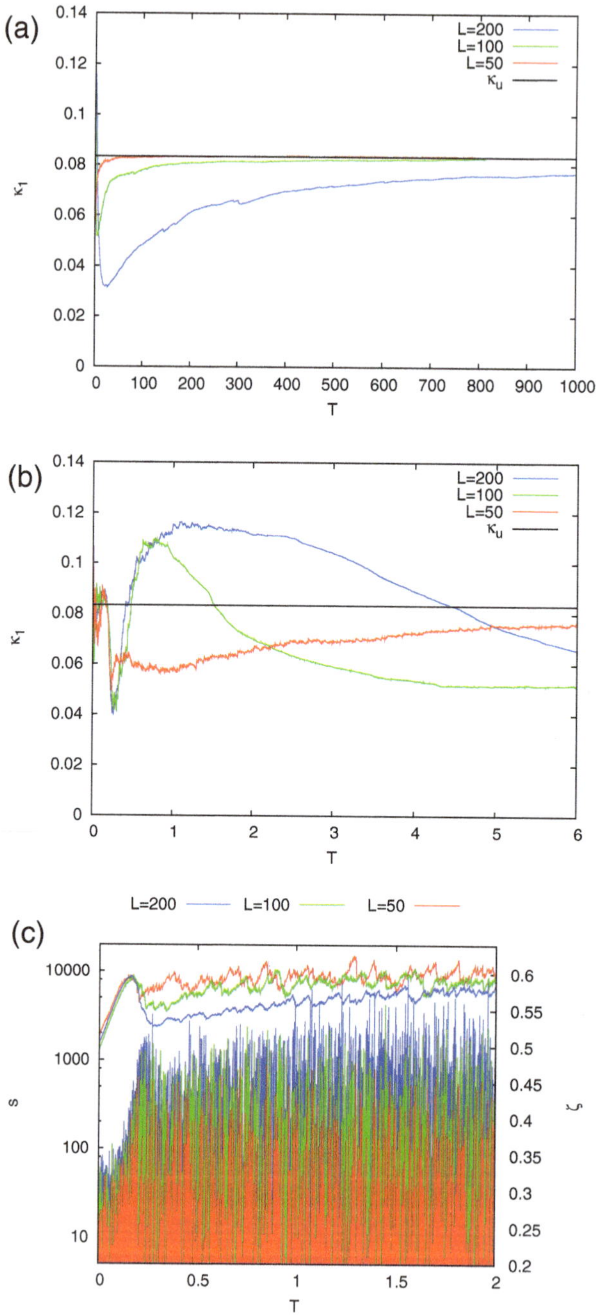

Fig. 8.7 The non-conservative ($\alpha = 0.24$) case of the OFC model for various L (= 50, 100, 200) and OBC: (a) κ_1 versus the 'time' T; (b) an excerpt of (a) showing the transient regime; (c) the avalanche size s (left scale) and the mean energy ζ (right scale) versus the 'time' T.

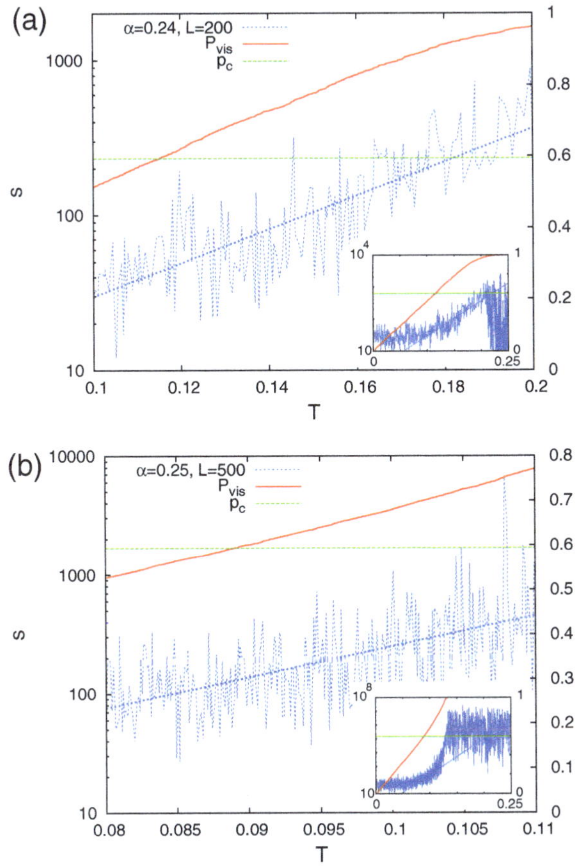

Fig. 8.8 (a): The (blue) dotted broken lines depict the maximum avalanche size observed every 20 avalanches versus T for the non-conservative case ($\alpha = 0.24$) for $L = 200$. The straight thick (blue) dotted line shows a linear fit in the log-linear plot for $T \in [0.1, 0.2]$. (b) is the same as (a), but every 30 avalanches for the conservative case ($\alpha = 0.25$) for $L = 500$ and $T \in [0.08, 0.11]$. Using the right scales in both (a) and (b), we also show the percentage P_{vis} (red solid lines) of the sites "visited" by the OFC toppling rule versus T together with the site-percolation threshold [49] $p_c = 0.59274598(4)$ (green dashed horizontal line) for the two-dimensional square lattice. The whole picture for $T \in [0, 0.25]$ is reproduced in the insets.

that an "unvisited" site of low or moderate random initial 'energy' $(z_{ij})_0$ will be toppled if it receives, apart from the overall increase of $\sum(\Delta F)(= T)$, enough energy to exceed the threshold due to the energy that has diffusively arrived at the site from another site (of possibly higher initial 'energy' $(z_{i'j'})_0$). During this stage it is reasonable to assume that the energy δz arriving at an "unvisited" site reaches it through a single path. Thus, the amount δz scales as α^n, where n is the (presumably small) number of the sites in the path. This amount is not significantly affected whether $\alpha = 0.24$ or $\alpha = 0.25$ and this is why the conservative and the non-conservative cases resemble each other. Later, as the visited sites cluster, the differences emerge dramatically: the energy loss in the non-conservative case occurs at all points of the lattice ($\propto L^2$) thus destroying finite-size scaling, whereas in the conservative case the clusters formed do not alter the finite-size scaling since the energy loss occurs only at the boundaries ($\propto L$) of the system. Returning, now to the κ_1 behavior, we observe the following:

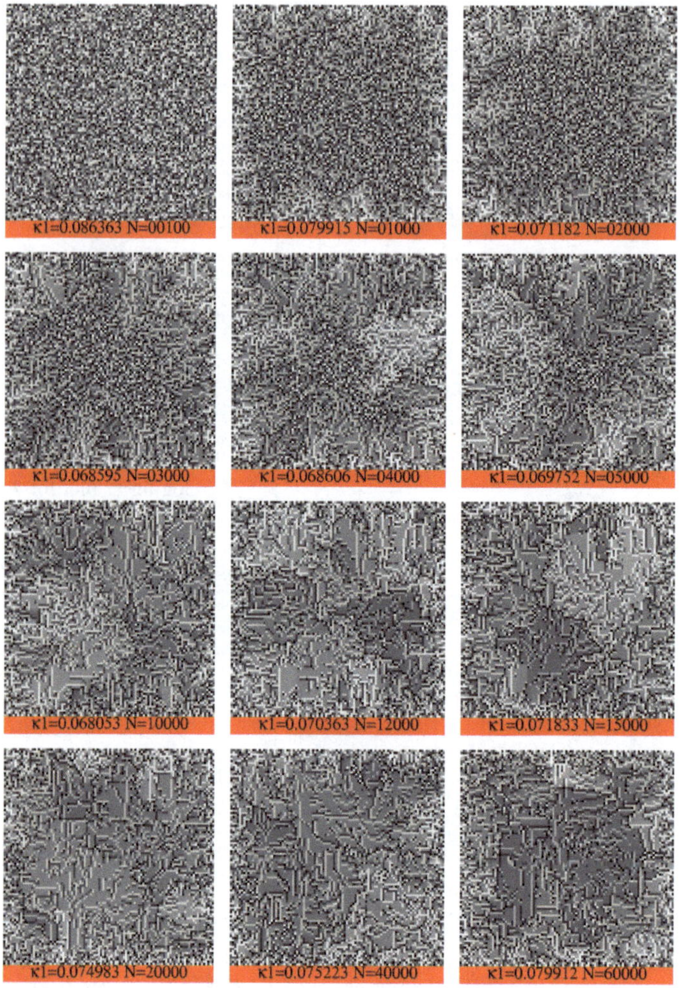

Fig. 8.9 Snapshot images of the strain field in the OFC model ($L = 100$ and $a = 0.22$, OBC), showing that the coherent structure formation is accompanied by values of κ_1 close to 0.070. Here only avalanches with $s > 10$ are considered in natural time analysis and N corresponds to the number of such avalanches. The grey scale (black to white) corresponds to the values of z_{ij} (zero to unity).

After the second minimum, κ_1 increases *slowly* up to the value κ_u. It is during this increase that a prolonged period exists in which coherent domains of the z_{ij} field (strain field) are developed in the non-conservative case; see Fig. 8.9.

Recall that Figs. 8.5 and 8.7 have been drawn by considering all avalanches generated, i.e, $Q_k = s_k \geq 1$. Similar natural time analysis, however, can be performed upon adopting an avalanche size threshold s_0 (i.e., an avalanche of size s is considered as an event in natural time only if $s > s_0$).

Selecting an appropriate threshold s_0 relative to the edge length L, we can find κ_1 values that scatter around 0.070 when the aforementioned coherent structures in the strain field start to appear in the non-conservative case. Such an example is shown in Fig. 8.9 in which a threshold $s_0 = 10$ was selected for $L = 100$ and $\alpha = 0.22$.

Recapitulating the aforementioned results in the transient regime, we see that when comparing the conservative and non-conservative cases, they exhibit considerable differences on how they move away from the initial random state. The question is raised, however, of whether some of these differences can shed light on which of these behaviors is critical and the other not. An answer can be provided on the basis of the following two key differences related to the curves κ_1 versus T and ζ versus T. First, let us consider the κ_1 versus T behavior. In the conservative case, when studying the system for various system sizes L and focusing on the behavior close to $\kappa_1 \approx 0$, we observe, as mentioned, that all the κ_1 vs T curves collapse onto a single curve (see Fig. 8.6).

In addition, in the conservative case, the value $\kappa_1 \approx 0.070$ (that occurs at $T = 0.119$) is accompanied by an abrupt exponential increase of the avalanche size s, which is indicative of the approach to a critical behavior; see the arrows in Fig. 8.6 that mark the cross-over points between the aforementioned log-linear fits (thick straight lines) of the (maximum) avalanche size.

It is this drastic increase of s – by several orders of magnitude – which leads to a κ_1 decrease down to $\kappa_1 \approx 0$ and then κ_1 gradually increases reaching the value $\kappa_u = 1/12$ in the stationary regime. On the other hand, in the non-conservative case the curves κ_1 vs T obtained upon increasing the system size L, do *not* collapse onto a single curve (see Fig. 8.7(b), where it is evident that the second part of the transient does not coincide for different L). Second, let us now consider the ζ vs T behavior: In the conservative case; Fig. 8.5(c) shows that the curves of the mean energy ζ upon studying different L collapse onto a single one after the exponential increase of s (subsequent to the appearance of $\kappa_1 \approx 0.070$ mentioned above). Such a collapse in the ζ vs T curves, however, is not observed in the non-conservative case, see Fig. 8.7(b) (note that in this case, Fig. 8.7(c), the s vs T curve exhibits an increase of only around one order of magnitude in contrast to the several orders of magnitude increase of the conservative case mentioned above).

We now summarize the behavior in the transient regime. *Only in the conservative case* when studying κ_1 vs T and/or ζ vs T (recall that the quantities κ_1, T and ζ are intensive and *not* extensive) the curves obtained upon varying the system size L collapse onto a single curve, as it should for a *critical* system. This is *not* observed in the non-conservative case, meaning that a larger system needs larger T to reach the 'steady-state' ζ value. This points to (the absence of a thermodynamic limit, and hence to) *non criticality*.

As for the stationary regime, for both the conservative and the non-conservative cases, the κ_1 value stabilizes at a value around $\kappa_u = 1/12$, see Figs. 8.5(a) and 8.7(a), which as mentioned reflects that the system has reached a steady state.

8.3.3 The predictability of the OFC model based either on the mean energy or on the interrelation between the κ_1 value and the exponent of the inverse Omori law

Here, in order to study the predictability of the OFC model which has been attributed to the occurrence of 'foreshocks' (note that 'aftershocks' have been also observed) in the non-conservative case of the model [39], we start with a prediction algorithm motivated by the one used by Zhang et al. [88]. This algorithm was inspired by an earlier one proposed by Keilis-Borok and coworkers [46, 45] and by Pepke and coworkers [62, 63]. In particular, we consider the mean 'energy' ζ which is a function of the 'time' T. For this function, the time increased probability (TIP) is turned on when $\zeta > \zeta_c$, where ζ_c is a given threshold in the prediction. If the size s of the next avalanche is greater than a target avalanche size threshold s_c, we have a successful prediction. For binary predictions, the prediction of events becomes a classification task with two type of errors: missing an event and giving a false alarm. We therefore choose, in a similar fashion as in § 6.4.2, the receiver operating characteristics (ROC) graph [30] to depict here the prediction quality. As an example, the ROC graph for $L = 100$ and $K = 2$ is shown in Fig. 8.10, where the various curves correspond to various values of s_c increasing from the bottom to the top. Recalling from § 6.4.2 that the diagonal line in such a plot corresponds to random predictions, and the points in each curve lie above it (meaningful predictions), we conclude that the precursory function ζ results in meaningful prediction which becomes very robust for larger values of s_c. We note, however, that the selection of the mean energy ζ as a precursory function suffers from the drawback that in the case of earthquakes the measurement of this quantity is dif-

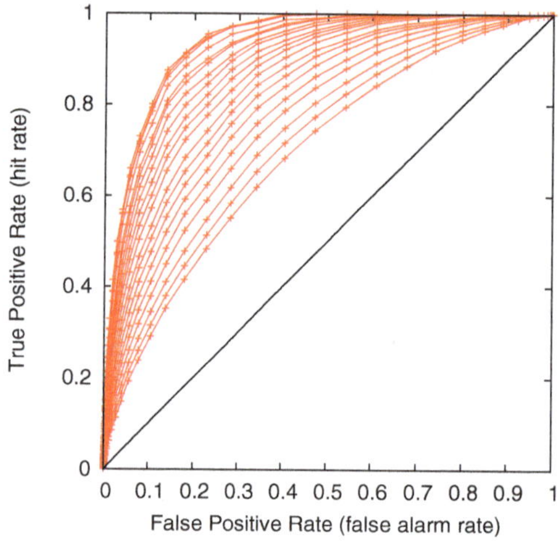

Fig. 8.10 Receiver operating characteristics graph for the OFC model with $L = 100$ and $K = 2$ when using the mean energy ζ as a predictor: the true positive rate (hit rate) versus the false positive rate (false alarm rate) for various s_c values increasing from the bottom ($s_c = 361$) to the top ($s_c = 1{,}938$) with constant steps.

ficult in practice and indirect (what can be measured by some techniques is the increment of stress or strain *not* the absolute values themselves [88]).

The occurrence of 'foreshocks' (f) as well as 'aftershocks' (a) in the OFC model has been exhaustively studied by Helmstetter et al. [36]. Here, we solely focus on the former (foreshocks) that are described by the so-called inverse Omori law [38, 36] which states that the average increase of seismicity observed at the time t before the occurrence time t_c of a mainshock is given by

$$N_f(t) = \frac{K_f}{(t_c - t + c)^{p_f}} \tag{8.2}$$

where the subscript "f" refers to the foreshocks and the quantities K_f and c are taken constants. The inverse Omori exponent p_f is usually close to or slightly *smaller* than the corresponding exponent p_a of the usual Omori law for aftershocks [37] (see also Section 6.1). Helmstetter et al. [36] defined as a "mainshock" (see their definition $d = 0$) any earthquake of magnitude m which was not preceded or followed by a larger earthquake in a time window of length $T(m)$ equal to 1% of the average return time of an earthquake of magnitude m. Foreshocks are then selected as all earthquakes occurring within the time $T(m)$ before a mainshock. The value of p_f has been found [37] by averaging the seismicity rate before a large number of mainshocks, because there are huge fluctuations of the rate of seismicity before individual mainshocks. Helmstetter et al. [36] generated synthetic catalogs with the OFC model and determined the p_f value using various lattice sizes L and K values, see their table I. They studied the cases for $K = 0.5$, 1, 2 and 4, i.e., for bulk $\alpha = 0.222$, 0.2, 0.167 and 0.125, and among the results presented in their table I, we only focus here on the larger lattice sizes, i.e., $L = 1024$ and $L = 2048$. The average value of these p_f exponents results equal to $p_f = 0.72$, if we consider all the relevant p_f values that correspond [36] to b values ranging from $b = 0.67$ to $b = 0.92$. If we restrict ourselves, to those p_f values corresponding to more reasonable values of b, e.g., $b \geq 0.76$, we find that the average p_f value increases somewhat to $p_f = 0.78$. These average p_f values (0.72 and 0.78) suggest that p_f may be considered to be around $p_f \approx 0.75$, which is just the p_f value given in their Fig. 2 for a synthetic catalog generated [36] with $L = 2048$ and $K = 2$. The power law form of the inverse Omori law, i.e., Eq. (8.2), implies that in natural time

$$p(\chi) = \frac{1 - p_f}{(1 - \chi)^{p_f}} \tag{8.3}$$

which reflects an increase of foreshocks as we approach the mainshock at $\chi = 1$. Equation (8.3), when substituted into Eq. (2.97) for the estimation of the variance κ_1 of natural time, leads to

$$\kappa_1 = \frac{2}{(2 - p_f)(3 - p_f)} - \frac{1}{(2 - p_f)^2} \tag{8.4}$$

The κ_1 values determined from Eq. (8.4) are plotted versus p_f in Fig. 8.11. At the same plot, the two κ_1 values that correspond to the aforementioned average p_f values (i.e., $p_f = 0.72$ and $p_f = 0.78$) are marked, which scatter around $\kappa_1 \approx 0.070$. This κ_1 value is comparable with the one ($\kappa_1 \approx 0.070$) determined when analyzing in natural time the small seismic events that occur after the initiation of a SES activity and before the occurrence of a mainshock (see Chapter 7). In other words:

The 'foreshocks' associated with the non-conservative OFC model give on the average a κ_1 value which is more or less comparable with that ($\kappa_1 \approx 0.070$) obtained from the analysis of the real seismic data that precede mainshocks.

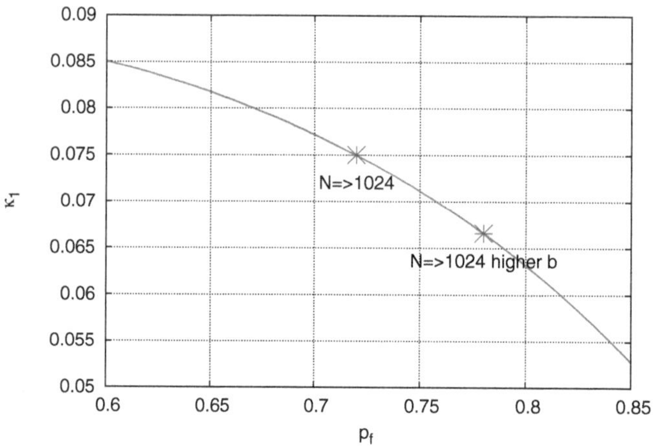

Fig. 8.11 The values of κ_1 vs the power law exponent p_f according to Eq. (8.4). The two points marked correspond to the two average p_f values discussed in the text.

We note, however, that this property of κ_1 is difficult to be used for the prediction of the avalanches in the OFC model in a way similar to that used for ζ. The reason is that the mean energy ζ solely depends on the current state of (the z_{ij} field of) the system, whereas κ_1 reflects the history of the system as it evolves starting from some (initial) state. Thus, the application of κ_1 for prediction purposes in the OFC model requires the real-time identification of the initiation of the foreshock time series. The latter is extremely difficult. Recall that in the field experiments, it is the initiation of the SES activity which signifies that the stress has reached a critical value (see § 1.6.2) in the preparatory volume; then, the identification of a "critical" time series, by employing the method explained in Chapter 7, allows the determination of the occurrence time of the impending strong earthquake.

8.3.4 The predictability of the OFC model on the basis of the change ΔS of the entropy in natural time under time reversal

Here, we focus on what happens before the occurrence time T_0 of a large avalanche during the stationary regime of the non-conservative case of the OFC model by employing the change ΔS of the entropy in natural time under time reversal (Section 3.5). In particular, for each large avalanche, we study the time evolution of ΔS_l (see § 3.5.1) obtained from the preceding avalanches time series. Following the study of foreshocks in Ref. [36] and in view of the fact that, as mentioned in § 8.3.3, there are huge fluctuations before individual large avalanches, our results have been found by averaging the values obtained before an appreciably high number of large avalanches. For example, Fig. 8.12 depicts the results for the average change ΔS (left scale) of the entropy in natural time under time reversal and the average value of the mean energy ζ (right scale) obtained by using the last 1,000 avalanches (irrespective of their size) before large avalanches of size $s \geq 100$ (red), 1,000 (green) and 2,000 (blue) in the non-conservative OFC model with $L = 100$ and $K = 2$. In the horizontal axis, the time is measured from the occurrence time T_0 of the large avalanche. We find that ΔS_l minimizes (note that. $|\Delta S_l|$ maximizes) *before* the impending large avalanche, thus signaling the imminent major event. The negative values of ΔS reflect, through Eq. (3.79), that the avalanche size tends to increase as the time approaches that of the large avalanche, "mainshock" (due to the foreshocks, mentioned in the previous subsection, that start to become discernable from the background "seismicity"). Furthermore, note that ΔS changes sign, becoming negative, when the parameter ζ almost starts to increase (recall that the quantity ζ, as shown in § 8.3.3, can be used as a predictor for the large avalanches).

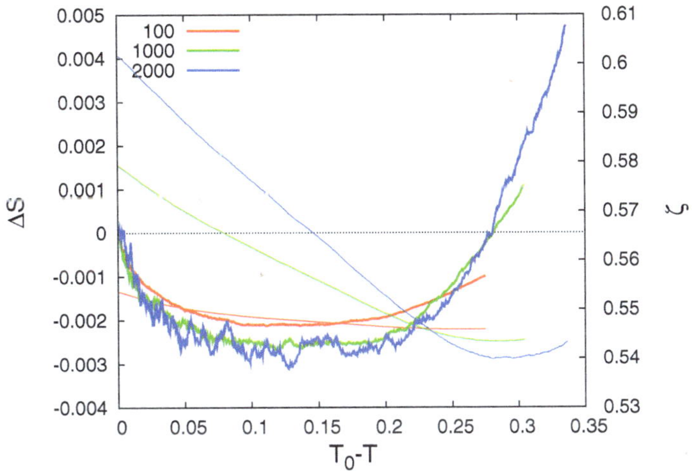

Fig. 8.12 Results from averaging the last 1,000 events before a large avalanche ($s \geq 100$, 1,000, 2,000 occurring at T_0) in the OFC model with $L = 100$ and $K = 2$: the change ΔS (left scale, thick lines) of the entropy in natural time under time reversal and the mean energy ζ (right scale, thin lines) as a function of the 'time' ($T_0 - T$) to the large avalanche. Note that ΔS minimizes *before* the occurrence time T_0 of the large avalanche, and changes sign when ζ almost starts to increase.

Proceeding one step further, Fig. 8.13 depicts the ROC graph when using $\overline{\Delta S}$ as a predictor. In this calculation, $\overline{\Delta S}$ was determined as the average value of ΔS_l using the (past) events that occurred within the time period $T_{now} - T = 0.05$ to 0.2, where T_{now} stands for the present time. This $\overline{\Delta S}$ is used as a predictor for the size of the next avalanche (in the sense described above in § 8.3.3; see also § 6.4.2). The results (red curves) lie above the diagonal and are statistically significant when compared with the cyan curves that correspond to the extrema of 100 trials obtained when performing the same calculation by using *randomly* shuffled $\overline{\Delta S}$ values. These results are certainly less impressive than those

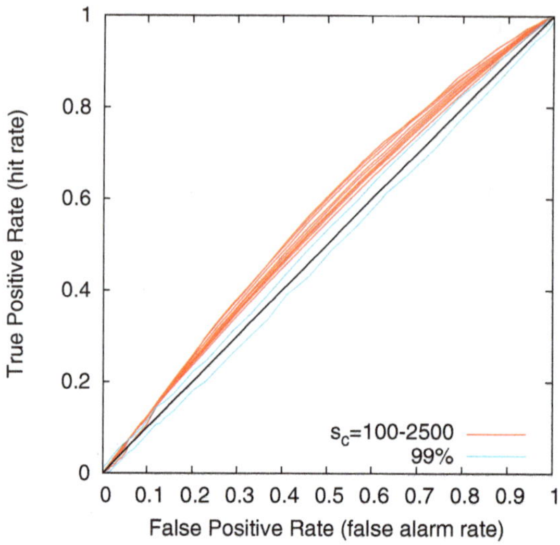

Fig. 8.13 Receiver operating characteristics graph (red) for the non-conservative OFC model with $L = 100$ and $K = 2$ when using $\overline{\Delta S}$ as a predictor (i.e., the TIP is on when $\overline{\Delta S}$ becomes less than some threshold): The true positive rate (hit rate) versus the false positive rate (false alarm rate) for various s_c values increasing from the bottom ($s_c = 100$) to the top ($s_c = 2{,}500$) with constant steps. The cyan curves depict the extrema obtained when repeating 100 times the same calculation using a random predictor (see the text).

in Fig. 8.10, but we emphasize that the predictor here is *solely* based on the sizes of the past avalanches.

8.3.5 Summary of the results

The main conclusions of the natural time analysis of the time series of avalanches in the OFC model could be summarized as follows.

First, concerning the transient period: the behavior is different depending on whether the model is conservative ($\alpha = 0.25$) or non-conservative ($\alpha < 0.25$). In the former case, there is a *single* transient which mainly consists of an abrupt decrease of the variance κ_1, down to a minimum $\kappa_1 \approx 0$ and then a gradual increase up to the value $\kappa_u = 1/12$ of the "uniform" distribution.

Before this minimum, the κ_1 vs T curves deduced for various system sizes ($L \geq 500$) *collapse* onto a single curve and when the κ_1 value reaches $\kappa_1 = 0.070$ (at around $T = 0.119$), an abrupt exponential increase of the avalanche size s occurs signaling the approach to the *critical* behavior.

It is this drastic increase of s which decreases the κ_1 value to $\kappa_1 \approx 0$ (the number of the generated avalanches corresponding to the minimum value $\kappa_1 \approx 0$ is found to scale with L^2). On the other hand, in the non-conservative case, the transient period splits into two parts. In the first part, the number of avalanches corresponding to the minimum κ_1 value does scale with L^2, but in the second part it increases much more quickly. During the second part, coherent domains of the strain field gradually develop. This coherent structure formation is accompanied by κ_1 values close to 0.070.

Second, the resulting power law exponent p_f of the inverse Omori law for the 'foreshocks' identified in the non-conservative cases for large lattice sizes ($L \geq 1024$), is shown to correspond to κ_1 values scattered around $\kappa_1 \approx 0.070$.

Third, there exists a nonzero change ΔS of the entropy in natural time under time reversal, thus signaling the breaking of the time symmetry and reflecting predictability in the OFC model.

8.4 Explanation of $\kappa_1 = 0.070$ for critical systems on the basis of the dynamic scaling hypothesis

We deal with time series of signals emitted from complex dynamical systems, i.e., systems consisting of interacting components that evolve with time. In natural time analysis, when the system is in thermodynamic equilibrium, it should produce stationary time series with probabilities p_k independent of χ_k. The situation is drastically different when the system is in non-equilibrium state. When the system approaches the critical state, clusters of the new phase are formed by enhanced fluctuations and their size increases as does the correlation length. But this happens not instantly, because long-range correlations develop gradually leading to the so-called dynamic phase transition (critical transition) (see § 1.5.2). Thus, the time series emitted in such a non-equilibrium process will be non-stationary and p_k, or the corresponding probability density function $p(\chi)$ will no longer be independent of χ.

Using $p(\chi)$, the normalized power spectrum of Eq. (2.31) can be re-written as

$$\Pi(\omega) = \int_0^1 \int_0^1 p(\chi)p(\chi') \cos\left[\omega(\chi - \chi')\right] \, d\chi \, d\chi' \tag{8.5}$$

A Taylor expansion of Eq. (8.5) around $\omega \to 0$ leads to the value

$$\kappa_1 = \frac{1}{2} \int_0^1 \int_0^1 p(\chi)p(\chi') \left(\chi - \chi'\right)^2 \, d\chi \, d\chi' \tag{8.6}$$

Since $p(\chi)$ is the normalized energy released at χ, for a dynamical system at criticality, it also characterizes the way energy is released during the evolution of the dynamic transition. Energy release may be caused by some existing external field coupled with the newly forming phase. The interaction energy *density* may comprise several terms the most significant of which is usually expected to be of the dipole type; for example the interaction Hamiltonian *density* $h_{int} = -\mathbf{E} \cdot \mathbf{p}$ ($h_{int} = -\mathbf{B} \cdot \mathbf{m}$) in the case of an electric $\mathbf{E}$ (magnetic $\mathbf{B}$) external field, where $\mathbf{p}$ ($\mathbf{m}$) stands for the electric (magnetic) dipole moment *density* of the new phase. This interaction energy is proportional to the linear dimension of the newly forming phase (the system volume is kept constant) and hence it is proportional to the correlation length ξ (this will be proven below). According to the *dynamic scaling hypothesis* (see Refs. [11, 73] and references therein), already explained in § 1.5.3, the time-dependent correlation length ξ at dynamic phase transitions scales as $\xi \propto t^{\frac{1}{z}}$, where z is the so-called dynamic critical exponent. The time t is usually measured in Monte Carlo steps, i.e., using the internal clock of the system. Assuming that the dynamic scaling hypothesis should also hold for $p(\chi)$ at *criticality*, we expect [81]:

$$p(\chi) = N_c \chi^{\frac{1}{\zeta}} \tag{8.7}$$

where ζ is another dynamic exponent (not to be confused with the mean energy ζ of Section 8.3), and N_c a normalization factor to make $\int_0^1 p(\chi)\, d\chi = 1$. In fact, Eq. (8.7) is plausible from the definition of p_k, i.e., it represents the normalized energy emitted during the k-th event and the energy at criticality has a power law distribution. By inserting Eq. (8.7) into Eq. (8.6), we obtain:

$$\kappa_1 = \frac{1+\zeta}{1+3\zeta} - \left(\frac{1+\zeta}{1+2\zeta}\right)^2 \tag{8.8}$$

Substituting the value of ζ by the dynamic critical exponent z for various universality classes of critical systems [58], we can obtain the values of κ_1 depicted in Fig. 8.14. Notice that for most universality classes, z varies in a region from $z = 2$ to $z = 2.4$ and thus (see Fig. 8.14) the value of κ_1 obtained by Eq. (8.8) are in the range 0.068 to 0.071. Especially for the two-dimensional (2D) Ising model, which is qualitatively similar to the process of SES emission (see § 8.4.1), one has $z = 2.165$ (see Ref. [42]) leading through Eq. (8.8) to $\kappa_1 = 0.0697 \approx 0.070$. These results seem to justify the substitution of ζ by z, strongly suggesting that they are the same dynamic exponent. This is not unreasonable since, in reality, the Monte Carlo steps used in the computation of z actually correspond to natural time steps.

Explanation of the statement that the interaction energy is proportional to the linear dimension of the newly formed phase. Following § 1.5.3, in the non-equilibrium scaling state, the equal-time correlation function $C(r,t)$ (see Eq. (1.37)) of an order parameter field

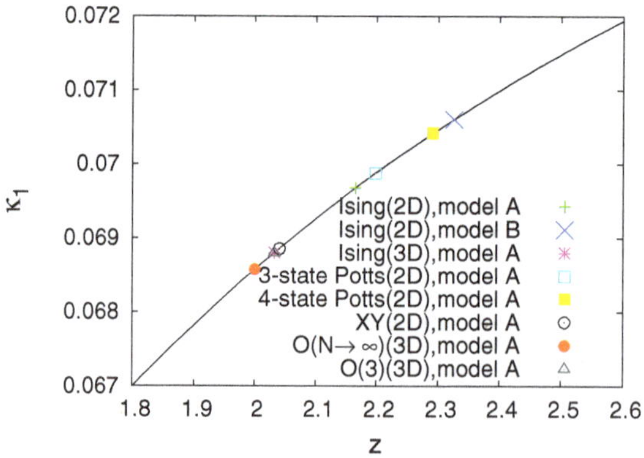

Fig. 8.14 The values of κ_1 as a function of dynamic critical exponent. Various dynamical universality classes are depicted according to their dynamic critical exponent value (see tables IV, VII, IX, XI of Ref. [58]). Models A and B correspond to non-conserved or conserved order parameter dynamics as defined by Hohenberg and Halperin [40]. Taken from Ref. [81]. Note that the value $\kappa_1 = 0.070$ corresponds to $z \approx 2.2$.

$\phi(\mathbf{x},t)$ in the space of D-dimension has the form [12] given by Eq. (1.38) that contains the usual critical exponent η.

It was stated above that the most significant part of the energy release is expected to be of the dipole type, caused by the interaction Hamiltonian $H_{\text{int}} = -\mathbf{E}\cdot\mathbf{P}(=\int_V h_{\text{int}}\,\mathrm{d}\mathbf{x})$, and that this interaction energy is proportional to the linear dimension of the newly forming phase. This statement, following a suggestion of Professor Hiroshi Ezawa, can be shown as follows. The dipole moment $\mathbf{P}$ is given by the integral of its density $\mathbf{p}$

$$\mathbf{P} = \int_V \mathbf{p}(\mathbf{x},t)\,\mathrm{d}\mathbf{x}. \tag{8.9}$$

Assuming $\mathbf{p}(\mathbf{x},t)$ fluctuating under a given electric field $\mathbf{E}$, one focuses on the magnitude P, considering that

$$P^2 = \int \mathrm{d}\mathbf{x} \int \mathrm{d}\mathbf{x}' \mathbf{p}(\mathbf{x},t)\cdot\mathbf{p}(\mathbf{x}',t),$$

which, on average, is related to the correlation function (see Eq. (1.37)) by

$$\begin{aligned}\langle P^2\rangle &= \int \mathrm{d}\mathbf{x} \int \mathrm{d}\mathbf{x}' \langle \mathbf{p}(\mathbf{x},t)\cdot\mathbf{p}(\mathbf{x}',t)\rangle \\ &= \int \mathrm{d}\mathbf{x} \int \mathrm{d}\mathbf{r} \langle \mathbf{p}(\mathbf{x},t)\cdot\mathbf{p}(\mathbf{x}+\mathbf{r},t)\rangle \\ &= D\int_V \mathrm{d}\mathbf{x} \int_0^R S_D r^{D-1} C(r,t)\,\mathrm{d}r, \end{aligned} \tag{8.10}$$

where V is D-dimensional and S_D is the surface area of the unit sphere in D dimensions, and $C(r,t)$ is given by Eq. (1.38), so that

$$\begin{aligned}\langle P^2\rangle &= DVS_D \int_0^R \frac{c}{r^{\eta-1}} f\left(\frac{r}{\xi(t)}\right)\mathrm{d}r \\ &= DVS_D \frac{c}{\xi(t)^{\eta-2}} \int_0^\infty \frac{c}{s^{\eta-1}} f(s)\,\mathrm{d}s \propto \frac{1}{\xi(t)^{\eta-2}} = \xi(t)^{2-\eta}, \end{aligned} \tag{8.11}$$

irrespective of the dimensionality D of V, where we have changed the variable of integration from r to $s = r/\xi(t)$ assuming $V \to \infty$. Since $\eta \approx 0$ irrespective of the dimensionality D of V (e.g., see Ref. [8]), Eq. (8.11) leads to

$$\sqrt{\langle P^2\rangle} \propto \xi(t), \tag{8.12}$$

which gives a linear growth of $P = \sqrt{\langle P^2\rangle}$ with $\xi(t)$.

8.4.1 Natural time analysis of the 2D Ising model quenched close to, but below, T_c. The qualitative similarity to the original SES generation model

Here we treat the case of quenching a ferromagnetic Ising system from a high temperature state to a temperature close to (but below) the critical temperature. This case is studied here since it is qualitatively similar to the pressure stimulated currents (PSC) generation mechanism of SES [78] (see § 1.6.2) in the following sense. In the focal region of a future earthquake, which contains ionic materials, the stress gradually changes before failure. In ionic solids containing aliovalent impurities, extrinsic defects are formed due to charge compensation, which are attracted by nearby aliovalent impurities, thus forming electric dipoles that can change their orientation in space through a defect migration. Stress variations may decrease the relaxation time of these dipoles and when the pressure, or the stress in general, reaches a critical value a *cooperative* orientation of these electric dipoles occurs, which results in the emission of a transient electric signal, which constitutes the SES. The amount of energy released during this relaxation is proportional to the electric dipole moment. This phenomenon may be considered as qualitatively similar to a rapid quench of a ferromagnetic Ising system from a high temperature state (corresponding to the initial random orientations of the magnetic dipoles) to a temperature close and below the critical temperature T_c. Of course, in the case of PSC and hence for the SES, it is not the temperature that changes, but it is the pressure. Pressure variations modify the coupling between the dipoles so that effectively the critical state is reached.

The calculations have been carried out as follows. A 2D Ising system (with Hamiltonian $H = -J\sum_{\langle ij\rangle} s_i s_j$, where $s_i = \pm 1$ and J stands for the coupling constant between the nearest neighbors s_i and s_j) in a square lattice of linear size L (with periodic boundary conditions) was prepared in a high-temperature state and then instantaneously quenched to a temperature (just) below T_c. The evolution of the magnetization per spin $M_k = \sum s_i/L^2$ was simulated by the standard Metropolis algorithm and studied as a function of the number k of Monte Carlo steps (MCS). The latter was set to zero when the system is quenched at a temperature close but below T_c, and increased by 1 after each Monte Carlo step, i.e., after all the spins of the system have been renewed following the standard Metropolis algorithm. For the purpose of the present simulation, k runs from $k = 1$ to 10^4 MCS. Figure 8.15(a) depicts the ensemble average $\langle |M_k| \rangle$ of $|M_k|$ obtained from 10^3 replicas for various sizes $L = 100$, 200, 400 and 1,000. It is observed in the figure that, due to the well-known phenomenon of critical slowing down [48], systems of larger linear size need larger number of MCS to finally reach the equilibrium magnetization. We now present in Fig. 8.15(b) a log-log plot of the values shown in Fig. 8.15(a). This reveals that, practically independent of L, the dynamics of $\langle |M_k| \rangle$ is a power law: $\langle |M_k| \rangle \propto k^{1/z}$ with the dynamic exponent z very close to [42] $z = 2.165$ (see the thick cyan straight line in Fig. 8.15(b)).

This dynamic model was then analyzed in natural time by setting $Q_k = |M_k|$. Figure 8.15(c), which depicts the results for κ_1 as a function of the number k of Monte Carlo steps that followed the instantaneous quench, clearly shows that $\kappa_1 \approx 0.070$.

This result coincides with the one obtained above (i.e., $\kappa_1 = 0.0697$) in § 8.4, when substituting in Eq. (8.8) ζ by the aforementioned value $z = 2.165$.

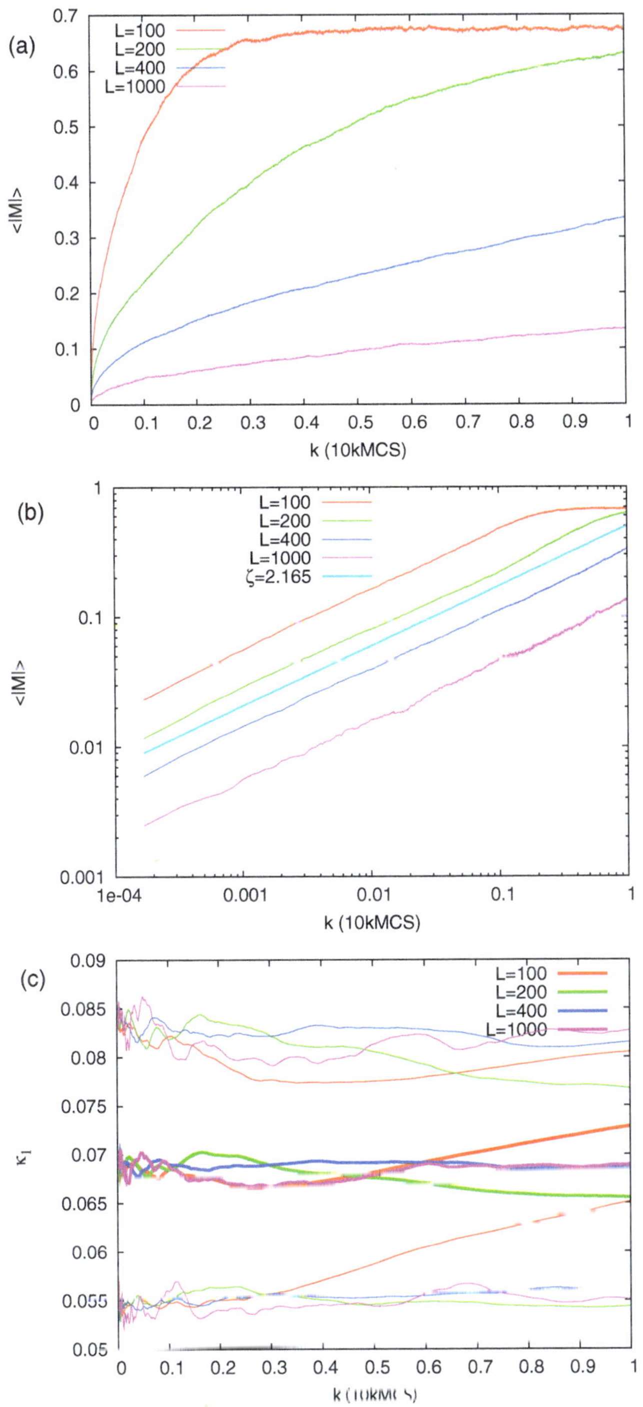

Fig. 8.15 (a), (b) Evolution of $\langle |M_k| \rangle$ as a function of the number k of MCS, after an abrupt quench close but below T_c, up to $k = 10^4$. The thick line in (b) corresponds to $z = 2.165$ (see Ref. [42]) and is drawn as a guide to the eye. (c) depicts the evolution of κ_1 as a function of k when $|M_k|$ is analyzed in natural time. The average value of κ_1 is drawn with the thick lines, whereas the thinner lines enclose the one standard deviation $(\mu \pm \sigma)$ confidence interval. The results were found by 10^3 runs of the model for various L. Taken from Ref. [81].

8.4.2 The original Bak–Tang–Wiesenfeld sandpile SOC model and its fully deterministic version. Natural time analysis

Here, we come back to the original archetypal "sandpile" automaton described in Ref. [6] (see also Section 8.1), hereafter called the Bak–Tang–Wiesenfeld (BTW) model. Let us consider the D-dimensional BTW model on a hypercubic lattice of linear size L in which integer variables $z_i \geq 0$ represent units of sand. We perturb the system by adding a unit of sand at a *randomly chosen* site $z_i \rightarrow z_i + 1$. If the corresponding z_i exceeds the critical value $2D$, the site is called unstable; an unstable site relaxes (topples): its z_i value is decreased by $2D$, and the amount of units of sand of its $2D$ nearest neighbors (nn) is increased by one:

$$z_i \rightarrow z_i - 2D \tag{8.13}$$

$$z_{\rm nn} \rightarrow z_{\rm nn} + 1 \tag{8.14}$$

Thus, the neighboring sites may be activated and an avalanche of relaxations may proceed. This avalanche stops when all sites are stable again. A relaxation event is characterized by its size s (total number of topplings), area a (number of distinct toppled sites), duration t (number of parallel update steps until stable configuration is reached), and its radius r (e.g., the maximal distance between the original and a toppled site). According to the basic hypothesis of Bak et al. [6], in the SOC state the probability distributions of values $x = s, a, t, r$ exhibit power law behavior

$$P_x(\mathrm{x}) \propto \mathrm{x}^{-\tau_x} \tag{8.15}$$

with $x \in \{s, a, t, r\}$. According to Ref. [47], Eq. (8.15) might *not* be in general true for complete avalanches but it does hold for waves of topplings. Specifically, waves represent relaxation processes in which any site topples at most once and hence do not contain multiple toppling events in the origin of the avalanche (note that the latter, for $D \gtrsim 4$, are so rare that they can be neglected). Ktitarev et al. [47] proved analytically that the upper critical dimension of the BTW model is $D_u = 4$, showing that previously observed deviations from mean field behavior at $D = 4$ are due to logarithmic corrections. For this case, $D = 4$, the scaling behavior of waves *and* avalanches is characterized by the same exponents and scaling functions.

In order to proceed to numerical simulations, we study a *fully* deterministic version of the BTW sandpile model, where the random site seeding is replaced by regular seeding at the central site of the hypercubic lattice, suggested by Wiesenfeld et al. [84]. They showed that despite this strict determinism, the system for $D = 2$ evolves into a SOC state. The natural time analysis (with initial condition $z_i = 0$) of the time series of avalanches lead to κ_1 values plotted in Fig. 8.16 for $D = 2$ to $D = 7$. Focusing on the aforementioned upper critical dimension (hence corresponding to the mean field case) $D = 4$, we see that the κ_1 value fluctuates close to 0.070.

The κ_1 values for various D plotted in Fig. 8.16 fluctuate around the value obtained from Eq. (8.8) for $\zeta = D/2$, i.e., 0.056 for $D = 2$, 0.064 for $D = 3$, 0.069 for $D = 4$, 0.071 for $D = 5$, 0.073 for $D = 6$ and 0.075 for $D = 7$. This result can be understood on the following grounds.

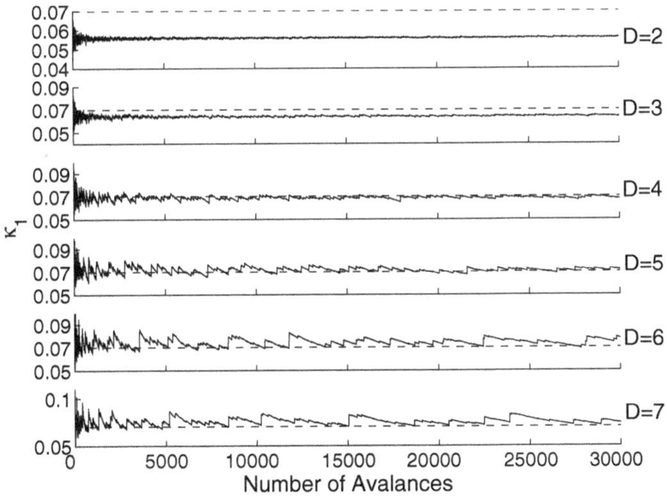

Fig. 8.16 The evolution of κ_1 values versus the number of consecutive avalanches for various D values, i.e., for $D = 2$ to $D = 7$, for the centrally fed sandpile. The initial condition is $z_i = 0$. For the sake of comparison, the broken horizontal line shows the value of $\kappa_1 = 0.070$. Taken from Ref. [81].

Since an avalanche occurs every $2D$ seeds are fed into the central site, the number of avalanches is equal to the number of seeds n fed divided by $2D$. Natural time increases by $1/N$ when an avalanche occurs, thus we have

$$\chi_k = \frac{k}{N}, \quad k = \left[\frac{n}{2D}\right]. \tag{8.16}$$

where the brackets $[\cdot]$ denote the integer part. The local conservation of the units of sand (i.e., sand particles can move only to nearest neighbors sites) expressed in Eqs. (8.13) and (8.14), leads to the fact that the expected number of toppling G_{ij} at site j, upon adding a particle at site i is characterized by [27]

$$G_{ij} \propto r_{ij}^{2-D}, \tag{8.17}$$

where r_{ij} is the distance between the sites i and j. Since we deal with a centrally fed sandpile, the total expected number of topplings $\langle s \rangle$ is found by integrating Eq. (8.17) in the hypersphere of radius l of the sandpile:

$$\langle s \rangle \propto \int_0^l G_{0j} S_D r_{0j}^{D-1} \, dr_{0j} \propto \int_0^l r_{0j} \, dr_{0j} \propto l^2. \tag{8.18}$$

With regard to l, recent mathematical studies [31] show that the linear dimension of the formed sandpile grows as

$$l \propto n^{1/D} \tag{8.19}$$

Inserting Eqs. (8.16) and (8.19) into Eq. (8.18), we obtain $\langle s \rangle \propto \chi^{2/D}$ which reflects (cf. Eq. (8.7)) that $\zeta = D/2$.

Notice that for the upper critical dimension, we have $D = D_u = 4$, and hence $\zeta = 2$ (which is equal to the *mean field dynamic exponent* $z = 2$, e.g. see Ref. [47]) which, in view of Eq. (8.8), leads to $\kappa_1 = 0.0686(\approx 0.070)$.

The fulfillment of the condition $\kappa_1 = 0.070$ could be alternatively seen as follows when considering some points discussed in Ref. [17]: The relaxation of a site can induce a number of other sites to relax in turn if, because of the particles received, they exceed the threshold. From the moment a site topples, the addition of particles stops until all sites have relaxed ($z_i < 2D$ for all i, see Eq. (8.13)). This condition assures that the driving force is *'slow'* being the driving time exceedingly longer than the characteristic time of toppling instances. The sequence of toppling events during this interval constitutes an avalanche. For conservative models, the number of transferred particles equals the number of particles lost by the relaxing site and dissipation occurs only at boundary, from which particles can escape the system. Under these conditions the system reaches a stationary state characterized by a sequence of avalanches. Since the SOC algorithm is implemented basically as a cellular automaton, the *cluster growth is intrinsically of diffusive nature*.

It is this *diffusive* nature of the cluster growth, which seems to lie behind the aforementioned result $\kappa_1 = 0.070$ in SOC models.

8.4.3 Natural time analysis of the mean field case

As mentioned above in Section 8.4 for most universality classes, z varies in a region from $z = 2$ to $z = 2.4$ and thus (see Fig. 8.14) the values of κ_1 obtained by Eq. (8.8) are in the range 0.068 to 0.071. Moreover, in the mean field case, e.g. of the *growing* centrally fed sandpile in § 8.4.2, we have

$$\zeta = 2. \tag{8.20}$$

By inserting Eq. (8.20) into Eq. (8.7), we get

$$p(\chi) = \frac{3}{2}\sqrt{\chi} \tag{8.21}$$

so that $\int_0^1 p(\chi)\,\mathrm{d}\chi = 1$. Upon using Eq. (8.21) for the estimation of the variance κ_1 of natural time (see Eq. (2.37))

$$\kappa_1 = \int_0^1 \chi^2 p(\chi)\,\mathrm{d}\chi - \left[\int_0^1 \chi p(\chi)\,\mathrm{d}\chi\right]^2, \tag{8.22}$$

we find and the corresponding κ_1 value is $\kappa_1 = 0.0686$. This value almost coincides with the value $\kappa_1 \approx 0.070$ found (see Chapter 7) from the natural time analysis of seismicity before large EQs.

The stability of the result $\kappa_1 \approx 0.070$ *if a single realization of the process is available.* The results of this investigation, depicted in Fig. 8.17, show the following:

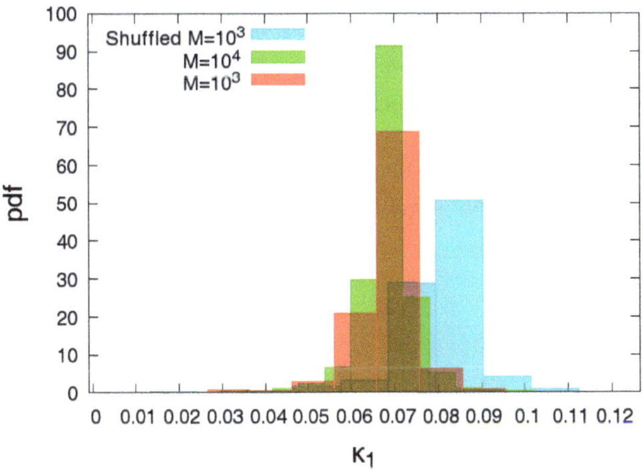

Fig. 8.17 The probability distribution of κ_1 obtained after randomly selecting $M = 10^3$ (red) or $M = 10^4$(green) subseries from a single realization of the process described by Eq. (8.21) using exponentially distributed Q_k (see the text). Both distributions are peaked close to $\kappa_1 = 0.070$. Once the events of the original realization are shuffled randomly and then $M = 10^3$ subseries are analyzed, the peak of the new distribution, shown in cyan, is displaced to the right

Even when using a single realization of the process described by Eq. (8.21) with Q_k exponentially distributed, i.e., $Q_k = r_k\sqrt{k}$ where r_k are exponential p.i.i.d. random variables, and select random subseries of the process to be analyzed in natural time, the pdf deduced for κ_1 maximizes at $\kappa_1 \approx 0.070$.

This is reminiscent of the *updated* procedure used in § 7.1.2. This so, because in that procedure we considered the time series of seismicity that occurs after the initiation of the SES activity in the area candidate to suffer a mainshock and then used the subseries corresponding to the seismicity in all possible subareas of the candidate area to construct the pdf of the resulting κ_1 values. It was then found that this pdf exhibits a maximum at $\kappa_1 \approx 0.070$ when approaching the occurrence time of the mainshock.

8.5 Natural time analysis of time series of avalanches observed in laboratory experiments

8.5.1 Time series of avalanches observed in ricepiles

Here, we consider the well-controlled experiment on three-dimensional ricepiles by Aegerter et al. [2, 1]. Since a genuine understanding of the nature of SOC can be gained only when the *approach* to the critical state is understood, Aegerter et al. studied the evolution of a three-dimensional pile of rice starting well away from the critical state and getting progressively closer. They found [1] that their experimental results are satisfactorily described by well-founded concepts proposed [60] in the context of extremal dynamics. In the latter context, Paczuski et al. [60] have derived an equation (predicting power law behavior), which they call the gap equation, describing the approach of the system to the critical state. Aegerter et al. [2] directly studied a measure of this gap given by the maximal local slope of the ricepile and hence could test various scaling relations of extremal dynamics. Furthermore, Aegerter et al. studied the evolution of avalanche sizes, as well as that of the avalanche distributions, which can be used as further tests of extremal dynamics aspects. Here we solely focus on the way the size ΔV of avalanches grow with time in the transient regime, which was measured directly.

Figure 8.18(a) depicts the time evolution of ΔV obtained in one experiment of Ref. [2]. Upon analyzing these data in natural time when assuming $Q_k = \Delta V_k$, we find the results shown in Fig. 8.18(b).

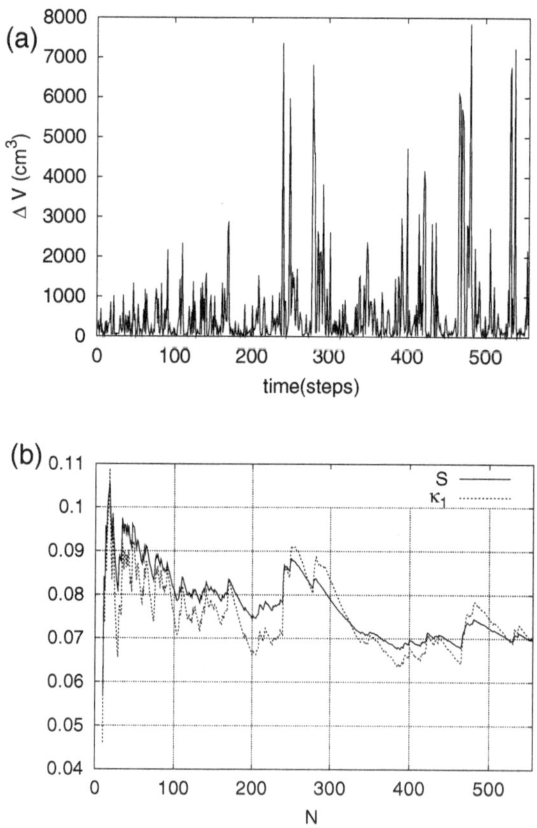

Fig. 8.18 (a): The evolution of the avalanche sizes, in the transient regime, for one of the experiments of Ref. [2] on ricepiles. (b): The results of the variance κ_1 (dotted) and the entropy S (solid) as they evolve event by event, when the data of (a) are analyzed in natural time. Taken from Ref. [70].

A careful inspection of these results in ricepiles reveals that actually at later times ($N \geq 350$) the κ_1 value scatters in the region around 0.070(10) (as well as that $S \approx 0.070(10) < S_u$).

We clarify that upon shuffling the data, which reflects that the values are put into random order and hence all correlations (memory) are destroyed [80] (see also § 2.5.2.1), we find that, for $N = 550$, $\text{Prob}[\kappa_1 \leq 0.070] < 2\%$ (and $\text{Prob}[S \leq 0.070] < 0.1\%$). This leads to the following conclusion:

The *sequential order* of the avalanches captured by the natural time analysis is of prominent importance [70] for establishing the SOC state and constitutes the basis for the observation of the result $\kappa_1 \approx 0.070$.

8.5.2 Time series of magnetic flux avalanches observed in high T$_c$ superconductors. A generalized stochastic directed SOC model

The archetypal example of SOC is, as mentioned (Section 8.1 and § 8.2.2), the growing of a sandpile [6, 7]. Furthermore, the critical state in superconductors has been proposed (e.g. see Ref. [87]) to be a SOC system. The strong analogy between these two systems, i.e., sandpiles and superconductors, as first pointed out by de Gennes (see p.83 of Ref. [34]), could be in principle understood as follows.

When a type II superconductor is put in a slowly ramped external field, magnetic vortices start to penetrate the sample from its edges. These vortices get pinned by crystallographic defects (e.g., dislocations [78]), leading to the build-up of a flux gradient which is only marginally stable in a similar fashion as is the slope in a slowly growing sandpile. Hence, it can happen that small changes in the applied field can result in large rearrangements of flux in the sample, known as flux avalanches [15, 4, 83].

We now proceed to the natural time analysis of the time series of the magnetic flux avalanches measured in a thin film of $YBa_2Cu_3O_{7-x}$. These measurements have been made by Aegerter et al. [3]. They studied the local changes in the magnetic flux over the whole central area of a sample via a highly sensitive magneto-optic setup, which allows that flux changes corresponding to $2.5\Phi_0$ can be resolved where $\Phi_0 = h/2e$ is the magnetic flux quantum (the flux of a single vortex). The pinning sites in the sample were uniformly distributed and consisted mostly of screw dislocations acting as point pins. For cuprate high T_c superconductors such as $YBa_2Cu_3O_{7-x}$ the coherence lengths are in the order of tens of Å, and thus atomic-scale structural inhomogeneities such as point defects and columnar defects can play an important role in flux-line pinning. (In these superconductors, Su et al. [77] found that Schottky defects formation energy increases almost linearly with $B\Omega$, where B is the isothermal bulk modulus and Ω the mean volume per atom, in striking agreement with an early model (termed the $cB\Omega$ model) proposed [79, 78] by one of the present authors.)

The data of Aegerter et al. [3] come from experimental runs consisting of 140 time steps. The size and shape of the avalanches was determined from the difference $\Delta B_z(x,y)$ of two consecutive images (50 μT increase between images), where $B_z(x,y)$ denotes the flux density at the surface of the sample measured. From these differences, the average increase in the applied magnetic field due to the stepwise field sweep, was subtracted in order to solely study the avalanches. Once the incremental field difference is determined, the size of an avalanche, corresponding to the displaced amount of flux $\Delta\Phi$, is estimated from ΔB_z by integrating over the whole area $\Delta\Phi = \frac{1}{2}\int \Delta B_z(x,y)\,\mathrm{d}x\,\mathrm{d}y$.

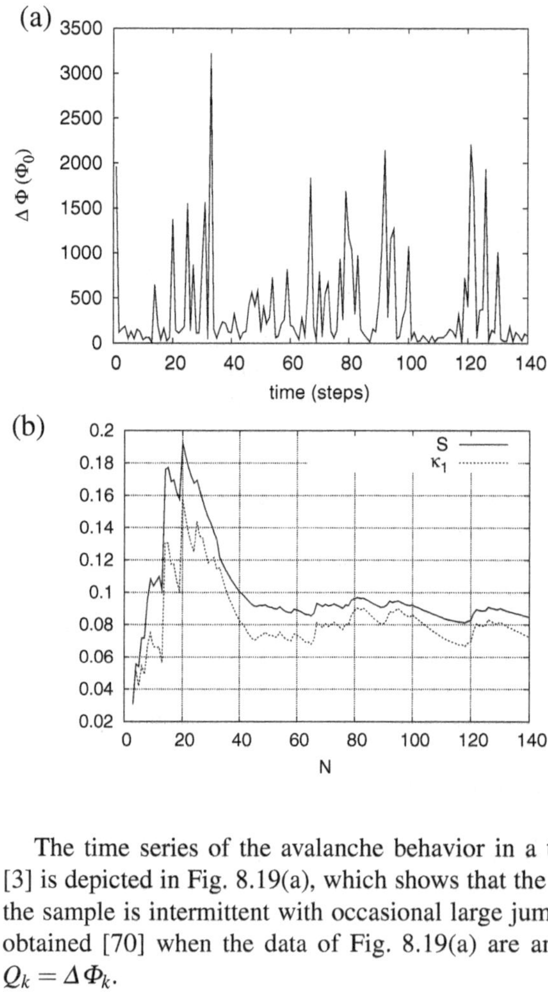

Fig. 8.19 (a): The time evolution of the magnetic flux in $YBa_2Cu_3O_{7-x}$ inside the sample over the 1st run of fig.2 of Ref. [3]. (b): The results of the variance κ_1 (dotted) and the entropy S (solid) as they evolve event by event, when the data of (a) are analyzed in the natural time-domain. Taken from Ref. [70].

The time series of the avalanche behavior in a typical experiment of Aegerter et al. [3] is depicted in Fig. 8.19(a), which shows that the evolution of the magnetic flux inside the sample is intermittent with occasional large jumps. Figure 8.19(b) depicts the results obtained [70] when the data of Fig. 8.19(a) are analyzed in natural time by assuming $Q_k = \Delta\Phi_k$.

An inspection of the latter figure shows that for $N = 140$ the κ_1 value is close to 0.070(5) (as well as that the S value is around 0.085, i.e., smaller than S_u).

The results obtained above have been compared [70] with those deduced from the natural time analysis of the numerical results from a generalized stochastic SOC model suggested by Carbone and Stanley [17]. It consists of a family of stochastic directed clusters generated by fractional Brownian paths with different correlation properties. Carbone and Stanley showed that the cluster area, length and duration exhibit the characteristic scaling behavior of SOC clusters. This model is considered [17] to be a generalized stochastic model, including the Dhar-Ramaswamy [28] directed sandpile model (which can be described assuming that the system is driven by particles added at the top layer $i = 0$ and removed from the bottom layer $i = L$) and the stochastic models as particular cases.

Carbone and Stanley [17] consider a generalized Brownian walk $y(i)$ defined by $y(i) \equiv \sum_{k=0}^{i-1} \xi_k$, where the steps ξ_k are taken from a discrete fGn (see § 1.5.1.1). The mean square displacement of $y(i)$ scales with Δi as $\langle y(i)^2 \rangle \propto (\Delta i)^{2H}$, where H is the Hurst exponent ($0 < H < 1$). The moving average function $\tilde{y}_n(i)$ is

$$\tilde{y}_n(i) \equiv \frac{1}{n} \sum_{k=0}^{n-1} y(i-k), \tag{8.23}$$

which is a linear operator whose output is still a generalized Brownian motion, but now with the high-frequency components of the signals averaged out [16] according to the window amplitude n. In order to characterize the clusters $\mathscr{C}$ corresponding to the regions bounded by $y(i)$ and $\tilde{y}_n(i)$ in terms of the characteristic exponents of SOC systems, they define – for each cluster – the cluster area s_j

$$s_j \equiv \sum_{i=i_c(j)}^{i_c(j+1)} |y(i) - \tilde{y}_n(i)| \Delta i, \tag{8.24}$$

where the index j refers to each cluster. The symbols $i_c(j)$ and $i_c(j+1)$ stand [17] for the values of the index i corresponding to two subsequent intersections of the "lines" defined by $\tilde{y}_n(i)$ and $y(i)$, and Δi is the elementary time interval corresponding to each step of the random walker. Then, the pdf $P(s)$ scales [17] as $P(s) \propto s^{-\tau}$ with $\tau = 2/(1+H)$. Considering that the exponent of the avalanche distribution reported from the data analysis of Aegerter et al. [3] is around $\tau = 1.3$, we find that it corresponds to $H \approx 0.5$. Thus, in Fig. 8.20, we plot $P(s)$ versus s calculated for various n values for $H = 0.5$.

We now turn to the comparison of the results of this model with the aforementioned experimental results in $YBa_2Cu_3O_{7-x}$ films. Taking into account that the maximum avalanche size s_{max} detected by Aegerter et al. [3] is of the order of 10^4, an inspection of Fig. 8.20 leads to $n \approx 200$. In Fig. 8.21, we plot with solid lines the pdfs of κ_1 and S that have been obtained from the model of Ref. [17] for $H = 0.5$, $n = 200$ and $N = 140$. An inspection of this figure shows that:

The maxima of the pdfs of κ_1 and S lie around $\kappa_1 = 0.070(10)$ and $S = 0.080(10)$, respectively. These are comparable with the corresponding κ_1 and S values obtained from the natural time analysis of the experimental data depicted in Fig. 8.19(b) (for $N = 140$).

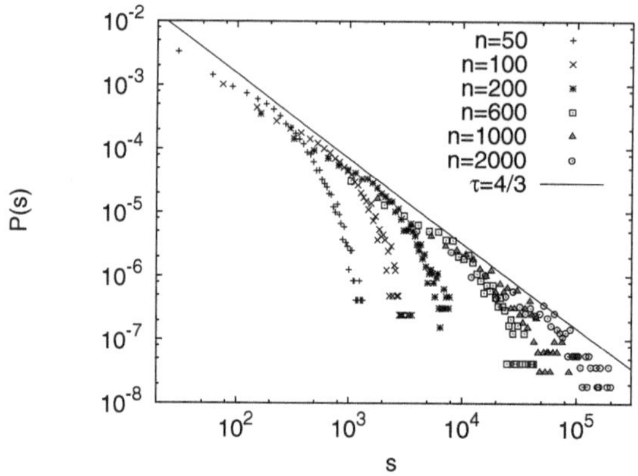

Fig. 8.20 The pdf $P(s)$ versus the cluster area s of the $\mathscr{C}$ clusters of Ref. [17], for $H = 0.5$ for various values of $n = 50$, 100, 200, 600, 1,000 and 2,000. The line $P(s) \propto s^{-\tau}$ with $\tau = 4/3$, as analytically found in Ref. [17] to describe $P(s)$, is also drawn as a guide to the eye. Taken from Ref. [70].

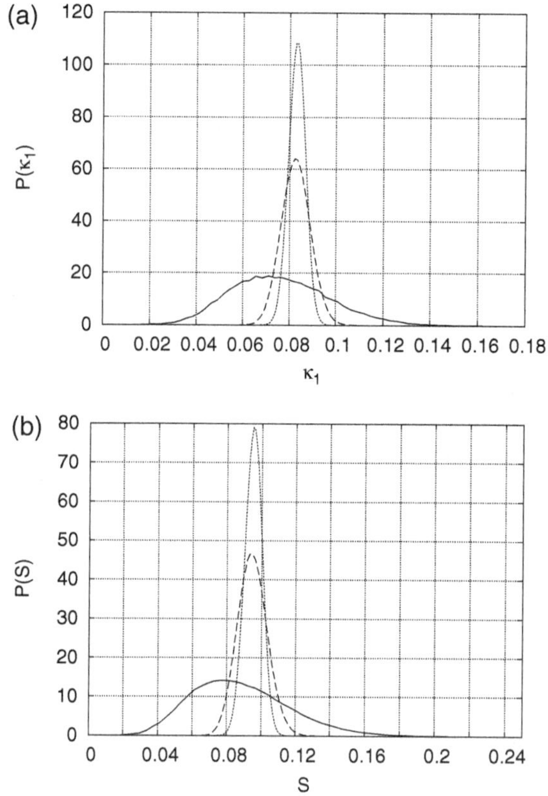

Fig. 8.21 The pdfs of κ_1 and S obtained from a Monte Carlo simulation for the generalized stochastic model of Ref. [17] for $H = 0.5$, $n = 200$ and $N = 140$. The dotted and dashed pdfs correspond to two different *non-critical* cases(see the text). Taken from Ref. [70].

For the sake of comparison, in Fig. 8.21, we also plot the corresponding pdfs for two *non-critical* cases associated with a "uniform" distribution (see § 2.1.3) i.e., (i) when Q_k are uniformly distributed in the range (0,1) (dotted) and (ii) when Q_k are exponentially distributed (broken) which corresponds to a dichotomous Markovian process (see Section 4.1), e.g, the case of the observed [25] RTS signals in metal-oxide-semiconductor transistors with tunneling oxide. The maxima of the latter two cases lie at $\kappa_1 \approx \kappa_u$ and $S \approx S_u$, which markedly differ from those deduced for $YBa_2Cu_3O_{7-x}$ in Fig. 8.19(b) (for $N = 140$).

By summarizing, the measurements of the penetration of magnetic flux into a thin film of $YBa_2Cu_3O_{7-x}$ have been analyzed in natural time. This analysis leads to a value of the variance $\kappa_1 = \langle\chi^2\rangle - \langle\chi\rangle^2$ equal to $\kappa_1 \approx 0.070$. The same κ_1 value is found from the natural time analysis of a generalized stochastic SOC model proposed by Carbone and Stanley [17].

Acknowledgments We express our sincere thanks Professor Rinke J. Wijngaarden, for sending us the $YBa_2Cu_3O_{7-x}$ and ricepile data discussed in § 8.5.2 and § 8.5.1, respectively.

References

1. Aegerter, C.M., Gunther, R., Wijngaarden, R.J.: Avalanche dynamics, surface roughening, and self-organized criticality: Experiments on a three-dimensional pile of rice. Phys. Rev. E **67**, 051306 (2003)
2. Aegerter, C.M., Lorincz, K.A., Welling, M.S., Wijngaarden, R.J.: Extremal dynamics and the approach to the critical state: Experiments on a three dimensional pile of rice. Phys. Rev. Lett. **92**, 058702 (2004)
3. Aegerter, C.M., Welling, M.S., Wijngaarden, R.J.: Self-organized criticality in the bean state in $YBa_2Cu_3O_{7-x}$ thin films. Europhys. Lett. **65**, 753–759 (2004)
4. Altshuler, E., Johansen, T.H.: Colloquium: Experiments in vortex avalanches. Rev. Mod. Phys. **76**, 471–487 (2004)
5. Bach, M., Wissel, F., Drossel, B.: Olami–Feder–Christensen model with quenched disorder. Phys. Rev. E **77**(6), 067101 (2008)
6. Bak, P., Tang, C., Wiesenfeld, K.: Self-organized criticality: An explanation of the 1/f noise. Phys. Rev. Lett. **59**(4), 381–384 (1987)
7. Bak, P., Tang, C., Wiesenfeld, K.: Self-organized criticality. Phys. Rev. A **38**, 364–374 (1988)
8. Binney, J.J., Dowrick, N.J., Fisher, A.J., Newman, M.E.J.: The Theory of Critical Phenomena – An Introduction to the Renormalization Group. Oxford University Press, UK (1992)
9. Bonachela, J.A., Mũnoz, M.A.: Self-organization without conservation: true or just apparent scale-invariance? J. Stat. Mech. P09009 (2009)
10. Boulter, C.J., Miller, G.: Nonuniversality and scaling breakdown in a nonconservative earthquake model. Phys. Rev. E **68**, 056108 (2003)
11. Bray, A.J.: Theory of phase-ordering kinetics. Adv. Phys. **43**, 357–459 (1994)
12. Bray, A.J., Briant, A.J., Jervis, D.K.: Breakdown of scaling in the nonequilibrium critical dynamics of the two-dimensional *XY* model. Phys. Rev. Lett. **84**, 1503–1506 (2000)
13. Bröker, H.M., Grassberger, P.: Random neighbor theory of the Olami–Feder–Christensen earthquake model. Phys. Rev. E **56**, 3944–3952 (1997)
14. Burridge, R., Knopoff, L.: Model and theoretical seismicity. Bull. Seismol. Soc. Am. **57**, 341–371 (1967)
15. Campbell, A.M., Evetts, J.E.: Flux vortices and transport currents in type II superconductors. Adv. Phys. **50**, 1249–1449 (2001)
16. Carbone, A., Castelli, G., Stanley, H.E.: Analysis of clusters formed by the moving average of a long-range correlated time series. Phys. Rev. E **69**, 026105 (2004)

17. Carbone, A., Stanley, H.E.: Directed self-organized critical patterns emerging from fractional Brownian paths. Physica A **340**, 544–551 (2004)
18. Carlson, J.M., Langer, J.S.: Properties of earthquakes generated by fault dynamics. Phys. Rev. Lett. **62**, 2632–2635 (1989)
19. Carlson, J.M., Langer, J.S., Shaw, B.E.: Dynamics of earthquake faults. Rev. Mod. Phys. **66**, 657–670 (1994)
20. Caruso, F., Pluchino, A., Latora, V., Vinciguerra, S., Rapisarda, A.: Analysis of self-organized criticality in the Olami–Feder–Christensen model and in real earthquakes. Phys. Rev. E **75**, 055101 (2007)
21. de Carvalho, J.X., Prado, C.P.C.: Self-organized criticality in the Olami–Feder–Christensen model. Phys. Rev. Lett. **84**, 4006–4009 (2000)
22. de Carvalho, J.X., Prado, C.P.C.: Dealing with transients in models with self-organized criticality. Physica A **321**, 519–528 (2003)
23. Ceva, H.: Influence of defects in a coupled map lattice modeling earthquakes. Phys. Rev. E **52**, 154–158 (1995)
24. Chabanol, M.L., Hakim, V.: Analysis of a dissipative model of self-organized criticality with random neighbors. Phys. Rev. E **56**, R2343–R2346 (1997)
25. Chen, M.J., Lu, M.P.: On-off switching of edge direct tunneling currents in metal-oxide-semiconductor field-effect transistors. Appl. Phys. Lett.. **81**, 3488–3490 (2002)
26. Davidsen, J., Paczuski, M.: $1/f^{\alpha}$ noise from correlations between avalanches in self-organized criticality. Phys. Rev. E **66**, 050101 (2002)
27. Dhar, D.: Theoretical studies of self-organized criticality. Physica A **369**, 29–70 (2006)
28. Dhar, D., Ramaswamy, R.: Exactly solved model of self-organized critical phenomena. Phys. Rev. Lett. **63**, 1659–1662 (1989)
29. Drossel, B.: Complex scaling behavior of nonconserved self-organized critical systems. Phys. Rev. Lett. **89**, 238701 (2002)
30. Fawcett, T.: An introduction to ROC analysis. Pattern Recogn. Lett. **27**, 861–874 (2006)
31. Fey, A., Levine, L., Peres, Y.: Growth rates and explosions in sandpiles. J. Stat. Phys. **138**, 143–159 (2010)
32. Garber, A., Hallerberg, S., Kantz, H.: Predicting extreme avalanches in self-organized critical sandpiles. Phys. Rev. E **80**, 026124 (2009)
33. Garber, A., Kantz, H.: Finite-size effects on the statistics of extreme events in the BTW model. Eur. Phys. J. B **67**, 437–443 (2009)
34. de Gennes, P.G.: Superconductivity of Metals and Alloys. Addison-Wesley, New York (1966)
35. Grassberger, P.: Efficient large-scale simulations of a uniformly driven system. Phys. Rev. E **49**, 2436–2444 (1994)
36. Helmstetter, A., Hergarten, S., Sornette, D.: Properties of foreshocks and aftershocks of the nonconservative self-organized critical Olami–Feder–Christensen model. Phys. Rev. E **70**, 046120 (2004)
37. Helmstetter, A., Sornette, D.: Foreshocks explained by cascades of triggered seismicity. J. Geophys. Res. **108(B10)**, 2457 (2003)
38. Helmstetter, A., Sornette, D., Grasso, J.R.: Mainshocks are aftershocks of conditional foreshocks: How do foreshock statistical properties emerge from aftershock laws. J. Geophys. Res. **108(B1)**, 2046 (2003)
39. Hergarten, S., Neugebauer, H.J.: Foreshocks and aftershocks in the Olami–Feder–Christensen model. Phys. Rev. Lett. **88**, 238501 (2002)
40. Hohenberg, P.C., Halperin, B.I.: Theory of dynamic critical phenomena. Rev. Mod. Phys. **49**, 435–479 (1977)
41. Huang, Y., Saleur, H., Sammis, C., Sornette, D.: Precursors, aftershocks, criticality and self-organized criticality. Europhys. Lett. **41**, 43–48 (1998)
42. Ito, N.: Non-equilibrium relaxation and interface energy of the Ising model. Physica A **196**, 591–614 (1993)
43. Jánosia, I.M., Kertész, J.: Self-organized criticality with and without conservation. Physica A **200**, 179–188 (1993)
44. Jensen, H.J.: Self-organized criticality: emergent complex behavior in physical and biological systems. Cambridge University Press, New York (1998)

45. Keilis-Borok, V.I., Kossobokov, V.G.: Premonitory activation of earthquake flow: algorithm M8. Phys. Earth Planet. Inter. **61**, 73–83 (1990)
46. Keilis-Borok, V.I., Rotwain, I.M.: Diagnosis of time of increased probability of strong earthquakes in different regions of the world: algorithm CN. Phys. Earth Planet. Inter. **61**, 57–72 (1990)
47. Ktitarev, D.V., Lübeck, S., Grassberger, P., B. Priezzhev, V.: Scaling of waves in the Bak–Tang–Wiesenfeld sandpile model. Phys. Rev. E **61**, 81–92 (2000)
48. Landau, D.P., Binder, K.: A Guide to Monte Carlo Simulations in Statistical Physics. Cambridge Univ. Press, New York (2005)
49. Lee, M.J.: Pseudo-random-number generators and the square site percolation threshold. Phys. Rev. E **78**, 031131 (2008)
50. Lippiello, E., de Arcangelis, L., Godano, C.: Influence of time and space correlations on earthquake magnitude. Phys. Rev. Lett. **100**, 038501 (2008)
51. Lippiello, E., Godano, C., de Arcangelis, L.: Dynamical scaling in branching models for seismicity. Phys. Rev. Lett. **98**, 098501 (2007)
52. Lise, S., Paczuski, M.: Self-organized criticality and universality in a nonconservative earthquake model. Phys. Rev. E **63**, 036111 (2001)
53. Middleton, A.A., Tang, C.: Self-organized criticality in nonconserved systems. Phys. Rev. Lett. **74**, 742–745 (1995)
54. Miller, G., Boulter, C.J.: Measurements of criticality in the Olami–Feder–Christensen model. Phys. Rev. E **66**, 016123 (2002)
55. Mori, T., Kawamura, H.: Simulation study of earthquakes based on the two-dimensional Burridge-Knopoff model with long-range interactions. Phys. Rev. E **77**, 051123 (2008)
56. Mousseau, N.: Synchronization by disorder in coupled systems. Phys. Rev. Lett. **77**, 968–971 (1996)
57. Nakanishi, H.: Statistical properties of the cellular-automaton model for earthquakes. Phys. Rev. A **43**, 6613–6621 (1991)
58. Ódor, G.: Universality classes in nonequilibrium lattice systems. Rev. Mod. Phys. **76**, 663–724 (2004)
59. Olami, Z., Feder, H.J.S., Christensen, K.: Self-organized criticality in a continuous, nonconservative cellular automaton modeling earthquakes. Phys. Rev. Lett. **68**, 1244–1247 (1992)
60. Paczuski, M., Maslov, S., Bak, P.: Avalanche dynamics in evolution, growth, and depinning models. Phys. Rev. E **53**, 414–443 (1996)
61. Peixoto, T.P., Davidsen, J.: Network of recurrent events for the Olami–Feder–Christensen model. Phys. Rev. E **77**, 066107 (2008)
62. Pepke, S.L., Carlson, J.M.: Predictability of self-organizing systems. Phys. Rev. E **50**, 236–242 (1994)
63. Pepke, S.L., Carlson, J.M., Shaw, B.E.: Prediction of large events on a dynamical model of a fault. J. Geophys. Res. **99(B4)**, 6769–6788 (1994)
64. Pérez, C.J., Corral, A., Díaz-Guilera, A., Christensen, K., Arenas, A.: On self-organized criticality and synchronization in lattice models of coupled dynamical systems. Int. J. Mod. Phys. B **10**, 1111–1151 (1996)
65. Ramos, O., Altshuler, E., Måløy, K.J.: Quasiperiodic events in an earthquake model. Phys. Rev. Lett. **96**, 098501 (2006)
66. Ramos, O., Altshuler, E., Måløy, K.J.: Avalanche prediction in a self-organized pile of beads. Phys. Rev. Lett. **102**, 078701 (2009)
67. Sammis, C.G., Smith, S.W.: Seismic cycles and the evolution of stress correlation in cellular automaton models of finite fault networks. Pure Appl. Geophys. **155**, 307–334 (1999)
68. Sarlis, N.V., Skordas, E.S., Varotsos, P.A.: Natural time analysis of the Olami–Feder–Christensen model. (to be published) (2011)
69. Sarlis, N.V., Skordas, E.S., Varotsos, P.A.: Nonextensivity and natural time: The case of seismicity. Phys. Rev. E **82**, 021110 (2010)
70. Sarlis, N.V., Varotsos, P.A., Skordas, E.S.: Flux avalanches in $YBa_2Cu_3O_{7-x}$ films and rice piles: Natural time domain analysis. Phys. Rev. B **73**, 054504 (2006)
71. Scholz, C.H.: Earthquakes and friction laws. Nature (London) **391**, 37–42 (1998)
72. Scholz, C.H.: The Mechanics of Earthquakes and Faulting, 2nd ed. Cambridge University Press, Cambridge U.K. (2002)

73. Sicilia, A., Arenzon, J.J., Bray, A.J., Cugliandolo, L.F.: Domain growth morphology in curvature-driven two-dimensional coarsening. Phys. Rev. E **76**, 061116 (2007)
74. Sornette, D.: Critical Phenomena in Natural Science, 2nd edn. Springer, Berlin (2004)
75. de Sousa Vieira, M.: Self-organized criticality in a deterministic mechanical model. Phys. Rev. A **46**, 6288–6293 (1992)
76. de Sousa Vieira, M.: Simple deterministic self-organized critical system. Phys. Rev. E **61**, R6056–R6059 (2000)
77. Su, H., Welch, D.O., Wong-Ng, W.: Strain effects on point defects and chain-oxygen order-disorder transition in 123 cuprate compounds. Phys. Rev. B **70**, 054517 (2004)
78. Varotsos, P., Alexopoulos, K.: Thermodynamics of Point Defects and their Relation with Bulk Properties. North Holland, Amsterdam (1986)
79. Varotsos, P., Ludwig, W., Alexopoulos, K.: Calculation of the formation volume of vacancies in solids. Phys. Rev. B **18**, 2683–2691 (1978)
80. Varotsos, P.A., Sarlis, N.V., Skordas, E.S., Lazaridou, M.S.: Entropy in natural time domain. Phys. Rev. E **70**, 011106 (2004)
81. Varotsos, P.A., Sarlis, N.V., Skordas, E.S., Uyeda, S., Kamogawa, M.: Natural time analysis of critical phenomena. under preparation (2011)
82. Varotsos, P.A., Sarlis, N.V., Skordas, E.S., Uyeda, S., Kamogawa, M.: Natural time analysis of critical phenomena. the case of seismicity. EPL **92**, 29002 (2010)
83. Welling, M.S., Aegerter, C.M., Wijngaarden, R.J.: Self-organized criticality induced by quenched disorder: Experiments on flux avalanches in NbH_x films. Phys. Rev. B **71**, 104515 (2005)
84. Wiesenfeld, K., Theiler, J., McNamara, B.: Self-organized criticality in a deterministic automaton. Phys. Rev. Lett. **65**, 949–952 (1990)
85. Wissel, F., Drossel, B.: Transient and stationary behavior of the Olami–Feder–Christensen model. Phys. Rev. E **74**, 066109 (2006)
86. Yang, X., Du, S., Ma, J.: Do earthquakes exhibit self-organized criticality? Phys. Rev. Lett. **92**, 228501 (2004)
87. Zaitsev, S.I.: Robin Hood as self-organized criticality. Physica A **189**, 411–416 (1992)
88. Zhang, S., Huang, Z., Ding, E.: Predictions of large events on a spring-block model. J. Phys. A: Math. Gen. **29**, 4445–4455 (1996)

9. Natural Time Analysis of Electrocardiograms

Abstract. Here, we present the results obtained from the natural time analysis of electrocardiograms. Considering that a general agreement about whether normal heart dynamics are chaotic or not is still lacking, and that a physiological time series may be due to a mixed process, stochastic and deterministic, we use here the concept of entropy which is equally applicable to deterministic as well as stochastic processes. Sudden cardiac death is a frequent cause of death and may occur even if the electrocardiogram seems to be strikingly similar to that of a healthy individual. Upon employing, however, the fluctuations of the entropy in natural time, when a time window of certain length is sliding each time by one "pulse" (heartbeat) through the whole time series, sudden cardiac death individuals (SD) can be clearly distinguished from the truly healthy individuals. Furthermore, by using the complexity measures introduced in § 3.6.1 to quantify the change of the natural entropy fluctuations either by changing the time window length scale or by shuffling the "pulses" randomly, we can achieve the classification of individuals into three categories: healthy, heart disease patients and SD. In addition, when considering the entropy change under time reversal, at certain time window length scales (which have a clear physical meaning), not only can the SD risk be identified, but also an estimate of the time of the impending cardiac arrest can be provided. In particular, after the maximization of the amplitude of ΔS at the scale of 13 heartbeats, ventricular fibrillation starts within $\approx$3 hours in 16 out of 18 SD. Finally, an $1/f$ model is proposed in natural time which leads to results that are consistent with the progressive modification of heart rate variability in healthy children and adolescents. The model results in complexity measures that separate healthy dynamics from heart disease patients as well as from SD.

9.1 Natural time analysis of the RR, QRS and QT time series

9.1.1 Introduction

The advantages of using the concept of the entropy in the analysis of a physiological time series in general, and of electrocardiograms (ECG) in particular, has been already explained in Section 3.1. In addition, it was explained there why the complexity measures associated with the entropy S defined in natural time (which is a *dynamic* entropy) have certain advantages compared to those based on *static* entropy (e.g. Shannon entropy). Earlier attempts in the ECG analysis have actually used measures related to dynamic entropy. For example, the so-called approximate entropy (AE) [48] or sample entropy (SE) [51] have been used earlier by other authors. Examples showing that the procedure developed here gives [63] better results than that based on AE or SE will be put forward later in § 9.2.3. Also, Costa et al. [11] introduced the multiscale entropy approach, the algorithm of which is based on AE or SE, calculating the entropy at different scales. As for the S, it differs essentially from the other entropies, because it is defined [61, 62] in an entirely different time-domain (see Fig. 9.1(b)). Moreover, as already mentioned (§ 4.8.3), in order to discriminate *similar-looking* electric signals emitted from systems of different dynamics, the following seems to hold [68]:

Signals that have S values more or less comparable to S_u (*which is the case of all ECG*, see Fig. 9.11 that will be discussed later) can be better classified by the complexity measures relevant to the fluctuations δS of the entropy.

If the S values differ *markedly* from S_u (which is usually the case for SES and AN), the classification of these signals should be preferably made by the use of the S values themselves (see Section 4.10). Hereafter, we focus on the case of ECG.

In a single sinus (normal) cycle of an ECG, the turning points are traditionally labeled with the letters Q, R, S, T; see Fig. 9.1(a). It has been clinically observed that the QT interval usually exhibits prolonged values before cardiac death (see Ref. [26] and references therein). In Fig. 9.1(b) we show how the QT interval time series can be read in natural time. By the same token, one can read in natural time the RR (beat-to-beat) interval time series (see Figs. 2.2(a) and 2.2(b)) as well as the QRS interval time series. The RR and

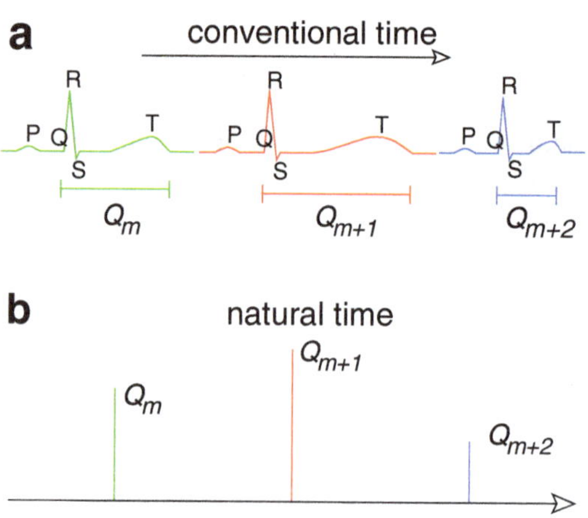

Fig. 9.1 (a) Schematic diagram (not in scale) of a three heartbeat excerpt of an ECG in the usual (conventional) time domain. Only the durations Q_m,Q_{m+1},Q_{m+2} of the QT interval (marked in each single cycle of the ECG corresponding to one heartbeat) are shown. (b) The QT interval time series of (a) read in natural time; the vertical bars are *equally* spaced and the length of each bar denotes the duration of the corresponding QT interval marked in (a). Taken from Ref. [66].

QRS intervals (mainly the RR) can be automatically detected [32, 30, 31, 22] more easily than the QT.

Sudden cardiac death, which is the primary cause of mortality in the industrialized world [7], may occur even if the ECG looks to be similar to that of truly healthy (H) humans.

Here, we present a surrogate data analysis which differentiates the ECG of H from those of sudden cardiac death individuals (SD) based on the fluctuations of the entropy S in natural time.

The fact that a system contains nonlinear components does not necessarily reflect that a specific signal we measure from the system also exhibits nonlinear features. Thus, before analyzing this signal by applying nonlinear techniques, we must first clarify if the use of such techniques is justified by the data available. The method of surrogate data has been extensively used to serve such a purpose (see Ref. [55] for a review). Surrogate data refer to data that preserve certain linear statistic properties of the experimental data, but are *random* otherwise [8, 57]. These data are prepared by various procedures, e.g., see Ref. [57]. Here, the surrogate data are obtained by shuffling the Q_k randomly and hence their distribution is conserved. Applying such a procedure, we do the following: consider the null hypothesis that the data consist of *independent* draws from a fixed probability distribution of the dwell times; if we find significantly different serial correlations in the data and their shuffles, we can reject the hypothesis of *independence* [55]. In other words, the tested null hypothesis is that Q_k are independent and identically distributed (i.i.d.) random variables, i.e., that there are no correlations between the lengths of consecutive intervals. If the original (continuous) time series is Markovian then the null hypothesis for the Q_k should hold, i.e., the Q_k are i.i.d. random variables. The terminology "Markovian" here always refers to the original time series.

Following § 3.6.1, as a measure of the natural time entropy S fluctuations we consider the standard deviation δS when we calculate the value of S for a number of consecutive pulses and study how S varies when sweeping this time-window through the whole time series. In all examples, we use here a sliding window of length 3 to 10 pulses, except otherwise stated. Concerning the symbols: we reserve δS *only* for the case when the calculation is made by a *single* time-window, e.g., 5 pulses. The symbol $\overline{\delta S}$ denotes the average of the δS values calculated for a sequence of single time-windows, e.g., 3, 4 and 5 pulses. Finally, $\langle \delta S \rangle$ stands for the δS values averaged over a group of individuals, e.g., the healthy subjects. The subscript "*shuf*" means that the relevant quantity refers to data obtained by shuffling Q_k randomly.

We used here the QT database from physiobank [14] (see also Ref. [31]), which is publicly accessible and consists of 105 fifteen-minute excerpts of Holter recordings as follows: 10 from MIT-BIH Normal Sinus Rhythm Database (i.e., healthy subjects, hereafter labeled H), 15 from MIT-BIH Arrhythmia Database (MIT), 13 from MIT-BIH Supraventricular Arrhythmia Database (MSV), 6 from MIT-BIH ST Change Database (MST), 33 from the European ST-T Database (EST), 4 from MIT-BIH Long-Term ECG Database (LT) and 24 from sudden cardiac death patients from BIH(SD) (BIH denotes the Beth Israel Hospital).

9.1.2 The quantities δS and δS_{shuf}. The non-Markovianity of electrocardiograms

We now investigate if the δS values alone can "recognize" the non-Markovianity in ECG [67]. In Fig. 9.2, we plot, for the QRS interval time series, the δS value averaged over each of the aforementioned seven groups versus the time-window length. Since all time series of these seven groups have $\approx 10^3$ intervals, we insert in the same figure the results calculated for a Markovian case of comparable length $\approx 10^3$. In particular, we consider a *dichotomous* Markovian time series, in which we recall (e.g. § 4.1.1 and § 4.1.3) that the dwell times (Q_k) are exponentially distributed. (Since in the calculation of S only ratios of Q_k are involved the result does not depend on the transition rates of the Markovian process.) An inspection of this figure shows that the Markovian case exhibits δS values that are roughly one order of magnitude larger than those of the seven groups of ECG, which clearly points to the non-Markovianity of *all* the signals in these groups. We emphasize that the same conclusions are drawn if we consider, instead of QRS, the time series of QT, or RR intervals.

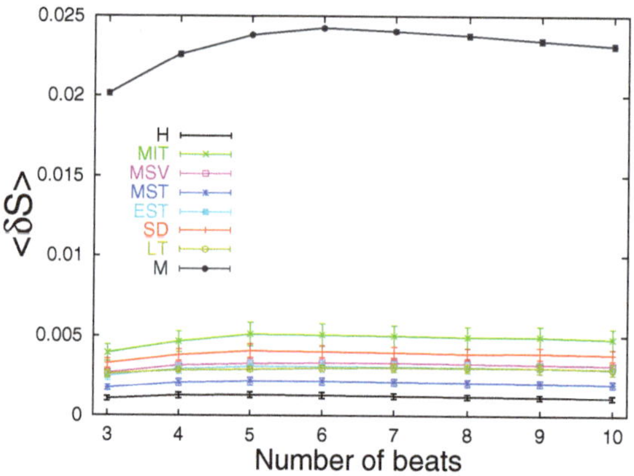

Fig. 9.2 The $\langle \delta S \rangle$ values for the QRS intervals (see the text) of the seven groups of ECG versus the time-window length. The corresponding values for a Markovian time series (10^3 pulses, labeled M) are also plotted. Taken from Ref. [67].

In summary, the δS value alone can recognize the non-Markovianity in ECG.

We now study δS_{shuf} (§ 3.6.1). Having in mind Eq. (3.63), in Fig. 9.3(a) we plot, for each of the 105 individuals, the value of σ/μ versus the corresponding value of $\overline{\delta S_{shuf}}$ (time-window range 3–10 beats) for the RR intervals. The same is repeated in Figs. 9.3(b) and 9.3(c) for the QT and QRS intervals, respectively. All these three plots, can be described by linear behavior and a least-squares fitting to a straight line passing through the origin leads to the following slopes: 38.6 ± 0.6, 36.8 ± 0.2 and 40.1 ± 0.4, for the RR, QT and QRS intervals, respectively. This points to the conclusion that δS_{shuf} provides a

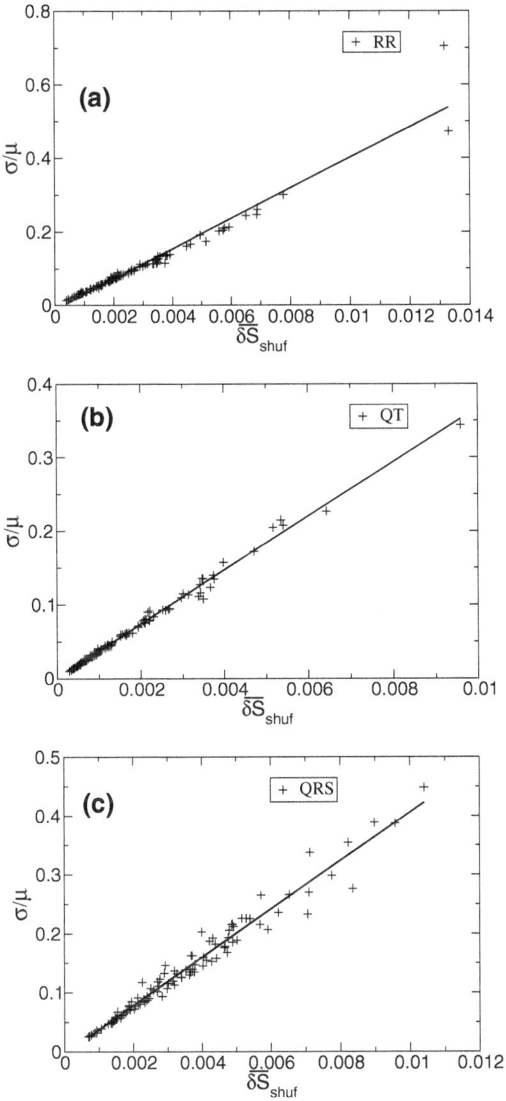

Fig. 9.3 The σ/μ value, for each of the 105 individuals, versus the corresponding $\overline{\delta S_{shuf}}$ value for the (a) RR, (b) QT and (c) QRS intervals. The identity of the individual associated with each point can be found in Ref. [64]. Taken from Ref. [67].

measure of σ/μ. Note that, although these three slopes are more or less comparable, they differ by amounts lying outside their standard error. Furthermore, if we study *altogether* the RR, QT and QRS intervals, for the 10 healthy humans *only* (Fig. 9.4), a good linearity of σ/μ versus $\overline{\delta S_{shuf}}$ results with a slope 37.5 $\pm$ 0.4. (note that if we study each of the three intervals separately, we find slopes that agree within the error margins, i.e., 37.5 $\pm$ 0.4, 37.1 $\pm$ 0.7 and 37.8 $\pm$ 0.1 for the RR, QT and QRS intervals, respectively). The origin of this *common* behavior in the *healthy* humans merits further investigation.

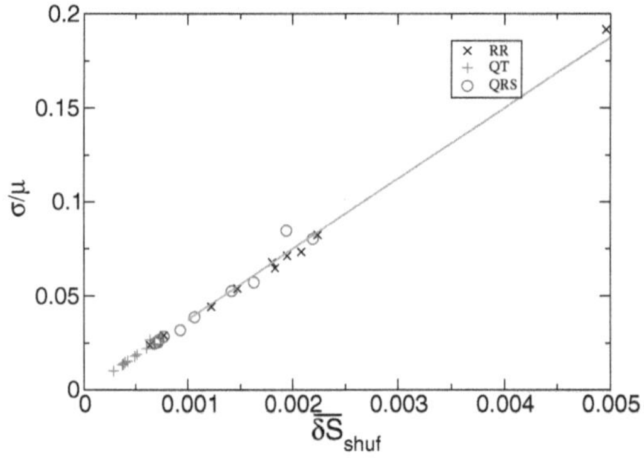

Fig. 9.4 The σ/μ value for RR, QT and QRS intervals of the ten H versus the corresponding $\overline{\delta S_{shuf}}$ value (time-window range 3–10 beats). The straight line results from a least-squares fit of all the thirty points. For the identity of the individual associated with each point see Ref. [64]. Taken from Ref. [67].

One could argue that Q_k may become i.i.d. upon their shuffling. In § 3.4.6, we showed that, when Q_k are i.i.d., δS is actually proportional to σ/μ, since the following relation holds (see Eq. (3.63)):

$$\delta S_{shuf} = \frac{\sigma}{\mu}\frac{1}{\sqrt{N-1}}\sqrt{\sum_{k=1}^{N}\left(\frac{k}{N}\ln\frac{k}{eN\overline{\chi}}\right)^2\frac{1}{N} - \left(\sum_{k=1}^{N}\frac{k}{N^2}\ln\frac{k}{eN\overline{\chi}}\right)^2} \tag{9.1}$$

where

$$\overline{\chi} = \sum_{k=1}^{N}\frac{k}{N^2} = \frac{1}{2} + \frac{1}{2N} \tag{9.2}$$

and e denotes, as usually, the base of the natural logarithms. The relation (9.1) reveals that δS_{shuf} versus σ/μ must be a straight line with a slope ranging from 34.2 to 40.4, for a time-window length 3 to 10. This result is comparable with the slopes determined above from the analysis of the ECG data.

We now proceed to compare δS_{shuf} with δS in ECG. We first point out that for a Markovian case we expect $\overline{\delta S_{shuf}} = \overline{\delta S}$ in view of the following:

Since, by definition, δS_{shuf} corresponds to the entropy fluctuations upon shuffling Q_k *randomly*, it is naturally expected that in a *Markovian* case the two quantities δS and δS_{shuf} *should* coincide. Note, however, that the reverse is *not* always true. The equality $\overline{\delta S_{shuf}} = \overline{\delta S}$ may also hold for *non*-Markovian time series, as will be demonstrated below with precise examples.

Figure 9.5(a) depicts the $\overline{\delta S}$ values, calculated for each of the 105 individuals, versus the corresponding $\overline{\delta S_{shuf}}$ for the RR intervals (time-window range 3–10 beats). The same is repeated in Figs. 9.5(b) and 9.5(c) for the QT and QRS intervals, respectively. In each case, we also plot the straight line $\overline{\delta S_{shuf}} = \overline{\delta S}$ to visualize that the vast majority of points fall below this line. The non-equality of $\overline{\delta S_{shuf}}$ and $\overline{\delta S}$ has been also verified by applying the Wilcoxon paired signed-rank test, which is recommended [42] to be fol-

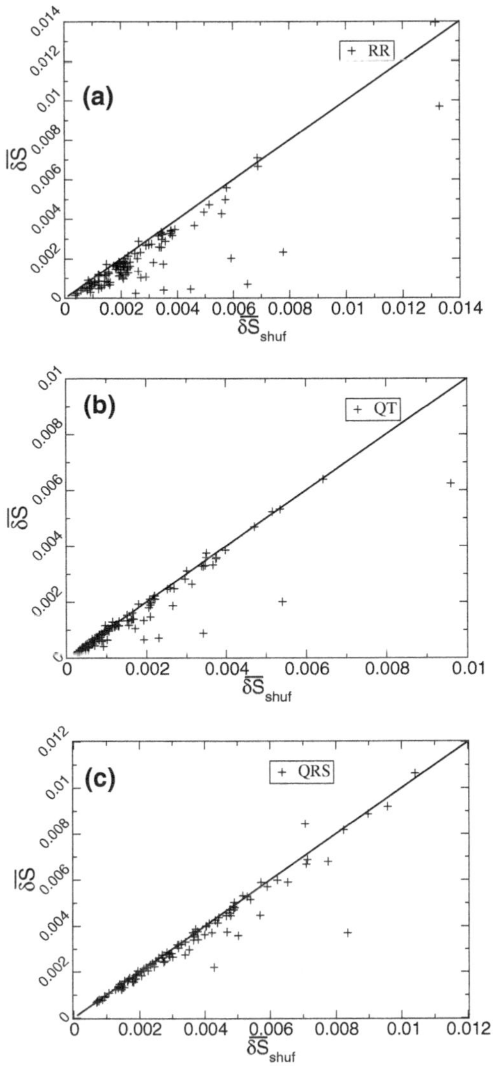

Fig. 9.5 The $\overline{\delta S}$ value, for each of the 105 individuals versus the corresponding $\overline{\delta S}_{shuf}$ value for (a) RR, (b) QT and (c) QRS intervals. The straight line, drawn in each case, corresponds to $\overline{\delta S}_{shuf} = \overline{\delta S}$. For the identity of the individual associated with each point see Ref. [64]. Taken from Ref. [67].

lowed for non-Gaussian paired data. The tested null hypothesis is that the means of $\overline{\delta S}_{shuf}$ and $\overline{\delta S}$ are the same and is rejected at a level of significance well below 0.01, since the data of Figs. 9.5(a),(b) and (c) lead to normally distributed variables $z = -8.29$, -6.81 and -6.32, respectively (note that the corresponding one-tailed asymptotic significance is given by $P(Z < z)$, i.e., the probability of obtaining a normally distributed variable obeying $N(0,1)$ that is smaller than z). Note that a least-squares fit to a straight line passing through the origin, results in the following expressions: $\delta S = 0.76(3)\delta S_{shuf}$, $\delta S = 0.85(2)\delta S_{shuf}$, $\delta S = 0.94(2)\delta S_{shuf}$ for the Figs. 9.5(a), 9.5(b), 9.5(c), respectively. The sampling rate f_{exp} in ECG is 250 Hz, thus the experimental error in their allocation is around $1/f_{exp} = 4$ ms.

This, if we take as an example the RR intervals, reflects in the calculation of δS and δS_{shuf} errors which are drastically smaller than those required to eventually justify a compatibility of the expression $\overline{\delta S} = 0.76(3)\overline{\delta S_{shuf}}$, obtained from Fig. 9.5(a), with a straight line of slope equal to unity, i.e., $\overline{\delta S} = \overline{\delta S_{shuf}}$.

The difference between δS and δS_{shuf} could be understood in the context that the former depends on the *sequential* order (of beats), while the latter does not.

Since short- (and long-) range correlations is a usual feature (see Ref. [16] and references therein) in heartbeat dynamics, which are possibly destroyed (or become weaker) upon randomizing the data, more "disorder" is intuitively expected to appear after randomization, thus reflecting $\delta S_{shuf} > \delta S$. Furthermore, note that in *all* plots of Fig. 9.5 there are some drastic deviations from the straight line $\overline{\delta S} = \overline{\delta S_{shuf}}$. The origin of some of these deviations will be discussed in Section 9.2.

Finally, by means of a precise example related to SD and H, we further clarify below the aforementioned point that the equality $\overline{\delta S} = \overline{\delta S_{shuf}}$ does *not* necessarily reflect Markovianity.

In Fig. 9.6, we plot for the QT intervals $\overline{\delta S_{shuf}}$ versus $\overline{\delta S}$ (for time-window range 3–10 beats) for SD and H. We see that there are several individuals (mainly SD, see also the next Section) whose values lie practically (i.e., within the error margins) on the straight line $\overline{\delta S} = \overline{\delta S_{shuf}}$. If we plot their δS- (or δS_{shuf}-) values versus the time-window (in a similar fashion as in Fig. 9.2), we find that they are distinctly smaller than those of the Markovian case (note that the δS values in Fig. 9.6 are smaller than 10^{-2}, while those of the Markovian case – depicted in the upper curve in Fig. 9.2 – are $\approx 2 \times 10^{-2}$ or larger). This makes clear that these individuals cannot be characterized as exhibiting Markovian behavior. (This non-Markovianity holds for *all* H and *all* SD.)

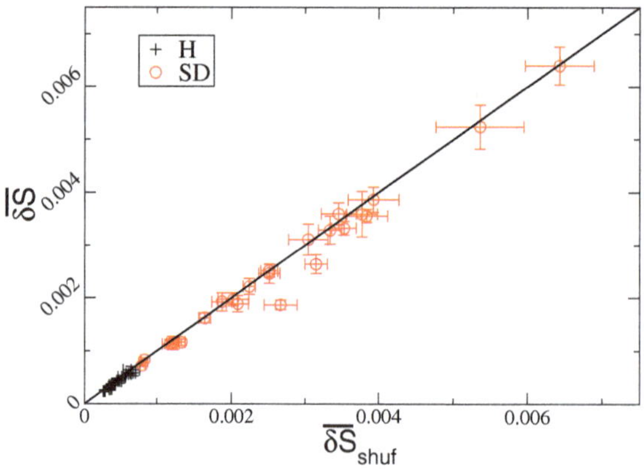

Fig. 9.6 The $\overline{\delta S}$ value, in each of the 10 H (black) and 24 SD (red), for the QT intervals versus $\overline{\delta S_{shuf}}$ (time-window range: 3–10 beats). Note that the values of the ordinates are appreciably smaller than the δS value ($\approx 2 \times 10^{-2}$) of the Markovian time series (10^3 events) depicted in Fig. 9.2. Taken from Ref. [67].

In addition, we note that in Ref. [67] (see § 4.8.3) the difference between δS and δS_{shuf} in the SES activities and "artificial" noises was also studied. It was found (see Table 4.5) that there is a systematic tendency pointing to a value of $\overline{\delta S_{shuf}}/\overline{\delta S}$ larger than unity either for the time-window range 3–5 or for the time-window range 3–10. This is consistent with the non-Markovianity of these signals, thus strengthening the conclusions of § 4.1.2 and § 4.1.3.

9.1.3 Distinction between healthy humans and sudden cardiac death ones by means of either δS(QT) or the ratio $\delta S_{shuf}/\delta S$ of the RR or QRS intervals

We emphasize that, in this subsection, we consider a set consisting *only* of two groups of ECG, namely H and SD. In other words, we are interested here in the distinction of the (otherwise healthy) SD from H, i.e., *if* the population under investigation does *not* include heart disease patients.

First, we point out that in *all* SD, the values of the quantities δS and δS_{shuf} themselves of the QT intervals exceed those of H, see Fig. 9.7. This important distinction between SD and H cannot be attributed (see Sec. VIII of Ref. [63]) to the allocation error of the QT interval.

We now turn to examine whether H and SD can also be distinguished by means of the ratio $\overline{\delta S_{shuf}}/\overline{\delta S}$, which is just the complexity measure ν introduced in § 3.6.1: we calculate this ratio, for each type of interval, at two ranges: (i) a short (s) range 3–4 beats and (ii) a longer (L) range 50–70 beats. By defining $\nu \equiv \overline{\delta S_{shuf}}/\overline{\delta S}$ (see Eq. (3.82)), hereafter the

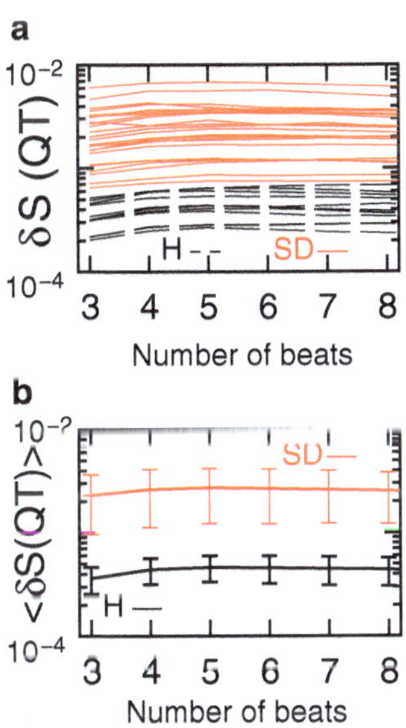

Fig. 9.7 (a) The $\delta S(QT)$ value for each of the 24 SD and 10 H (see Table 9.2) and (b) the average of the δS(QT) values – designated by $\langle \delta S(QT) \rangle$ – along with their standard deviation for each of the two groups SD and H versus the time-window length. Taken from Ref. [68].

following ratios are investigated: $\nu_s(\tau)$ and $\nu_L(\tau)$, where τ denotes the type of interval (i.e., τ = RR, QRS or QT) and s, L refer to the range studied (i.e., $s = 3$–4 beats and $L = 50$–70 beats).

The calculated values for $\nu_s(\tau)$ and $\nu_L(\tau)$ for the three types of intervals are given, for all H and SD, in Table 9.1. The minima $\min_H[\nu_\kappa(\tau)]$ and maxima $\max_H[\nu_\kappa(\tau)]$ (where κ denotes either the short, $\kappa = s$, or the longer, $\kappa = L$, range) among the healthy subjects are also inserted in two separate rows, for each type of interval and each range studied. These minima and maxima are labeled H_{min} and H_{max}, respectively. The cases of SD which have smaller and larger values than H_{min} and H_{max} (reported in each column) are marked with superscripts "*)" and "**)", respectively.

A careful inspection of Table 9.1 leads to the following main conclusion: *all* SD violate one or more *H-limits* (i.e., they have values that are smaller than H_{min} or larger than H_{max}). We intentionally emphasize that this conclusion is also drawn *even* when disregarding the results for the QT intervals. Concerning the latter intervals: Only 5 SD out of 24 violate the *H-limits*; however, in *all* SD, their δS values themselves, as mentioned, are larger than those in H, see also Figs. 9.6 and 9.7. The usefulness of this difference will be discussed later in Section 9.2.

In other words, when focusing our investigation *solely* on the RR and QRS intervals, *all* SD violate one or more of the four *H-limits* related to $\nu_s(RR)$, $\nu_L(RR)$, $\nu_s(QRS)$ and $\nu_L(QRS)$.

This is of profound importance from practical point of view, because the RR and QRS intervals can be detected more easily (and accurately) than the QT by means of an automatic threshold based detector (e.g., see Ref. [22] that evaluated the results of a detector that has been forwarded in Refs. [32] and [30] to determine automatically the waveform limits in Holter ECG).

A further inspection of Table 9.1 leads to the following additional comment:

When investigating the RR intervals *alone* (which can be detected automatically more easily and precisely than the other intervals), i.e., studying $\nu_s(RR)$ and $\nu_L(RR)$, the vast majority of SD (22 out of 24 cases) can be distinguished from H. Only two SD, i.e., sel30 and sel47, obey the corresponding *H-limits*.

Specifically, concerning $\nu_s(RR)$, fifteen SD have values smaller than $H_{min} = 1.18$, while only one SD (i.e., sel43) has a value exceeding $H_{max} = 2.25$. As for $\nu_L(RR)$, eighteen SD exceed $H_{max} = 0.77$, while only 2 SD (i.e., sel34 and sel42) have values smaller than $H_{min} = 0.44$.

9.1.3.1 Tentative physical interpretation of the above results

The main feature of the aforementioned results focuses on the fact that most SD *simultaneously* have $\nu_s(RR)$ values smaller than $H_{min}(= 1.18)$ *and* $\nu_L(RR)$ values exceeding

Table 9.1 The values of the ratios $\overline{\delta S}_{shuf}/\overline{\delta S}$ in the short (s) range 3–4 (ν_s) or in the longer (L) range 50–70 beats (ν_L) in H (sel16265 to sel17453) and SD (sel30 to sel17152) for the RR, QRS and QT intervals. Taken from Ref. [67].

	ν_s, 3–4 beats			ν_L, 50–70 beats		
Individual	RR	QRS	QT	RR	QRS	QT
sel16265	1.82	1.00	1.24	0.48	1.02	0.76
sel16272	1.74	0.99	0.98	0.77	1.08	1.11
sel16273	2.21	1.00	1.48	0.50	0.88	0.71
sel16420	1.55	0.98	1.08	0.53	1.09	0.90
sel16483	2.25	1.02	1.14	0.52	1.16	0.92
sel16539	1.42	1.06	1.25	0.50	1.08	0.65
sel16773	1.94	1.00	0.99	0.44	1.05	0.96
sel16786	1.42	1.00	1.19	0.56	1.04	0.77
sel16795	1.18	0.98	1.08	0.73	0.96	0.99
sel17453	1.38	1.01	1.02	0.56	0.98	0.81
H_{min}	1.18	0.98	0.98	0.44	0.88	0.65
H_{max}	2.25	1.06	1.48	0.77	1.16	1.11
sel30	1.29	1.11**)	1.09	0.65	0.72*)	1.09
sel31	0.96*)	1.08**)	1.17	1.23**)	0.94	0.62*)
sel32	1.39	1.14**)	1.12	1.02**)	0.69*)	0.90
sel33	1.05*)	0.99	1.00	0.86**)	0.82*)	0.99
sel34	2.11	1.29**)	1.11	0.42*)	0.78*)	0.67
sel35	1.00*)	1.00	0.96*)	1.01**)	1.05	1.08
sel36	1.02*)	1.02	1.04	0.92**)	1.00	0.88
sel37	1.07*)	1.18**)	1.07	0.55	0.75*)	0.65
sel38	0.99*)	1.09**)	1.13	1.37**)	0.89	1.04
sel39	0.96*)	1.02	1.06	2.93**)	0.92	0.90
sel40	1.01*)	1.00	0.93*)	0.78**)	0.93	1.29**)
sel41	1.07*)	1.04	1.02	1.07**)	0.84*)	0.96
sel42	1.63	1.08**)	1.23	0.42*)	1.06	0.67
sel43	2.71**)	1.11**)	1.05	0.56	0.76*)	0.89
sel44	0.91*)	0.95*)	0.88*)	2.24**)	1.46**)	1.32**)
sel45	0.98*)	1.24**)	1.29	0.98**)	0.86*)	0.79
sel46	1.03*)	1.01	1.03	1.00**)	0.84*)	1.01
sel47	1.56	0.97*)	1.03	0.45	0.97	1.01
sel48	0.82*)	1.18**)	1.44	1.48**)	0.68*)	0.73
sel49	0.93*)	1.11**)	0.96*)	1.22**)	0.70*)	1.14**)
sel50	1.05*)	0.98	0.98	0.93**)	1.23**)	1.50**)
sel51	1.25	1.01	0.97*)	1.05**)	1.24**)	0.91
sel52	1.50	1.16**)	1.22	1.00**)	0.73*)	0.68
sel17152	1.64	1.01	1.04	0.90**)	1.01	0.97

*) These values are smaller than the minimum (H_{min}) value of $\overline{\delta S}_{shuf}/\overline{\delta S}$ in H for each range.
**) These values are larger than the maximum (H_{max}) value of $\overline{\delta S}_{shuf}/\overline{\delta S}$ in H for each range.

$H_{max}(=0.77)$. The RR time series of healthy subjects are characterized by high complexity (e.g., see Refs. [18, 16]); this, if we recall that in a Markovian series we intuitively expect $\delta S_{shuf}/\delta S = 1$ (and hence $\nu_s = 1$ and $\nu_L = 1$), is compatible with the fact that in *all* H *both* $\nu_s(RR)$ and $\nu_L(RR)$ distinctly differ from unity (see Table 9.1).

We now turn to SD by considering that for individuals at high risk of sudden cardiac death the fractal physiological organization (long-range correlations) breaks down and this is often accompanied by emergence of *uncorrelated randomness*, see Ref. [16] and references therein; see also § 9.2.1.

It is therefore naturally expected that in SD the values of $\nu_s(RR)$ and $\nu_L(RR)$ become closer to the Markovian value (i.e., unity) compared to H. Hence, in SD, $\nu_s(RR)$ naturally becomes smaller than the value 1.18 (the corresponding H_{min}-limit) and $\nu_L(RR)$ larger than 0.77 (the corresponding H_{max}-limit).

We now focus on the following important property of H: although *both* $\nu_s(RR)$ and $\nu_L(RR)$ differ from unity, as mentioned, they systematically behave *differently*, i.e., $\nu_s(RR) > 1$ while $\nu_L(RR) < 1$. The exact origin of the latter difference has not yet been identified with certainty, but the following comments might be relevant: First, in the frame of the frequency-domain characteristics of heart rate variability (e.g., Refs. [38, 49]), we may state that $\nu_s(RR)$ and $\nu_L(RR)$ are associated with the *high-frequency* (HF, 0.15–0.4 Hz) and *low-frequency* (VLF: 0.015–0.04 Hz, LF: 0.04–0.15 Hz) *range in the RR* tachogram ("instantaneous" heart rate, i.e., 1/RR, see also § 9.4.3 and § 9.5.1). An important difference on the effect of the sympathetic and parasympathetic modulation of the RR intervals has been noticed (e.g., see Ref. [38] and references therein): Sympathetic tone is believed to influence the VLF-LF component whereas *both* sympathetic and parasympathetic activity have an effect on the HF component (recall that our results show $\nu_s(RR) > \nu_L(RR)$). Second, at short time-scales (high frequencies), it has been suggested [46] that we have relatively *smooth* heartbeat oscillations associated with respiration (e.g., 15 breaths per minute corresponds to a 4 sec oscillation with a peak in the power spectrum at 0.25 Hz, see Ref. [38]); this is lost upon randomizing the consecutive intervals Q_k, thus probably leading to (larger variations – compared to the original experimental data – between the durations of consecutive intervals and hence to) δS_{shuf} values larger than δS, i.e., a $\nu_s(RR)$ value larger than unity. Such an argument, if true, cannot be applied, of course, in the longer range 50–70 beats and hence explain why the opposite behavior, i.e., $\delta S_{shuf} < \delta S$, then holds. The latter finding must be inherently connected to the nature itself of the long-range correlations. The existence of the latter is evident from the fact that (in this range also) the RR-intervals result in δS values ($\approx 10^{-3}$) which significantly differ from the Markovian δS value ($\approx 10^{-2}$), compare Fig. 9.5(a) with the upper curve in Fig. 9.2.

A simplified interpretation of the results of Fig. 9.6, and in particular the reason why for the QT intervals the quantity δS is larger for the SD than for the H, could be attempted if we consider that: (i) S could be thought as a measure of the "disorder" (in the consecutive intervals) (ii) the essence of the natural time analysis is built on the variation of the durations of consecutive pulses, and (iii) it has been clinically observed (e.g., see Ref. [26]) that the QT interval (which corresponds to the time in which the heart in each beat "recovers" – electrically speaking – from the previous excitation) exhibits frequent prolonged values before cardiac death. Thus, when a time-window is sliding on an ECG of H, it is intuitively expected to find, more or less, the same S values (when sweeping through various parts of the ECG) and hence a small δS value is envisaged. By the same token, in an ECG of SD, we expect that, in view of the short–long–short sequences of the QT intervals, the corresponding S values will be much different (compared to H), thus leading to a larger δS value (note that in the same frame we may also understand why the σ/μ values – and hence δS_{shuf}, see Eq. (9.1) – are larger in SD than those in H, as shown in Fig. 9.6).

9.2 Complexity measures of the RR, QRS and QT intervals in natural time to classify sudden cardiac death individuals, heart disease patients and truly healthy ones

9.2.1 Introduction

In complex systems operating far from equilibrium like the case of heart dynamics [16], long-range correlations play an important role (such correlations are also of prominent importance in equilibrium systems when approaching a critical point, e.g., the "critical" temperature T_c, i.e., $T \to T_c$; see Section 1.5). Specifically the existence of long-range correlations in the heart rate variability has been independently established by several applications of DFA, e.g., see Refs. [46, 16] and references therein. Additional studies [21, 18] showed that healthy dynamics exhibits even higher complexity characterized by a broad multifractal spectrum (note that both methods for its determination, i.e., MF-DFA and wavelet transform, see Sections 4.5 and 4.6, respectively, have been employed). This high complexity breaks down in illness and is usually associated with increased mortality in cardiac patients (for more details see § 9.5.1). Thus, in ECG it is advisable that *both* correlations (i.e., short- and long-range), in general, be studied carefully and hence appropriate complexity measures should be envisaged. This is, in simple terms, the physics underlying the procedure that is followed in this Section.

In particular, here we employ the complexity measures introduced in § 3.6.1 to quantify the change of the natural entropy fluctuations at different length scales in time series emitted from systems operating far from equilibrium. Along these lines, we use in ECG the ratios $\delta S_i(RR)/\delta S_j(RR)$, $\delta S_i(QRS)/\delta S_j(QRS)$ and $\delta S_i(QT)/\delta S_j(QT)$ for the RR, QRS and QT intervals, respectively, where i, j denote the time-window length used in the calculation of δS. Assuming that $j < i$, these three ratios provide measures of the *δS-variability* when a scale i changes to a scale j. We select as a common scale (for all RR, QRS and QT) the *smallest* j value reasonable for natural time analysis, i.e., $j = 3$ beats, and for the short-range (s) $i = 5$, while for the longer (L) $i = 60$ beats.

Thus, in accordance to § 3.6.1, the following ratios are studied: $\lambda_s(\tau) \equiv \delta S_5(\tau)/\delta S_3(\tau)$ and $\lambda_L(\tau) \equiv \delta S_{60}(\tau)/\delta S_3(\tau)$, where τ denotes the type of interval, i.e., $\tau =$ RR, QRS or QT.

We also define [68] the ratios

$$\rho_i(\tau) = \delta S_i(RR)/\delta S_i(\tau), \tag{9.3}$$

which provide a *relative* measure of the δS values of the RR intervals compared to either QRS or QT (for the *same* number of beats i). Here, we will use for the short-range $\rho_s(\tau) \equiv \rho_3(\tau)$ and for the long-range $\rho_L(\tau) \equiv \rho_{60}(\tau)$.

Thus, we have 10 complexity measures related to λ and ρ in total: six variability measures, i.e., $\lambda_s(RR)$, $\lambda_L(RR)$, $\lambda_s(QRS)$, $\lambda_L(QRS)$, $\lambda_s(QT)$, $\lambda_L(QT)$, and four relative ones, i.e., $\rho_s(QRS)$, $\rho_L(QRS)$, $\rho_s(QT)$, $\rho_L(QT)$.

We shall show below that these complexity measures identify SD by analyzing fifteen-minute electrocardiograms and comparing them to those of truly healthy humans. In addition, these measures seem to be *complementary* to the ones employed in § 9.1.3, and

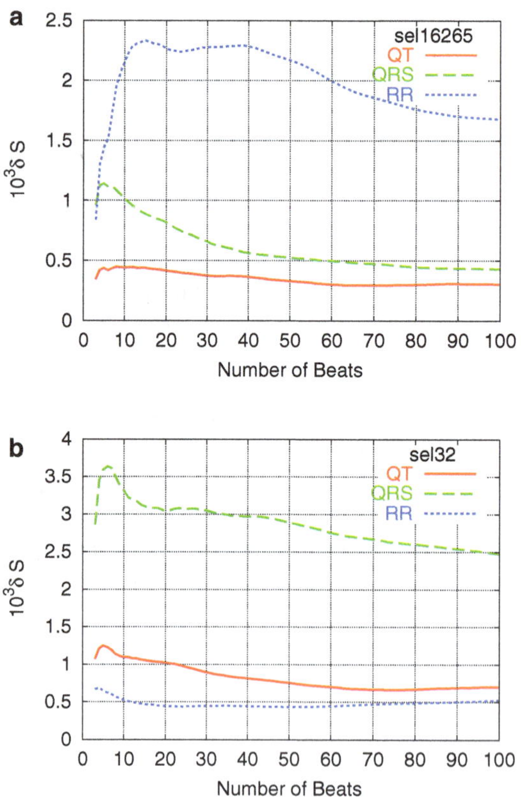

Fig. 9.8 The δS value versus the time-window length for one H (a) and one SD (b). Intervals: QT (solid red), QRS (broken green), and RR (dotted blue). Taken from Ref. [68].

altogether enable the classification of individuals into three categories: H, heart disease patients and SD. We use here the QT-Database of physiobank mentioned in Section 9.1 by considering, beyond the 10H and 24 SD, four groups of heart disease patients, i.e., 15 MIT, 13 MSV, 33 EST and 6 MST. Thus, 101 individuals out of 105 have been investigated (note that the group LT consisting of 4 individuals was discarded in view of its small population). Examples of the δS values, calculated for the RR, QRS and QT intervals in the range 3 to 100 beats are plotted in Figs. 9.8(a) and (b) for one H and one SD, respectively. As for the symbols, we use the same as those mentioned in § 9.1.1.

9.2.2 Distinction of sudden cardiac death individuals (SD) from truly healthy ones (H)

Here, as in § 9.1.3, we consider a set consisting *only* of two groups of ECG, namely H and SD. Thus, we focus here on the distinction of the (otherwise healthy) SD from H, i.e., *if* the population under investigation does *not* include heart disease patients.

The calculated values for the complexity measures $\lambda_\kappa, \rho_\kappa$ (where κ denotes either the short, $\kappa = s$, or the longer, $\kappa = L$, range) are given, for all H and SD, in Table 9.2. The minima $\min_H[\lambda_\kappa(\tau)]$ and maxima $\max_H[\lambda_\kappa(\tau)]$ among the healthy individuals for the RR ($\tau =$ RR) and QRS ($\tau =$ QRS) intervals are also inserted in this Table. We also include the corresponding minima $\min_H[\rho_\kappa(\tau)]$ and maxima $\max_H[\rho_\kappa(\tau)]$ for (the relative δS-variability measure) ρ. For the sake of simplicity, they are labeled H_{min} and H_{max}, respectively, and jointly named *H-limits*. The superscripts 'a' and 'b' show the cases of SD which have smaller and larger values than H_{min} and H_{max}, respectively. In two individuals, i.e., sel41 and sel51, it is uncertain whether their measure $\lambda_s(QRS)$ violates the value $H_{min} = 1.15$.

Table 9.2 reveals that *all* SD violate one or more *H-limits* of the four complexity measures $\lambda_s(RR)$, $\lambda_L(RR)$, $\rho_s(QRS)$ and $\rho_L(QRS)$, and hence can be distinguished from H.

In other words, the δS-variability measures of the RR-intervals, together with their relative ones with respect to the QRS (i.e., four parameters in total), seem to achieve a distinction between SD and H.

Note that $\lambda_\kappa(RR)$ *alone* can classify the vast majority of SD, i.e, all SD except sel47. Furthermore, attention is drawn to the point that if we also consider the $\lambda_\kappa(\tau)$ values calculated (*not* in the original, but) in the randomized ("shuffled") sequence of Q_m, we find that *all* SD violate one or more *H-limits* of $\lambda_\kappa(RR)$ and $\lambda_{\kappa,shuf}(RR)$ (see Table 9.2 and table VII of Ref. [63], respectively). This allows using again four parameters in total the distinction of *all* SD from H by using the RR intervals *only*.

Thus, we found that among the 10 parameters defined in the original time series extracted from each ECG (or 20 parameters, in total, if we also account for the correspond-

Table 9.2 The variability measures (λ), the relative ones (ρ), and the ratios $\nu \equiv \overline{\delta S_{shuf}}/\overline{\delta S}$ in the short (s) range and in the longer (L) range in H (sel16265 to sel17453) and SD (sel30 to sel17152) along with their $\overline{\delta S_{3-4}}(QT)$ values. Taken from Ref. [68].

Individual	RR		QRS		QT		RR over QRS	
	$\lambda_s(RR)$	$\lambda_L(RR)$	$\lambda_s(QRS)$	$\lambda_L(QRS)$	$\lambda_s(QT)$	$\lambda_L(QT)$	$\rho_s(QRS)$	$\rho_L(QRS)$
sel16265	1.72	2.38	1.19	0.52	1.27	0.88	0.88	4.01
sel16272	1.69	1.35	1.29	0.61	1.21	0.50	0.18	0.40
sel16273	1.61	2.69	1.16	0.59	1.30	1.11	1.11	5.05
sel16420	1.51	1.74	1.22	0.48	1.37	0.66	0.96	3.46
sel16483	1.43	2.37	1.23	0.49	1.31	0.68	0.25	1.22
sel16539	2.00	1.94	1.26	0.50	1.41	1.08	1.85	7.10
sel16773	1.92	2.61	1.21	0.49	1.31	0.70	0.90	4.84
sel16786	1.71	1.57	1.19	0.51	1.31	0.84	1.16	3.56
sel16795	1.77	0.99	1.24	0.55	1.16	0.56	0.77	1.37
sel17453	1.87	1.67	1.26	0.54	1.22	0.68	1.49	4.59
H_{min}	1.43	0.99	1.16	0.48	1.16	0.50	0.18	0.40
H_{max}	2.00	2.69	1.29	0.61	1.41	1.11	1.85	7.10
sel30	1.11[a)]	0.89[a)]	1.20	1.05[b)]	1.28	0.56	0.51	0.43
sel31	0.96[a)]	0.34[a)]	1.39[b)]	0.89[b)]	1.30	0.84	1.10	0.42
sel32	0.96[a)]	0.67[a)]	1.26	0.96[b)]	1.16	0.65	0.23	0.16[a)]
sel33	1.14[a)]	0.77[a)]	0.96[a)]	0.52	1.21	0.53	0.79	1.17
sel34	1.87	3.04[b)]	1.33[b)]	1.22[b)]	1.15[a)]	0.85	0.40	1.00
sel35	1.12[a)]	0.52[a)]	1.24	0.66[b)]	1.12[a)]	0.44[a)]	1.72	
sel36	1.31[a)]	0.62[a)]	1.12[a)]	0.51	1.26	0.60	2.35[b)]	2.88
sel37	0.92[a)]	0.71[a)]	1.26	0.87[b)]	1.11[a)]	0.78	0.71	0.58
sel38	0.91[a)]	0.34[a)]	1.27	0.65[b)]	1.03[a)]	0.50	0.65	0.34[a)]
sel39	0.81[a)]	0.11[a)]	1.23	0.72[b)]	1.17	0.58	0.80	0.12[a)]
sel40	1.66	0.81[a)]	1.14[a)]	0.55	1.19	0.43[a)]	0.12[a)]	0.18[a)]
sel41	1.14[a)]	0.48[a)]	1.18	0.70[b)]	1.22	0.56	0.21	0.15[a)]
sel42	1.10[a)]	1.81	1.16	0.51	1.31	1.01	0.95	3.40
sel43	1.69	3.04[b)]	1.24	0.77[b)]	1.26	0.68	0.06[a)]	0.23[a)]
sel44	1.18[a)]	0.18[a)]	1.52[b)]	0.43[a)]	1.02[a)]	0.34[a)]	0.59	0.25[a)]
sel45	0.92[a)]	0.42[a)]	1.16	0.73[b)]	1.37	0.68	1.46	0.85
sel46	0.94[a)]	0.43[a)]	1.05[a)]	0.71[b)]	1.12[a)]	0.55	1.35	0.82
sel47	1.54	2.07	1.19	0.54	1.36	0.57	0.16[a)]	0.63
sel48	0.84[a)]	0.30[a)]	1.23	1.08[b)]	1.14[a)]	1.00	0.91	0.26[a)]
sel49	0.93[a)]	0.33[a)]	1.17	0.83[b)]	1.16	0.50	1.27	0.50
sel50	1.32[a)]	0.59[a)]	1.28	0.46[a)]	1.21	0.32[a)]	1.78	2.31
sel51	1.83	0.72[a)]	1.14[a)]	0.42[a)]	1.24	0.66	0.16[a)]	0.27[a)]
sel52	1.40[a)]	0.73	1.32[b)]	1.02[b)]	1.29	1.01	0.14[a)]	0.10[a)]
sel17152	1.06[a)]	0.93[a)]	1.31[b)]	0.58	1.13[a)]	0.54	0.06[a)]	0.10[a)]
min	0.81	0.11	0.96	0.42	1.02	0.32	0.06	0.10
max	1.87	3.04	1.52	1.22	1.37	1.01	2.35	3.40

Table 9.2 Continued

RR over QT		3–4 beats (ν_s)[c]			50–70 beats (ν_L)[c]			
$\rho_s(QT)$	$\rho_L(QT)$	RR	QRS	QT	RR	QRS	QT	$\overline{\delta S_{3-4}(QT)} \times 10^3$
2.44	6.62	1.87	0.98	1.29	0.48	1.02	0.75	0.38
0.67	1.79	1.65	0.88	0.94	0.77	1.10	1.07	0.48
3.17	7.65	2.18	0.99	1.46	0.50	0.88	0.71	0.24
1.97	5.21	1.60	0.99	1.07	0.53	1.09	0.90	0.36
0.96	3.37	2.27	0.99	1.17	0.52	1.15	0.92	0.35
5.57	10.04	1.43	1.07	1.27	0.50	1.08	0.65	0.52
1.49	5.54	1.85	1.01	0.91	0.44	1.05	0.97	0.55
3.97	7.43	1.39	1.01	1.19	0.55	1.04	0.77	0.23
2.87	5.08	1.10	0.98	1.05	0.74	0.95	1.00	0.56
2.91	7.12	1.46	1.01	1.02	0.57	0.98	0.81	0.34
0.67	1.79	1.10	0.88	0.91	0.44	0.88	0.65	0.23
5.57	10.04	2.27	1.07	1.46	0.77	1.15	1.07	0.56
1.73	2.73	1.15	1.08[b]	1.13	0.66	0.71[a]	1.10[b]	1.04[b]
0.80	0.32[a]	0.90[a]	1.06	1.15	1.23[b]	0.97	0.63[a]	3.01[b]
0.63[a]	0.64[a]	1.31	1.11[b]	1.13	1.02[b]	0.69[a]	0.90	1.14[b]
2.41	3.50	1.07[a]	1.00	1.08	0.85[b]	0.83[a]	1.00	0.76[b]
1.16	4.12	2.13	1.11[b]	1.12	0.41[a]	0.77[a]	0.67	0.69[b]
0.83	0.99[a]	1.02[a]	0.97	0.97	1.02[b]	1.05	1.07	6.45[b]
1.45	1.52[a]	1.03[a]	1.01	1.08	0.93[b]	0.99	0.89	2.08[b]
1.19	1.07[a]	1.11	1.17[b]	1.07	0.56	0.75[a]	0.64[a]	3.30[b]
0.37[a]	0.25[a]	1.15	1.08	1.12	1.33[b]	0.89	1.03	2.71[b]
1.53	0.28[a]	0.97[a]	0.97	0.99	2.93[b]	0.93	0.89	2.44[b]
0.20[a]	0.38[a]	1.03[a]	1.01	0.93	0.79[b]	0.94	1.30[b]	3.43[b]
0.80	0.68[a]	0.91[a]	1.04	1.06	1.05[b]	0.84[a]	0.96	1.53[b]
1.62	2.89	1.63	1.09[b]	1.26	0.43[a]	1.06	0.66	0.95[b]
0.11	0.48[a]	2.79[b]	1.12[b]	1.08	0.56	0.77[a]	0.89	2.23[b]
1.08	0.58[a]	0.91[a]	0.92	0.90[a]	2.25[b]	1.46[b]	1.33[b]	4.12[b]
1.14	0.71[a]	0.97[a]	1.05	1.11	0.98[b]	0.88	0.79	1.71[b]
1.59	1.26[a]	1.01[a]	0.99	1.01	0.99[b]	0.85[a]	1.01	3.44[b]
0.14[a]	0.49[a]	1.60	0.97	0.97	0.45	0.96	1.02	2.85[b]
1.36	0.41[a]	0.84[a]	1.24[b]	1.42	1.49[b]	0.68[a]	0.74	1.75[b]
1.08	0.71[a]	0.86[a]	1.15[b]	0.96	1.21[b]	0.71[a]	1.11[b]	3.96[b]
1.21	2.26	1.07[a]	1.00	0.91	0.93[b]	1.20[b]	1.62[b]	5.21[b]
0.30[a]	0.33[a]	1.30	1.04	1.00	1.05[b]	1.24[b]	0.90	1.83[b]
0.42[a]	0.31[a]	1.51	1.13[b]	1.17	1.02[b]	0.73[a]	0.67	1.66[b]
0.23[a]	0.40[a]	1.68	1.01	1.03	0.91[b]	1.01	0.97	1.15[b]
0.11	0.25	0.84	0.92	0.90	0.41	0.68	0.63	0.69
2.41	4.12	2.79	1.24	1.42	2.93	1.46	1.62	6.45

a) These values are smaller than the H_{min} given in each column.
b) These values are larger than the H_{max} given in each column.
c) These values do not fully coincide with those given in Ref. [67] for the reasons discussed in § 9.2.7.

ing parameters defined in the time series obtained after shuffling the Q_m randomly), *only* four are required for the distinction between SD and H. We clarify that this seems to be extremely difficult to be achieved by chance. In order to visualize it, if we assume (for the sake of convenience only) independent and identically distributed (i.i.d.) values of the parameters for one subject, we find that the probability that *all* 4 parameters are within the bounds (minima and maxima) set by 10 other subjects (i.e., the healthy ones) is $(1-2/11)^4 \approx 0.448$. Thus, the probability that all 24 additional subjects are classified as SD by pure chance is $(1-0.448)^{24} \approx 6.4 \times 10^{-7}$, i.e., extremely small (note that only if one decides which parameters one wants to use *before* the calculation of the values is this probability valid; this is the reason why blind evaluation – defining all methods, parameters and criteria studying one set of data, and *then* testing the significance using an additional set of independent data – is considered very important in medical applications and/or publications). If one just picks 4 parameters out of the original 20 as in our case, the above probability should be multiplied by the possible combinations of selecting 4 objects among 20, i.e., $20!/(4!16!) = 4{,}845$, leading to a value 0.31% of achieving our result by chance.

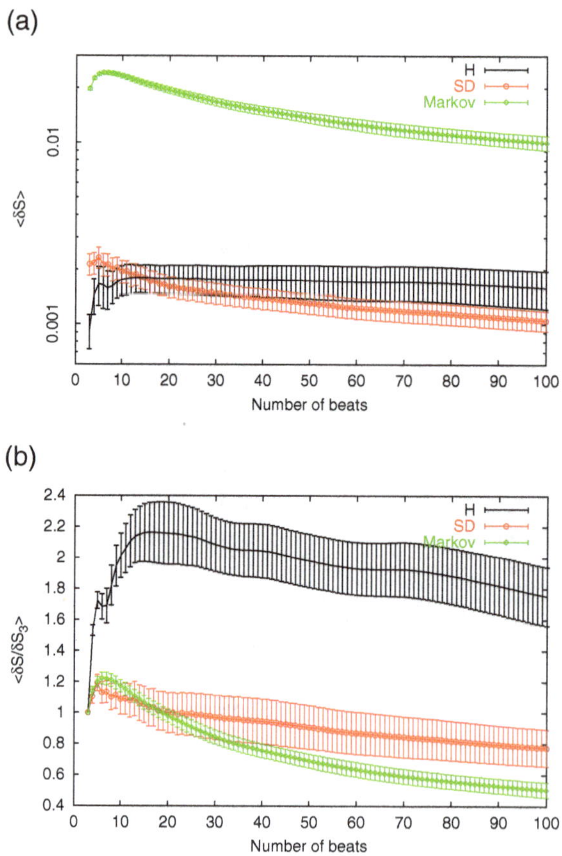

Fig. 9.9 The average (denoted by the brackets) values of (a): the $\delta S_i(RR)$ and (b): $\delta S_i(RR)/\delta S_3(RR)$ for the SD (solid black) and H (red circles) versus the time-window length; the bars correspond to the standard error of the mean. The results for a Markovian time series are also plotted (green squares), but the bars here denote the standard deviation. Taken from Ref. [68].

9.2.2.1 Physical interpretation of the aforementioned results in § 9.2.2

The main feature of these results focuses on the fact that both ratios $\lambda_s(RR)$ and $\lambda_L(RR)$ become smaller in the vast majority of SD, compared to H.

Recall that the $\delta S_i(RR)$ values themselves cannot distinguish SD from H, see Fig. 9.9(a), in contrast to the ratios $\delta S_i(RR)/\delta S_3(RR)$, see Fig. 9.9(b).

We now consider that for individuals at high risk of sudden cardiac death, the fractal organization (long-range correlations) that characterizes the healthy subjects breaks down (see Refs. [18, 15] and references therein; see also § 9.2.1 and § 9.5.1). This breakdown is often accompanied by emergence of *uncorrelated randomness* (as already mentioned in § 9.1.3.1) or *excessive* order (e.g., periodic oscillations appear in the heart rate recordings of "frequency" $\approx$ 1/min, which are associated with Cheyne–Stokes breathing) [15].

Let us now calculate [67] the δS values in a (dichotomous) Markovian (hereafter labeled $\mathscr{M}$) time series (exponentially distributed durations), see § 9.1.2, hereafter labeled $\delta S_i(\mathscr{M})$, for a total number of $N = 10^3$ pulses (i.e., length comparable to that of the ECG analyzed here). These values are plotted – along with those for SD and H – in green in Fig. 9.9(a) and show that the corresponding λ_s and λ_L variability measures are

$$\lambda_s(\mathscr{M}) = 1.20 \pm 0.03 \text{ and } \lambda_L(\mathscr{M}) = 0.64 \pm 0.05; \qquad (9.4)$$

see Fig. 9.9(b). Three comments are now in order:

First, the $\delta S_i(\mathscr{M})$ values differ drastically, see Fig. 9.9(a), from the $\delta S_i(RR)$ values themselves of *both* SD and H, which indicates that the RR intervals (both in SD and H) exhibit non-Markovian behavior, as mentioned in § 9.1.2.

This is consistent with the aspects that bodily rhythms, such as the heartbeat, show complex dynamics, e.g., Refs. [18, 15].

Second, the fact that $\lambda_s(RR)$ in SD becomes smaller than in H can now be understood as follows: Since H exhibit a high degree of complexity, it is expected that (even) their H_{min} value (= 1.43) should markedly exceed $\lambda_s(\mathscr{M})$. On the other hand, in SD this high complexity breaks down and hence their $\lambda_s(RR)$ values naturally approach $\lambda_s(\mathscr{M})$, thus becoming smaller.

This is strengthened by the fact that the SD average value of $\lambda_s(RR)$ in Table 9.2 is 1.19, which almost coincides with $\lambda_s(\mathscr{M})(= 1.20)$.

The latter coincidence also occurs for the QRS intervals in *both* H and SD, which agrees with the observations [26] mentioned above (§ 9.1.3.1) that the prolonged QT intervals in SD mainly originate from enlarged ST values, while their QRS intervals may remain the *same*.

Third, we now turn to the interpretation of the results related to $\lambda_L(RR)$. In H, it is expected that (in view of the RR long-range correlations [15]) the corresponding values must be appreciably larger than $\lambda_L(\mathscr{M}) = 0.64 \pm 0.05$. We now examine the SD: If, in SD "*uncorrelated randomness*" appears, this reflects that their $\lambda_L(RR)$ values naturally approach $\lambda_L(\mathscr{M})$, thus becoming smaller (compared to H); this actually occurs in the vast majority of SD in Table 9.2.

If in SD the aforementioned periodicities (associated with Cheyne–Stokes breathing) appear, it is naturally expected (as shown below in § 9.2.2.2) to find *large* δS values when a time-window of length around 60 beats, or so (i.e., related to the aforementioned "frequency" $\approx$ 1/min) sweeps through the RR time series. This for SD, results in δS values even larger than those in H, since in H no such periodicities appear, as actually observed in the *two* cases marked with superscript 'b' (i.e., those exceeding H_{max}) in Table 9.2.

The plausibility of the above interpretation is considerably strengthened by the following remarks. Recall that the H_{min} values for $\lambda_s(RR)$ and $\lambda_L(RR)$ have been determined *empirically* by selecting the smallest values among the 10 H. We may overcome this empirical selection, however, as follows. We divide each ECG in equal and non-overlapping segments of length (l) significantly larger than the time-window of 60 beats (e.g., $l = 180$ or 120 beats; see Tables 9.3 and 9.4, respectively) and calculate the corresponding measures $[\lambda_s(RR)]_l$ and $[\lambda_L(RR)]_l$ for the various segments labeled by l. The mean values $\langle\lambda_\kappa(RR)\rangle_l$ for each individual, agree more or less with the values that have been obtained above, when the time-window swept through the whole record and their standard deviations provide a measure of the variability of each of these two complexity measures among the various segments studied in each record. Comparing the values of $\min\{[\lambda_s(RR)]_l\}$ and $\min\{[\lambda_L(RR)]_l\}$ (see the Tables 9.3 and 9.4) to $\lambda_s(\mathscr{M})$ and $\lambda_L(\mathscr{M})$, respectively, we find the following. In H , the values of $\min\{[\lambda_\kappa(RR)]_l\}$ significantly exceed $\lambda_\kappa(\mathscr{M})$ for $\kappa = s$ or L, as they should (with a possible exception of $\min\{[\lambda_L(RR)]_l\}$ for sel16795, which might be due to the fact that the ECG of this individual has the smallest length, i.e., 760 beats, among the H). On the other hand, most SD (e.g., in Table 9.3 those marked with 'c' and 'd') exhibit $\min\{[\lambda_\kappa(RR)]_l\}$ values which are smaller than (or equal to) $\lambda_\kappa(\mathscr{M})$ for $\kappa = s$ or L (the values in bold, in both Tables 9.3 and 9.4, indicate the minority of cases of SD in which the resulting $\min\{[\lambda_\kappa(RR)]_l\}$ values exceed $\lambda_\kappa(\mathscr{M})$). Interestingly, all these (21 or 22 out of 24) SD cases coincide with those already marked with 'a' in Table 9.2 on the basis of the empirically determined *H-limits* of $\lambda_s(RR)$ and $\lambda_L(RR)$. Thus, the essence of our findings could be summarized as follows:

When a time-window sweeps through the *whole* record available, the vast majority of SD exhibits $\lambda_s(RR)$ and $\lambda_L(RR)$ values which are significantly smaller than those in H and hence SD are distinguished from H. This finding stems from the fact that some segments of the SD records exhibit values of these measures that are comparable with those of a Markovian behavior (see Fig. 9.9(b)).

Table 9.3 The resulting values of the *variability* measures $\lambda_s(RR)$ and $\lambda_L(RR)$ when using segments of length $l = 180$ beats and then calculating their mean and minimum values. Taken from Ref. [63].

Signal	$\lambda_s(RR)$				$\lambda_L(RR)$			
	$\lambda_s(RR)^{a)}$	$\lambda_s(RR)^{b)}$	$\langle\lambda_s(RR)\rangle_l$	$\min\{[\lambda_s(RR)]_l\}$	$\lambda_L(RR)^{a)}$	$\lambda_L(RR)^{b)}$	$\langle\lambda_L(RR)\rangle_l$	$\min\{[\lambda_L(RR)]_l\}$
sel16265	1.72	1.73	1.69	1.52	2.38	2.40	1.78	0.92
sel16272	1.69	1.66	1.67	1.56	1.35	1.44	1.31	1.12
sel16273	1.61	1.60	1.60	1.52	2.69	2.67	2.50	1.11
sel16420	1.51	1.54	1.50	1.43	1.74	1.80	1.80	1.37
sel16483	1.43	1.38	1.40	1.30	2.37	2.51	2.19	1.44
sel16539	2.00	2.10	2.02	1.73	1.94	2.08	1.92	1.03
sel16773	1.92	1.93	1.90	1.66	2.61	2.64	2.26	1.52
sel16786	1.71	1.78	1.76	1.54	1.57	1.70	1.51	0.95
sel16795	1.77	1.81	1.77	1.67	0.99	1.10	0.82	$0.41^{e)}$
sel17453	1.87	1.91	1.90	1.85	1.67	1.73	1.68	0.93
sel30	$1.11^{c)}$	1.12	1.17	1.03	0.89	1.06	1.38	**1.21**
sel31	$0.96^{c)}$	0.96	0.97	0.88	$0.34^{d)}$	0.34	0.35	0.28
sel32	$0.96^{c)}$	1.12	1.28	0.93	$0.67^{d)}$	0.95	1.32	0.39
sel33	$1.14^{c)}$	0.90	1.07	0.92	0.77	0.74	0.87	**0.77**
sel34	1.87	2.07	1.99	**1.50**	3.04	3.48	2.82	**1.32**
sel35	$1.12^{c)}$	1.13	1.14	1.07	$0.52^{d)}$	0.58	0.56	0.44
sel36	$1.31^{c)}$	1.30	1.33	1.16	$0.62^{d)}$	0.63	0.64	0.48
sel37	$0.92^{c)}$	0.91	0.94	0.75	$0.71^{d)}$	0.78	0.69	0.51
sel38	$0.91^{c)}$	0.81	1.09	0.79	$0.34^{d)}$	0.12	0.36	0.08
sel39	$0.81^{c)}$	0.81	0.81	0.79	$0.11^{d)}$	0.11	0.10	0.07
sel40	1.66	1.16	1.65	**1.60**	$0.81^{d)}$	0.82	0.67	0.35
sel41	$1.14^{c)}$	1.13	1.31	0.91	$0.48^{d)}$	0.44	0.63	0.10
sel42	$1.10^{c)}$	1.22	1.31	0.87	$1.81^{d)}$	2.13	2.59	0.69
sel43	1.69	1.55	1.63	**1.52**	3.04	3.85	3.24	**1.65**
sel44	$1.18^{c)}$	1.17	1.19	1.17	$0.18^{d)}$	0.18	0.17	0.13
sel45	$0.92^{c)}$	0.92	1.12	0.82	$0.42^{d)}$	0.42	0.65	0.11
sel46	$0.94^{c)}$	0.96	0.94	0.88	$0.43^{d)}$	0.46	0.41	0.30
sel47	1.54	1.54	1.54	**1.37**	2.07	2.16	2.32	**1.81**
sel48	$0.84^{c)}$	0.84	0.93	0.84	$0.30^{d)}$	0.30	0.79	0.14
sel49	$0.93^{c)}$	0.89	0.93	0.87	$0.33^{d)}$	0.37	0.32	0.20
sel50	$1.32^{c)}$	1.33	1.33	1.16	$0.59^{d)}$	0.73	0.61	0.49
sel51	1.83	1.87	1.79	**1.63**	$0.72^{d)}$	0.75	0.77	0.66
sel52	$1.40^{c)}$	1.41	1.13	0.99	$0.73^{d)}$	0.74	0.69	0.49
sel17152	$1.06^{c)}$	0.94	1.00	0.87	$0.93^{d)}$	0.98	1.12	0.51

a) They come from Table 9.2.
b) These values, for the sake of comparison, are obtained after applying a detection algorithm which excludes the "outliers"; this algorithm is analogous to the one used by Ivanov et al. [21].
c) These individuals have $\min\{[\lambda_s(RR)]_l\}$ values which are equal to or smaller than the value $\lambda_s(\mathscr{M}) = 1.20 \pm 0.03$ discussed in the text.
d) These individuals have $\min\{[\lambda_L(RR)]_l\}$ values which are equal to or smaller than the value $\lambda_L(\mathscr{M}) = 0.64 \pm 0.05$ discussed in the text.
e) The ECG of this individual has the smallest length (760 beats) among the H, which might be one of the reasons why this case only deviates from the other H.

Table 9.4 The resulting values of the *variability* measures $\lambda_s(RR)$ and $\lambda_L(RR)$ when using segments of length $l = 120$ beats and then calculating their mean and minimum values. Taken from Ref. [63].

Signal	$\lambda_s(RR)$		$\lambda_L(RR)$	
	$\langle\lambda_s(RR)\rangle_l$	$\min\{[\lambda_s(RR)]_l\}$	$\langle\lambda_L(RR)\rangle_l$	$\min\{[\lambda_L(RR)]_l\}$
sel16265	1.70	1.46	1.87	0.98
sel16272	1.66	1.46	1.20	0.82
sel16273	1.59	1.47	1.95	0.79
sel16420	1.51	1.39	1.57	0.86
sel16483	1.42	1.23	2.45	0.90
sel16539	2.04	1.67	1.50	0.90
sel16773	1.91	1.67	2.41	0.77
sel16786	1.78	1.49	1.18	0.69
sel16795	1.77	1.68	0.68	0.44[e)]
sel17453	1.93	1.77	1.33	0.77
sel30	1.09	0.93	1.02	0.68
sel31	0.99	0.87	0.31	0.19
sel32	1.34	0.92	1.82	0.27
sel33	1.13	0.91	0.70	0.46
sel34	2.01	**1.39**	2.92	**1.26**
sel35	1.15	1.03	0.45	0.35
sel36	1.33	1.21	0.64	0.36
sel37	0.96	0.75	0.53	0.33
sel38	1.11	0.78	0.34	0.07
sel39	0.81	0.78	0.10	0.06
sel40	1.66	**1.58**	0.64	0.23
sel41	1.32	0.88	0.58	0.18
sel42	1.43	0.81	2.31	0.48
sel43	1.62	**1.42**	3.39	**1.11**
sel44	1.19	1.13	0.16	0.09
sel45	1.17	0.81	0.69	0.19
sel46	0.94	0.85	0.41	0.29
sel47	1.55	**1.34**	1.83	**1.28**
sel48	0.98	0.77	1.64	0.14
sel49	0.91	0.86	0.25	0.08
sel50	1.32	1.09	0.51	0.34
sel51	1.80	**1.60**	0.63	0.57
sel52	1.11	0.94	0.72	0.29
sel17152	0.99	0.79	1.16	0.40

[e)] The ECG of this individual has the smallest length (760 beats) among the H, which might be one of the reasons why this case only deviates from the other H.

The same conclusions are drawn irrespective of whether we use a detection algorithm to exclude 'outliers' from the records. In the third column (labeled with a superscript 'b') of Table 9.3, we present the values obtained after applying such a detection algorithm. More precisely a moving window average filter was applied. For each set of five contiguous intervals, a local mean was computed, excluding the central interval. If the value of the central interval exceeded the local average by a factor 1.5 or larger, it was considered to be an outlier and excluded from the interval series. This algorithm is analogous to the one used by Ivanov et al. [21].

9.2.2.2 Study of the δS values for time series with a "sinusoidal" background

In Fig. 9.10, we show the δS value calculated when a time-window of length 3–100 beats is sliding through the time series given by

$$x_k = a + b\sin(2\pi k/T), \tag{9.5}$$

or

$$y_k = \mu + \sigma\sin(2\pi k/T)\eta, \tag{9.6}$$

where η is an exponentially distributed random variable of unit mean and standard deviation. The amplitude of the "oscillation" b or σ is comparable to the standard deviation of the RR intervals in ECG and the "period" T is 60 beats, i.e., comparable to that of the periodic oscillations in the heart rate recordings which are associated with Cheyne–Stokes breathing [15] mentioned above in § 9.2.2.1. The main result of Fig. 9.10 could be summarized as follows:

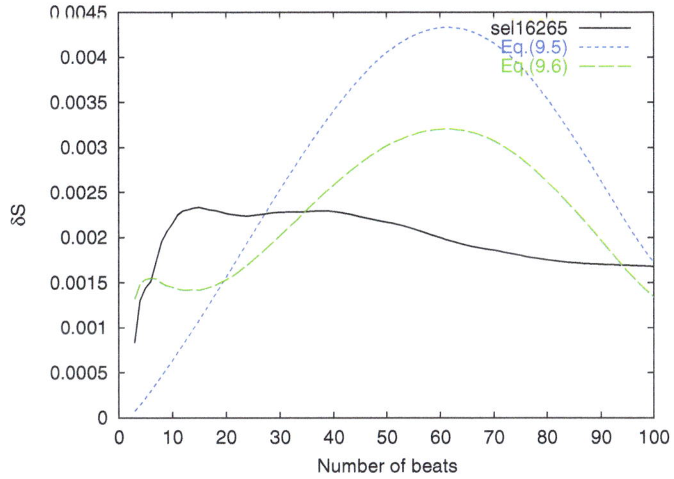

Fig. 9.10 The δS values versus the time-window length for one H (sel16265) together with those obtained using Eq. (9.5) (dotted blue) or Eq. (9.6) (broken green). Note that *no* maximum at around 60 beats appears in the case of H. Taken from Ref. [63].

When the length of the sliding time-window becomes equal to the "period" ($T = 60$ beats) of the "oscillating" background, the δS value becomes maximum.

Note that the window length corresponding to the maximum amplitude is practically equal to that observed *if* the "oscillating" background were *solely* present; the latter case for the sake of comparison is also plotted in dotted blue in Fig. 9.10.

9.2.3 Comparison of the present results in natural time with those deduced from the Approximate Entropy (AE) or the Sample Entropy (SE) to distinguish SD from H

In § 9.1.1, it was mentioned that two other dynamic entropies, i.e., AE or SE, have been applied to ECG analysis. Here, we compare [63] the results of these two entropies to distinguish SD from H with those achieved above in § 9.2.2 by means of the complexity measures in natural time.

AE and SE are based on two input parameters: the sequence length m and the tolerance level r. The smallest values of entropy correspond to perfectly regular sequences, since the output of these algorithms provides a likelihood measure that two sequences (within tolerance level r) remain close at the next point. Note that as r decreases both AE and SE increase, because the criterion for sequence matching becomes more stringent [51].

In Fig. 9.11, we plot the values of AE calculated for $r = 0.2$STD and $m = 2$ (as recommended in the program `apen` [25]) and SE, again for $m = 2$, and $r = 0.2$STD (by means of the program `sampen` [33]) along with the values of the entropy S in natural time for SD and H.

Note that no distinction of *all* individuals can be achieved by means of either AE or SE (note that this still holds if we calculate AE for $r = 0.65$STD as recommended in Ref. [44]), although the average values of the two groups actually turn out to be different. This shows the necessity of using the complexity measures based on the fluctuations δS of the entropy S in natural time in order to obtain the distinction of *all* SD from H as in § 9.1.3 and § 9.2.2. Such a distinction cannot be achieved by means of the S values themselves (which are close to S_u, see Fig. 9.11) as already emphasized in § 9.1.1.

9.2.4 The procedure for identifying SD among other individuals that include healthy ones and heart disease patients

We first address the question of distinguishing all SD from the other individuals (heart disease patients and H).

We use here the 101 individuals mentioned in § 9.2.1.

The values of all the complexity measures in natural time: λ, ρ, ν,$\overline{\delta S_{3-4}}$(QT), λ_{shuf}, ρ_{shuf} and $\overline{\delta S_{3-4,shuf}}$(QT) for each one of the 101 ECG can be found in Table 9.2 and in tables III to VII of Ref. [63] which are freely accessible.

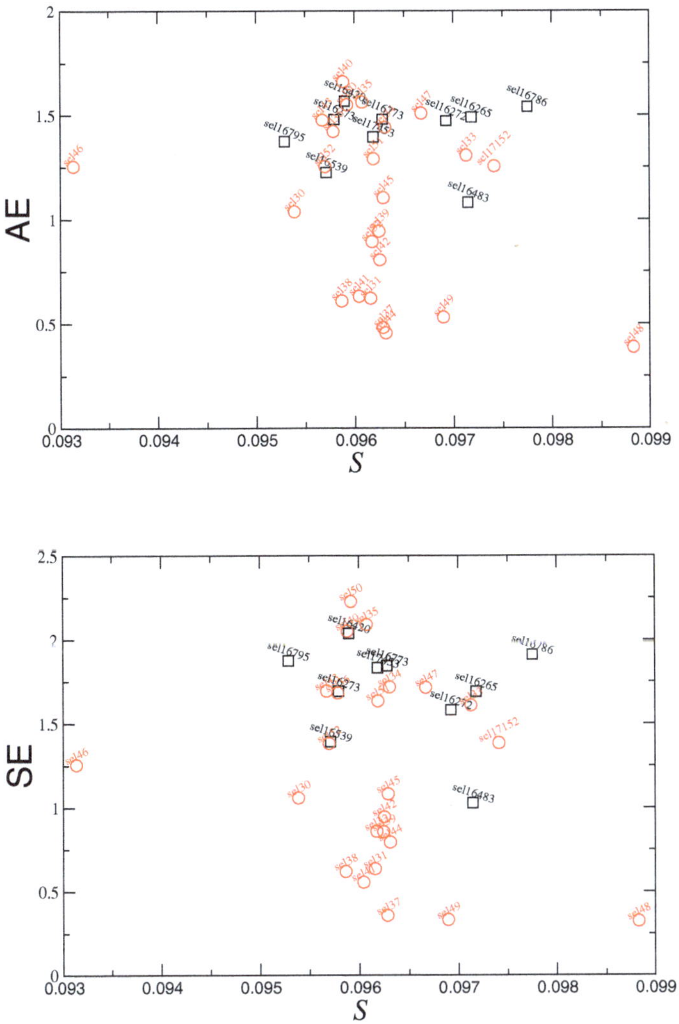

Fig. 9.11 The values (for $m = 2$, $r = 0.2$STD) of AE (upper panel) or SE (lower panel) versus the entropy S in natural time calculated for SD and H. Taken from Ref. [63].

In addition, the quality of ECG data was discussed in Ref. [63] with the following results: Among the 101 individuals investigated, five patients have been identified as "outliers". The appearance of such "outliers" is not surprising (see below) when using (as we did) an automatic threshold detector [31, 22, 32, 30] for the allocation of the intervals. More precisely, their recognition was made as follows: four individuals, i.e., two MIT (sel230 and sel231) and two EST (sele0612 and sele0704), have been identified as "outliers", because they exhibit $\nu_s(QRS)$ values which are *unusually* larger than unity (a simple statistical test – by means of the STATIST [39] – of the 101 $\nu_s(QRS)$ values, immediately shows that these four cases can be considered as "outliers"). The fifth individual identified as "outlier", i.e., sele0136, has a $\rho_L(QRS)$ value drastically larger than the corresponding values of *all* other patients.

An inspection of the measures λ, ρ, ν shows three facts. First, *all* SD and *all* patients violate one or more *H-limits*. Second, *none* of the measures λ, ρ, ν alone, or a combination of two of them, can effectively differentiate the SD from the patients. Third, if we consider the three measures λ, ρ, ν (i.e., 16 parameters consisting of the 10 parameters explained in § 9.2.1 and the 6 parameters of ν_s and ν_L related to the RR, QRS and QT intervals, e.g., see Table 9.1) altogether, we find that 20 SD out of 24 violate some of the limits of both patients and H, thus allowing in principle a distinction of the *vast* majority of SD from the other individuals.

Thus, in summary, the consideration of the quantities (λ, ρ, ν) *only*, does not lead to a distinction of *all* SD from the patients. The same conclusion is drawn if we alternatively consider the quantities (λ, λ_{shuf},ρ) *only*.

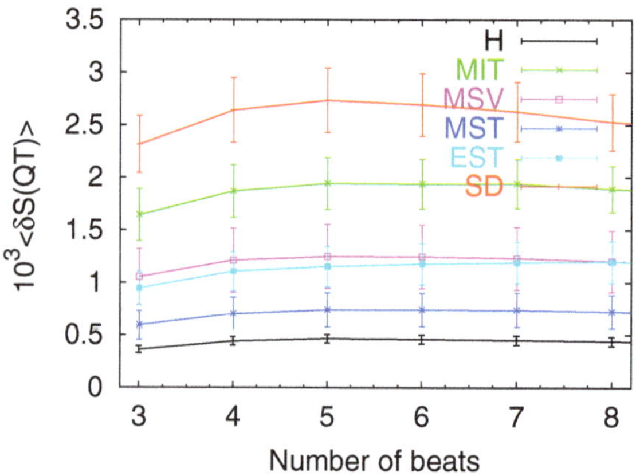

Fig. 9.12 The average of the $\delta S(QT)$ values – labeled $\langle \delta S(QT) \rangle$ – for each of the six groups labeled H, MIT, MSV, MST, EST and SD versus the time-window length. The bars denote the standard error of the mean. (The corresponding standard deviations overlap considerably and hence are not shown for the sake of clarity.) The lowermost curve and the uppermost curve correspond to H and SD, respectively and hence coincide with the two curves depicted in Fig. 9.7(b). Taken from Ref. [68].

We now turn to the investigation of the δS(QT) values, which as shown in Fig 9.7(a) allows the distinction of *all* SD from H. In Fig. 9.12, the average value $\langle \delta S \rangle$(QT) for each group is plotted versus the time-window length. It is intriguing that the results of the four groups (MIT, MSV, MST, EST) of patients are located between H (the lowermost curve) and SD (the uppermost curve). We emphasize, however, that if we plot the curves for each one of the 101 individuals (in a way similar to that of Fig. 9.7(a)), we find that there are some patients whose results overlap with either SD or H. We now restrict ourselves to $\overline{\delta S}_{3-4}$(QT) which for the sake of simplicity will be hereafter simply denoted δS(QT). Let us consider only the limiting cases – i.e., the values corresponding to the lowermost and the uppermost curve, to be called hereafter $\delta S(\mathrm{QT})_{min,\xi}$ and $\delta S(\mathrm{QT})_{max,\xi}$, respectively – obtained in each group ξ of heart disease patients, i.e., ξ = MIT, MSV, EST or MST. In order to distinguish SD from heart disease patients, we must appropriately discriminate the overlap which refers to those of the patients that lie above the uppermost δS(QT) of H;

the latter from now on will be called $\delta S(\mathrm{QT})_{max,H}$. Thus, the limits of the patients we are currently interested in, do not extend from $\delta S(\mathrm{QT})_{min,\xi}$ to $\delta S(\mathrm{QT})_{max,\xi}$, since they must exceed $\delta S(\mathrm{QT})_{max,H}$, i.e.,

$$\delta S(QT) > \delta S(QT)_{max,H}. \tag{9.7}$$

The curve which corresponds to the one of the patients that has $\delta S(\mathrm{QT})$ lying just above the $\delta S(QT)_{max,H}$ corresponds to a value, which will be hereafter labeled $\delta S(\mathrm{QT})_{min',\xi}$ (e.g. see fig.3 of Ref. [63]). Thus, if we apply to each group ξ of patients the condition

$$\delta S(QT)_{min',\xi} \leq \delta S(QT) \leq \delta S(QT)_{max,\xi} \tag{9.8}$$

we are left only with those of the patients of the group ξ that actually overlap with SD.

We now recall that the measures λ, ρ, ν altogether, which are in fact ratios of δS values, enable the discrimination of the *vast* majority of SD from all the others (i.e., heart disease patients and H), while the $\delta S(\mathrm{QT})$ values themselves efficiently distinguish, as mentioned (see Fig. 9.7), *all* SD from H. This motivates us to investigate whether a proper combination of these two facts can serve our purpose, which refers to the identification of all SD among the other individuals (heart disease patients and H). Thus, we now compare the quantities λ, ρ, ν, $\delta S(\mathrm{QT})$ altogether of each SD to the corresponding parameters of only those among the patients that happen to have $\delta S(\mathrm{QT})$ values exceeding the corresponding values of H, i.e., obey the condition (9.7), or preferably the more accurate condition (9.8).

Such a comparison reveals that some of the 17 parameters of λ, ρ, ν, $\delta S(\mathrm{QT})$, in all SD, lie outside the limits of these patients (cf. the same happens, of course, if we compare each SD to the limits of H). These results point to the conclusion that all 24 SD are distinguished from the patients (and H). The same conclusion is drawn if we consider instead, the 17 parameters λ, λ_{shuf}, ρ, $\delta S(\mathrm{QT})$.

We emphasize, however, that the study of the estimation errors (see § 9.2.7 and Section 9.3; see also the Appendix of Ref. [68]) reveals that:

The confidence level for the distinction of all SD from the patients becomes appreciably larger if we combine all the measures λ, λ_{shuf}, ρ ρ_{shuf}, ν (of all intervals) with the condition (9.8) applied to both $\delta S(\mathrm{QT})$ and $\delta S_{shuf}(\mathrm{QT})$ (i.e., in reality, we then consider the limits of those patients whom *both* $\delta S(\mathrm{QT})$ and $\delta S_{shuf}(\mathrm{QT})$ values are larger than those in H which are shown in Fig. 9.6).

A compilation of the limits of each of the complexity measures λ, ρ,λ_{shuf}, ρ_{shuf}, ν along with those of $\overline{\delta S}_{3-4}(QT)$ and $\overline{\delta S}_{3-4,shuf}(QT)$ in healthy humans (H) and in four groups (MIT, MSV, EST, MST) of heart disease patients is given in Table 9.5.

Table 9.5 Compilation of the limits of each of the complexity measures λ, ρ, λ_{shuf}, ρ_{shuf}, ν along with those of $\overline{\delta S_{3-4}}(QT)$ and $\overline{\delta S_{3-4,shuf}}(QT)$ in healthy humans (H) and in four groups (MIT, MSV, EST, MST) of heart disease patients. In parenthesis we put the limits which change when considering only the patients who have $\overline{\delta S_{3-4}}(QT)$ values larger than those in H, see § 9.2.4. The modified estimation errors, discussed in § 9.2.7, of the various parameters investigated are also shown. For the sake of brevity, the subscript "*sh*" stands for "*shuf*".

Parameter	H		MIT		MSV		EST		MST		$\varepsilon_m(\%)$
	min	max	min	max	min	max	min	max	min	max	
$\lambda_s(RR)$	1.43	2.00	0.80	1.96	0.85	1.22	0.79(0.94)	2.33	1.01	1.72(1.42)	11.66
$\lambda_L(RR)$	0.99	2.69	0.17	1.74	0.16	1.87	0.48(0.61)	6.66	0.45	3.75(2.68)	14.62
$\lambda_s(QRS)$	1.16	1.29	0.99	1.63	1.09	1.36	0.99	1.48	1.20	1.30(1.28)	10.53
$\lambda_L(QRS)$	0.48	0.61	0.46	0.71	0.47	1.06	0.42	1.63(0.90)	0.58	0.71(0.70)	11.19
$\lambda_s(QT)$	1.16	1.41	0.99	1.77	1.10	1.32	1.03(1.09)	1.71(1.36)	1.18	1.38	32.92
$\lambda_L(QT)$	0.50	1.11	0.33	0.96	0.50	0.74	0.43	2.82(1.73)	0.67	1.43(1.36)	41.37
$\rho_s(QRS)$	0.18	1.85	0.21	2.04	0.24	2.81	0.03	1.54	0.17	0.99(0.96)	18.23
$\rho_L(QRS)$	0.40	7.10	0.09	3.45	0.38	1.85	0.13	2.86(2.61)	0.51	2.62(1.25)	18.93
$\rho_s(QT)$	0.67	5.57	0.45	5.96	0.87	6.76	0.15	10.04(4.63)	0.53	4.98(1.97)	53.56
$\rho_L(QT)$	1.79	10.04	0.48	11.03	0.29	5.35	0.51	8.93(5.12)	1.11	8.70(2.07)	50.92
$\nu_s(RR)$ [a]	1.10	2.27	0.86	1.64	0.84	1.94	1.01	11.11	1.76(2.30)	11.65	13.47
$\nu_L(RR)$ [a]	0.44	0.77	0.59	2.48	0.51	2.75	0.31	1.90	0.44(1.93)	3.13	12.73
$\nu_s(QRS)$ [a]	0.88	1.07	0.92	1.23	0.95	1.18	0.92(0.93)	1.25(1.19)	0.95	1.17	10.97
$\nu_L(QRS)$ [a]	0.88	1.15	0.86	1.38	0.76	1.25	0.67(0.78)	1.25	0.83(0.97)	1.04	11.33
$\nu_s(QT)$ [a]	0.91	1.46	0.95	1.20	0.97	1.58	0.96	3.04(1.82)	1.17	4.14	36.96
$\nu_L(QT)$ [a]	0.65	1.07	0.74	1.38	0.76	1.31	0.50	1.75	0.59(0.95)	2.22	37.33
$\overline{\delta S_{3-4}}(QT)\times 10^3$	0.23	0.56	0.79	3.71	0.45(0.61)	4.59	0.20(0.56)	5.75	0.22(0.63)	1.33	28.51
$\lambda_{s,sh}(RR)$	1.04	1.27	0.92	1.34	0.96(1.05)	1.24	1.00	1.26	1.11(1.16)	1.21	11.00
$\lambda_{L,sh}(RR)$	0.48	0.57	0.37	0.92	0.45	0.54	0.43(0.46)	0.58	0.50(0.51)	0.56	10.86
$\lambda_{s,sh}(QRS)$	1.10	1.44	1.13	1.46	1.15	1.37	1.10(1.11)	1.47(1.43)	1.10	1.43	11.34
$\lambda_{L,sh}(QRS)$	0.46	0.73	0.47	0.75	0.49	0.80	0.51	0.96(0.77)	0.50(0.51)	0.69	12.27
$\lambda_{s,sh}(QT)$	1.12	1.32	0.89	1.26	0.85	1.23	1.10	1.56	1.17	1.28(1.24)	32.46
$\lambda_{L,sh}(QT)$	0.49	0.70	0.37	0.57	0.43	0.60	0.49	0.70	0.52	0.57	35.29
$\rho_{s,sh}(QRS)$	0.39	3.02	0.29	2.35	0.27	2.54	0.08	2.47(1.78)	1.07(1.75)	3.34	17.71
$\rho_{L,sh}(QRS)$	0.30	3.21	0.20	4.22	0.27	1.64	0.06	2.52(1.67)	0.97(1.48)	3.66	18.29
$\rho_{s,sh}(QT)$	1.25	9.63	0.46	7.01	0.75	6.13	0.57	10.50(6.71)	4.31	9.95(8.54)	53.11
$\rho_{L,sh}(QT)$	1.34	7.30	0.56	13.04	0.62	5.07	0.54	6.81(3.26)	1.80	7.03(3.62)	50.69
$\overline{\delta S_{3-4,sh}}(QT)\times 10^3$	0.27	0.66	0.79	3.84	0.46(0.67)	4.57	0.32(0.63)	9.07	0.31(1.57)	3.12	28.50

a)The values of these quantities do not fully coincide with those given in Ref. [66] for the reasons discussed in § 9.2.7.

Table 9.6 The number of SD and patients that can be distinguished from H when using λ_κ(RR) or $\lambda_{\kappa,shuf}$(RR) *alone*.

Group	Total number of individuals	λ_κ(RR)	$\lambda_{\kappa,shuf}$(RR)	λ_κ(RR) and $\lambda_{\kappa,shuf}$(RR)
SD	24	23	10	24
MIT	15	14	6	14
MSV	13	13	2	13
EST	33	29	8	29
MST	6	5	0	5

We now comment on two points.

First, since it is known that heart rate variability depends strongly on age, it is highly recommended that when comparing values of the aforementioned complexity measures, the corresponding limits should be taken from subjects (heart disease patients and H) of comparable age [66].

Second, we now focus on the importance of the sequential order of Q_m on the aforementioned complexity measures. We prefer to deal with the results related to the RR intervals since it is known that the healthy heart beats irregularly and that the RR intervals fluctuate widely, following complicated patterns [9]. Let us investigate, for example, the possibility of using λ_κ(RR) alone to distinguish the SD as well as the four groups of patients from H, i.e., examine whether the λ_κ(RR) values of each individual violate one (at least) of the relevant *H-limits*.

The results show (see Table 9.6) that the vast majority of SD and of each group of patients is well distinguished from H by means of λ_κ(RR) *alone*.

The situation drastically changes, however, if we use, instead of λ_κ(RR), the $\lambda_{\kappa,shuf}$ values (see the tables V to VII in Ref. [63]): only the minority of SD and of each group of patients can be differentiated from H. Since the calculation of the λ_κ(RR) values takes into account the sequential order of Q_m, while the $\lambda_{\kappa,shuf}$(RR) values do not, this points to the following conclusion:

It is the *sequential* order of beats that contains the primary information which enables the distinction between the SD and heart disease patients, on the one hand, and the H, on the other.

This might explain why procedures based on the entropy in natural time (which is dynamic entropy, affected by the sequential order [67]) – and hence consider the complexity measures mentioned in § 9.2.1 – can achieve such a distinction, while static entropy (e.g., Shannon entropy, see Ref. [67]) cannot.

9.2.5 Distinction of heart disease patients from H

This distinction can be made by identifying as heart disease patients the individuals whom one or more of the parameters associated with λ, ρ, ν (of RR, QRS, QT) and $\delta S(QT)$ violate the *H-limits provided* that the distinction of the SD has been *preceded* by the procedure described above in § 9.2.4.

Furthermore, comparing each of the tables in Ref. [63] that present the aforementioned parameters for each group of heart disease patients to (the H in) Table 9.2, we also find that:

In *all* heart disease patients, at least one of their four λ parameters associated with RR and QRS, i.e., $\lambda_s(RR)$, $\lambda_L(RR)$, $\lambda_s(QRS)$ and $\lambda_L(QRS)$, violates one of the corresponding *H-limits*, thus allowing again a distinction between patients and H. In other words, only four parameters are needed to distinguish heart disease patients from H.

A further inspection reveals that among the limits of these four λ parameters most of the heart disease patients violate the ones of $\lambda_s(RR)$ and/or $\lambda_L(RR)$.

Thus, in a future population consisting of all three categories SD, heart disease patients and H, in order to separate the last two, we may work as follows. By considering the limits given in Table 9.5, we first apply the procedure to identify the SD (as described in § 9.2.4) among the other individuals, thus only heart disease patients and H remain. It seems then that, in the latter population, the λ parameters of the RR and QRS can efficiently distinguish heart disease patients from H (this can be further strengthened by the additional use of the corresponding ν parameters, which differentiate most of the heart disease patients – but *not* all of them – from the H). In other words, any (explicit) information on the QT may not be prerequisite to distinguish between heart disease patients and H. This is consistent with the aforementioned (§ 9.1.1) clinical observations that the prolongation of the QT (due to the lengthening of the ST interval) is mainly a characteristic of the SD.

9.2.6 Complementarity of the complexity measures for identifying sudden cardiac death individuals (SD)

We first discuss the complementarity of the two procedures described above in § 9.1.3 and § 9.2.2 for the distinction of the (otherwise healthy) SD from H, i.e., *if* the population under investigation does *not* include heart disease patients.

Recall that in § 9.1.3 entropy fluctuations – deduced from the original and the "shuffled" time series – on *fixed* time-scales have been employed, while in § 9.2.2 entropy fluctuations on *different* time-scales have been considered.

This complementarity holds in the following sense: if in the frame of the one procedure an ambiguity emerges in the distinction between SD and H, the other procedure gives a clear answer.

We now study, as an example, the following two procedures: i.e., the one that uses δS(QT) (see § 9.1.3) and the other which combines the measures λ, ρ (see § 9.2.2). The $\overline{\delta S_{3-4}}$(QT) values of SD and H given in the last column of Table 9.2 are classified into two classes: the larger values correspond to SD, and the lower ones correspond to H (see also Figs. 9.7 and 9.12). Let us focus on the two lowermost SD values and the uppermost H value. The former two correspond to sel33 and sel34 ($\overline{\delta S_{3-4}}(\mathrm{QT}) = 0.00076$ and 0.00069, respectively) and the latter one to sel16795 ($\overline{\delta S_{3-4}}(\mathrm{QT}) = 0.00056$). In view of their $\overline{\delta S_{3-4}}$(QT) values proximity, one may wonder whether these two SD could be confused with H. This ambiguity can be resolved in the light of the other procedure (i.e., λ, ρ), as follows. Table 9.2 reveals that sel33 markedly violates both the H_{min}*-limit* for λ_s(QRS) and H_{min} for λ_s(RR) (the latter can be visualized in Fig. 9.13). As for sel34, the H_{max}*-limit* of λ_L(QRS) is strongly violated. We now turn to an alternative example, i.e., sel47, which, by means of the method using the complexity measures λ, ρ (of the RR and QRS intervals, see § 9.2.2) could be confused with H, because a deviation of only around 12% from the H_{min}*-limit* of $\min_H[\rho_s(QRS)] = 0.18$ is noticed. This ambiguity can be resolved by means of the procedure using δS(QT) (§ 9.1.3) as follows: sel47 has $\overline{\delta S_{3-4}}(\mathrm{QT}) = 0.0029$, which exceeds significantly, i.e., by a factor 5, the corresponding value of sel16795, who has the largest $\overline{\delta S_{3-4}}(\mathrm{QT}) = 0.00056$ value among the H.

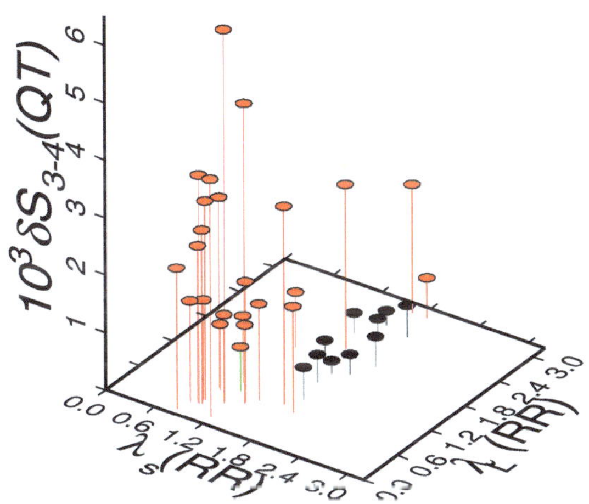

Fig. 9.13 The $\overline{\delta S_{3-4}}$(QT) values along with those of λ_s(RR) and λ_L(RR) for SD (red) and H (black). The individual sel33 is marked with a green column. Taken from Ref. [68].

We now turn to the investigation (for details see Ref. [63]) of the complementarity of the *four* quantities λ, ρ, ν and $\delta S(QT)$ on differentiating *all* SD from the others (i.e., heart disease patients and H). This can be judged from an inspection of Table 9.7, which contains the results to distinguish the SD among 101 individuals, for all possible combinations, upon considering only *three* of these quantities (i.e., see the cases in Table 9.7 except of the upper two where all four quantities are used). For example, the combination λ, ρ, ν cannot differentiate four SD (i.e., sel30, sel32, sel34, sel37) from the heart disease patients. As a

Table 9.7 Results of the distinction of 24 SD among 101 individuals upon using combinations of the measures λ, ρ, ν along with $\overline{\delta S_{3-4}}(QT)$. Taken from Ref. [63]

Measures combined[a)]	The non-differentiated SD[b)]	Number of SD distinguished
λ,ρ,ν and relation (9.8)	**None**	24 **(all)**
λ,ρ,ν and relation (9.7)	**One:** sel35(MIT)	23
λ,ρ,ν	**Four:** sel30(EST), sel32(EST), sel34(EST), sel37(EST)	20
λ,ρ and relation (9.8)	**Four:** sel30(MSV), sel41(MIT), sel46(MIT), sel49(MSV)	20
ρ,ν and relation (9.8)	**Three:** sel33(MSV,EST), sel45(MIT,MSV), sel46(MSV,EST)	21
λ,ν and relation (9.8)	**Seven:** sel36(MIT,EST), sel38(MIT), sel41(MSV), sel42(EST), sel47(EST), sel51(EST), sel17152(MSV,EST)	17
λ, ρ, ν of RR and QRS only	**Twelve:** sel30(EST), sel32(EST), sel34(EST), sel35(MIT,MSV), sel37(EST), sel38(MIT), sel40(EST), sel43(EST), sel45(MSV), sel47(EST), sel50(MIT), sel51(EST)	12

a) In all cases the data of the five heart disease patients sel230, sel231, sele0612, sele0704, sele0136 have been excluded (see § 9.2.4).
b) In parenthesis we mark the group(s) of heart disease patients in which the corresponding SD is mislocated.

second example, the combination ρ, ν and $\delta S(QT)$ cannot identify three SD (i.e., sel33, sel45, sel46), who are *different* from the four that could *not* be discriminated by the former combination λ, ρ, ν. By the same token, we find that each of the remaining combinations fails to identify certain SD, who can be distinguished by another combination(s).

Therefore, we conclude that each of the four quantities λ, ρ, ν, $\delta S(QT)$ seems to *complement* the others in identifying all SD (note that the same conclusion is drawn if we alternatively use the four quantities λ, λ_{shuf}, ρ and δS(QT); see table XIII of Ref. [63]).

In general, measures that employ entropy fluctuations of the original and shuffled time series on fixed time-scales, seem to complement those that take into account entropy fluctuations on different time-scales.

This might be understood in the context that each of these quantities, as already mentioned, presumably captures certain "elements" of heart dynamics only. As for the *necessity* of using all these quantities, it might stem from the following fact. The database we used, consists of SD individuals in which different physiological processes might have led to sudden cardiac death. The selection of such a *heterogeneous* database was intentionally made, because it was our aim to find, if possible, a general procedure for identifying SD. If a study of "homogeneous" SD databases (in the sense that the same physiological processes preceded the sudden cardiac death) is made, it may happen that a smaller number of parameters are necessary to distinguish *all* SD. Until the completion of such studies, however, it is recommended to use *all* the parameters associated with the aforementioned quantities, as described in Ref. [68].

9.2.7 The estimation errors in the procedure for identifying SD

Beyond the error introduced by the use of an automatic threshold detector for the allocation of the corresponding intervals which is largest for the QT and smallest for the RR intervals, the following two sources of errors must be considered [67, 68]: First, an estimation error emerges when analyzing – instead of the original time series of length $l \approx 10^3$ heartbeats – smaller lengths l' (e.g., see Table 9.3), which, however, still significantly exceed the time-window lengths used, for example $l' \approx 2 \times 10^2$ (obviously the errors associated with the measures in the short-range, s, are smaller from those corresponding to the longer range, L, because for the latter range the l/l' values – due to the restricted length of the records available – are small, thus not allowing more reliable statistics). Second, a source of (statistical) error in the results emerges when considering the ratio(s) $\delta S_{shuf}/\delta S$ (i.e., when dealing with ν and λ_{shuf}) instead of δS itself. While δS may be considered to have a *unique* value for a (given) original Q_m time series, the value of δS_{shuf} depends on the randomly shuffled Q_m series each time selected (note that such differences are well known [23] when dealing with randomized series of *finite* length). This is why the ν values given in Ref. [67] for SD and H do not fully coincide with those tabulated in Ref. [68]. To account roughly for the extent of this statistical error, we averaged here the δS_{shuf} values calculated over a number (e.g. 20) of randomly shuffled Q_m-series generated from the *same* original series and the corresponding standard deviation was estimated.

The final results of the above sources, could be summarized as follows [68]: The (percentage) estimation error was found to be on the average $\approx$10% for the complexity measures λ, λ_{shuf}, ρ, ρ_{shuf}, ν associated with the RR and QRS intervals. Furthermore, since the error in the δS(QT) may reach 20%, the estimation error in those of the complexity measures that involve δS(QT) may be as high as $\approx$30%. Upon considering such error-levels, hereafter called "*plausible estimation errors*" ε_p, a study of each of the methods for the distinction of SD was made. The study was repeated by assuming larger (percentage) estimation errors, hereafter labeled "*modified estimation errors*" ε_m, calculated for each parameter from

$$\varepsilon_m = \varepsilon_p \left(1 + \frac{H_{max} - H_{min}}{H_{max} + H_{min}} \right), \tag{9.9}$$

see the last column in Table 9.5. Both studies led, more or less, to the same results, e.g., those obtained when using ε_m, which are tabulated in columns 5–7 in Table 9.8. The calculation, in each study, was made as follows. Each parameter was assumed to be equal to its value (initially estimated from the original time series available) multiplied by a number randomly selected in the range $1 \pm \varepsilon_p$ or $1 \pm \varepsilon_m$, respectively) and then each of the methods for the distinction of SD was applied. This application was repeated, for each method, 10^3 times via Monte Carlo.

Table 9.8 The confidence levels to distinguish SD from either H or heart disease patients when considering the estimation errors ε_m of Eq. (9.9) discussed in § 9.2.7 and given in Table 9.5. Taken from Ref. [68].

Method employed				Confidence levels to distinguish SD						
				Using the limits from the data analyzed			Using broader limits[c)]			
Aim	Measures	Type of intervals	No. of para-meters	All SD %	All but one SD %	All but two SD[d)] %	All SD %	All but one SD %	All but two SD %	All but five SD[d)] %
Distinction of SD from H	λ, ρ	RR, QRS, QT	10	>99	>99	>99	88	99	>99	>99
	λ, ρ	RR, QRS	4	63	95	>99	8	43	90	>99
	λ, λ_{shuf}	RR	4	49	90	99	1	11	36	97
	ν	RR,QRS	4	32	74	96	<0.5	1	8	60
	$\delta S_{3-4}(QT)$	QT	1	59	93	>99	11	39	77	>99
	$\lambda, \rho, \lambda_{sh}, \rho_{sh}, \nu, \delta S_{3-4}(QT), \delta S_{sh,3-4}(QT)$	RR, QRS, QT	28	>99	>99	>99	>99	> 99	>99	>99
Distinction of SD from heart disease patients	$\lambda, \rho, \nu, \delta S_{3-4}(QT)^{a)}$	RR, QRS, QT	17	51	83	95	<0.1	<0.1	<0.1	1
	$\lambda, \rho, \lambda_{sh}, \delta S_{3-4}(QT)^{a)}$	RR, QRS, QT	17	62	91	98	<0.1	<0.1	<0.1	1
	$\lambda, \rho, \lambda_{sh}, \rho_{sh}, \nu, \delta S_{3-4}(QT), \delta S_{sh,3-4}(QT)^{b)}$	RR, QRS, QT	28	95	>99	>99	16	41	68	98

a) Considering the limits of those patients that have $\delta S_{3-4}(QT)$ larger than those in H.
b) Considering the limits of those patients that have *both* $\delta S_{3-4}(QT)$ and $\delta S_{sh,3-4}(QT)$ larger than those in H.
c) By amounts ε_m of Eq. (9.9) discussed in § 9.2.7 and given in Table 9.5.
d) When stating, e.g., "All but one", it means that when allowing *at the most*, one SD – out of 24 – to be misinterpreted as being H or heart disease patient, respectively.

The extent to which these conclusions hold, was also investigated in the following *extreme* case: the limits of the parameters of H (and patients), which are automatically adjusted for each "random" selection of the values described above, have been assumed to *additionally* relax by (extra) amounts equal to ε_p or ε_m. Such a "relaxation" faces the *extreme* possibility that the populations of H and heart disease patients analyzed here are not considered large enough to allow a precise determination of their limits, and hence future increased populations' studies could somehow broaden these limits by *extra* amounts as large as ε_p or ε_m. The corresponding confidence levels to distinguish SD from either H or heart disease patients can be found in the last four columns of Table 9.8.

9.3 Summarizing the conclusions for identifying sudden cardiac death individuals (SD) upon considering the error levels

As already mentioned in § 9.1.1, sudden cardiac death may occur *even if* the ECG looks similar to that of truly healthy humans. In other words, we are interested here in the distinction of the (otherwise healthy) SD from H, i.e., *if* the population under investigation does *not* include heart disease patients. To distinguish such cases, i.e., when we consider a set consisting *only* of two groups of ECG, namely H and SD, the conclusions drawn from the procedures developed in § 9.1.3 and § 9.2.2 above, are summarized below in § 9.3.1 and the relevant confidence levels are compiled in Table 9.8 under the Aim "Distinction of SD from H". As for the procedures developed to identify SD in a population that includes H as well as heart disease patients (§ 9.2.4) that led to the limits compiled in Table 9.5, the conclusions are summarized in § 9.3.2 and the corresponding confidence levels are given in Table 9.8 against the Aim "Distinction of SD from heart disease patients".

9.3.1 Summary of the conclusions for distinguishing SD from H

Among the four methods suggested (i.e., two in § 9.1.3 and two in § 9.2.2), the one that uses the measures λ, ρ (associated, however, with *all* three types of intervals, i.e., 10 parameters in total, see first row in Table 9.8) seems to be robust [68] in the following sense:

(i) When assuming the error-levels (see § 9.2.7) deduced from the data analyzed here (the relevant results are inserted in Table 9.8 under the heading "Using the limits from the data analyzed"):

The use of λ, ρ related to *all* intervals, thus 10 parameters in total, allows the distinction of *all* SD from H with a confidence level above 99%.

The confidence level decreases to 63%, 49%, 32% and 59% when using either four parameters or one parameter only as follows: first: λ_κ(RR) and ρ_κ(QRS); second: λ_κ(RR) and $\lambda_{\kappa,shuf}$(RR); third: ν_κ(RR) and ν_κ(QRS); fourth: δS_{3-4}(QT), respectively.

(ii) If we investigate the extreme case of the additional "relaxation" of the *H-limits* mentioned in § 9.2.7 (the relevant results in Table 9.8 are under the heading "Using broader limits"), the capability for the distinction of *all* SD still remains with the following results:

In the case of using solely λ, ρ for all intervals, the confidence level in distinguishing *all* SD is 88%. It becomes *appreciably higher*, i.e., larger than 99%, if we use the quantities λ, ρ, λ_{shuf}, ρ_{shuf}, ν, $\overline{\delta S_{3-4}}(QT)$, $\overline{\delta S_{3-4,shuf}}(QT)$ *altogether*.

When using, however, four parameters only in the first three combinations mentioned above, the confidence level decreases to 90%, 36% and 8%, respectively (and to 77% when using $\overline{\delta S_{3-4}}$(QT)), even when allowing two at the most SD – out of 24 – to be misinterpreted as being H.

9.3.2 Summary of the conclusions for identifying SD among individuals that also include heart disease patients and H

The corresponding conclusions related to the distinction of SD from heart disease patients can be drawn on the basis of the values given in the lower part of Table 9.8.

In summary, the study of the estimation errors reveals [68] that *if* the limits of the parameters that have been deduced from the ECG data analyzed here will *not* be broadened by future investigations:

We can satisfactorily distinguish the *totality* of SD from H as well as discriminate the totality of SD from heart disease patients, upon employing the quantities λ, λ_{shuf}, ρ, ρ_{shuf}, ν,$\overline{\delta S_{3-4}}$(QT), $\overline{\delta S_{3-4,shuf}}$(QT) *altogether*, i.e., the sixth and the last method in Table 9.8, respectively.

These quantities also allow the distinction of the *totality* of SD from H (as well as distinguish the *vast majority* of SD from the heart disease patients) *even if* their limits will be eventually broadened (by ε_m of Eq. (9.9), see § 9.2.7).

Concerning the number of parameters required to achieve the desired distinction [68]: In reality, only twelve *independent* quantities, (i.e., the six: $\delta S_\kappa(\tau)$ and the six $\delta S_{\kappa,shuf}(\tau)$, where $\kappa = s, L$ and $\tau =$ RR, QRS, QT) are extracted from the experimental data. Thus, for example, beyond $\overline{\delta S_{3-4}}(QT)$ *or* $\overline{\delta S_{3-4,shuf}}(QT)$, eleven additional parameters (out of 26) of the ratios: λ, λ_{shuf}, ρ, ρ_{shuf}, ν are in principle required to be used for the distinction. These twelve quantities, however, should *not* be fortuitously selected, but the following points must be carefully considered: (i) priority should be given to the eight parameters associated with λ values and λ_{shuf} (or ν) values of RR and QRS, (ii) using, at least, one ρ-parameter (involving $\overline{\delta S_{3-4}}(QT)$ or $\overline{\delta S_{3-4,shuf}}(QT)$), and (iii) examining whether the totality of the parameters used can actually reproduce the aforementioned twelve δS values determined directly from the data. However, in order to avoid the difficulty arising from the completeness (or not) of the aforementioned selection, at the present stage (i.e., until an appreciably larger number of H and heart disease patients will be analyzed to allow a better precision in the determination of the corresponding limits, see § 9.2.7), it is recommended to use – instead of twelve – *all* the 28 parameters associated with the quantities λ, λ_{shuf}, ρ, ρ_{shuf}, ν, $\overline{\delta S_{3-4}}(QT)$ and $\overline{\delta S_{3-4,shuf}}(QT)$.

9.4 The change ΔS of the entropy in natural time under time reversal: identifying the sudden cardiac death risk and specifying its occurrence time

9.4.1 Specifying the occurrence time of the impending cardiac arrest by means of ΔS

Here, we make use of the Definition 3.2 of ΔS (see Eq. (3.64)) and the points developed in § 3.5.1.

In particular, a window of length l is sliding, each time by one pulse, through the whole time series. The entropies S and S_-, and therefrom their difference ΔS_l, are calculated each time. Thus, we form a new time series consisting of successive ΔS_l values.

We will show and that the determination of the occurrence time of the impending cardiac arrest can be obtained [69] from the time evolution of ΔS_l deduced from the RR time series.

9.4.1.1 The ECG data analyzed in natural time

These are 159 long-lasting (from several hours to around 24 h) ECG recordings, which come from databases [14], containing: (i) 72 healthy subjects, (ii) 44 patients with congestive heart failure (CHF) (iii) 25 subjects with atrial fibrillation (AF) and (iv) 18 individuals suffered sudden cardiac death. In particular (see Ref. [65]), these data come from the following databases [14]: (i) the MIT-BIH Normal Sinus Rhythm Database (nsrdb) containing 18 H digitized with frequency $f_{exp} = 128$ Hz, (ii) the Normal Sinus Rhythm RR Interval Database (nsr2db) containing 54 H, $f_{exp} = 128$ Hz (iii) the Congestive Heart Failure RR Interval Database (chf2db) containing 29 subjects with congestive heart failure, $f_{exp} = 128$ Hz, (iv) the BIDMC Congestive Heart Failure Database (chfdb) with 15 subjects with severe congestive heart failure, $f_{exp} = 250$ Hz (v) the MIT-BIH Atrial Fibrillation Database (afdb) with 25 subjects with atrial fibrillation (AF) mostly paroxysmal, $f_{exp} = 250$ Hz and (vi) the Sudden Cardiac Death Holter Database (sddb), $f_{exp} = 250$ Hz. The latter contains 24 SD among which 12 had ECG with audited annotations. Here, beyond these 12 individuals, we studied six more (i.e., "33", "37", "44", "47", "48", "50") whose ECG could be analyzed with confidence. Thus, we consider 18 (out of 24) SD individuals of the sddb.

The results presented in this Section refer to the RR intervals (see Fig. 2.2), i.e., $Q_m = \mathrm{RR}_m$. For reasons that will be explained later, the study will be extended (in all these 159 individuals except the 25 AF for which NN annotations were not available) to the so-called NN intervals, i.e., $Q_m = \mathrm{NN}_m$. These are intervals obtained from ECG annotation files by using the option [41] "-c -PN pN", which yields only intervals between consecutive *normal* beats, while intervals between pairs of *normal* beats surrounding an ectopic beat are discarded. In both the RR and NN time series, in order to exclude "outliers" from the records, the detection algorithm proposed in Ref. [21] has been applied, i.e., for each set of five contiguous intervals, if the local mean, excluding the central interval, is larger than twice the central interval then this interval is excluded from further analysis. In Fig. 9.15(a) one H out of 72, i.e., the one labeled 16539, has been discarded because the resulting $\sigma[\Delta S_3]$(NN) value was unusually high compared to that in other H of nsrdb (see table 2 of Ref. [65]). Furthermore, in Fig. 9.15(b), three H out of 72 (i.e., 16539, nsr024 and nsr044) have been also discarded since they have $\sigma[\Delta S_3]$(RR) value unusually higher than that in other H (see table 2 of Ref. [65]). For more details on the annotators used see Ref. [65].

Table 9.9 Results of the application of the complexity measure ΔS_l to the *RR* time series: the extrema $\max(\Delta S_{13})$ and $\min(\Delta S_{13})$ in SD along with the time of their occurrence, i.e., T_{max} and T_{min}, respectively. The latter time is measured from the time of the VF onset (except for "49", who paced with no VF). In the last column, the total duration of the record T_{total} measured from the time of the VF onset is also inserted. Taken from Ref. [65].

Individual	$\max(\Delta S_{13})$	$T_{\max}$(s)	$\min(\Delta S_{13})$	$T_{\min}$(s)	T_{total} (s)
30	0.0129	28,150.65	−0.0107	6,000.90	28,470.75
31	0.0182	1,497.47	−0.0174	1,492.78	49,341.89
32	0.0069	59,754.38	−0.0047	59,746.80	60,315.61
33	0.0168	3,021.60	−0.0237	11,212.63	17,176.40
34	0.0102	10,642.46	−0.0097	7,408.24	23,743.42
35	0.0214	22,674.56	−0.0220	7,872.32	86,398.19
36	0.0218	5,603.68	−0.0197	5,598.33	68,338.58
37	0.0355	5,361.32	−0.0569	5,370.84	5,470.82
41	0.0240	3,303.27	−0.0212	3,060.47	10,762.66
44	0.0146	7,993.19	−0.0123	34,421.23	70,723.33
45	0.0157	62,992.88	−0.0145	62,985.09	65,354.88
46	0.0184	13.38	−0.0166	5,244.22	13,304.91
47	0.0241	13,282.90	−0.0230	8,481.94	22,378.26
48	0.0146	8,921.66	−0.0150	8,930.64	8,978.57
49	0.0145	5,677.80	−0.0140	1,805.06	84,528.44
50	0.0353	1,349.73	−0.0347	4,349.58	42,339.39
51	0.0151	53,067.89	−0.0161	1,957.63	82,701.48
52	0.0293	2,552.97	−0.0252	2,567.82	9,158.85

9.4.1.2 Presentation of the ΔS_l results

In Fig. 9.14(a), we give as an example the time series of ΔS_{13} for one SD, i.e., the one labeled "30". In the horizontal axis the time is measured from the ventricular fibrillation (VF) onset. The time of the VF initiation for each SD (except for the individual "49", who

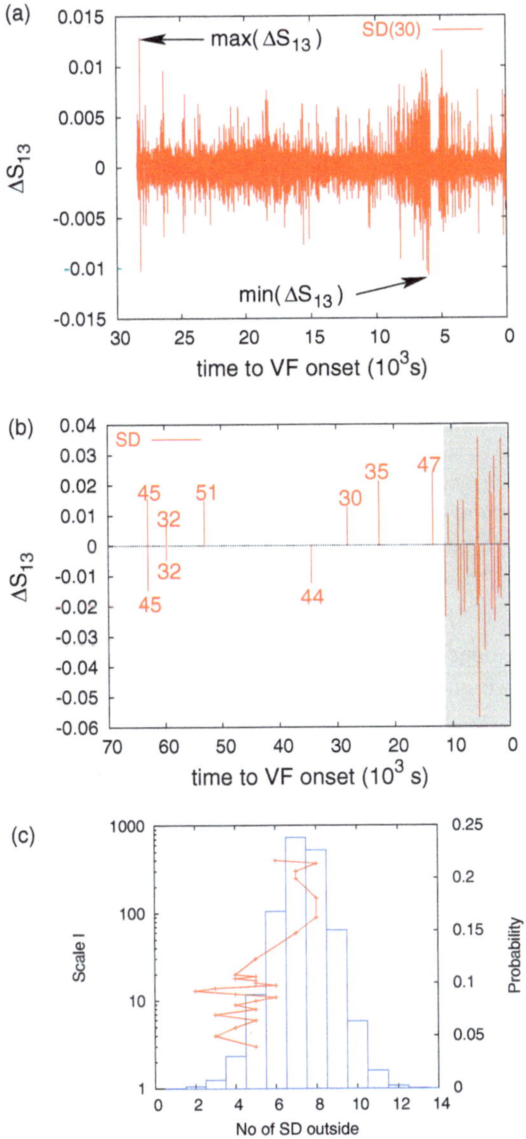

Fig. 9.14 Results from the analysis of the RR time series: (a) Plot of the quantity ΔS_{13} versus the time to the VF onset for one SD, i.e., "30". The quantities max[ΔS_{13}] and min[ΔS_{13}] are shown by arrows. (b) For each of the 18 SD (each bar corresponds to each individual), we plot the max[ΔS_{13}] value – in the upper part (i.e., positive ΔS_{13} axis) – and the value min[ΔS_{13}] – in the lower part (i.e., negative ΔS_{13} axis) – versus the time it appeared *before* the VF onset. The shaded part indicates the last 3 h *before* the VF onset. (c) The red curve shows the number of SD that violate *both* conditions $T_{max} \leq 3$ h and $T_{min} \leq 3$ h as a function of scale l. The probability achieving by chance the relevant number of SD is drawn by blue bars (right vertical scale). Reprinted with permission from Ref. [69]. Copyright (2007), American Institute of Physics.

paced with no VF) is given in the database used [14]. The VF initiation remains one of the leading immediate causes of sudden cardiac death [1]. The maximum and the minimum values of ΔS_{13} will be labeled max[ΔS_{13}] and min[ΔS_{13}], respectively. The time of their appearances are designated T_{max} and T_{min}, respectively. An inspection of Fig. 9.14(a) in conjunction to Table 9.9, reveals that $T_{max} \approx 28{,}150$ s and $T_{min} \approx 6{,}000$ s (before the VF onset). The corresponding values for all the other SD studied, are also given in the same Table, which presents the extrema of ΔS_{13} along with the time of their appearance. These values, which are depicted in Fig. 9.14(b), reveal that interestingly in the vast majority of SD (i.e., in all the 18 SD except the individuals "32" and "45", the latter having a history of ventricular ectopy) they are smaller than around 3 hours. In other words, *only* for two individuals (i.e., "32" and "45") out of eighteen, *both* $T_{\max}$ and $T_{\min}$ are larger than around 3 hours. The results for a variety of other length scales are summarized in Fig. 9.14(c), where we plot in red the number of SD that violate *both* conditions, i.e., $T_{\max} \leq 3$ h and $T_{\min} \leq 3$ h, at various scales. The probability having such a result by chance is also shown in the right vertical scale. This probability has been found by Monte Carlo calculation, in which the observation times for *both* extrema, i.e., $T_{\max}$ and $T_{\min}$, were assumed to be uniformly distributed within the total duration T_{total} of the record for each individual (see Table 9.9). We observe that for small scales ($l < 30$) the observed number of SD differs significantly from the one expected by chance. Especially, the probability to find by chance the result obtained at $l = 13$ is smaller than 0.2%.

In other words, an optimum length scale (i.e., $l = 13$ heartbeats) exists, at which the magnitude of ΔS_l (deduced from the RR time series, alone) maximizes (in 16 out of 18 cases) ≈ 3 hours at the most before the VF onset, thus signaling the imminent cardiac death risk.

Since many SD experience arrhythmia (consisting of one or more types including premature ventricular contractions (PVCs), AF and non-sustained tachycardia), it has been confirmed (through a direct inspection of the ECG) that the extreme values of ΔS_{13} in Fig. 9.14(b) mainly come from trains of occurrences of PVCs. We emphasize, however, that beyond the PVCs, the method of ΔS_l captures *additional* elements of cardiac dynamics that distinguish SD from other individuals as will be discussed in § 9.4.2.

9.4.2 Identifying the sudden cardiac death risk by means of complexity measures based on ΔS

We now make use of the points treated in § 3.5.1 and § 3.6.2. In particular, following § 3.5.1, we recall that when we form the new time series consisting of successive ΔS_l values, the standard deviation of these values is denoted by $\sigma[\Delta S_l]$. Upon shuffling the Q_m randomly (thus destroying any information hidden in the *ordering* of the events), the ΔS_l values turn to a sequence of different values labeled ΔS_l^{shuf} whose standard deviation is designated by $\sigma[\Delta S_l^{\text{shuf}}]$ (its theoretical estimation was given in § 3.5.2). The complexity measure $N_l \equiv \sigma[\Delta S_l^{\text{shuf}}]/\sigma[\Delta S_l]$ (see Eq. (3.83)), which quantifies the extent to which the ordering of the heartbeats contributes to the ΔS_l values (being unity for a *random* process), is also computed.

In Fig. 9.15(a), we plot the quantities N_3(NN) versus $\sigma[\Delta S_7]$(NN) deduced from the analysis of the NN time series of all individuals except of the 25 AF (since for the latter, relevant NN annotations were not available).

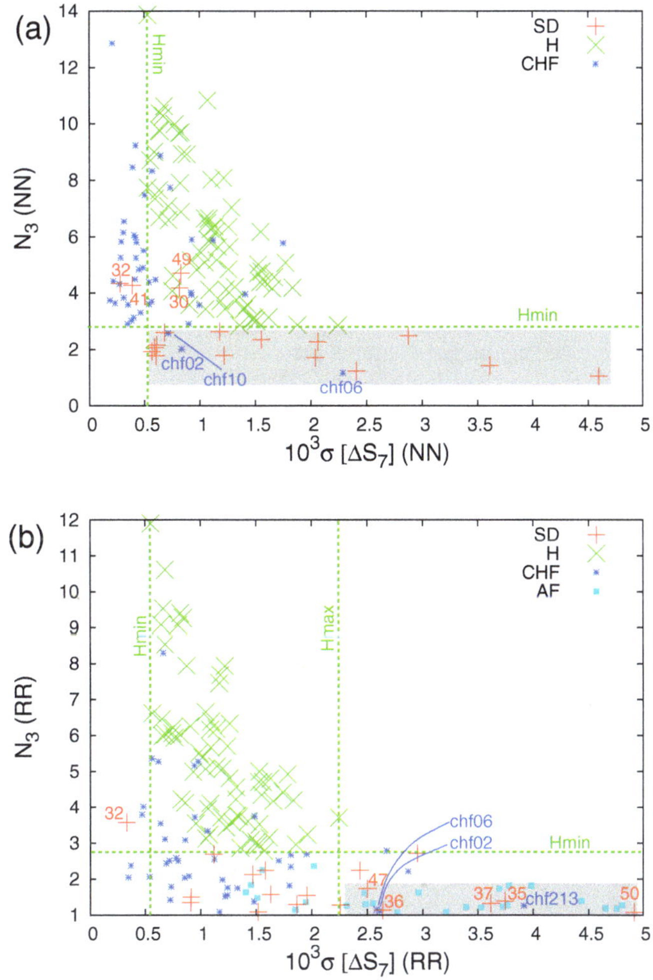

Fig. 9.15 The quantity N_3 versus $\sigma[\Delta S_7]$ for (a) the NN and (b) the RR time series. The green horizontal line corresponds to the minimum N_3 value computed in H. Reprinted with permission from Ref. [69]. Copyright (2007), American Institute of Physics.

Thus, when using the NN time series alone, an inspection of Fig. 9.15(a) reveals the major importance of the measure N_3(NN) in two respects.

First, the vast majority of SD (i.e., 14 out of 18, lying in the shaded region) exhibit N_3(NN) values that are *smaller* than the minimum N_3(NN) value computed among the H which is labeled $H_{\min}$ and marked with a horizontal green line in Fig. 9.15(a).

Second, the vast majority of CHF have N_3(NN) values *larger* than $H_{\min}$, thus allowing in principle a distinction between CHF and SD.

In Fig. 9.15(b), we plot N_3(RR) versus $\sigma[\Delta S_7]$(RR) deduced from the RR time series. This figure shows that the distinction between CHF and SD achieved in Fig. 9.15(a) is now lost. This is understood in the context that frequent PVCs influence the RR time series (but *not* the NN) of *both* CHF and SD.

Thus, when using the RR time series alone, a closer inspection of Fig. 9.15(b) reveals two important points:

First, almost *all* SD (i.e., except "32") exhibit N_3(RR) values that are smaller (hence high complexity breaks down) than the minimum value $H_{\min}$ computed in H, thus emphasizing again the importance of the scale $l = 3$.

Second, the shaded region that contains the vast majority of AF (18 out of 25) lies to the right of the maximum value of $\sigma[\Delta S_7]$(RR) observed in H, labeled $H_{\max}$ (see the rightmost vertical green line). Four out of the five SD (i.e., except "47") located in this region, suffered from atrial fibrillation, thus this shaded region seems to separate AF from the others.

Thus, in short, the aforementioned method *not only* identifies the sudden cardiac death risk *but also* provides a distinction of congestive heart failure patients from SD when NN annotations are available.

9.4.3 Summary of the findings based on ΔS and their tentative explanation

In order to understand the physical origin of the findings in § 9.4.1 and § 9.4.2 we resort to the neural influences on cardiovascular variability. Let us recall that:

Physiologically, the origin of the complex dynamics of heart rate has been attributed to antagonistic activity of the two branches of the autonomic nervous system, i.e., the parasympathetic and the sympathetic nervous systems, respectively, decreasing and increasing heart rate [47, 29, 20, 2]. Their net result is what seems to be actually captured by ΔS_l, as shown in § 3.5.3.

A variety of research has now established [35], as already mentioned in § 9.1.3.1, two clear frequency bands in heart rate and blood pressure with autonomic involvement. (i) A higher frequency (HF) band, which lies in [6, 49] the range 0.15 to 0.40 Hz and is [29] "indicative of the presence of respiratory modulation of the heart rate" or reflects [6] "modulation of vagal activity, primarily by breathing". (ii) A lower frequency (LF) band from 0.04 to 0.15Hz (i.e., at around 0.1 Hz), which is usually described as corresponding to [49] "the process of slow regulation of blood pressure and heart rate" or that [6] "it reflects modulation of sympathetic or parasympathetic activity by baroflex mechanisms" due to [29] "the emergence of a limit cycle caused by the vascular sympathetic delay" (note that its exact explanation, however, is still strongly debated [38]). The aforementioned scale $l = 13$ (see ΔS_{13} in Fig. 9.14(b)) corresponds to the LF band, while the scale $l = 3$ (see N_3 in Fig. 9.15), to the HF band. Thus, the magnitude of ΔS_l, when calculated for length scales corresponding to the HF and LF bands, quantifies the extent to which the processes: "modulation of vagal activity primarily by breathing" and the "slow regulation of blood pressure and heart rate" are "disorganized", respectively.

An alternative way of understanding intuitively the aforementioned findings is the following. If we consider [67] that S could be thought of as a measure of the "disorder" (in successive intervals) and that the essence of the natural time analysis is built on the variation of the durations of consecutive pulses, we may say the following: when approaching sudden cardiac death, the difference between the "disorder" looking in the (immediate) future, i.e., S, and that in the (immediate) past, i.e., S_-, becomes in SD of profound importance when compared to the corresponding difference under truly healthy conditions.

In summary, the complexity measure N_3, based on the entropy change ΔS_l under time reversal at the scale $l = 3$ heartbeats, identifies the sudden cardiac death risk and distinguishes SD from truly healthy individuals as well as from those with the life-threatening congestive heart failure. Furthermore, the study of ΔS_l at the scale $l = 13$ heartbeats provides an estimate of the occurrence time of the impending VF onset in those classified as SD.

The importance of the aforementioned scale of $l = 13$ heartbeats also emerges from studies on the correlation properties of the magnitude and the sign of the increments in the intervals between successive heartbeats during daytime activity as well as during sleep stages. Interestingly, it was found [24, 19] that the correlation behavior of the heartbeat increments and their signs and magnitudes during daytime activity is similar to the behavior in REM (rapid eye-movement) sleep, but significantly different from the behavior in deep sleep. It has been empirically observed [24, 19] by DFA that the most significant differences between the different sleep stages occur in the following ranges: $8 \leq l \leq 13$ and $11 \leq l \leq 150$ heartbeats for the sign-series and magnitude-series respectively. It is challenging that the scale $l = 13$ is just in the verge of these two important ranges. This coincidence cannot be fortuitous, but might stem from the reasons (LF-band, etc.) discussed above.

9.5 Heart rate variability (HRV) and 1/*f* "noise". A model in natural time that exhibits 1/*f* behavior

9.5.1 The 1/*f* "noise". Background

Among the different features that characterize complex physical systems, the most ubiquitous is the presence of $1/f^a$ noise in fluctuating physical variables [36]. This means that the Fourier power spectrum $S(f)$ of fluctuations scales with frequency f as $S(f) \propto 1/f^a$, as already mentioned in § 1.4.2 (see also § 1.5.1.1). The power law behavior often persists over several orders of magnitude with cutoffs present at both high and low frequencies. Typical values of the exponent a approximately range between 0.8 and 4 (e.g., see Ref. [4] and references therein), but in a loose terminology all these systems are said to exhibit $1/f$

"noise". Such a "noise" is found in a large variety of systems, e.g., condensed matter systems(e.g. Ref. [70]), granular flow [43], DNA sequence [45], ionic current fluctuations in membrane channels [40], the number of stocks traded daily [34], chaotic quantum systems [17, 50, 52, 53], human cognition [13] and coordination [72], burst errors in communication systems [5], electrical measurements [28], the electric noise in carbon nanotubes [10] and in nanoparticle films [27], the SES activities (see § 1.4.3), etc. In some of these systems, the exponent a was reported to be very close to 1, but good quality data supporting such a value exist in a few of them [70]. As an example we refer to the voltage fluctuations when current flows through a resistor [71]. As a second example we recall the case of SES activities discussed in § 1.4.3 in which we concluded that $\alpha \approx 1$. As a third example, we mention the case of heart rate variability to which we now turn.

Various tests of time variation have been applied to heart rate variability to show that in healthy subjects heart rate fluctuations display $1/f$ noise and fractal dynamics with long-range correlations, e.g., see Ref. [47]. These initial studies indicated rich dynamics with differences between normal individuals and patients [15]. In particular, it has been found (see Ref. [19] and references therein) that at scales above $\approx$1 min ($l > 60$ heartbeats) the data during waking hours display long-range power law correlations over two decades with average exponents $\alpha_{wake} \approx 1.05$ for the healthy group and $\alpha_{wake} \approx 1.2$ for congestive heart failure patients. These values change to a smaller exponent $\alpha_{sleep} \approx 0.85$ for the healthy group and $\alpha_{sleep} \approx 0.95$ for the heart failure group for the sleep data. Heart rate variability (HRV) is a useful tool that might provide indices of autonomic modulation of the sinus node [58] and its reduced value is a sign of autonomic imbalance. Later findings (e.g., Refs. [21, 18]) showed that healthy heartbeat dynamics exhibits even higher complexity, which is characterized by a broad multifractal spectrum as already mentioned in § 9.2.1 (concerning the distinction between monofractals and multifractals, see § 4.5.1). This high complexity breaks down in illness associated with altered cardiovascular autonomic regulation (e.g., Refs. [29, 19] and references therein). In particular, the heart rate in healthy subjects is a multifractal signal while for subjects with a pathological condition, e.g. congestive heart failure, it shows a clear loss of multifractality [18, 21]. In other words, for the heart failure subjects the multifractal spectrum is nonzero only over a very narrow range of exponents indicating an almost monofractal behavior.

The $1/f^a$ behavior has been well understood on the basis of dynamic scaling observed at *equilibrium* critical points (e.g., § 1.5.3) where the power law correlations in time stem from the infinite-range correlations in space (see Ref. [4] and references therein). Most of the observations mentioned above, however, refer to *non-equilibrium* phenomena for which – despite some challenging theoretical attempts [3, 12] – possible *generic* mechanisms leading to scale-invariant fluctuations have not yet been identified.

In other words, despite its ubiquity, there is no yet universal explanation about the phenomenon of the $1/f^a$ behavior.

9.5.2 An evolution model in natural time that exhibits 1/f behavior

We describe here a simple evolutionary model which, in the frame of natural time, leads to $1/f^a$ behavior with an exponent a close to unity.

This model [54] considers the following simple evolution picture. As the number of generations n increases by one, a new species – whose ability to survive is characterized by a number η_n – appears. The new species competes and eliminates *only* the existing species that have a lesser ability to survive. We show below that the number of species ε_n, if considered as a function of the number of generations n, exhibits an $1/f$ behavior and that it increases very slowly with n, actually logarithmically, thus very few species survive in this competitive process.

The mathematical description of the model, in terms of set theory, is as follows. Let us consider the cardinality ε_n (see § 2.7.1) of the family of sets E_n of *successive extrema* obtained from a given probability distribution function (pdf); E_0 equals the empty set. Each E_n is obtained by following the procedure described below for n times. Select a random number η_n from a given pdf (here, we use the exponential pdf, i.e., $p(\eta_n) = \exp(-\eta_n)$) and compare it with all the members of E_{n-1}. In order to construct the set E_n, we *discard* from the set E_{n-1} all its members that are smaller than η_n and furthermore *include* η_n. Thus, $\mathrm{E}_n \neq \emptyset$ for *all* $n > 0$ and E_n is a finite set of real numbers whose members are always larger or equal to η_n. Moreover, $\max[\mathrm{E}_n] \geq \max[\mathrm{E}_{n-1}]$. The increase of the cardinality $\varepsilon_n \equiv |\mathrm{E}_n|$ of these sets is at the most 1, but its decrease may be as large as $\varepsilon_{n-1} - 1$. This reflects an asymmetry if ε_n is considered as time series with respect to the *natural* number n. An example of ε_n vs n is shown in Fig. 9.16(a). The cardinality ε_n exhibits $1/f^a$ noise with a very close to unity; see Fig. 9.16(b). The mathematical model described above, the analytical properties of which has been discussed in detail in Ref. [60], corresponds to an asymptotically non-stationary process, since $\langle \varepsilon_n \rangle \propto \ln n$ with a variance $\langle (\varepsilon_n - \langle \varepsilon_n \rangle)^2 \rangle \propto \ln n$ (see Fig. 9.16(c)). In particular, it has been shown analytically in Ref. [60] that:

$$\langle \varepsilon_n \rangle = \sum_{k=1}^{n} \frac{1}{k}, \tag{9.10}$$

$$\langle (\varepsilon_n - \langle \varepsilon_n \rangle)^2 \rangle = \sum_{k=1}^{n} \left(\frac{1}{k} - \frac{1}{k^2} \right). \tag{9.11}$$

Equations (9.10) and (9.11) reveal that both the average value $\mu \equiv \langle \varepsilon_n \rangle$ and the variance $\sigma^2 \equiv \langle (\varepsilon_n - \langle \varepsilon_n \rangle)^2 \rangle$ diverge logarithmically as n tends to infinity. The point probabilities $p(\varepsilon_n = m)$, however, remain localized around $\mu = \langle \varepsilon_n \rangle \propto \ln n$ since $\sigma/\mu \propto 1/\sqrt{\ln n}$.

Thus, in short, the model suggests that the cardinality ε_n of the family of sets E_n of successive extrema exhibits a logarithmic creep and the $1/f^a$ behavior when considered as time series with respect to the natural (time) number n.

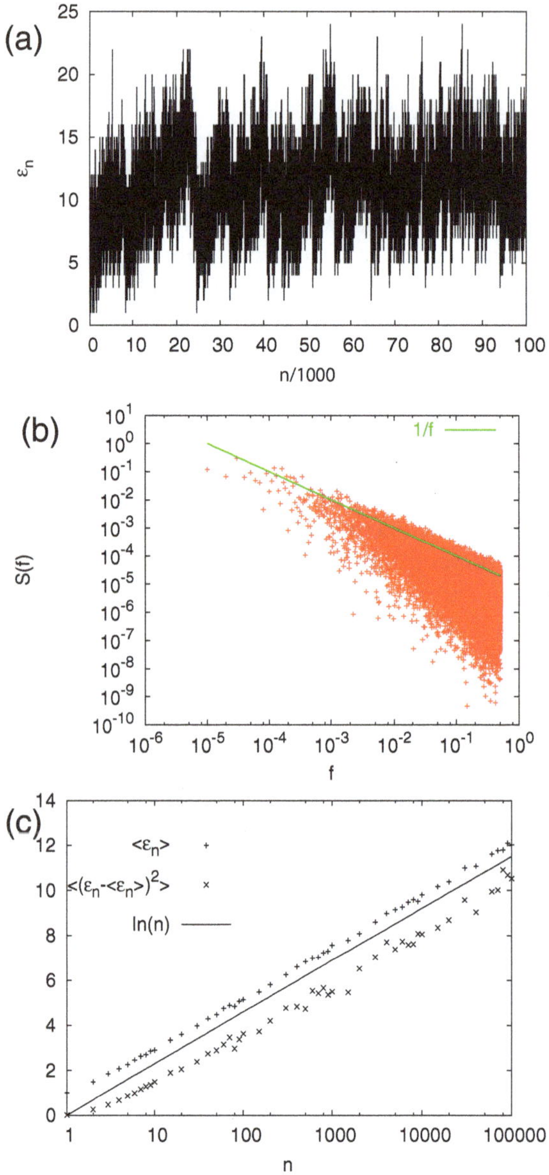

Fig. 9.16 (a): Example of the evolution of ε_n versus the number of generations n, i.e., in natural time. An exponential pdf has been considered for the selection of η_n. (b): The Fourier power spectrum of (a); the (green) solid line corresponds to $1/f$ and was drawn as a guide to the eye. (c): Properties of the distribution of ε_n. The average value $\langle\varepsilon_n\rangle$ (plus) and the variance $\langle(\varepsilon_n - \langle\varepsilon_n\rangle)^2\rangle$ (crosses) as a function of n. The straight solid line depicts $\ln(n)$ and was drawn for the reader's convenience. Taken from Ref. [54].

Note that an interconnection between $1/f^a$ noise and extreme value statistics has been proposed as providing a new angle at the generic aspect of the phenomena [3].

In order to check the stability of the results of Fig. 9.16, we present in Fig. 9.17(a) the average power spectrum obtained from 10^4 runs of the model. A sharp $1/f$ behavior is observed. Moreover, in Fig. 9.17(b), we present the results of the corresponding average values of F_{DFA-l} of the DFA obtained for various orders l, i.e., when detrending with a

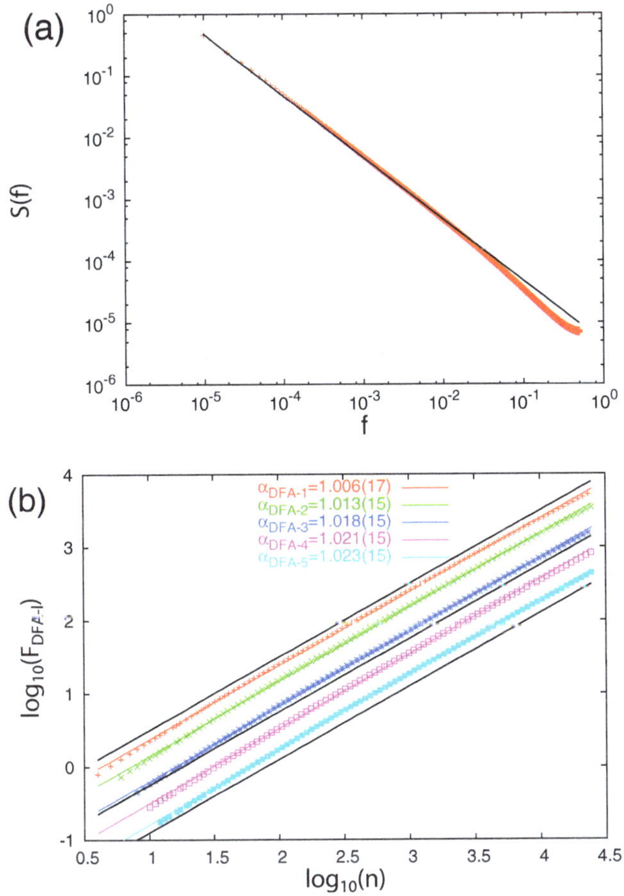

Fig. 9.17 Results from 10^4 runs of the model presented in Fig. 9.16: (a) the average power spectrum, (b) detrended fluctuation analyses of order l (DFA-l). The black solid line in (a) corresponds to $1/f$ spectrum and was drawn as a guide to the eye. For the same reason in (b), the black solid lines correspond to $\alpha_{DFA} = 1$. In (b), the colored solid lines correspond to the least squares fit of the average F_{DFA-l} depicted by symbols of the same color. The numbers in parentheses denote the standard deviation of α_{DFA-l} obtained from the 10^4 runs of the model. The various F_{DFA-l} have been displaced vertically for the sake of clarity. Taken from Ref. [54].

polynomial of order l, see § 1.4.2. Figure 9.17(b) indicates that α_{DFA-l} is close to unity, thus being compatible with the $1/f$ power spectrum depicted in Figs. 9.16(b) and 9.17(a).

We recall that in the aforementioned example of Fig. 9.16(a) showing the evolution of ε_n versus the number of generations n (i.e., in natural time), an exponential pdf has been considered. After investigating several different distributions of η_n, we conclude that the resulting spectral density depends only very weakly – if at all – on the pdf of η_n.

We find that, in order to obtain $\alpha \approx 1$, the only essential condition to be fulfilled is that the corresponding pdf should be bounded from below (note that this is a reasonable assumption if η_n is to be considered a measure of the ability to survive; a negative measure would correspond to a species that is unable to survive).

This holds, of course, under the assumption that η_n come from the *same* pdf, i.e., they are independent and identically distributed variables. Let us now investigate the case when

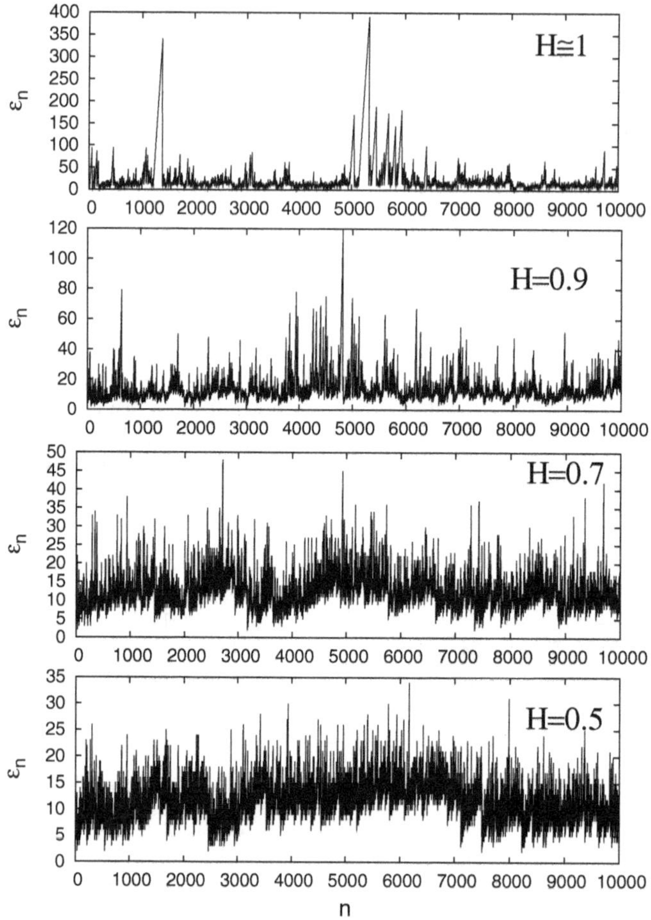

Fig. 9.18 Time series of ε_n when η_n come from fGn for various values of H (increasing from the bottom to the top). Taken from Ref. [54].

η_n come from a stationary but long-range (time) correlated process, for example from fractional Gaussian noise (fGn) (see § 1.5.1.1). To this end, several values of the H exponent have been considered and indicative results are depicted in Fig. 9.18 for $H = 0.5, 0.7, 0.9$ and ≈ 1. A noticeable difference can be visualized in this figure upon increasing H: for $H = 1$, which corresponds, for example, to the case of SES activities (see § 1.4.3, § 4.3.2, § 4.4.2 and Section 4.10) the results differ greatly from those corresponding to smaller exponents, e.g., $H = 0.5$–0.7, which are occasionally found in the analysis of electric signal time series emitted from "artificial" (man-made) electrical sources (see § 4.4.2).

This model, beyond its applicability to HRV (see below in § 9.5.3), may be useful in other disciplines as well. For example, in the frame of a formal similarity between the discrete spectrum of quantum systems and a discrete time series [50], the following striking similarity is noticed. The fact that $a \approx 1$ together with the behavior $\langle(\varepsilon_n - \langle\varepsilon_n\rangle)^2\rangle \propto \ln n$ of the present model, is reminiscent of the power law exponent and the $\langle\delta_n^2\rangle$ statistic in chaotic quantum systems [50, 52].

Furthermore, ε_n may be considered as equivalent to the dimensionality of the thresholds distribution in the so-called coherent noise model (e.g. see Ref. [59] and references therein).

9.5.3 The 1/f model proposed and the progressive modification of HRV in healthy children and adolescents

The model described above in § 9.5.2 amounts to a sort of shot noise in a process showing logarithmic creep, a non-stationary process. We now compare this prediction of the model with the heart rate variability data in healthy children and adolescents versus age.

We consider here the HRV data in healthy children and adolescents presented by Silvetti et al. [56]. In particular, the following two standard 24 h time-domain measures, among others, were computed: SDNN (standard deviation of all normal sinus RR intervals over 24 h) and SDANN (standard deviation of the averaged normal sinus RR intervals for all 5-min segments). They evaluated 103 subjects (57 males and 46 females, aged 1–20 years) and found that SDNN and SDANN, overall HRV measures, increased with age and were gender-related. These data demonstrate that in healthy children and adolescents there is a progressive modification of HRV that may reflect a progressive evolution of the autonomic nervous system.

Using the results of Silvetti et al. [56], we plot in Fig. 9.19(a) SDNN vs age in a semilogarithmic plot. An inspection of this figure reveals that, for ages below 14 yr, in both male (blue) and female (red) subjects an almost logarithmic creep is present, a property also exhibited by the model.

This logarithmic creep can also emerge from the results of Ref. [37] where the SDNN versus age (A) was fitted by a power law, i.e., SDNN $= 97.2 \times \mathrm{A}^{0.20}$ [ms], for the period from infancy to adolescence.

In particular, in Fig. 9.19(b), drawn on the basis of the data presented in fig. 1 of Ref. [37] by using averages every one year of age, a logarithmic creep seems to provide a better description for SDNN from early childhood to adolescence.

This behavior could be, in principle, understood in the following context. The present model may simulate the variation of RR intervals around a mean value determined by the sinoatrial node, thus leading to the logarithmic creep of SDNN visualized in Fig. 9.19. We note that the model intrinsically represents a competitive evolution which is also present

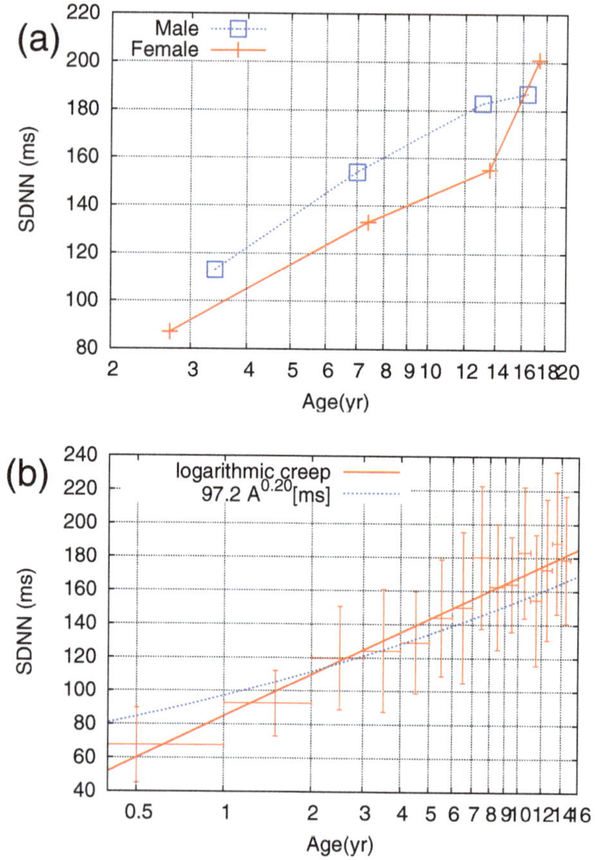

Fig. 9.19 (a): The mean values of SDNN for male (blue) and female (red) subjects as a function of their age. The data come from table 1 of Ref. [56]. The x-axis is in logarithmic scale. (b): Variation of SDNN with respect to age. The data come from fig. 4 of Ref. [37] and are binned every year of age. The vertical error bars stand for $\pm$ one standard deviation. The dotted (blue) curve corresponds to the power law fit suggested in Ref. [37] whereas the solid (red) line corresponds to a logarithmic creep predicted by the model of § 9.5.2. Taken from Ref. [54].

during the period of childhood. The complexity of heart rate dynamics is high in children and illustrates [37]: "an increase of cholinergic and a decrease of adrenergic modulation of heart rate variability with age, confirming the progressive maturation of the autonomic nervous system." In other words, in order to shed light on the underlying connection between the presented model and the development of heartbeat regulation we could say the following. As already mentioned in § 9.4.3, the origin of the complex dynamics of heart rate has been attributed to the *antagonistic* activity of the parasympathetic and sympathetic nervous system:

It is this *antagonistic* activity which seems to be captured by the model since its basic spirit stems from a competitive evolution process.

9.5.4 The complexity measures obtained from the 1/f model and their comparison with HRV data

We now compare the results of the model in natural time with the HRV data – actually the RR time series – of heart disease patients and healthy subjects that have been already analyzed in natural time in Section 9.4. Recall that those data came from long time ECG recordings [14] containing on average $N \simeq 10^5$ heartbeats for each record. Thus, in order to compare with the results already presented in Fig. 9.15(b) on HRV, we consider only mature models with $n \simeq 10^6$ and examine their evolution, i.e, the time series ε_n, for the later 10^5 generations (cf. this is the order of magnitude of heartbeats in a 24 h ECG recording). The proposed model results in $N_3 = 2.52 \pm 0.19$ and $\sigma[\Delta S_7] = (2.46 \pm 0.25) \times 10^{-3}$ shown by the (black) square in Fig. 9.20. This figure just reproduces Fig. 9.15(b) to which the calculated values of the model (as well as those from the INAGS model, see below) are now added. Concerning the calculated value of N_3, this is close to (but below) the minimum value H_{min} observed in H and larger than the N_3 values in the vast majority of SD (where high complexity breaks down). As for the calculated $\sigma[\Delta S_7]$ value, it lies to the right of the maximum value of $\sigma[\Delta S_7]$ observed in H as well as in the vast majority of CHF located outside the shaded region which seems to separate AF from the others. This is consistent with the fact that the (black) square corresponds to an $1/f$ behavior, while healthy heartbeat dynamics exhibits even higher complexity [21, 18] as mentioned in § 9.5.1.

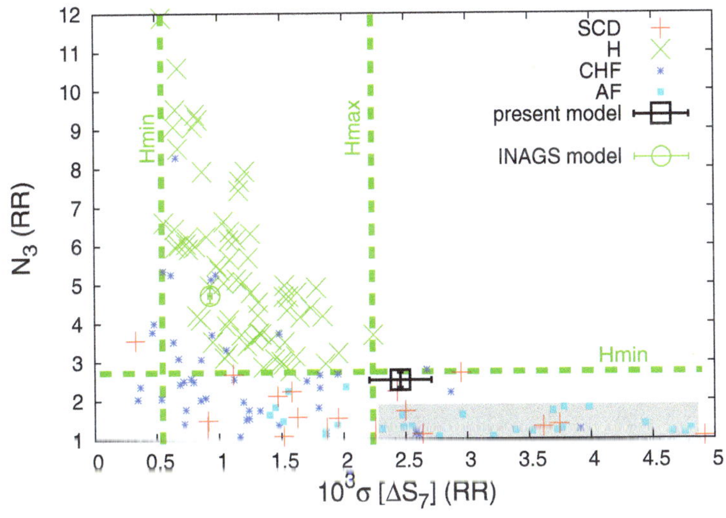

Fig. 9.20 The complexity measure N_3 vs $\sigma[\Delta S_7]$ for the RR time series. This figure is the same as Fig. 9.15(b) to which the complexity measures obtained from the present $1/f$ model as well as those deduced from the model of Ref. [20] have been added, marked with (black) square and (green) circle, respectively. Taken from Ref. [54].

Indeed, let us consider the stochastic feedback model proposed by Ivanov, Nunes Amaral, Goldberger and Stanley (INAGS) in Ref. [20] which describes the *healthy* regulation of biological rhythms with a clear relation to the physiology of the heart; the effects of the sinoatrial node along with the parasympathetic and the sympathetic influences were taken into account. The INAGS model leads [20] to an approximately $1/f^{1.1}$ behavior and generates complex dynamics that account for the functional form and scaling of the distribution of variations of RR. The aforementioned complexity measures in natural time that correspond to this model (by using the same parameters as those mentioned in fig. 2 of Ref. [20]) have been calculated [54] and the results are depicted by the (green) circle in Fig. 9.20. Interestingly, this point lies within the *H-limits*, as it should.

Summarizing, using the concept of natural time, a simple competitive evolution model has been proposed that exhibits $1/f^a$ behavior with a close to unity. The model amounts to a sort of shot noise in a process showing logarithmic creep (non-stationary process), a behavior which is similar to the fact that the standard deviation of all normal sinus RR intervals over 24 h exhibits a logarithmic creep with age for children and adolescents. The model predicts complexity measures (see the black square in Fig. 9.20) that separate healthy dynamics from heart disease patients and SD, as intuitively expected since it corresponds to a simple $1/f$ behavior.

References

1. Allexandre, D., Otani, N.F.: Preventing alternans-induced spiral wave breakup in cardiac tissue: An ion-channel-based approach. Phys. Rev. E **70**, 061903 (2004)
2. Amaral, L.A.N., Ivanov, P.C., Aoyagi, N., Hidaka, I., Tomono, T., Goldberger, A.L., Stanley, H.E., Yamamoto, Y.: Behavioral-independent features of complex heartbeat dynamics. Phys. Rev. Lett. **86**, 6026–6029 (2001)
3. Antal, T., Droz, M., Györgyi, G., Rácz, Z.: 1/f noise and extreme value statistics. Phys. Rev. Lett **87**, 240601 (2001)
4. Antal, T., Droz, M., Györgyi, G., Rácz, Z.: Roughness distributions for $1/f^{\alpha}$ signals. Phys. Rev. E **65**, 046140 (2002)
5. Berger, J.M., Mandelbrot, B.B.: A new model for the clustering of errors on telephone circuits. IBM J. Res. Dev. **7**, 224–236 (1963)
6. Bigger J. Thomas, J., Fleiss, J.L., Steinman, R.C., Rolnitzky, L.M., Schneider, W.J., Stein, P.K.: RR variability in healthy, middle-aged persons compared with patients with chronic coronary heart disease or recent acute myocardial infarction. Circulation **91**, 1936–1943 (1995)
7. Bray, M.A.P.: Visualization and analysis of electrodynamic behavior during cardiac arrhythmias. Med. Phys. **30**, 3045 (2003)
8. Chang, T., Sauer, T., Schiff, S.J.: Tests for nonlinearity in short stationary time-series. CHAOS **5**, 118–126 (1995)
9. Chialvo, D.R.: Physiology – unhealthy surprises. Nature **419**, 263 (2002)
10. Collins, P.G., Fuhrer, M.S., Zettl, A.: 1/f noise in carbon nanotubes. Appl. Phys. Lett. **76**, 894–896 (2000)
11. Costa, M., Goldberger, A.L., Peng, C.K.: Multiscale entropy analysis of complex physiologic time series. Phys. Rev. Lett. **89**, 068102 (2002)
12. Davidsen, J., Schuster, H.G.: Simple model for $1/f^{\alpha}$ noise. Phys. Rev. E **65**, 026120 (2002)

13. Gilder, D.L., Thornton, T., Mallon, M.W.: 1/f noise in human cognition. Science **267**, 1837–1839 (1995)
14. Goldberger, A.L., Amaral, L.A.N., Glass, L., Hausdorff, J.M., Ivanov, P.C., Mark, R.G., Mictus, J.E., Moody, G.B., Peng, C.K., Stanley, H.E.: Physiobank, physiotoolkit, and physionet – components of a new research resource for complex physiologic signals. Circulation **101**, E215 (see also www.physionet.org) (2000)
15. Goldberger, A.L., Amaral, L.A.N., Hausdorff, J.M., Ivanov, P.C., Peng, C.K., Stanley, H.E.: Fractal dynamics in physiology: Alterations with disease and aging. Proc. Natl. Acad. Sci. USA **99**, 2466–2472 (2002)
16. Goldberger, A.L., Amaral, L.A.N., Hausdorff, J.M., Ivanov, P.C., Peng, C.K., Stanley, H.E.: Self-organized complexity in the physical, biological, and social sciences: Fractal dynamics in physiology: Alterations with disease and aging. Proc. Natl. Acad. Sci. USA **99**, 2466–2472 (2002)
17. Gómez, J.M.G., Relaño, A., Retamosa, J., Faleiro, E., Salasnich, L., Vraničar, M., Robnik, M.: $1/f^{\alpha}$ noise in spectral fluctuations of quantum systems. Phys. Rev. Lett **94**, 084101 (2005)
18. Ivanov, P.C., Amaral, L.A.N., Goldberger, A.L., Halvin, S., Rosenblum, M.G., Stanley, H.E., Struzik, Z.R.: From 1/f noise to multifractal cascades in heartbeat dynamics. CHAOS **11**, 641–652 (2001)
19. Ivanov, P.C., Chen, Z., Hu, K., Stanley, H.E.: Multiscale aspects of cardiac control. Physica A **344**, 685–704 (2004)
20. Ivanov, P.C., Nunes Amaral, L.A., Goldberger, A.L., Stanley, H.E.: Stochastic feedback and the regulation of biological rhythms. Europhys. Lett. **43**, 363–368 (1998)
21. Ivanov, P.C., Rosenblum, M.G., Peng, C.K., Mietus, J., Havlin, S., Stanley, H.E., Goldberger, A.L.: Multifractality in human heartbeat dynamics. Nature **399**, 461–465 (1999)
22. Jané, R., Blasi, A., J. Garcia, J., Laguna, P.: Evaluation of an automatic threshold based detector of waveform limits in Holter ECG with the QT database. In: Computers in Cardiology, Vol. 24, p. 295. IEEE Computer Society Press, Piscataway, NJ (1997)
23. Kantelhardt, J., Zschiegner, S.A., Koscienly-Bunde, E., Bunde, A., Havlin, S., Stanley, H.E.: Multifractal detrended fluctuation analysis of nonstationary time series. Physica A **316**, 87–114 (2002)
24. Kantelhardt, J.W., Ashkenazy, Y., Ivanov, P.C., Bunde, A., Havlin, S., Penzel, T., Peter, J.H., Stanley, H.E.: Characterization of sleep stages by correlations in the magnitude and sign of heartbeat increments. Phys. Rev. E **65**, 051908 (2002)
25. Kaplan, D., Staffin, P.: (2004). Computer codes `Software for Heart Rate Variability` available from http://www.macalester.edu/~kaplan/hrv/doc/
26. Khan, I.A.: Long QT syndrome: Diagnosis and management. Am. Heart J. **143**, 7–14 (2002)
27. Kiss, L.B., Klein, U., Muirhead, C.M., Smithyman, J., Gingl, Z.: Diffusive fluctuations, long-time and short-time cross-correlations in the motion of vortice-pancakes in different layers of YBCO/PBCO superlattices. Solid State Commun. **101**, 51–56 (1997)
28. Kogan, S.: Electronic Noise and Fluctuations in Solids. Cambridge Univrsity Press, Cambridge (1996)
29. Kotani, K., Struzik, Z.R., Takamasu, K., Stanley, H.E., Yamamoto, Y.: Model for complex heart rate dynamics in health and diseases. Phys. Rev. E **72**, 041904 (2005)
30. Laguna, P., Jané, R., Caminal, P.: Automatic detection of wave boundaries in multilead ECG signals: Validation with the CSE database. Computers and Biomedical Research **27**, 45–60 (1994)
31. Laguna, P., Mark, R.G., Goldberger, A., Moody, G.B.: A database for evaluation of algorithms for measurement of QT and other waveform intervals in the ECG. In: Computers in Cardiology, Vol. 24, p. 673. IEEE Computer Society Press, Piscataway, NJ (1997)
32. Laguna, P., Thakor, N.V., Caminal, P., Jané, R., Yoon, H.R.: New algorithm for QT interval analysis in 24-hour Holter ECG - performance and applications. Medical & Biological Engineering & Computing **28**, 67–73 (1990)
33. Lake, D.K., Moorman, J.R., Hanqing, C.: sampen. Computer code `sampen` available from http://www.physionet.org/physiotools/sampen/ (2004)
34. Lillo, F., Mantegna, R.N.: Variety and volatility in financial markets. Phys. Rev. E **62**, 6126–6134 (2000)
35. Malpas, S.C.: Neural influences on cardiovascular variability: possibilities and pitfalls. Am. J. Physiol. Heart. Circ. Physiol. **282**, H6–H20 (2002)
36. Mandelbrot, B.B.: Multifractals and 1/f Noise. Springer-Verlag, New York (1999)

37. Massin, M., von Bernouth, G.: Normal ranges of heart rate variability during infancy and childhood. Pediatr. Cardiol. **18**, 297–302 (1997)
38. McSharry, P.E., Clifford, G.D., Tarassenko, L., Smith, L.A.: A dynamical model for generating synthetic electrocardiogram signals. IEEE Trans. Biomed. Eng. **550**, 289–294 (2003)
39. Melcher, D.: Statist. Computer code STATIST available from `http://www.usf.uos.de/~breiter/tools/statist/index.en.html` (2001)
40. Mercik, S., Weron, K., Siwy, Z.: Statistical analysis of ionic current fluctuations in membrane channels. Phys. Rev. E **60**, 7343–7348 (1999)
41. Moody, G.B.: Computer code `ann2rr` available from `http://www.physionet.org/physiotools/wag/ann2rr-1.htm`.
42. Motulsky, H.: Intuitive Biostatistics. Oxford University Press, New York (1995)
43. Nakahara, A., Isoda, T.: $1/f^{\alpha}$ density fluctuations at the slugging transition point of granular flows through a pipe. Phys. Rev. E **55**, 4264–4273 (1997)
44. Nikolopoulos, S., Alexandridi, A., Nikolakeas, S., Manis, G.: Experimental analysis of heart rate variability of long-recording electrocardiograms in normal subjects and patients with coronary artery disease and normal left ventricular function. J. Biomed. Informatics **36**, 202–217 (2003)
45. Peng, C.K., Buldyrev, S., Goldberger, A., Havlin, S., Sciortino, F., Simons, M., Stanley, H.E.: Long-range correlations in nucleotide sequences. Nature **356**, 168–170 (1992)
46. Peng, C.K., Havlin, S., Stanley, H.E., Goldberger, A.L.: Quantification of scaling exponents and crossover phenomena in nonstationary heartbeat time series. CHAOS **5**, 82–87 (1995)
47. Peng, C.K., Mietus, J., Hausdorff, J.M., Havlin, S., Stanley, H.E., Goldberger, A.L.: Long-range anticorrelations and non-Gaussian behavior of the heartbeat. Phys. Rev. Lett. **70**, 1343–1346 (1993)
48. Pincus, S.M.: Approximate entropy as a measure of system complexity. Proc. Natl. Acad. Sci. USA **88**, 2297–2301 (1991)
49. Prokhorov, M.D., Ponomarenko, V.I., Gridnev, V.I., Bodrov, M.B., Bespyatov, A.B.: Synchronization between main rhythmic processes in the human cardiovascular system. Phys. Rev. E **68**, 041913 (2003)
50. Relaño, A., Gómez, J.M.G., Molina, R.A., Retamosa, J., Faleiro, E.: Quantum chaos and $1/f$ noise. Phys. Rev. Lett. **89**, 244102 (2002)
51. Richman, J.S., Moorman, J.R.: Physiological time-series analysis using approximate entropy and sample entropy. Am. J. Physiol. Heart Circ. Physiol., **278**, H2039–H2049 (2000)
52. Santhanam, M.S., Bandyopadhyay, J.N.: Spectral fluctuations and 1/f noise in the order-chaos transition regime. Phys. Rev. Lett. **95**, 114101 (2005)
53. Santhanam, M.S., Bandyopadhyay, J.N., Angom, D.: Quantum spectrum as a time series: Fluctuation measures. Phys. Rev. E **73**, 015201(R) (2006)
54. Sarlis, N.V., Skordas, E.S., Varotsos, P.A.: Heart rate variability in natural time and $1/f$ "noise". EPL **87**, 18003 (2009)
55. Schreiber, T., Schmitz, A.: Surrogate time series. Physica D **142**, 346–382 (2000)
56. Silvetti, M.S., Drago, F., Ragonese, P.: Heart rate variability in healthy children and adolescents is partially related to age and gender. Int. J. Cardiol. **81**, 169–174 (2001)
57. Siwy, Z., Mercik, S., Weron, K., Ausloos, M.: Application of dwell-time series in studies of long-range correlation in single channel ion transport: analysis of ion current through a big conductance locust potassium channel. Physica A **297**, 79–96 (2001)
58. Taskforce ESC/NASPE: Heart rate variability, standards of measurement, physiological interpretation, and clinical use. Circulation **93**, 1043–1065 (1996)
59. Tirnakli, U., Abe, S.: Aging in coherent noise models and natural time. Phys. Rev. E **70**, 056120 (2004)
60. Varotsos, P.A., Sarlis, N.V., Skordas, E.S.: Seismic electric signals and $1/f$ "noise" in natural time. arXiv:0711.3766v3 [cond-mat.stat-mech] (1 February 2008)
61. Varotsos, P.A., Sarlis, N.V., Skordas, E.S.: Spatio-temporal complexity aspects on the interrelation between seismic electric signals and seismicity. Practica of Athens Academy **76**, 294–321 (2001)
62. Varotsos, P.A., Sarlis, N.V., Skordas, E.S.: Attempt to distinguish electric signals of a dichotomous nature. Phys. Rev. E **68**, 031106 (2003)

63. Varotsos, P.A., Sarlis, N.V., Skordas, E.S., Lazaridou, M.S.: See (the freely available) EPAPS Document No. E-PLEEE8-71-134501 originally from P.A. Varotsos, N.V. Sarlis, E.S. Skordas and M.S. Lazaridou *Phys. Rev. E* **71**, 011110 (2005). For more information on EPAPS, see http://www.aip.org/pubservs/epaps.html.
64. Varotsos, P.A., Sarlis, N.V., Skordas, E.S., Lazaridou, M.S.: See (the freely available) EPAPS Document No. E-PLEEE8-69-107405 originally from P.A. Varotsos, N.V. Sarlis, E.S. Skordas and M.S. Lazaridou *Phys. Rev. E* **70**, 011106 (2004). For more information on EPAPS, see http://www.aip.org/pubservs/epaps.html.
65. Varotsos, P.A., Sarlis, N.V., Skordas, E.S., Lazaridou, M.S.: See (the freely available) EPAPS Document No. E-APPLAB-91-062732 originally from P.A. Varotsos, N.V. Sarlis, E.S. Skordas, and M.S. Lazaridou, *Appl. Phys. Lett.* **91**, 064106 (2007). For more information on EPAPS, see http://www.aip.org/pubservs/epaps.html.
66. Varotsos, P.A., Sarlis, N.V., Skordas, E.S., Lazaridou, M.S.: The use of the entropy in the natural time-domain to distinguish electric signals. Practica of Athens Academy **78**, 281–298 (2003)
67. Varotsos, P.A., Sarlis, N.V., Skordas, E.S., Lazaridou, M.S.: Entropy in natural time domain. Phys. Rev. E **70**, 011106 (2004)
68. Varotsos, P.A., Sarlis, N.V., Skordas, E.S., Lazaridou, M.S.: Natural entropy fluctuations discriminate similar-looking electric signals emitted from systems of different dynamics. Phys. Rev. E **71**, 011110 (2005)
69. Varotsos, P.A., Sarlis, N.V., Skordas, E.S., Lazaridou, M.S.: Identifying sudden cardiac death risk and specifying its occurrence time by analyzing electrocardiograms in natural time. Appl. Phys. Lett. **91**, 064106 (2007)
70. Weissman, M.B.: 1/f noise and other slow, nonexponential kinetics in condensed matter. Rev. Mod. Phys. **60**, 537–571 (1988)
71. Yakimov, A.V., Hooge, F.N.: A simple test of the Gaussian character of noise. Physica B **291**, 97–104 (2000)
72. Yoshinaga, H., Miyazima, S., Mitake, S.: Fluctuation of biological rhythm in finger tapping. Physica A **280**, 582–586 (2000)

Index

Symbols

A

E

I

J

K

L

M

N

T

U

V

W

Z